Complexity
and
Approximation

Springer-Verlag Berlin Heidelberg GmbH

G. Ausiello P. Crescenzi G. Gambosi
V. Kann A. Marchetti-Spaccamela M. Protasi

Complexity and Approximation

Combinatorial Optimization Problems
and Their Approximability Properties

With 69 Figures and 4 Tables

Springer

Giorgio Ausiello
Alberto Marchetti-Spaccamela
Dipartimento di Informatica
e Sistemistica
Università di Roma "La Sapienza"
Via Salaria 113, I-00198 Rome, Italy

Giorgio Gambosi
Marco Protasi†
Dipartimento di Matematica
Università di Roma "Tor Vergata"
Via della Ricerca Scientifica
I-00133 Rome, Italy

Pierluigi Crescenzi
Dipartimento di Sistemi e Informatica
Università degli Studi di Firenze
Via C. Lombroso 6/17
I-50134 Florence, Italy

Viggo Kann
NADA, Department of Numerical
Analysis and Computing Science
KTH, Royal Institute of Technology
SE-10044 Stockholm, Sweden

Cover picture "What happened in the night" by J. Nešetřil and J. Načeradský

Second corrected printing 2003

Library of Congress Cataloging-in-Publication Data
Complexity and approximation: combinatorial optimization problems and
 their approximability properties/G. Ausiello... [et al.].
  p. cm.
Includes bibliographical references and index.
ISBN 978-3-642-63581-6   ISBN 978-3-642-58412-1 (eBook)
DOI 10.1007/978-3-642-58412-1
1. Combinatorial optimization. 2. Computational complexity.
3. Computer algorithms. I. Ausiello, G. (Giorgio), 1941–
QA402.5.C555   1999
519.3—dc21   99-40936
                                                                              CIP

ACM Subject Classification (1998): F.2, G.1.2, G.1.6, G.2, G.3, G.4
1991 Mathematics Subject Classification: 05-01, 90-01
Additional material to this book can be downloaded from http://extras.springer.com
ISBN 978-3-642-63581-6

This work is subject to copyright. All rights are reserved, whether the whole or part of the material is concerned, specifically the rights of translation, reprinting, reuse of illustrations, recitation, broadcasting, reproduction on microfilm or in any other way, and storage in data banks. Duplication of this publication or parts thereof is permitted only under the provisions of the German Copyright Law of September 9, 1965, in its current version, and permission for use must always be obtained from Springer-Verlag. Violations are liable for prosecution under the German Copyright Law.

© Springer-Verlag Berlin Heidelberg 1999
Softcover reprint of the hardcover 1st edition 1999

The use of general descriptive names, trademarks, etc. in this publication does not imply, even in the absence of a specific statement, that such names are exempt from the relevant protective laws and regulations and therefore free for general use.

Typesetting: Camera-ready by the authors
Design: design + production GmbH, Heidelberg
Printed on acid-free paper   SPIN 10885020   06/3142SR - 5 4 3 2 1 0

To our dear colleague
and friend Marco Protasi, *in memoriam.*

*...And soonest our best men with thee doe goe,
rest of their bones, and soules deliverie.*
                    JOHN DONNE

G.A.
P.L.C.
G.G.
V.K.
A.M.-S.

To Gabriele, Igor, Irene, and Sara
To Giorgia and Nicole
To Benedetta
To Elon
To Salvatore
To Davide and Sara

*I will tell you plainly
all that you would like to know,
not weaving riddles,
but in simple language,
since it is right
to speak openly to friends.*

AESCHYLUS,
Prometheus bound, 609-611

Contents

1 The Complexity of Optimization Problems   **1**
    1.1 Analysis of algorithms and complexity of problems   2
         1.1.1 Complexity analysis of computer programs   3
         1.1.2 Upper and lower bounds on the complexity of problems   8
    1.2 Complexity classes of decision problems   9
         1.2.1 The class NP   12
    1.3 Reducibility among problems   17
         1.3.1 Karp and Turing reducibility   17
         1.3.2 NP-complete problems   21
    1.4 Complexity of optimization problems   22
         1.4.1 Optimization problems   22
         1.4.2 PO and NPO problems   26
         1.4.3 NP-hard optimization problems   29
         1.4.4 Optimization problems and evaluation problems .   31
    1.5 Exercises   33
    1.6 Bibliographical notes   36

2 Design Techniques for Approximation Algorithms   **39**
    2.1 The greedy method   40
         2.1.1 Greedy algorithm for the knapsack problem   41
         2.1.2 Greedy algorithm for the independent set problem   43
         2.1.3 Greedy algorithm for the salesperson problem . . .   47

Table of contents

    2.2 Sequential algorithms for partitioning problems 50
        2.2.1 Scheduling jobs on identical machines 51
        2.2.2 Sequential algorithms for bin packing 54
        2.2.3 Sequential algorithms for the graph coloring problem 58
    2.3 Local search 61
        2.3.1 Local search algorithms for the cut problem 62
        2.3.2 Local search algorithms for the salesperson problem 64
    2.4 Linear programming based algorithms 65
        2.4.1 Rounding the solution of a linear program 66
        2.4.2 Primal-dual algorithms 67
    2.5 Dynamic programming 69
    2.6 Randomized algorithms 74
    2.7 Approaches to the approximate solution of problems . . . 76
        2.7.1 Performance guarantee: chapters 3 and 4 76
        2.7.2 Randomized algorithms: chapter 5 77
        2.7.3 Probabilistic analysis: chapter 9 77
        2.7.4 Heuristics: chapter 10 78
        2.7.5 Final remarks 79
    2.8 Exercises 79
    2.9 Bibliographical notes 83

3 Approximation Classes   **87**
    3.1 Approximate solutions with guaranteed performance . . . 88
        3.1.1 Absolute approximation 88
        3.1.2 Relative approximation 90
        3.1.3 Approximability and non-approximability of TSP . 94
        3.1.4 Limits to approximability: The gap technique . . . 100
    3.2 Polynomial-time approximation schemes 102
        3.2.1 The class PTAS 105
        3.2.2 APX versus PTAS 110
    3.3 Fully polynomial-time approximation schemes 111
        3.3.1 The class FPTAS 111
        3.3.2 The variable partitioning technique 112
        3.3.3 Negative results for the class FPTAS 113
        3.3.4 Strong NP-completeness and pseudo-polynomiality 114
    3.4 Exercises 116
    3.5 Bibliographical notes 119

4 Input-Dependent and Asymptotic Approximation   **123**
    4.1 Between APX and NPO 124
        4.1.1 Approximating the set cover problem 124
        4.1.2 Approximating the graph coloring problem 127

	4.1.3	Approximating the minimum multi-cut problem . .	129
4.2	Between APX and PTAS		139
	4.2.1	Approximating the edge coloring problem	139
	4.2.2	Approximating the bin packing problem	143
4.3	Exercises		148
4.4	Bibliographical notes		150

5 Approximation through Randomization — 153

- 5.1 Randomized algorithms for weighted vertex cover 154
- 5.2 Randomized algorithms for weighted satisfiability 157
  - 5.2.1 A new randomized approximation algorithm . . . 157
  - 5.2.2 A 4/3-approximation randomized algorithm 160
- 5.3 Algorithms based on semidefinite programming 162
  - 5.3.1 Improved algorithms for weighted 2-satisfiability . 167
- 5.4 The method of the conditional probabilities 168
- 5.5 Exercises 171
- 5.6 Bibliographical notes 173

6 NP, PCP and Non-approximability Results — 175

- 6.1 Formal complexity theory 175
  - 6.1.1 Turing machines 175
  - 6.1.2 Deterministic Turing machines 178
  - 6.1.3 Nondeterministic Turing machines 180
  - 6.1.4 Time and space complexity 181
  - 6.1.5 NP-completeness and Cook-Levin theorem 184
- 6.2 Oracles 188
  - 6.2.1 Oracle Turing machines 189
- 6.3 The PCP model 190
  - 6.3.1 Membership proofs 190
  - 6.3.2 Probabilistic Turing machines 191
  - 6.3.3 Verifiers and PCP 193
  - 6.3.4 A different view of NP 194
- 6.4 Using PCP to prove non-approximability results 195
  - 6.4.1 The maximum satisfiability problem 196
  - 6.4.2 The maximum clique problem 198
- 6.5 Exercises 200
- 6.6 Bibliographical notes 204

7 The PCP theorem — 207

- 7.1 Transparent long proofs 208
  - 7.1.1 Linear functions 210
  - 7.1.2 Arithmetization 214

Table of contents

			7.1.3	The first PCP result	218
	7.2	Almost transparent short proofs			221
		7.2.1		Low-degree polynomials	222
		7.2.2		Arithmetization (revisited)	231
		7.2.3		The second PCP result	238
	7.3	The final proof			239
		7.3.1		Normal form verifiers	241
		7.3.2		The composition lemma	245
	7.4	Exercises			248
	7.5	Bibliographical notes			249

8 Approximation Preserving Reductions — 253

	8.1	The World of NPO Problems	254
	8.2	AP-reducibility	256
		8.2.1 Complete problems	261
	8.3	NPO-completeness	261
		8.3.1 Other NPO-complete problems	265
		8.3.2 Completeness in exp-APX	265
	8.4	APX-completeness	266
		8.4.1 Other APX-complete problems	270
	8.5	Exercises	281
	8.6	Bibliographical notes	283

9 Probabilistic analysis of approximation algorithms — 287

	9.1	Introduction	288
		9.1.1 Goals of probabilistic analysis	289
	9.2	Techniques for the probabilistic analysis of algorithms . .	291
		9.2.1 Conditioning in the analysis of algorithms	291
		9.2.2 The first and the second moment methods	293
		9.2.3 Convergence of random variables	294
	9.3	Probabilistic analysis and multiprocessor scheduling . . .	296
	9.4	Probabilistic analysis and bin packing	298
	9.5	Probabilistic analysis and maximum clique	302
	9.6	Probabilistic analysis and graph coloring	311
	9.7	Probabilistic analysis and Euclidean TSP	312
	9.8	Exercises	316
	9.9	Bibliographical notes	318

10 Heuristic methods — 321

	10.1	Types of heuristics	322
	10.2	Construction heuristics	325
	10.3	Local search heuristics	329

		10.3.1 Fixed-depth local search heuristics	330
		10.3.2 Variable-depth local search heuristics	336
	10.4	Heuristics based on local search	341
		10.4.1 Simulated annealing	341
		10.4.2 Genetic algorithms	344
		10.4.3 Tabu search	347
	10.5	Exercises	349
	10.6	Bibliographical notes	350
A	**Mathematical preliminaries**		**353**
	A.1	Sets	353
		A.1.1 Sequences, tuples and matrices	354
	A.2	Functions and relations	355
	A.3	Graphs	356
	A.4	Strings and languages	357
	A.5	Boolean logic	357
	A.6	Probability	358
		A.6.1 Random variables	359
	A.7	Linear programming	361
	A.8	Two famous formulas	365
B	**A List of NP Optimization Problems**		**367**
	Bibliography		**471**
	Index		**515**

Preface

IN COMPUTER applications we are used to live with approximation. Various notions of approximation appear, in fact, in many circumstances. One notable example is the type of approximation that arises in numerical analysis or in computational geometry from the fact that we cannot perform computations with arbitrary precision and we have to truncate the representation of real numbers. In other cases, we use to approximate complex mathematical objects by simpler ones: for example, we sometimes represent non-linear functions by means of piecewise linear ones.

The need to solve difficult optimization problems is another reason that forces us to deal with approximation. In particular, when a problem is computationally hard (i.e., the only way we know to solve it is by making use of an algorithm that runs in exponential time), it may be practically unfeasible to try to compute the exact solution, because it might require months or years of machine time, even with the help of powerful parallel computers. In such cases, we may decide to restrict ourselves to compute a solution that, though not being an optimal one, nevertheless is close to the optimum and may be determined in polynomial time. We call this type of solution an *approximate solution* and the corresponding algorithm a *polynomial-time approximation algorithm*.

Most combinatorial optimization problems of great practical relevance are, indeed, computationally intractable in the above sense. In formal terms, they are classified as NP-hard optimization problems. Even if until now no proof is known that NP-hard problems cannot be solved by means of algorithms that run in polynomial time, there is a strong evidence that such algorithms may not exist and we have to make use of approximation algorithms.

This happens for several well known problems: task scheduling problems, routing problems, covering and packing problems, and, more generally, problems that can be formulated as integer or Boolean programming problems. Many of these problems are paradigmatic. Consider, for example, the traveling salesperson problem, the minimum graph coloring problem, the minimum bin packing problem, or the maximum satisfiability problem. Research has always concentrated on such paradigmatic problems, because knowing their computational properties and studying the design of approximation algorithms for them allows us to understand how to solve, in an approximate way, hundreds of different, though related, optimization problems that are met in everyday applications.

This book is about the approximate solution of NP-hard combinatorial optimization problems, one of the domains in computer science whose scientific developments have considerably enriched our understanding of problem complexity and our ability to cope with difficult applications.

Heuristic techniques for solving optimization problems in an approximate way have always been used throughout the history of computing and, in particular, since the early development of operations research. Nevertheless, no precise notion of approximation had been proposed until the late sixties and early seventies, when researchers introduced the notion of approximation with guaranteed *performance ratio*, in order to characterize those situations in which the approximate solution is provably close to an optimal one (e.g., within a factor bounded by a constant or by a slowly growing function of the input size).

Historically, the problems for which the first examples of this kind were proposed have been multiprocessor scheduling and bin packing problems, both related to computer resource management. Since then thousands of approximation algorithms, for all kinds of optimization problems, have been designed and published in the literature. However, two aspects continued to remain unsolved until recently. First of all, while, in some cases, one was able to devise algorithms that could provide approximate solutions that were arbitrarily close to the optimum, in many cases the quality of the approximation was very poor (allowing, for example, a relative error of 50%) and it was not clear whether this was due to our lack in ability to design good approximation algorithms or to some intrinsic, structural properties of the problems. Second, some other problems were resisting to any type of attack and even very weak approximate solutions seemed to be impervious.

Recent results allow us to have now a more clear picture with respect to such aspects. On one side, in fact, extremely good approximation algorithms have been designed for some of the paradigmatic problems (such as the maximum satisfiability and the maximum cut problems), good both with respect to the theoretical performance bounds and with respect to the practical applicability. At the same time a powerful (and surprising, we should say) technique, based on the notion of probabilistically checkable proofs, has been introduced for studying bounds in approximability of problems. By means of such techniques, it has been shown that some of the classical problems, such as the minimum graph coloring and the maximum clique problems, do not allow any approximation algorithm with constant performance ratio, while, for other problems, precise bounds to approximability have been found. In some cases, these bounds meet the degree of approximability that can actually be achieved and in such, unfortunately rare, cases (one of them is, indeed, the maximum satisfiability problem restricted to formulas with exactly three literals per clause), we may say that we possess the best possible approximation algorithm.

The content of the book is organized as follows. In the first chapter, after surveying classical topics concerning the complexity of decision problems, we introduce the basic notions of NP-hard optimization problems, and we give the motivation for studying approximation algorithms for such problems. The second chapter is devoted to basic techniques for the design of approximation algorithms and to the presentation of the first examples of such algorithms. In particular, algorithms based on general approaches such as the greedy approach, local search, and dynamic programming are considered, together with algorithms based on classical methods of discrete optimization, such as linear programming. The examples shown in this chapter testify that the quality of approximation that we may achieve in various cases may be quite different, ranging from constant to polynomial factors.

The third, fourth, and fifth chapters form the main part of the book in terms of positive results. In fact, they contain a large body of approximation algorithms and approximability results for several paradigmatic optimization problems. In Chap. 3, we concentrate on problems that belong to the so-called class APX, i.e., problems that allow polynomial-time approximation algorithms with a performance ratio bounded by a constant. After showing some examples of this kind of problems, we turn to the particular subclasses of APX consisting of problems for which any desired performance ratio can be obtained (i.e., the classes PTAS and FPTAS). In the same chapter, we give the first results that show that, unless P=NP, the approximability classes FPTAS, PTAS, and APX are strictly contained one into the other. Chapter 4 is devoted to problems that cannot be suitably classified in the above seen approximability classes. First, we consider problems that do not belong to APX and for which we can guarantee only a non-constant approximation ratio. Then, we consider problems that, while being outside of PTAS, indeed allow polynomial-time approximation schemes under a somewhat weaker notion of approximability, i.e., the asymptotic approximability. In Chap. 5, more advanced design techniques for approximation algorithms, based on linear and quadratic programming, are presented. Essentially, they are characterized by the use of randomization in rounding real solution to integer ones but, in all cases, we will see that the resulting algorithms can be transformed into deterministic ones without any loss in the approximation performance. Such techniques are, indeed, quite interesting from the application point of view, because their discovery led to very good approximation algorithms for some relevant optimization problems.

The sixth, seventh, and eighth chapters are devoted to the theoretical background and to the proof techniques that are needed to derive non-approximability results. First, in Chap. 6, after reviewing the formal basis of NP-completeness theory, the notion of probabilistically checkable proof is introduced and it is shown how, from the characterization of the class NP in terms of such proofs, a bound to the approximability of the maximum satisfiability problem and the non-approximability of the maximum clique problem can be derived. The proof of the above mentioned characterization (known as the *PCP theorem*) is given in Chap. 7. The result is quite involved and the proofs of some lemmas can be skipped in a first reading, but various ingredients of the main proof, such as the arithmetization of Boolean formulas, the linear function and the polynomial coding schemes, and the notion of transparent proof are important results *per se* and, in our opinion, deserve being known by any researcher working in this field. Finally, in Chap. 8, a formal study of the approximability classes is performed. The chapter is based on the notion of approximation preserving reducibility that, as usual, can be applied both for inferring inapproximability results and for deriving new approximation algorithms.

The last two chapters extend the subject of approximate solution of NP-hard optimization problems in two directions. On one direction, Chap. 9 provides techniques and results for analyzing the expected performance of approximation algorithms. Such performance, which may often be substantially better than the worst case performance analyzed in the previous chapters, is determined by comparing, in probabilistic terms, the quality of the solution achieved by the algorithms with the expected value of the optimal solutions of the given combinatorial problem. On the other direction, Chap. 10 is intended to provide pointers to the literature concerning various popular techniques used in practice in the solution of discrete optimization

Preface

problems. Despite the fact that they do not have guaranteed performance, such techniques may in practice behave quite well. Most of them are elaborated versions of local search (such in the case of tabu search, simulated annealing, and genetic algorithms). In other cases (i.e., branch-and-bound and branch-and-cut), they are based on classical operations research approaches.

One peculiar aspect of this book, that characterizes it at the same time as a text book and a reference book (similarly to [Garey and Johnson, 1979]) is the rich compendium containing information on the approximability of more than 200 problems. The compendium is structured into twelve categories according to subject matters and, for each problem, it contains the best approximation positive result (i.e., the approximability upper bound) and the worst approximation negative result (i.e., the approximability lower bound) for the problem. Besides, each problem entry includes a section of additional comments where approximability results for variations of the problem are mentioned. Finally, a reference to the closest problem appearing in the list published in Garey and Johnson's book is given (whenever it is possible).

Each chapter contains several exercises which range from easy to very difficult ones: the degree of difficulty of an exercise is represented by the number of stars in front of it. Finally, the book is supplemented with an extensive bibliography with cross references towards the compendium entries.

The book assumes that the reader is familiar with basic notions of (a) design and analysis of computer algorithms, (b) discrete mathematics, and (c) linear programming. However, for sake of completeness, the material presented in the various chapters is supplemented by the content of an appendix that is devoted to basic mathematical concepts and notations and to essential notions of linear programming.

The reading of the material presented in the book may be organized along four main threads. The first reading thread regards the design and performance analysis of basic approximation algorithms (BAA) for NP-hard optimization problems and develops through Chaps. 1 to 4: it may be taught in a senior undergraduate course within a curriculum in computer science or in applied mathematics and operations research. The second thread regards advanced techniques for the design and analysis of approximation algorithms (AAA) and develops through Chaps. 5, 9 and 10 with Chaps. 3 and 4 as prerequisites. The third reading thread regards the theoretical study of basic notions of approximation complexity (BAC). It develops through Chaps. 1, 2, 3, and 6 and it is meant for a senior undergraduate audience in computer science and mathematics. Finally, the fourth thread

Preface

is devoted to advanced concepts of approximation complexity and inapproximability results (AAC). It develops through Chaps. 6, 7, 8 and it is meant for a graduate audience in computer science and mathematics with the preceding material (at least Chap. 3) as a prerequisite. In Table 1 the above four reading threads are summarized.

Threads	Chapters									
	1	2	3	4	5	6	7	8	9	10
BAA	X	X	X	X						
AAA			X	X	X				X	X
BAC	X	X	X			X				
AAC			X			X	X	X		

Table 1 Four possible reading threads

This book has benefited from the collaboration of several students and friends who helped us in making the presentation as clean and clear as possible. In particular, we want to thank Paola Alimonti, Gunnar Andersson, Andrea Clementi, Alberto Del Lungo, Gianluca Della Vedova, Lars Engebretsen, Guido Fiorino, Manica Govekar, Roberto Grossi, Stefano Leonardi, Alessandro Panconesi, Gianluca Rossi, Massimo Santini, Riccardo Silvestri, Luca Trevisan. An incredible amount of people supported the compilation of the compendium. The following list certainly may not cover all contributors and we apologize for all omissions: Christoph Albrecht, Edoardo Amaldi, Yossi Azar, Reuven Bar-Yehuda, Bo Chen, Pascal Cherrier, Janka Chlebikova, Keith Edwards, Igor Egorov, Uriel Feige, Michel Goemans, Joachim Gudmundsson, Magnús M. Halldórsson, Refael Hassin, Cor Hurkens, Arun Jagota, David Johnson, Marek Karpinski, Owen Kaser, Sanjeev Khanna, Samir Khuller, Jon Kleinberg, Goran Konjevod, Guy Kortsarz, Matthias Krause, Sven O. Krumke, Ming Li, B. Ma, Prabhu Manyem, Madhav Marathe, Tarcisio Maugeri, Petra Mutzel, Nagi Nahas, Claudio Nogueira de Menezes, Alessandro Panconesi, Dan Pehoushek, Erez Petrank, Jan Plesník, Yuval Rabani, Satish Rao, Eilon Reshef, Günter Rote, Jean-Pieree Seifert, Norbert Sensen, Hadas Shachnai, Martin Skutella, Roberto Solis-Oba, Arie Tamir, Rakesh Vohra, Jürgen Wirtgen, Gerhard Woeginger, Bang Ye Wu, Mihalis Yannakakis. Finally, we are grateful to the following persons who have reported errors of the first printing of the book: Ari Freund, Joachim Gudmundsson, Jonas Holmerin, Jan Johannsen, Amit Kagan, Gunnar W. Klau, Jochen Konemann, Jan Korst, Jan van Leeuwen, Daniel Lehmann, Elena Lodi, Fredrik Manne, Klaus Meer, Takao Nishizeki, Sander van Rijnswou, Ali Tofigh, Heribert Vollmer, Osamu Watanabe.

Preface

Giorgio Ausiello, Giorgio Gambosi, and Alberto Marchetti-Spaccamela are grateful for the support received by the European Union through the ESPRIT LTR Project 20244 *Algorithms and Complexity in Information Technology (ALCOM-IT)*.

We also wish to thank our colleague and friend Jarik Nešetřil for giving us the permission to choose for our front cover one of the most beautiful paintings that he has produced with his coauthor Jiři Načeradskỳ. And above all, we wish to thank Angela, Anna, Linda, Paola, and Susanna for their patience, support, and encouragement.

Finally we wish to remember our beloved friend and colleague Marco Protasi, who died at the age of 48, without having the pleasure to see this work, to which he devoted a considerable effort in the recent years, completed.

Europe, July 1999
G. A.
P.L. C.
G. G.
V. K.
A. M.-S.

Chapter 1

The Complexity of Optimization Problems

IN INTRODUCTORY computer programming courses we learn that computers are used to execute algorithms for the solution of *problems*. Actually, the problems we want to solve by computer may have quite varying characteristics. In general, we are able to express our problem in terms of some *relation* $P \subseteq I \times S$, where I is the set of *problem instances* and S is the set of *problem solutions*. As an alternative view, we can also consider a predicate $p(x,y)$ which is true if and only if $(x,y) \in P$. If we want to analyze the properties of the computations to be performed, it is necessary to consider the characteristics of the sets I, S and of the relation P (or of the predicate p) more closely.

In some cases, we just want to determine if an instance $x \in I$ satisfies a given condition, i.e., whether $\pi(x)$ is verified, where π is a specified (unary) predicate. This happens, for example, when we want to check if a program is syntactically correct, or a certain number is prime, or when we use a theorem prover to decide if a logical formula is a theorem in a given theory. In all these cases, relation P reduces to a function $f : I \mapsto S$, where S is the binary set $S = \{\text{YES}, \text{NO}\}$ (or $S = \{0,1\}$), and we denote our problem as a *decision* (or *recognition*) *problem*. We may also consider *search* problems, where, for any instance $x \in I$, a solution $y \in S$ has to be returned such that $(x,y) \in P$ is verified. This includes such problems as, for example, finding a path in a graph between two given nodes or determining the (unique) factorization of an integer. In other cases, given an instance $x \in I$, we are interested in finding the "best" solution y^* (according to some measure) among all solutions $y \in S$ such that $(x,y) \in P$ is verified. This

Chapter 1

THE COMPLEXITY OF OPTIMIZATION PROBLEMS

is the case when, given a point q and a set of points Q in the plane, we want to determine the point $q' \in Q$ which is nearest to q or when, given a weighted graph, we want to find a Hamiltonian cycle, if any, of minimum cost. Problems of this kind are called *optimization problems* and have occurred frequently in most human activities, since the beginning of the history of mathematics.

The aim of this book is to discuss how and under what conditions optimization problems which are computationally hard to solve can be efficiently approached by means of algorithms which only return "good" (and possibly not the best) solutions. In order to approach this topic, we first have to recall how the efficiency of the algorithms and the computational complexity of problems are measured. The goal of this chapter is, indeed, to introduce the reader to the basic concepts related to the complexity of optimization problems. Since the whole setting of complexity theory is built up in terms of decision problems, we will first discuss these concepts by showing the main results concerning decision problems. In particular, we will introduce the two complexity classes P and NP, whose properties have been deeply investigated in the last decades, and we will briefly discuss their relationship. Subsequently, we will shift our attention to the optimization problem context and we will show how the previously introduced concepts can be adapted to deal with this new framework.

1.1 Analysis of algorithms and complexity of problems

THE MOST direct way to define the efficiency of an algorithm would be to consider how much time (or memory), for any instance, the algorithm takes to run and output a result on a given machine. Such a point of view depends on the structural and technological characteristics of the machines and of their system software. It is possible to see that, as a matter of fact, the cost of the same algorithm on two different machines will differ only by no more than a multiplicative constant, thus making cost evaluations for a certain machine significant also in different computing environments.

As a consequence, the algorithms presented throughout the book will be usually written in a Pascal-like language and will be hypothetically run on an abstract Pascal machine, composed by a control and processing unit, able to execute Pascal statements, and a set of memory locations identified by all variable and constant identifiers defined in the algorithm. Unless otherwise specified, all statements concerning the efficiency of algorithms and the complexity of problem solutions will refer to such a computation model.

1.1.1 Complexity analysis of computer programs

The simplest way to measure the running time of a program in the model chosen above is the *uniform cost* measure and consists in determining the overall number of instructions executed by the algorithm before halting.

This approach to computing the execution cost of a program is natural, but, since it assumes that any operation can be executed in constant time on operands of any size (even arbitrarily large), it may lead to serious anomalies if we consider the possibility that arbitrarily large values can be represented in memory locations. Thus, in practice, the uniform cost model may be applied in all cases where it is implicitly assumed that all memory locations have the same given size and the values involved in any execution of the algorithm are not greater than that size, i.e., any value to be represented during an execution of the algorithm can be stored in a memory location. In all other cases (for example, where a bound on the size of the values involved cannot be assumed) a different cost model, known as *logarithmic cost* model should be used.

The logarithmic cost model is obtained by assigning to all instructions a cost which is a function of the number of bits (or, equivalently for positive integers, of the logarithm) of all values involved. In particular, basic instructions such as additions, comparisons, and assignments are assumed to have cost proportional to the number of bits of the operands, while a cost $O(n \log n)$ for multiplying or dividing two n-bit integers may be adopted.

For example, in the execution of an assignment instruction such as $a := b + 5$, we consider the execution cost equal to $\log |b| + \log 5$. Notice that we did not specify the base of the logarithm, since logarithms in different bases differ only by a multiplicative constant.

Such a cost model avoids the anomalies mentioned above and corresponds to determining the number of operations required to perform arithmetical operations with arbitrary precision on a real computer.

Both approaches can also be applied to evaluate the execution cost in terms of the amount of memory (space) used by the algorithm. In this framework, the uniform cost model will take into account the overall number of distinct memory locations accessed by the algorithm during the computation. On the other hand, according to the logarithmic cost model it should be necessary to consider the number of bits of all values contained in such locations during the computation.

◀ Example 1.1

Consider Program 1.1 for computing $r = x^y$, where $x, y \in \mathbf{N}$. In the uniform cost model, this algorithm has time cost $2 + 3y$, while in the logarithmic cost model, it presents a time cost bounded by the expression $ay \log y + by^2 \log x (\log y + \log \log x) + c$, where a, b, c are suitable constants (see Exercise 1.1). Concern-

Chapter 1

THE COMPLEXITY OF OPTIMIZATION PROBLEMS

Program 1.1: Exponentiation

input Nonnegative integers x, y;
output Nonnegative integer $r = x^y$;
begin
  $r := 1$;
  while $y \neq 0$ **do**
  begin
    $r := r * x$;
    $y := y - 1$
  end;
  return r
end.

ing the amount of memory used, the resulting uniform cost is 3 (the variables x, y, r), while the logarithmic cost is $\log y + (y+1) \log x$.

In the following, we will refer to the uniform cost model in all (time and space) complexity evaluations performed. This is justified by the fact that the numbers used by the algorithms described in this book will be sufficiently small with respect to the length of their inputs. Before going into greater detail, we need to make clear some aspects of cost analysis that may help us to make the evaluation easier and more expressive and to introduce a suitable notation.

Worst case analysis. The first aspect we have to make clear is that, for most applications, at least as a first step of the analysis, we are not interested in determining the precise running time or space of an algorithm for all particular values of the input. Our concern is rather more in determining the behavior of these execution costs as the input size grows. However, instances of the same size may anyway result in extremely different execution costs. It is well known, for example, that sorting algorithms may have a different behavior, from the point of view of the running time, if they have to sort an array whose elements are randomly chosen or an array of partially sorted items. For such reasons, in order to specify the performance of an algorithm on inputs of size n, we determine the cost of applying the algorithm on the *worst case* instance of that size, that is on the instance with highest execution cost.

In contrast with worst case analysis, in several situations a different approach may be taken (known as *average case* analysis), consisting of determining the average cost of running the algorithm on all instances of size n, assuming that such instances occur with a specified (usually uniform) probability distribution.

Worst case analysis is widely used since it provides us with the certainty that, in any case, the given algorithm will perform its task within the established time bound. Besides, it is often easier to obtain, while the average case analysis may require complex mathematical calculations and, moreover, has validity limited by the probabilistic assumptions that have been made on the input distribution.

Input size. As stated above, we are interested in expressing the execution costs as a (growing) function of the size of the instance of the problem to be solved. But, how to measure that size? In Example 1.1 the running time of the algorithm was provided as a function of the input values x, y. Even if this seemed a reasonable choice in the case of the computation of an integer function, conventionally, a different *input size measure* is adopted: the *length* or *size* of the input, that is the number of digits (possibly bits) needed to present the specific instance of the problem. This conventional choice is motivated by the fact that problems may be defined over quite heterogeneous data: integers or sequences of integers, sets, graphs, geometric objects, arrays and matrices, etc. In all cases, in order to be submitted as input to a computer algorithm, the data will have to be presented in the form of a character string over a suitable finite alphabet (e.g., the binary alphabet $\{0,1\}$). It is therefore natural to adopt the length of such a string as a universal input size measure good for all kinds of problems.

Thus we assume there exists an *encoding scheme* which is used to describe any problem instance in a string of characters over some alphabet. Even if different encoding schemes usually result in different strings (and input sizes) for the same instance, it is important to observe that for a wide class of *natural* encoding schemes (i.e., encoding schemes that do not introduce an artificially redundant description) the input sizes of the same instance are not too different from each other. This is expressed by saying that for any pair of natural encoding schemes e_1, e_2 and for any problem instance x, the resulting strings are polynomially related, that is there exist two polynomials p_1, p_2 such that $|e_1(x)| \leq p_1(|e_2(x)|)$ and $|e_2(x)| \leq p_2(|e_1(x)|)$, where $|e_i(x)|$ denotes the length of string $e_i(x)$, for $i = 1, 2$.

In the following, for any problem instance x, we will denote with the same symbol x the string resulting by any natural encoding of the instance itself. In general, we will denote with the same symbol both an object and its encoding under any encoding scheme.

Section 1.1
ANALYSIS OF ALGORITHMS AND COMPLEXITY OF PROBLEMS

◀ Example 1.2

Let us consider the problem of determining if an integer $x \in Z^+$ is a prime number. A trivial algorithm which tries all possible divisors d, $1 \leq d \leq \sqrt{x}$, can perform (if x is prime) a number of steps proportional to $\sqrt{x}$. If we use the natural encoding scheme which represents integer x as a binary string of length $n = |x| \approx \log x$, the

Chapter 1

THE COMPLEXITY OF OPTIMIZATION PROBLEMS

execution cost is instead proportional to $2^{n/2}$. Even if the two values are equal, the second one gives greater evidence of the complexity of the algorithm, which grows exponentially with the size of the input. Actually, at the current state of knowledge, it is not yet known whether any algorithm for primality testing with a polynomial number of steps exists. Notice that using a unary base encoding (an unnatural scheme) would result in an evaluation of the execution cost as the square root of the instance size.

Asymptotic analysis of algorithms. If we go back to the analysis of the running time of Program 1.1, we realize the following two facts. First, the expressions contain constants, which represent both the cost of some instructions (such as during the initialization phase of an algorithm) whose execution does not depend on the particular instance and the fact that single instructions, on real machines, present execution costs which are not unitary but depend on technological characteristics of the machine (and system software) and would have been extremely difficult to determine precisely. Second, the expressions themselves are upper bounds and not precise evaluations of the computation costs.

For the above two reasons, by making use of the standard *O-notation* (see Appendix A), we may as well describe the execution costs of the algorithm by saying that its execution time *asymptotically* grows no more than $O(y)$ in the uniform cost model and $O(y^2 \log x(\log y + \log \log x))$ in the logarithmic cost model.

More generally, let us see how the notation used to express the asymptotic behavior of functions can be used to describe the running time (or space) of an algorithm.

Let us denote as $\hat{t}_\mathcal{A}(x)$ the running time of algorithm $\mathcal{A}$ on input x. The *worst case running time* of $\mathcal{A}$ is then given by

$$t_\mathcal{A}(n) = \max\{\hat{t}_\mathcal{A}(x) \mid \forall x : |x| \leq n\}.$$

Similarly, let us denote as $\hat{s}_\mathcal{A}(x)$ the running space of algorithm $\mathcal{A}$ on input x. The *worst case running space* of $\mathcal{A}$ is then given by

$$s_\mathcal{A}(n) = \max\{\hat{s}_\mathcal{A}(x) \mid \forall x : |x| \leq n\}.$$

Definition 1.1 ▶ *Algorithm complexity bounds*

We say that algorithm $\mathcal{A}$

1. *has complexity (upper bound) $O(g(n))$ if $t_\mathcal{A}(n)$ is $O(g(n))$;*

2. *has complexity (lower bound) $\Omega(g(n))$ if $t_\mathcal{A}(n)$ is $\Omega(g(n))$;*

3. *has complexity (exactly) $\Theta(g(n))$ if $t_\mathcal{A}(n)$ is $\Theta(g(n))$.*

Similar definitions can be introduced for the space complexity.

By using the preceding notation, if an algorithm performs a number of steps bounded by $an^2 + b\log n$ to process an input of size n, we say that its complexity is $O(n^2)$. If, moreover, we are able to prove that for any n sufficiently large there exists an input instance of size n on which the algorithm performs at least cn^2 steps, we say that its complexity is also $\Omega(n^2)$. In such a case we also say that the complexity of the algorithm is $\Theta(n^2)$.

◂ Example 1.3

Let us again consider the analysis of Program 1.1 in Example 1.1. As we have already seen, the running time of this algorithm under logarithmic cost measure is $ay\log y + by^2\log x(\log y + \log\log x) + c$. As noticed above, the term $by^2\log x(\log y + \log\log x)$ dominates all others. Let $n = \log x + \log y$ be the overall input size: a proper expression for the asymptotic cost analysis as a function of the input size would be $O(n^2 2^{2n})$.

The typical procedure to perform a complexity analysis of an algorithm requires deciding, case by case, on the following issues:

1. How to measure the execution cost. Usually, in the case of evaluating the running time, this is done by determining a *dominant* operation, that is an operation which is executed at least as often as any other operation in the algorithm, thus characterizing the asymptotic complexity of the algorithm. More formally, if the asymptotic complexity of the algorithm is $\Theta(g(n))$, a dominant operation is any operation whose contribution to the cost is $\Theta(g(n))$.

   Typically, in a sorting algorithm a dominant operation is the comparison of two elements, in matrix multiplication a dominant operation may be assumed to be the multiplication of elements, etc.

2. Which cost measure we want to adopt. As said above, the logarithmic cost model is more precise, but the simpler uniform cost model can be applied if we make the assumption that all values involved in the algorithm execution are upper bounded by some value (usually some function of the input size).

3. How to measure the input size, that is what characteristic parameter of the input instance is the one whose growth towards infinity determines the asymptotic growth of the computation cost. In this case, typical examples are the number of graph nodes and edges in graph algorithms, the number of rows and columns of matrices in algebraic algorithms, etc.

◂ Example 1.4

Let us consider a simple version of the well known *Insertion Sort* algorithm (see Program 1.2). Suppose that an array of size n is given and that we want to sort

Section 1.1

ANALYSIS OF ALGORITHMS AND COMPLEXITY OF PROBLEMS

Chapter 1

THE COMPLEXITY OF OPTIMIZATION PROBLEMS

Program 1.2: Insertion Sort

input Array $A[1,\ldots,n]$ of integers;
output Array A sorted in increasing order;
begin
    for $i := 1$ **to** $n-1$ **do**
    begin
        $j := i+1$;
        while $j \geq 2$ **and** $A[j] < A[j-1]$ **do**
        begin
            swap $A[j]$ and $A[j-1]$;
            $j := j-1$
        end
    end;
    return A
end.

it in increasing order. Without loss of generality let us assume that the range of the elements is the finite interval $\{1,\ldots,M\}$, where M is a constant. Here the input size is characterized by the number n of elements in the array. In the uniform model we may simply take into consideration comparison operations. Since the algorithm contains two loops each of which is repeated at most n times, the running time is clearly bounded by $O(n^2)$. On the other side, it is not difficult to observe that the number of comparisons performed by the algorithm is indeed $n(n-1)/2$ in the worst case, that is when the array is ordered in decreasing order. Hence, the algorithm also has a lower bound $\Omega(n^2)$ on the running time.

Regarding the running space of the algorithm, it is immediate to see that an $\Omega(n)$ lower bound holds, due to the need for accessing all values in the input instance. The algorithm accesses such values plus a constant number of additional locations, thus resulting in a $O(n)$ upper bound.

1.1.2 Upper and lower bounds on the complexity of problems

Suppose we have to solve a problem and we have an algorithm whose time complexity is $O(g(n))$. According to the following definition, we say that $O(g(n))$ is an upper bound to the time complexity of the given problem.

Definition 1.2 ▶ *Problem complexity upper bound*
Given a problem $\mathcal{P}$ we say that the time complexity upper bound of $\mathcal{P}$ is $O(g(n))$ if there exists an algorithm for $\mathcal{P}$ whose running time is $O(g(n))$.

In other words, the complexity upper bound for a problem provides information on the amount of time which is asymptotically sufficient to solve the problem. Knowing the upper bound does not mean that a precise knowledge of the complexity of the problem is available. In fact, other more efficient algorithms with smaller running times may exist, and we may just be unaware of their existence.

A more precise characterization of the time complexity of a problem is achieved when we are also able to find a time complexity lower bound, that is we are able to establish how much time is anyhow needed (asymptotically) to solve the problem, no matter what algorithms are used. Knowing a lower bound for a problem provides important information on the intrinsic difficulty of solving the problem, especially in the case that the lower bound is sufficiently close to (or even coincides with) the upper bound.

◀ **Definition 1.3**
Problem complexity bounds

Given a problem $\mathcal{P}$ we say that the time complexity lower bound of $\mathcal{P}$ is $\Omega(g(n))$ if any algorithm for $\mathcal{P}$ has a running time $\Omega(g(n))$, and that the time complexity of $\mathcal{P}$ is $\Theta(g(n))$ if its upper bound is $O(g(n))$ and its lower bound is $\Omega(g(n))$.

Establishing a complexity lower bound for a problem is a hard task, since it requires stating (and proving) a property that has to hold for all (known and unknown) algorithms that solve the problem.

In the case of sorting, by means of an information theoretic argument, it can be proved that, in terms of comparisons, the time complexity lower bound is $\Omega(n \log n)$, that is any algorithm for sorting that uses only comparisons requires $\Omega(n \log n)$ comparisons. Since the time complexity upper bound for sorting is also $O(n \log n)$, this means that, in terms of comparisons, the time complexity of sorting is exactly determined to be $\Theta(n \log n)$.

Unfortunately, such a precise characterization is not achieved in all cases. For most problems of great practical interest (the satisfiability of a Boolean formula in the propositional calculus, the multiplication of two matrices, the decomposition of a number into prime factors, etc.) the time complexity has not presently been precisely determined. In particular, this is the case for all optimization problems which form the subject of this book. Despite this difficulty, a classification of (both decision and optimization) problems according to their complexity can be established in formal terms. Before examining how the complexity of optimization problems can be defined, let us recall some basic notions about the complexity of decision problems.

1.2 Complexity classes of decision problems

A PROBLEM $\mathcal{P}$ is called a *decision problem* if the set $I_\mathcal{P}$ of all instances of $\mathcal{P}$ is partitioned into a set $Y_\mathcal{P}$ of *positive instances* and a set $N_\mathcal{P}$ of

Chapter 1

THE COMPLEXITY OF OPTIMIZATION PROBLEMS

negative instances and the problem asks, for any instance $x \in I_\mathcal{P}$, to verify whether $x \in Y_\mathcal{P}$.

Actually, since any algorithm for $\mathcal{P}$ also can receive inputs that do not correspond to legal instances of $\mathcal{P}$, the set of inputs to such an algorithm is partitioned in $Y_\mathcal{P}$, $N_\mathcal{P}$ and $D_\mathcal{P}$, the set of inputs not corresponding to instances of $\mathcal{P}$.

We assume that any algorithm for a decision problem is able to return a YES or NO answer, for example by printing it (see Fig. 1.1).

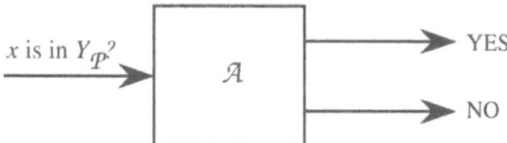

Figure 1.1
Solving a decision problem

Definition 1.4 ▶ *A decision problem $\mathcal{P}$ is solved by an algorithm $\mathcal{A}$ if the algorithm halts for every instance $x \in I_\mathcal{P}$, and returns YES if and only if $x \in Y_\mathcal{P}$. We also say that set $Y_\mathcal{P}$ is recognized by $\mathcal{A}$. Moreover, we say that $\mathcal{P}$ is solved in time $t(n)$ (space $s(n)$) if the time (space) complexity of $\mathcal{A}$ is $t(n)$ ($s(n)$).*

Problem solution

Notice that, if an input string belongs to $N_\mathcal{P} \cup D_\mathcal{P}$, then $\mathcal{A}$ may either halt returning NO, halt without returning anything, or never halt.

There are several reasons why the main concepts of computational complexity have been stated in terms of decision problems instead of computation of functions.

First of all, given an encoding scheme, any decision problem can be seen, independently from its specific kind of instances (graphs, numbers, strings, etc.), as the problem of discriminating among two sets of strings: encodings of instances in $Y_\mathcal{P}$ and strings in $N_\mathcal{P} \cup D_\mathcal{P}$. This allows a more formal and unifying treatment of all of them as language recognition problems. In general, for any decision problem $\mathcal{P}$, we will denote the corresponding language as $L_\mathcal{P}$.

Secondly, decision problems have a YES or NO answer. This means that in the complexity analysis we do not have to pay attention to the cost of producing the result. All the costs are strictly of a computational nature and they have nothing to do with the size of the output or the time needed to return it (for example by printing it).

The main aim of the theory of computational complexity is the characterization of collections of problems with respect to the computing resources (time, space) needed to solve them: such collections are denoted as *complexity classes*.

Let us now define some important complexity classes (it is intended that Defs. 1.2 and 1.3 can be applied –*mutatis mutandis*– to space complexity).

> **Problem 1.1: Satisfying truth assignment**
> INSTANCE: CNF formula $\mathcal{F}$ on a set V of Boolean variables, truth assignment on V, $f : V \mapsto \{\text{TRUE}, \text{FALSE}\}$.
> QUESTION: Does f satisfy $\mathcal{F}$?

Section 1.2
COMPLEXITY CLASSES OF DECISION PROBLEMS

For any function $f(n)$, let TIME$(f(n))$ (SPACE$(f(n))$) *be the collection of decision problems which can be solved with a time (space) complexity $O(f(n))$.*

◀ Definition 1.5
Complexity classes

Often, we are more interested in classifying problems according to less refined classes. Thus, we may define the following complexity classes:

1. the class of all problems solvable in time proportional to a polynomial of the input size: $\text{P} = \cup_{k=0}^{\infty} \text{TIME}(n^k)$;

2. the class of all problems solvable in space proportional to a polynomial of the input size: $\text{PSPACE} = \cup_{k=0}^{\infty} \text{SPACE}(n^k)$;

3. the class of all problems solvable in time proportional to an exponential of the input size: $\text{EXPTIME} = \cup_{k=0}^{\infty} \text{TIME}(2^{n^k})$.

Clearly, P $\subseteq$ PSPACE: indeed, any algorithm cannot access more memory locations than the number of computing steps performed. Moreover, under suitable assumptions on the computation model, it is also easy to prove that PSPACE $\subseteq$ EXPTIME (see, for example, Exercise 6.11 which refers to the Turing machine model of computation). Whether these inclusions are strict (that is, P $\subset$ PSPACE and PSPACE $\subset$ EXPTIME), are open problems in complexity theory. The only separation result already known concerning the above defined classes is P $\subset$ EXPTIME.

The class P is traditionally considered as a reasonable threshold between tractable and intractable problems.

Let us consider SATISFYING TRUTH ASSIGNMENT, that is, Problem 1.1. This problem is in P. In fact, there exists a trivial decision algorithm which first plugs in the truth values given by f and then checks whether each clause is satisfied. The time complexity of this algorithm is clearly linear in the input size.

◀ Example 1.5

Let us consider SATISFIABILITY, that is, Problem 1.2. It is easy to show that this problem belongs to PSPACE. Consider the algorithm that tries every possible truth assignment on V and for each assignment computes the value of $\mathcal{F}$. As soon as the value of $\mathcal{F}$ is TRUE, the algorithm returns YES. If $\mathcal{F}$ is false for all truth assignments, the algorithm returns NO. The space needed by the algorithm to check one truth assignment is clearly polynomial in the size of the input. Since

◀ Example 1.6

11

Chapter 1

THE COMPLEXITY
OF OPTIMIZATION
PROBLEMS

> **Problem 1.2: Satisfiability**
>
> INSTANCE: CNF formula $\mathcal{F}$ on a set V of Boolean variables.
>
> QUESTION: Is $\mathcal{F}$ satisfiable, i.e., does there exist a truth assignment $f: V \mapsto \{\text{TRUE}, \text{FALSE}\}$ which satisfies $\mathcal{F}$?

this space can be recycled and at most polynomial space is required for the overall control, this algorithm has indeed polynomial-space complexity.

Example 1.7 ▶ Given a sequence of Boolean variables $V = \{v_1, v_2, \ldots, v_n\}$ and a CNF Boolean formula $\mathcal{F}$ on V, let the corresponding quantified Boolean formula be defined as

$$Q_1 v_1 Q_2 v_2 Q_3 v_3 \ldots Q_n v_n \mathcal{F}(v_1, v_2, \ldots, v_n),$$

where, for any $i = 1, \ldots, n$, $Q_i = \exists$ if i is odd and $Q_i = \forall$ if i is even. We can see that QUANTIFIED BOOLEAN FORMULAS, that is Problem 1.3, belongs to PSPACE by modifying the PSPACE algorithm for SATISFIABILITY above. We try all truth assignments on V by assigning values to the variables from left to right, and after each assignment computing the value of the expression to the right. For example, assume that $v_1 \ldots v_{i-1}$ have been assigned values $z_1, \ldots, z_{i-1}$, respectively. Besides, let r_1 be the value of the expression

$$Q_{i+1} v_{i+1} \ldots Q_n v_n \mathcal{F}(z_1, z_2, \ldots, z_{i-1}, \text{TRUE}, v_{i+1}, \ldots, v_n)$$

and r_0 the value of the expression

$$Q_{i+1} v_{i+1} \ldots Q_n v_n \mathcal{F}(z_1, z_2, \ldots, z_{i-1}, \text{FALSE}, v_{i+1}, \ldots, v_n).$$

Then the value of expression

$$Q_i v_i Q_{i+1} v_{i+1} \ldots Q_n v_n \mathcal{F}(z_1, z_2, \ldots, z_{i-1}, v_i, v_{i+1}, \ldots, v_n)$$

is $r_0 \wedge r_1$ if i is even (i.e., $Q_i = \forall$) and $r_0 \vee r_1$ if i is odd (i.e., $Q_i = \exists$). This algorithm can easily be implemented in order to compute the value of the whole expression using space polynomial in the input size (see Exercise 1.6).

Note that no polynomial-time algorithm is known for the problems considered in the last two examples; indeed, it is widely believed that no such algorithm can exist. However, as we will see in the next subsection, these problems present rather different complexity properties, that make them paradigmatic problems in different complexity classes.

1.2.1 The class NP

The classification of problems that we have provided so far is indeed too coarse with respect to the need for characterizing the complexity of several problems of great practical interest. For example, even though we have

Section 1.2

COMPLEXITY CLASSES OF DECISION PROBLEMS

> **Problem 1.3: Quantified Boolean formulas**
> INSTANCE: CNF Boolean formula $\mathcal{F}$ on a set $V = \{v_1, v_2, \ldots, v_n\}$ of Boolean variables.
>
> QUESTION: Is the quantified Boolean formula
>
> $$\exists v_1 \forall v_2 \exists v_3 \ldots Q v_n \, \mathcal{F}(v_1, v_2, \ldots, v_n)$$
>
> true, where $Q = \exists$ if n is odd, otherwise $Q = \forall$?

shown that both QUANTIFIED BOOLEAN FORMULAS and SATISFIABILITY belong to PSPACE, it is possible to see that these two problems present quite different characteristics if we consider their behavior when we analyze, instead of the cost of *solving* a problem, the cost of *verifying* that a given mathematical object is indeed the solution of a problem instance.

According to the definition of decision problem given in the preceding section, given an instance x of a decision problem $\mathcal{P}$ (encoded as a string of symbols), the corresponding solution consists of a single binary value 1 or 0 (YES or NO, TRUE or FALSE), expressing the fact that x belongs to the set $Y_\mathcal{P}$ of positive instances or not. Actually, in most cases, determining that a problem instance x belongs to $Y_\mathcal{P}$ should also be supported by deriving some object $y(x)$ (string, set, graph, array, etc.) which depends on x, whose characteristics are stated in the problem instance and whose existence is what is asked for in the decision problem. In this case we would call $y(x)$ a *constructive solution* of $\mathcal{P}$.

◀ Example 1.8

Let us consider again SATISFIABILITY. Any positive instance of this problem has (at least) one associated solution $y(x)$ represented by a truth assignment f which satisfies the Boolean formula $\mathcal{F}$.

Given a decision problem $\mathcal{P}$ and an instance $x \in Y_\mathcal{P}$, verifying whether a string of characters σ is the description of a constructive solution $y(x)$ of $\mathcal{P}$ is often a much easier task than solving the decision problem itself. In the case of SATISFIABILITY, whereas deciding whether there exists a truth assignment which satisfies a CNF Boolean formula is a complex task (no algorithm is known which solves this problem in time polynomial in the length of the input), verifying whether a string represents a satisfying truth assignment can be done in time linear in the size of the instance (see Example 1.5).

The cost of verifying problem solutions can be used as an additional measure to characterize problem complexity. In order to do this, the concept of a *nondeterministic algorithm* has been introduced to provide a com-

13

Chapter 1

THE COMPLEXITY
OF OPTIMIZATION
PROBLEMS

Program 1.3: Nondeterministic SAT

input CNF Boolean formula $\mathcal{F}$ over a set V of Boolean variables;
output YES if $\mathcal{F}$ is satisfiable;
begin
  for each v in V **do**
  begin
    guess $y \in \{0,1\}$;
    if $y = 0$ **then** $f(v) :=$ FALSE **else** $f(v) :=$ TRUE
  end;
  if f satisfies $\mathcal{F}$ **then return** YES **else return** NO
end.

putational framework coherent to the one assumed when problem solving is considered.

A nondeterministic algorithm is an algorithm which, apart from all usual constructs, can execute commands of the type "**guess** $y \in \{0,1\}$". Such an instruction means that y can take as its value either 0 or 1. Essentially, a nondeterministic algorithm has the additional option of "guessing" a value, in particular of guessing a continuation (in a finite set of possible continuations) of the computation performed so far. Thus, while the computations performed by the (deterministic) algorithms considered so far have a linear structure (since at any time there is only one possible next step of the computation), nondeterministic computations can be described by a tree-like structure (called *computation tree*), where **guess** instructions correspond to branching points.

Notice also that a deterministic algorithm presents one outcome for each input instance, while a nondeterministic one will have many different outcomes, in correspondence to different sequences of guesses. If, in particular, we consider nondeterministic algorithms for acceptance or rejection of strings, some sequences of guesses will lead the algorithm to return YES, some to return NO.

Definition 1.6 ▶ *Given a decision problem $\mathcal{P}$, a nondeterministic algorithm $\mathcal{A}$ solves $\mathcal{P}$ if,*
Nondeterministic problem *for any instance $x \in I_\mathcal{P}$, $\mathcal{A}$ halts for any possible guess sequence and $x \in Y_\mathcal{P}$*
solution *if and only if there exists at least one sequence of guesses which leads the algorithm to return the value* YES.

Example 1.9 ▶ Program 1.3 is a nondeterministic algorithm for the SATISFIABILITY problem. As one can see, the algorithm essentially guesses a candidate truth assignment among the $2^{|V|}$ possible ones and then verifies whether the guess has been successful. The behavior of the algorithm with input the Boolean formula containing the two clauses $v_1 \vee v_2 \vee \bar{v}_3$ and $\bar{v}_1 \vee \bar{v}_2 \vee v_3$ is shown in Fig. 1.2 where the computation tree is represented. Each path in this tree denotes a possible sequence of guesses of the algorithm: the input is accepted if and only if at least one path produces a truth assignment that satisfies the formula. Notice that, since the verification phase can be performed in polynomial time (see Example 1.5) and the sequence of guesses is polynomially long, the nondeterministic algorithm itself takes polynomial time.

Section 1.2

COMPLEXITY CLASSES OF DECISION PROBLEMS

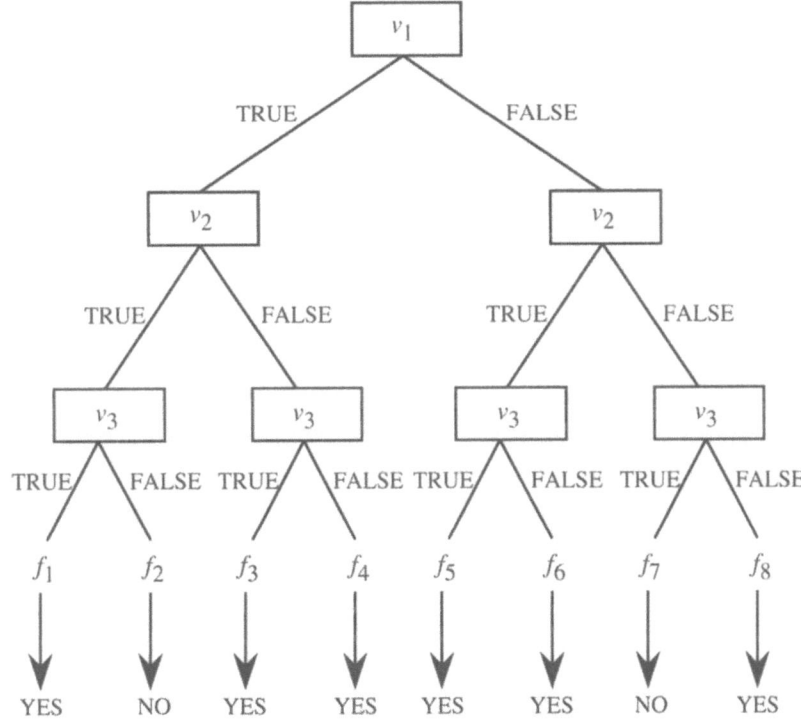

Figure 1.2
A nondeterministic algorithm for SATISFIABILITY with input $v_1 \vee v_2 \vee \bar{v}_3, \bar{v}_1 \vee \bar{v}_2 \vee v_3$

A nondeterministic algorithm $\mathcal{A}$ solves a decision problem $\mathcal{P}$ in time complexity $t(n)$ if, for any instance $x \in I_\mathcal{P}$ with $|x| = n$, $\mathcal{A}$ halts for any possible guess sequence and $x \in Y_\mathcal{P}$ if and only if there exists at least one sequence of guesses which leads the algorithm to return the value YES in time at most $t(n)$.

◀ Definition 1.7
Complexity of nondeterministic algorithm

Without loss of generality, we may assume that all (accepting and non-accepting) computations performed by a nondeterministic algorithm that solves a decision problem in time $t(n)$ halt: we may in fact assume that the algorithm is equipped with a step counter which halts the execution when more than $t(n)$ steps have been performed.

Moreover, without loss of generality, we may also consider a simplified model of a polynomial-time nondeterministic algorithm which performs a polynomially long sequence of **guess** operations at the beginning of its

Chapter 1

THE COMPLEXITY OF OPTIMIZATION PROBLEMS

execution. Such a sequence can be seen as a unique overall guess of all the outcomes of the at most polynomial number of guesses in any computation, that is, we assume that all nondeterministic choices to be performed during any computation are guessed at the beginning of the computation.

Definition 1.8 ▶ *For any function $f(n)$, let $\text{NTIME}(f(n))$ be the collection of decision problems which can be solved by a nondeterministic algorithm in time $O(f(n))$.*

Nondeterministic complexity classes

We may then define the class NP as the class of all decision problems which can be solved in time proportional to a polynomial of the input size by a nondeterministic algorithm, i.e., $\text{NP} = \cup_{k=0}^{\infty} \text{NTIME}(n^k)$. By the above considerations, this is equivalent to the class of all decision problems whose constructive solutions can be verified in time polynomial in the input size.

It is easy to see that $\text{P} \subseteq \text{NP}$, since a conventional (deterministic) algorithm is just a special case of a nondeterministic one, in which no guess is performed. Moreover, several problems in EXPTIME (and probably not belonging to P) are also in NP. For all such problems a solution to the corresponding constructive problem, if any, has to be found in an exponentially large search space. Therefore, their complexity arises from the need to generate and search such an exponentially large space, whereas, given any element of the search space, checking whether it is a solution or not is a relatively simpler task. A paradigmatic example of such problems is SATISFIABILITY (see Example 1.5).

On the other side, in the case of QUANTIFIED BOOLEAN FORMULAS, there is no known method of exploiting nondeterminism for obtaining a solution in polynomial time. In fact, this problem is believed not to belong to NP, and that, therefore, $\text{NP} \subset \text{PSPACE}$. Observe that in the case of QUANTIFIED BOOLEAN FORMULAS the natural constructive solution for a positive instance would be a subtree of the tree of all truth assignments: this subtree alternatively contains nodes with branching factor 1 and nodes with branching factor 2. Then, this solution has size exponential in the input length and cannot be guessed in polynomial time.

Before closing this section we point out that nondeterministic algorithms present an asymmetry between instances in $Y_\mathcal{P}$ and instances in $N_\mathcal{P} \cup D_\mathcal{P}$. In fact, while in order for an instance to be recognized as belonging to $Y_\mathcal{P}$ there must exist *at least one* accepting computation, the same instance is recognized to be in $N_\mathcal{P} \cup D_\mathcal{P}$ if and only if *all* computations are non-accepting. Even in the case that all computations halt, this makes the conditions to be verified inherently different, since they correspond to checking an existential quantifier in one case and a universal quantifier in the other case.

> **Problem 1.4: Falsity**
>
> INSTANCE: A CNF formula $\mathcal{F}$ on a set V of Boolean variables.
>
> QUESTION: Is it true that there does not exist any truth assignment $f : V \mapsto \{\text{TRUE}, \text{FALSE}\}$ which satisfies $\mathcal{F}$?

Let us consider the FALSITY problem, that is, Problem 1.4. Clearly, this problem has the same set of instances of SATISFIABILITY, but its set of YES instances corresponds to the set of NO instances of SATISFIABILITY, and vice versa. Checking whether an instance is a YES instance is a process that, in the case of SATISFIABILITY, can stop as soon as a satisfying truth assignment has been found, while, in the case of FALSITY, it has to analyze every truth assignment in order to be sure that no satisfying one exists. The relation between SATISFIABILITY and FALSITY is an example of complementarity between decision problems.

For any decision problem $\mathcal{P}$ the complementary problem $\mathcal{P}^c$ is the decision problem with $I_{\mathcal{P}^c} = I_{\mathcal{P}}$, $Y_{\mathcal{P}^c} = N_{\mathcal{P}}$, and $N_{\mathcal{P}^c} = Y_{\mathcal{P}}$. ◀ Definition 1.9
Complementary problem

The class of all decision problems which are complementary to decision problems in NP is called co–NP (e.g., FALSITY belongs to co–NP).

1.3 Reducibility among problems

REDUCTIONS ARE a basic tool for establishing relations among the complexities of different problems. Basically, a reduction from a problem $\mathcal{P}_1$ to a problem $\mathcal{P}_2$ presents a method for solving $\mathcal{P}_1$ using an algorithm for $\mathcal{P}_2$. Notice that, broadly speaking, this means that $\mathcal{P}_2$ is at least as difficult as $\mathcal{P}_1$ (since we can solve the latter if we are able to solve the former), provided the reduction itself only involves "simple enough" calculations.

1.3.1 Karp and Turing reducibility

Different types of reducibilities can be defined, as a consequence of the assumption on how a solution to problem $\mathcal{P}_2$ can be used to solve $\mathcal{P}_1$.

A decision problem $\mathcal{P}_1$ is said to be Karp-reducible *(or many-to-one reducible) to a decision problem $\mathcal{P}_2$ if there exists an algorithm $\mathcal{R}$ which* ◀ Definition 1.10
Karp reducibility

Chapter 1

THE COMPLEXITY OF OPTIMIZATION PROBLEMS

given any instance $x \in I_{\mathcal{P}_1}$ of $\mathcal{P}_1$, transforms it into an instance $y \in I_{\mathcal{P}_2}$ of $\mathcal{P}_2$ in such a way that $x \in Y_{\mathcal{P}_1}$ if and only if $y \in Y_{\mathcal{P}_2}$. In such a case, $\mathcal{R}$ is said to be a Karp-reduction *from* $\mathcal{P}_1$ *to* $\mathcal{P}_2$ and we write $\mathcal{P}_1 \leq_m \mathcal{P}_2$. If both $\mathcal{P}_1 \leq_m \mathcal{P}_2$ and $\mathcal{P}_2 \leq_m \mathcal{P}_1$ we say that $\mathcal{P}_1$ and $\mathcal{P}_2$ are Karp-equivalent (*in symbols* $\mathcal{P}_1 \equiv_m \mathcal{P}_2$).

As a consequence of this definition, if a decision problem $\mathcal{P}_1$ is Karp-reducible to a decision problem $\mathcal{P}_2$, then for any instance x of $\mathcal{P}_1$, x is a positive instance if and only if the transformed instance y is a positive instance for $\mathcal{P}_2$.

It should then be clear that, if a reduction $\mathcal{R}$ from $\mathcal{P}_1$ to $\mathcal{P}_2$ exists and if an algorithm $\mathcal{A}_2$ is known for $\mathcal{P}_2$, an algorithm $\mathcal{A}_1$ for $\mathcal{P}_1$ can be obtained as follows (see Fig. 1.3):

1. given $x \in I_{\mathcal{P}_1}$, apply $\mathcal{R}$ to x and obtain $y \in I_{\mathcal{P}_2}$;

2. apply $\mathcal{A}_2$ to y: if $\mathcal{A}_2$ returns YES then return YES, otherwise ($\mathcal{A}_2$ returns NO) return NO.

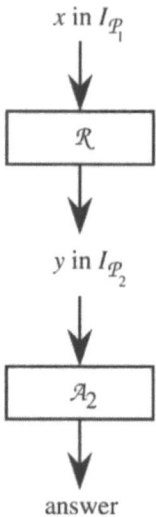

Figure 1.3
How to use a Karp reduction

As a particular case, we define the *polynomial-time Karp-reducibility* by saying that $\mathcal{P}_1$ is polynomially Karp-reducible to $\mathcal{P}_2$ if and only if $\mathcal{P}_1$ is Karp-reducible to $\mathcal{P}_2$ and the corresponding reduction $\mathcal{R}$ is a polynomial-time algorithm. In such a case we write $\mathcal{P}_1 \leq_m^p \mathcal{P}_2$ (again, if both $\mathcal{P}_1 \leq_m^p \mathcal{P}_2$ and $\mathcal{P}_2 \leq_m^p \mathcal{P}_1$ then $\mathcal{P}_1 \equiv_m^p \mathcal{P}_2$).

In the case $\mathcal{P}_1 \leq_m^p \mathcal{P}_2$, efficiently solving $\mathcal{P}_2$, i.e. in polynomial time, implies that also $\mathcal{P}_1$ can be solved efficiently, that is, $\mathcal{P}_2 \in P$ implies $\mathcal{P}_1 \in P$. On the contrary, if it is proved that $\mathcal{P}_1$ cannot be solved in polynomial time, then also $\mathcal{P}_2$ cannot be solved efficiently, that is, $\mathcal{P}_1 \notin P$ implies $\mathcal{P}_2 \notin P$.

> **Problem 1.5: {0,1}-Linear Programming**
>
> INSTANCE: Set of variables $Z = \{z_1, \ldots, z_n\}$ with domain $\{0,1\}$, set I of linear inequalities on Z with integer coefficients.
>
> QUESTION: Does there exist any solution of I, that is, any assignment of values to variables in Z such that all inequalities are verified?

◀ Example 1.10

Let us consider {0,1}-LINEAR PROGRAMMING, that is, Problem 1.5. It is not hard to see that SATISFIABILITY$\leq_m^p$ {0,1}-LINEAR PROGRAMMING. In fact, any instance $x_{SAT} = (V, \mathcal{F})$ of SATISFIABILITY can be reduced to some instance $x_{LP} = (Z, I)$ of {0,1}-LINEAR PROGRAMMING as follows. Let $l_{j_1} \vee l_{j_2} \vee \cdots \vee l_{j_{n_j}}$ be the j-th clause in $\mathcal{F}$: a corresponding inequality $\zeta_{j_1} + \zeta_{j_2} + \cdots + \zeta_{j_{n_j}} \geq 1$ is derived for x_{LP}, where $\zeta_{j_k} = z_i$ if $t_{j_k} = v_i$ (for some $v_i \in V$) and $\zeta_{j_k} = (1 - z_i)$ if $t_{j_k} = \bar{v}_i$ (for some $v_i \in V$). It is easy to see that any truth assignment $f : V \mapsto \{\text{TRUE}, \text{FALSE}\}$ satisfies $\mathcal{F}$ if and only if all inequalities in I are verified by the value assignment $f' : Z \mapsto \{0,1\}$ such that $f'(z_i) = 1$ if and only if $f(v_i) = \text{TRUE}$. Moreover, the reduction is clearly polynomial-time computable.

Let us now introduce a different, more general, type of reducibility, which can be applied also to problems which are not decision problems and that, basically, models the possibility, in writing a program to solve a problem, to use subprograms for solving another problem as many times as it is required. Such a situation is formalized in complexity theory by the use of *oracles*.

Let $\mathcal{P}$ be the problem of computing a (possibly multivalued) function $f : I_{\mathcal{P}} \mapsto S_{\mathcal{P}}$. An oracle *for problem $\mathcal{P}$ is an abstract device which, for any $x \in I_{\mathcal{P}}$, returns a value $f(x) \in S_{\mathcal{P}}$. It is assumed that the oracle may return the value in just one computation step.*

◀ Definition 1.11
Oracle

Let $\mathcal{P}_1$ be the problem of computing a (possibly multivalued) function $g : I_{\mathcal{P}_1} \mapsto S_{\mathcal{P}_1}$. Problem $\mathcal{P}_1$ is said to be Turing-reducible *(see Fig. 1.4) to a problem $\mathcal{P}_2$ if there exists an algorithm $\mathcal{R}$ for $\mathcal{P}_1$ which queries an oracle for $\mathcal{P}_2$. In such a case, $\mathcal{R}$ is said to be a* Turing-reduction *from $\mathcal{P}_1$ to $\mathcal{P}_2$ and we write $\mathcal{P}_1 \leq_T \mathcal{P}_2$.*

◀ Definition 1.12
Turing reducibility

Again, if $\mathcal{P}_1 \leq_T \mathcal{P}_2$ and $\mathcal{P}_2 \leq_T \mathcal{P}_1$ we have that $\mathcal{P}_1$ and $\mathcal{P}_2$ are Turing-equivalent ($\mathcal{P}_1 \equiv_T \mathcal{P}_2$).

Clearly, Karp-reducibility is just a particular case of Turing-reducibility, corresponding to the case when $\mathcal{P}_1$ and $\mathcal{P}_2$ are both decision problems, the oracle for $\mathcal{P}_2$ is queried just once, and $\mathcal{R}$ returns the same value answered by the oracle. In general, Karp-reducibility is weaker than Turing-

Chapter 1
THE COMPLEXITY OF OPTIMIZATION PROBLEMS

reducibility. For example, for any decision problem $\mathcal{P}$, $\mathcal{P}$ is always Turing-reducible to $\mathcal{P}^c$ (and vice versa), while the same does not hold for Karp-reducibility (see Bibliographical notes).

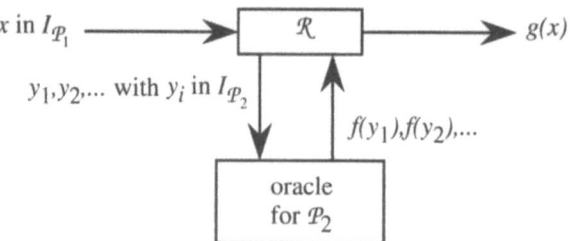

Figure 1.4
Turing-reduction from $\mathcal{P}_1$ to $\mathcal{P}_2$

As in the case of Karp-reducibility, we may also introduce a polynomial-time Turing-reducibility $\leq_T^p$, by saying that $\mathcal{P}_1 \leq_T^p \mathcal{P}_2$ if and only if $\mathcal{P}_1 \leq_T \mathcal{P}_2$ and the Turing-reduction $\mathcal{R}$ is polynomial-time computable with respect to the input size.

Again, in the case $\mathcal{P}_1 \leq_T^p \mathcal{P}_2$, efficiently solving $\mathcal{P}_2$ (in polynomial time) implies that also $\mathcal{P}_1$ can be solved efficiently, that is, $\mathcal{P}_2 \in P$ implies $\mathcal{P}_1 \in P$. On the contrary, if it is proved that $\mathcal{P}_1$ cannot be solved in polynomial time, then also $\mathcal{P}_2$ cannot be solved efficiently, that is, $\mathcal{P}_1 \notin P$ implies $\mathcal{P}_2 \notin P$.

Example 1.11 ▶ Given two CNF Boolean formulas $\mathcal{F}_1$ and $\mathcal{F}_2$, the EQUIVALENT FORMULAS problem consists of deciding whether $\mathcal{F}_1$ is *equivalent* to $\mathcal{F}_2$, that is, if, for any assignment of values, $\mathcal{F}_1$ is satisfied if and only if $\mathcal{F}_2$ is satisfied.

It is easy to verify that SATISFIABILITY can be solved in polynomial time by a deterministic Turing machine with oracle EQUIVALENT FORMULAS.

Indeed, let $\mathcal{F}$ be a Boolean formula in conjunctive normal form. To decide whether $\mathcal{F}$ is satisfiable it is sufficient to check whether $\mathcal{F}$ is equivalent to $\mathcal{F}_{\bar{f}} = x \wedge \bar{x}$ (observe that $\mathcal{F}_{\bar{f}}$ cannot be satisfied by any assignment of values). If this is the case, $\mathcal{F}$ is not satisfiable, otherwise it is satisfiable. Clearly, this check can be done by querying the oracle EQUIVALENT FORMULAS about the instance formed by $\mathcal{F}$ and $\mathcal{F}_{\bar{f}}$. If the oracle answers positively, $\mathcal{F}$ is not satisfiable, otherwise it is satisfiable.

Let us now introduce two definitions valid for any complexity class and type of reducibility.

Definition 1.13 ▶ A complexity class C is said to be closed *with respect to a reducibility* $\leq_r$
Complexity class closure   *if, for any pair of decision problems* $\mathcal{P}_1$, $\mathcal{P}_2$ *such that* $\mathcal{P}_1 \leq_r \mathcal{P}_2$, $\mathcal{P}_2 \in C$ *implies* $\mathcal{P}_1 \in C$.

Definition 1.14 ▶ For any complexity class C, a decision problem $\mathcal{P} \in C$ is said to be com-
Complete problem   plete *in* C *(equivalently, C-complete) with respect to a reducibility* $\leq_r$, *if, for any other decision problem* $\mathcal{P}_1 \in C$, $\mathcal{P}_1 \leq_r \mathcal{P}$.

The above definition immediately implies that, for any two problems $\mathcal{P}_1$ and $\mathcal{P}_2$ which are C-complete with respect to a reducibility $\leq_r$, $\mathcal{P}_1 \equiv_r \mathcal{P}_2$.

It is easy to see that, for any pair of complexity classes C_1 and C_2 such that $C_1 \subset C_2$ and C_1 is closed with respect to a reducibility $\leq_r$, any C_2-complete problem $\mathcal{P}$ belongs to $C_2 - C_1$. In fact, for any problem $\mathcal{P}_1 \in C_2 - C_1$, $\mathcal{P}_1 \leq_r \mathcal{P}$ by the completeness of $\mathcal{P}$. Then, by the closure of C_1, $\mathcal{P} \in C_1$ would imply $\mathcal{P}_1 \in C_1$, which is a contradiction. This property suggests that if $C_1 \subseteq C_2$ then the best approach to determining whether $C_1 \subset C_2$ or $C_1 = C_2$ is to study the complexity of C_2-complete problems. In fact, if for any C_2-complete problem $\mathcal{P}$ we prove that $\mathcal{P} \in C_1$, then we have that $C_1 = C_2$; conversely, if we have that $\mathcal{P} \notin C_1$ then $C_1 \subset C_2$.

Section 1.3

REDUCIBILITY AMONG PROBLEMS

1.3.2 NP-complete problems

The above definitions turn out to be particularly relevant when applied to the class NP in order to define complete problems in NP with respect to Karp-reductions.

A decision problem $\mathcal{P}$ is said to be NP-complete *if it is complete in* NP *with respect to $\leq_m^p$, that is, $\mathcal{P} \in$* NP *and, for any decision problem $\mathcal{P}_1 \in$* NP*, $\mathcal{P}_1 \leq_m^p \mathcal{P}$.*

◀ Definition 1.15
NP-complete problem

Since $P \subseteq NP$, by the closure of NP with respect to $\leq_m^p$ (see Exercise 1.10), $P \neq NP$ if and only if for each NP-complete problem $\mathcal{P}$, $\mathcal{P} \notin P$. At the moment, it is not known whether polynomial-time algorithms exist for NP-complete problems and only superpolynomial time algorithms are currently available, even if no superpolynomial lower bound on the time complexity has been proved for any of these problems.

SATISFIABILITY is NP-complete. We have already seen in Example 1.9 that this problem is in NP. Showing its NP-completeness is quite a harder task, which we will present in Chap. 6.

◀ Example 1.12

By the definition of NP-completeness and by the transitive property of polynomial-time Karp reductions (see Exercise 1.7), any problem $\mathcal{P}$ can be proved to be NP-complete by first showing a polynomial-time nondeterministic algorithm for $\mathcal{P}$ (which proves that $\mathcal{P} \in$ NP) and then providing a polynomial-time Karp-reduction from some other problem $\mathcal{P}'$, already known to be NP-complete, to $\mathcal{P}$.

{0,1}-LINEAR PROGRAMMING is NP-complete. Indeed, this problem can be shown to be in NP by considering a polynomial-time nondeterministic algorithm similar to the one considered for SATISFIABILITY (that is, Program 1.3).

◀ Example 1.13

Chapter 1

THE COMPLEXITY OF OPTIMIZATION PROBLEMS

The NP-completeness can be shown by providing a polynomial-time Karp-reduction $\mathcal{R}$ from any other problem which is known to be NP-complete. In particular, a polynomial-time Karp-reduction from SATISFIABILITY has been shown in Example 1.10.

As said above, it is not known whether NP-complete problems can be solved efficiently or not. However, due to the fact that even after a wide effort from the whole computer science community, polynomial-time algorithms for such problems (now numbering in the thousands) are still lacking, NP-completeness is considered as a sign that the considered problem is intractable.

1.4 Complexity of optimization problems

LET US now turn our attention to optimization problems. As already observed, most of the basic concepts of complexity theory have been introduced with reference to decision problems. In order to extend the theoretical setting to optimization problems we need to reexamine such concepts and consider how they apply to optimization.

1.4.1 Optimization problems

The study of the cost of solving optimization problems is probably one of the most relevant practical aspects of complexity theory, due to the importance of such problems in many application areas.

In order to extend complexity theory from decision problems to optimization problems, suitable definitions have to be introduced. Also, the relationships between the complexity of optimization problems and the complexity of decision problems have to be discussed.

First, let us provide a formal definition of an optimization problem.

Definition 1.16 ▶ *An optimization problem $\mathcal{P}$ is characterized by the following quadruple of objects $(I_\mathcal{P}, \text{SOL}_\mathcal{P}, m_\mathcal{P}, \text{goal}_\mathcal{P})$, where:*
Optimization problem

1. *$I_\mathcal{P}$ is the set of instances of $\mathcal{P}$;*

2. *$\text{SOL}_\mathcal{P}$ is a function that associates to any input instance $x \in I_\mathcal{P}$ the set of feasible solutions of x;*

3. *$m_\mathcal{P}$ is the measure function, defined for pairs (x,y) such that $x \in I_\mathcal{P}$ and $y \in \text{SOL}_\mathcal{P}(x)$. For every such pair (x,y), $m_\mathcal{P}(x,y)$ provides a*

> **Problem 1.6:** Minimum Vertex Cover
> **INSTANCE:** Graph $G = (V, E)$.
> **SOLUTION:** A subset of nodes $U \subseteq V$ such that $\forall (v_i, v_j) \in E$, $v_i \in U$ or $v_j \in U$.
> **MEASURE:** $|U|$.

Section 1.4

COMPLEXITY OF OPTIMIZATION PROBLEMS

positive integer which is the value of the feasible solution y;[1]

4. $\text{goal}_\mathcal{P} \in \{\text{MIN}, \text{MAX}\}$ *specifies whether $\mathcal{P}$ is a maximization or a minimization problem.*

Given an input instance x, we denote by $\text{SOL}^*_\mathcal{P}(x)$ the set of *optimal solutions* of x, that is the set of solutions whose value is optimal (minimum or maximum depending on whether $\text{goal}_\mathcal{P} = \text{MIN}$ or $\text{goal}_\mathcal{P} = \text{MAX}$). More formally, for every $y^*(x)$ such that $y^*(x) \in \text{SOL}^*_\mathcal{P}(x)$:

$$m_\mathcal{P}(x, y^*(x)) = \text{goal}_\mathcal{P}\{v \mid v = m_\mathcal{P}(x, z) \land z \in \text{SOL}_\mathcal{P}(x)\}.$$

The value of any optimal solution $y^*(x)$ of x will be denoted as $m^*_\mathcal{P}(x)$. In the following, whenever the problem we are referring to is clear from the context we will not use the subscript that explicitly refers to the problem $\mathcal{P}$.

Given a graph $G = (V, E)$, the MINIMUM VERTEX COVER problem is to find a vertex cover of minimum size, that is a minimum set U of nodes such that, for each edge $(v_i, v_j) \in E$, either $v_i \in U$ or $v_j \in U$ (that is to say, at least one among v_i and v_j must belong to U). Formally, this problem is defined as follows:

◄ Example 1.14

1. $I = \{G = (V, E) \mid G \text{ is a graph}\}$;
2. $\text{SOL}(G) = \{U \subseteq V \mid \forall (v_i, v_j) \in E [v_i \in U \lor v_j \in U]\}$;
3. $m(G, U) = |U|$;
4. $\text{goal} = \text{MIN}$.

In the following, however, whenever a new problem will be introduced, we will make use of the more intuitive notation shown in Problem 1.6.

It is important to notice that any optimization problem $\mathcal{P}$ has an associated decision problem $\mathcal{P}_D$. In the case that $\mathcal{P}$ is a minimization problem, $\mathcal{P}_D$

---

[1] Notice that, in practice, for several problems the measure function is defined to have values in $\mathbb{Q}$. It is however possible to transform any such optimization problem into an equivalent one satisfying our definition.

Chapter 1

THE COMPLEXITY
OF OPTIMIZATION
PROBLEMS

Problem 1.7: Vertex cover

INSTANCE: Graph $G = (V, E)$, $K \in \mathbf{N}$.

QUESTION: Does there exist a vertex cover on G of size $\leq K$, that is a subset $U \subseteq V$ such that $|U| \leq K$ and $\forall (u, v) \in E$, $u \in U$ or $v \in U$?

---

asks, for some $K > 0$, for the existence of a feasible solution y of instance x with value $m(x, y) \leq K$. Analogously, if $\mathcal{P}$ is a maximization problem, the associated decision problem asks, given $K > 0$, for the existence of a feasible solution y of x with $m(x, y) \geq K$. Moreover, an evaluation problem $\mathcal{P}_E$ can also be associated with $\mathcal{P}$, which asks for the value of an optimal solution of $\mathcal{P}$.

More precisely, we can say that the definition of an optimization problem $\mathcal{P}$ naturally leads to the following three different problems, corresponding to different ways of approaching its solution.

Constructive Problem ($\mathcal{P}_C$) – Given an instance $x \in I$, derive an optimal solution $y^*(x) \in \text{SOL}^*(x)$ and its measure $m^*(x)$.

Evaluation Problem ($\mathcal{P}_E$) – Given an instance $x \in I$, derive the value $m^*(x)$.

Decision Problem ($\mathcal{P}_D$) – Given an instance $x \in I$ and a positive integer $K \in Z^+$, decide whether $m^*(x) \geq K$ (if goal = MAX) or whether $m^*(x) \leq K$ (if goal = MIN). If goal = MAX, the set $\{(x, K) \mid x \in I \land m^*(x) \geq K\}$ (or $\{(x, K) \mid x \in I \land m^*(x) \leq K\}$ if goal = MIN) is called the *underlying language* of $\mathcal{P}$.

Example 1.15 ▶ The decision problem relative to MINIMUM VERTEX COVER is Problem 1.7. This problem can be shown to be NP-complete (see Exercise 1.12).

Notice that, for any optimization problem $\mathcal{P}$, the corresponding decision problem $\mathcal{P}_D$ is not harder than the constructive problem $\mathcal{P}_C$. In fact, to answer $\mathcal{P}_D$ on instance x it is sufficient to run some algorithm for $\mathcal{P}_C$, thus obtaining the optimal solution $y^*(x)$ together with its value $m(x, y^*(x))$; then, it is sufficient to check whether $m(x, y^*(x)) \leq K$, in the minimization case, or whether $m(x, y^*(x)) \geq K$, in the maximization case.

Example 1.16 ▶ Let us consider MINIMUM TRAVELING SALESPERSON (TSP), that is, Problem 1.8. An instance of this problem can also be represented by a complete graph $G = (V, E)$ of n vertices with positive weights on the edges (the vertices represent the cities and the weight of an edge is equal to the distance between the corresponding pair of cities). Feasible solutions of the problem are then subsets I of

> **Problem 1.8:** Minimum Traveling Salesperson
>
> INSTANCE: Set of cities $\{c_1,\ldots,c_n\}$, $n \times n$ matrix D of distances in Z^+.
>
> SOLUTION: A (traveling salesperson) tour of all cities, that is, a permutation $\{c_{i_1},\ldots,c_{i_n}\}$.
>
> MEASURE: $\sum_{k=1}^{n-1} D(i_k, i_{k+1}) + D(i_n, i_1)$.

> **Problem 1.9:** Minimum Graph Coloring
>
> INSTANCE: Graph $G = (V, E)$.
>
> SOLUTION: An assignment $f : V \mapsto \{1,\ldots,K\}$ of K colors to the vertices of G such that $\forall (u,v) \in E$, $f(u) \neq f(v)$.
>
> MEASURE: K.

Section 1.4

COMPLEXITY OF OPTIMIZATION PROBLEMS

edges such that the graph (V, I) is a cycle. In Fig. 1.5 it is shown an instance of MINIMUM TRAVELING SALESPERSON in both ways (observe that in this case the distance matrix is symmetric): the thick edges in the graph denote the optimal tour. Since the problem is an optimization problem it cannot belong to NP, but it has a corresponding decision problem that is NP-complete (see Bibliographical notes).

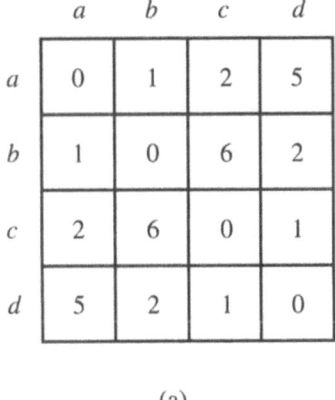

(a)

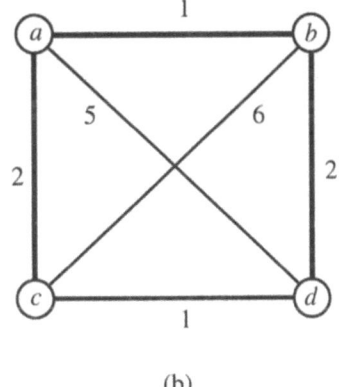

(b)

Figure 1.5
An instance of MINIMUM TRAVELING SALESPERSON represented (a) as a matrix and (b) as a graph

The problem MINIMUM GRAPH COLORING is defined as follows (see Problem 1.9): given a graph $G = (V, E)$, find a vertex coloring with a minimum number of colors, that is a partition of V in a minimum number of classes $\{V_1,\ldots,V_K\}$ such that for any edge $(u, v) \in E$, u and v are in different classes. For example, in the left side of Fig. 1.6 a sample graph is given with a coloring using 4 colors.

◂ Example 1.17

Note that such a coloring is not optimal: a minimum coloring for the same graph is shown in the right side of the same figure. As above, the problem cannot belong to NP, but the corresponding decision problem can be shown to be NP-complete (see Exercise 1.14).

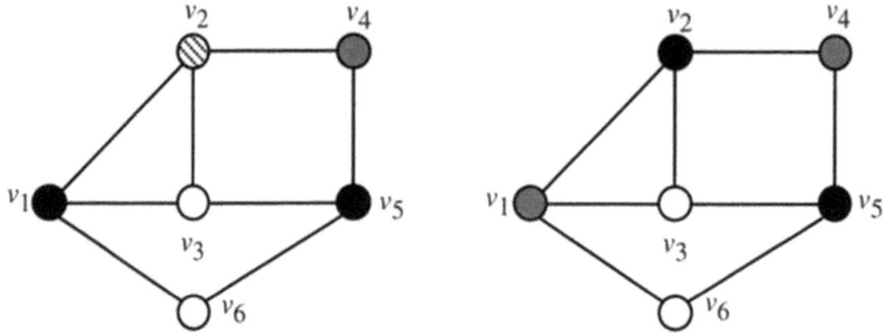

Figure 1.6
A coloring of a graph with 4 colors and an optimal coloring requiring 3 colors

1.4.2 PO and NPO problems

In order to characterize the complexity of optimization problems and to classify such problems accordingly, various approaches may be followed.

The most direct point of view is to consider the time needed for solving the given problem and to extend to optimization problems the theory developed for decision problems.

Of course, the most relevant issue is to characterize optimization problems $\mathcal{P}$ which can be considered tractable, i.e. such that there exists a polynomial-time computable algorithm $\mathcal{A}$ that, for any instance $x \in I$ returns an optimal solution $y \in \text{SOL}^*(x)$, together with its value $m^*(x)$. This means that our main concern will be the study of constructive versions of optimization problems: this will indeed be the approach that will be followed throughout the book (even though we will usually avoid to explicitly return the measure of the computed solution).

Example 1.18 ▶ The problem MINIMUM PATH is defined as follows (see Problem 1.10): given a graph $G = (V, E)$ and a pair of nodes v_s and v_d, derive the shortest path from v_s to v_d. This problem can be proved to be tractable by showing the polynomial-time Program 1.4 which constructs all minimum paths from all nodes in V to v_d. The algorithm visits the graph in a breadth-first order and constructs a tree with root v_d by setting in any node a pointer π to its parent node. An example of the behavior of the algorithm is shown in Fig. 1.7 where the lower part represents the resulting tree obtained with input the graph depicted in the upper part and $v_d = v_1$: the arrows denote the pointers to the parents while the values in the square brackets represent the visiting order.

> **Problem 1.10: Minimum Path**
> INSTANCE: Graph $G = (V, E)$, two nodes $v_s, v_d \in V$.
> SOLUTION: A path $(v_s = v_{i_1}, v_{i_2}, \ldots, v_{i_k} = v_d)$ from v_s to v_d.
> MEASURE: The number k of nodes in the path.

> **Program 1.4: Shortest path by breadth-first search**

input Graph $G = (V, E)$ and two nodes $v_s, v_d \in V$;
output Minimum path from v_s to v_d in G;
begin
   Enqueue(v_d, Q);
   Mark node v_d as visited;
   $\pi[v_d] := $ **nil**;
   while Q is not empty **do**
   begin
     $v := $ Dequeue(Q);
     for each node u adjacent to v **do**
       if u has not been already visited **then**
       begin
         Enqueue(u, Q);
         Mark u as visited;
         $\pi[u] := v$;
       end
   end;
   if v_s has been visited **then**
     return the path from v_s to v_d by following pointers π;
end.

Section 1.4
COMPLEXITY OF OPTIMIZATION PROBLEMS

In this book we are mainly interested in those optimization problems which stand on the borderline between tractability and intractability and which, by analogy with the NP decision problems, are called NPO problems.

An optimization problem $\mathcal{P} = (I, \text{SOL}, m, \text{goal})$ belongs to the class NPO *if the following holds:* ◂ **Definition 1.17** *Class NPO*

1. *the set of instances I is recognizable in polynomial time;*

2. *there exists a polynomial q such that, given an instance $x \in I$, for any $y \in \text{SOL}(x)$, $|y| \leq q(|x|)$ and, besides, for any y such that $|y| \leq q(|x|)$, it is decidable in polynomial time whether $y \in \text{SOL}(x)$;*

3. *the measure function m is computable in polynomial time.*

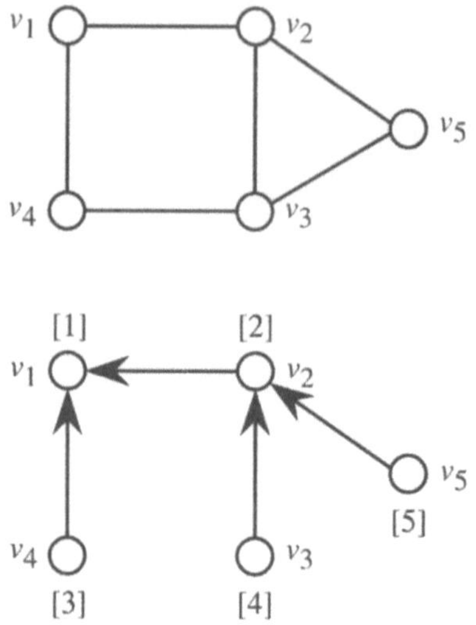

Chapter 1

THE COMPLEXITY
OF OPTIMIZATION
PROBLEMS

Figure 1.7
An example of application of
Program 1.4

Example 1.19 ▶ MINIMUM VERTEX COVER belongs to NPO since:

1. the set of instances (any undirected graph) is clearly recognizable in polynomial time;

2. since any feasible solution is a subset of the set of nodes its size is smaller than the size of the instance; moreover, testing that a subset $U \subseteq V$ is a feasible solution requires testing whether any edge in E is incident to at least one node in U, which can be clearly performed in polynomial time;

3. given a feasible solution U, the measure function (size of U) is trivially computable in polynomial time.

Even if not explicitly introduced, underlying the definition of NPO problem there is a nondeterministic computation model. This is formally stated by the following result which basically shows that the class NPO is the natural optimization counterpart of the class NP.

Theorem 1.1 ▶ *For any optimization problem $\mathcal{P}$ in* NPO, *the corresponding decision problem $\mathcal{P}_D$ belongs to* NP.

PROOF Assume that $\mathcal{P}$ is a maximization problem (the proof in the minimization case is similar). Given an instance $x \in I$ and an integer K, we can solve $\mathcal{P}_D$ by performing the following nondeterministic algorithm. In time $q(|x|)$, where q is a polynomial, any string y such that $|y| \leq q(|x|)$ is guessed. Afterwards the string is tested for membership in SOL(x) in polynomial time. If the test is positive, $m(x,y)$ is computed (again in polynomial time) and if $m(x,y) \geq K$ the answer YES is returned. Otherwise (i.e., either y is not a feasible solution or $m(x,y) < K$), the answer NO is returned.

QED

The relationship between classes NP and NPO, which holds in the case of nondeterministic computations, can be translated, in the case of deterministic algorithms, by the following definition, that introduces the class of NPO problems whose constructive versions are efficiently solvable .

Section 1.4

COMPLEXITY OF OPTIMIZATION PROBLEMS

An optimization problem $\mathcal{P}$ belongs to the class PO *if it is in* NPO *and there exists a polynomial-time computable algorithm $\mathcal{A}$ that, for any instance $x \in I$, returns an optimal solution $y \in \text{SOL}^*(x)$, together with its value $m^*(x)$.*

◀ Definition 1.18
Class PO

MINIMUM PATH belongs to PO. Indeed, it is easy to see that this problem satisfies the three conditions of Def. 1.17. Moreover, as we have seen in Example 1.18, MINIMUM PATH is solvable in polynomial time.

◀ Example 1.20

Practically all interesting optimization problems belong to the class NPO: in addition to all optimization problems in PO (such as MINIMUM PATH), several other problems of great practical relevance belong to NPO. Beside MINIMUM VERTEX COVER, MINIMUM TRAVELING SALESPERSON, and MINIMUM GRAPH COLORING, many other graph optimization problems, most packing and scheduling problems, and the general formulations of integer and binary linear programming belong to NPO but are not known to be in PO since no polynomial-time algorithm for them is known (and it is likely that none exists). For some of them, not only does the exact solution appear to be extremely complex to obtain but, as we will see, even to achieve good approximate solutions may be computationally hard.

Indeed, for all the optimization problems in NPO − PO the intrinsic complexity is not precisely known. Just as with the decision problems in NP − P, no polynomial-time algorithms for them have been found but no proof of intractability is known either. In fact, the question "PO = NPO?" is strictly related to the question "P = NP?" since it can be proved that the two questions are equivalent in the sense that a positive answer to the first would imply a positive answer to the second and vice versa. In order to establish such relationships between the two questions we have to make more precise how the complexity of decision problems may be related to the complexity of optimization problems.

1.4.3 NP-hard optimization problems

In order to assess the intrinsic complexity of optimization problems we may think of proceeding as in the case of decision problems. In that case the relative complexity of problems was established by making use

Chapter 1

THE COMPLEXITY
OF OPTIMIZATION
PROBLEMS

of polynomial-time Karp reductions and of the related notion of NP-completeness. Unfortunately, Karp reductions are defined for decision problems and cannot be applied to optimization problems. In this case, instead, we can make use of the polynomial-time Turing reducibility (see Sect. 1.3).

Definition 1.19 ▶ *An optimization problem $\mathcal{P}$ is called NP-hard if, for every decision problem $\mathcal{P}' \in \text{NP}$, $\mathcal{P}' \leq_T^p \mathcal{P}$, that is, $\mathcal{P}'$ can be solved in polynomial time by an algorithm which uses an oracle that, for any instance $x \in I_\mathcal{P}$, returns an optimal solution $y^*(x)$ of x along with its value $m^*_\mathcal{P}(x)$.*[2]

NP-hard problem

Thus, a problem is NP-hard if it is at least as difficult to solve (in terms of time complexity and apart from a polynomial-time reduction) as any problem in NP. As a consequence of the definition of NP-completeness, in order to prove that an optimization problem $\mathcal{P}$ is NP-hard it is enough to show that $\mathcal{P}' \leq_T^p \mathcal{P}$ for an NP-complete problem $\mathcal{P}'$. Besides, if a problem $\mathcal{P}$ is NP-hard, $\mathcal{P} \in \text{PO}$ implies P=NP.

Indeed, many interesting problems are NP-hard. For example, this happens with all problems in NPO whose underlying language is NP-complete.

Theorem 1.2 ▶ *Let a problem $\mathcal{P} \in \text{NPO}$ be given; if the underlying language of $\mathcal{P}$ is NP-complete then $\mathcal{P}$ is NP-hard.*

PROOF   Clearly, the solution of the decision problem could be obtained for free if an oracle would give us the solution of the constructive optimization problem.

QED

From the preceding result we may easily derive a first consequence concerning the complexity of NPO problems. In fact it turns out that if we could solve the problem $\mathcal{P}$ of Theorem 1.2 in polynomial time we could also solve its underlying decision problem in polynomial time. Hence, unless P = NP, no problem in NPO whose underlying language is NP-complete (e.g., MINIMUM TRAVELING SALESPERSON) can belong to PO and the next result follows.

Corollary 1.3 ▶ *If P $\neq$ NP then PO $\neq$ NPO.*

The fact that an NPO problem $\mathcal{P}$ is NP-hard naturally places $\mathcal{P}$ at the highest level of complexity in the class NPO, like what happens in NP with the NP-complete problems.

Example 1.21 ▶ As a consequence of Exercises 1.12 and 1.14 and of Example 1.16, we have that MINIMUM VERTEX COVER, MINIMUM GRAPH COLORING, and MINIMUM TRAVELING SALESPERSON are NP-hard.

---

[2] More formally, and according to the definition of Turing reducibility (i.e., Def. 1.12), we should write $\mathcal{P}' \leq_T^p \text{SOL}^*_\mathcal{P}$ instead of $\mathcal{P}' \leq_T^p \mathcal{P}$.

1.4.4 Optimization problems and evaluation problems

As we have seen, the results of the preceding section have provided us with some information on the relative complexity of decision problems and optimization problems in some particular cases.

Let us now see the mutual relations existing among the various versions of optimization problems (decision, evaluation, and constructive problems) in a more systematic manner.

First of all, let us state more formally three results that have already been mentioned in the preceding section.

For any problem $\mathcal{P} \in \text{NPO}$, $\mathcal{P}_D \equiv_T^p \mathcal{P}_E \leq_T^p \mathcal{P}_C$. ◀ Theorem 1.4

The proofs of $\mathcal{P}_D \leq_T^p \mathcal{P}_E$ and of $\mathcal{P}_E \leq_T^p \mathcal{P}_C$ are immediate. Regarding $\mathcal{P}_E \leq_T^p \mathcal{P}_D$, due to the properties of NPO problems we have that, for any $x \in I$ and for any $y \in \text{SOL}(x)$, the range of possible values of $m(x,y)$ is bounded by $M = 2^{p(|x|)}$ for some polynomial p. Hence, by applying binary search, the evaluation problem can be solved by at most $\log M = p(|x|)$ queries to the oracle $\mathcal{P}_D$.

PROOF

QED

Section 1.4

COMPLEXITY OF OPTIMIZATION PROBLEMS

Less clear is the possibility of deriving the constructive solution when knowing only the solution of the evaluation problem.

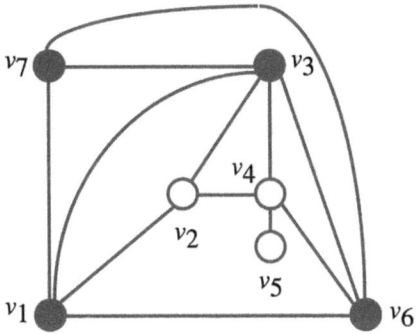

Figure 1.8
A clique of size 4

Let us consider MAXIMUM CLIQUE, that is, Problem 1.11. An example of a clique of size 4 is shown in Fig. 1.8. Given the solution of the evaluation problem for MAXIMUM CLIQUE we can construct an optimal solution in polynomial time as described by Program 1.5 where, given a node v, $G(v)$ denotes the subgraph induced by v and the set of nodes adjacent to v and $G^-(v)$ denotes the subgraph induced by the set of nodes adjacent to v.

◀ Example 1.22

The cost of running Program 1.5 is given by the following recurrence relation as a function T of the number of nodes in the graph:

1. $T(1)$ is $O(1)$,

Chapter 1

THE COMPLEXITY
OF OPTIMIZATION
PROBLEMS

> **Problem 1.11: Maximum Clique**
>
> INSTANCE: Graph $G = (V,E)$.
>
> SOLUTION: A clique in G, i.e., a subset of nodes $U \subseteq V$ such that $\forall (v_i, v_j) \in U \times U, (v_i, v_j) \in E$ or $v_i = v_j$.
>
> MEASURE: $|U|$.

> **Program 1.5: Maxclique**

input Graph $G = (V,E)$;
output Maximum clique in G;
begin
    $k :=$ MAXIMUM CLIQUE$_E(G)$;
    if $k = 1$ **then return** any node in V;
    Find node v such that MAXIMUM CLIQUE$_E(G(v)) = k$;
    return $\{v\} \cup$ Maxclique$(G^-(v))$
end.

2. $T(n) = (n+1) + T(n-1)$

(recall that querying the oracle MAXIMUM CLIQUE$_E$ costs only one step). The solution of the recurrence relation is $O(n^2)$.

Unfortunately, the straightforward construction of the previous example cannot be applied in general. Intuitively it may be that the constructive problem is indeed more complex than the evaluation problem since it yields additional information.

The next result, however, shows that whenever the decision problem is NP-complete the constructive, evaluation, and decision problems are equivalent.

Theorem 1.5 ▶ *Let $\mathcal{P}$ be an NPO problem whose corresponding decision problem $\mathcal{P}_D$ is NP-complete. Then $\mathcal{P}_C \leq^p_T \mathcal{P}_D$.*

PROOF   Let us assume, without loss of generality, that $\mathcal{P}$ is a maximization problem. To prove the theorem we will derive an NPO problem $\mathcal{P}'$ such that $\mathcal{P}_C \leq^p_T \mathcal{P}'_D$, and then use the fact that, since $\mathcal{P}_D$ is NP-complete, $\mathcal{P}'_D \leq^p_T \mathcal{P}_D$ (actually, $\mathcal{P}'_D \leq^p_m \mathcal{P}_D$).

Problem $\mathcal{P}'$ has the same definition of $\mathcal{P}$ except for the measure function $m_{\mathcal{P}'}$, which is defined as follows. Let p be a polynomial which bounds the length of the solutions of $\mathcal{P}$ with respect to the length of the corresponding instances and, for any solution $y \in \text{SOL}_\mathcal{P}$, let $\lambda(y)$ denote the rank of y

in the lexicographic order. Then, for any instance $x \in I_{\mathcal{P}'} = I_{\mathcal{P}}$ and for any solution $y \in \text{SOL}_{\mathcal{P}'}(x) = \text{SOL}_{\mathcal{P}}(x)$, let $m_{\mathcal{P}'}(x,y) = 2^{p(|x|)+1} m_{\mathcal{P}}(x,y) + \lambda(y)$.

Notice that, for all instances x of $\mathcal{P}'$ and for any pair $y_1, y_2 \in \text{SOL}_{\mathcal{P}'}(x)$, we have that $m_{\mathcal{P}'}(x,y_1) \neq m_{\mathcal{P}'}(x,y_2)$. Therefore there exists a unique optimal feasible solution $y^*_{\mathcal{P}'}(x)$ in $\text{SOL}^*_{\mathcal{P}'}(x)$. Note also that, by definition, if $m_{\mathcal{P}'}(x,y_1) > m_{\mathcal{P}'}(x,y_2)$ then $m_{\mathcal{P}}(x,y_1) \geq m_{\mathcal{P}}(x,y_2)$, thus implying that $y^*_{\mathcal{P}'}(x) \in \text{SOL}^*_{\mathcal{P}}(x)$.

The optimal solution $y^*_{\mathcal{P}'}(x)$ can be easily derived in polynomial time by means of an oracle for $\mathcal{P}'_E$, since, given $m^*_{\mathcal{P}'}(x)$, the position of $y^*_{\mathcal{P}'}(x)$ in the lexicographic order can be derived by computing the remainder of the division between $m^*_{\mathcal{P}'}(x)$ and $2^{p(|x|)+1}$.

We know that an oracle for $\mathcal{P}'_D$ can be used to simulate $\mathcal{P}'_E$ in polynomial time. Thus we can construct an optimal solution of $\mathcal{P}$ in polynomial time using an oracle for $\mathcal{P}'_D$, and since $\mathcal{P}'_D \in \text{NP}$ and $\mathcal{P}_D$ is NP-complete, an oracle for $\mathcal{P}_D$ can be used to simulate the oracle for $\mathcal{P}'_D$. QED

The following question still remains open: Is there an NPO problem $\mathcal{P}$ whose corresponding constructive problem is harder than the evaluation problem $\mathcal{P}_E$? Indeed, there is some evidence that the answer to this question may be affirmative (see Bibliographical notes).

1.5 Exercises

Exercise 1.1 Prove the cost evaluations given in Example 1.1.

Exercise 1.2 Program 1.6 is the Euclidean algorithm for the greatest common divisor of two integers $x, y \in Z$. Determine its execution cost under the uniform cost model and the logarithmic cost model. Moreover, show which are the dominant operations and derive the asymptotic execution cost.

Exercise 1.3 Given a set P of points in the plane, the convex hull of P is defined as the minimal size convex polygon including all points in P. It can be easily proved that the set of vertices of the convex hull is a subset of P. Prove that the time complexity of the problem of computing the convex hull of a set of n points is:

1. $O(n^2)$;
2. $\Omega(n \log n)$ (by reduction from sorting);
3. (*) $\Theta(n \log n)$.

Chapter 1

THE COMPLEXITY
OF OPTIMIZATION
PROBLEMS

Program 1.6: Euclid

input Integers x, y;
output Greatest common divisor z of x and y;
begin
  while $x > 0$ **do**
    if $x > y$ **then**
      $x := x - y$
    else if $x < y$ **then**
      $y := y - x$
    else begin
      $z := x$;
      $x := 0$
    end;
  return z
end.

Problem 1.12: Maximum Path in a Tree
INSTANCE: Tree T, integer $K > 0$.
QUESTION: Is the length of the longest path in T less than K?

Exercise 1.4 Prove that SATISFYING TRUTH ASSIGNMENT is solvable (with respect to the logarithmic cost model) in *work space* proportional to the logarithm of the input size where the work space is the number of memory locations used for performing the computation, not considering the space required for the initial description of the problem instance (for example, we may assume that the problem instance is represented on a read-only memory device, whose size is not considered in the space complexity evaluation).

Exercise 1.5 Prove the same as the previous exercise for Problem 1.12.

Exercise 1.6 Derive a polynomial-space algorithm solving QUANTIFIED BOOLEAN FORMULAS.

Exercise 1.7 Show that polynomial-time Karp reductions have the transitive property, that is, if $\mathcal{P}_1 \leq_m^p \mathcal{P}_2$ and $\mathcal{P}_2 \leq_m^p \mathcal{P}_3$, then $\mathcal{P}_1 \leq_m^p \mathcal{P}_3$.

Exercise 1.8 Define a polynomial-time nondeterministic algorithm for the Problem 1.13.

Exercise 1.9 Recall that a *disjunctive normal form* (DNF) formula is defined as a collection of conjunctions $C = \{c_1, c_2, \ldots, c_m\}$ and is assumed to

Section 1.5
EXERCISES

Problem 1.13: Subset Sum

INSTANCE: Set $S = \{s_1, s_2, \ldots, s_n\}$, weight function $w : S \mapsto \mathbf{N}$, $K \in \mathbf{N}$.

QUESTION: Does there exist any $S' \subseteq S$ such that $\sum_{s_i \in S'} w(s_i) = K$?

Problem 1.14: Tautology

INSTANCE: Boolean formula $\mathcal{F}$ in DNF.

QUESTION: Is it true that every truth assignment on $\mathcal{F}$ is a satisfying assignment?

be satisfied by a truth assignment f if and only if at least one conjunction c_i, with $1 \leq i \leq m$, is satisfied by f. Show that Problem 1.14 is in co–NP.

Exercise 1.10 Show that NP is closed with respect to polynomial-time Karp-reducibility. Is NP closed also with respect to polynomial-time Turing reducibility?

Exercise 1.11 Prove that if there exists a problem $\mathcal{P}$ such that both $\mathcal{P}$ and $\mathcal{P}^c$ are NP-complete then NP = co–NP.

Exercise 1.12 (*) Prove that VERTEX COVER is NP-complete. (Hint: use the NP-completeness of SATISFIABILITY.)

Exercise 1.13 Prove that the decision problem corresponding to the MAXIMUM CLIQUE problem is NP-complete. (Hint: use the previous exercise).

Exercise 1.14 (*) Prove that the decision problem corresponding to the MINIMUM GRAPH COLORING problem is NP-complete. (Hint: use the NP-completeness of SATISFIABILITY.)

Exercise 1.15 Prove that the problem of deciding whether a graph is colorable with two colors is in P. (Hint: assign a color to a vertex and compute the consequences of this assignment.)

Exercise 1.16 Prove that the optimization problem corresponding to Problem 1.15 is in NPO.

Exercise 1.17 A problem $\mathcal{P}$ is called NP-easy if there exists a decision problem $\mathcal{P}' \in $ NP such that $\mathcal{P} \leq_T^p \mathcal{P}'$. Show that MINIMUM TRAVELING SALESPERSON is NP-easy. (Hint: Use the language $\{(G, p, k) \mid G$ is a graph, p is a path that can be extended to a Hamiltonian tour of cost $\leq k\}$ as the problem $\mathcal{P}'$ in the definition.)

Chapter 1

THE COMPLEXITY OF OPTIMIZATION PROBLEMS

> **Problem 1.15**: 4-th vertex cover
>
> INSTANCE: Graph $G = (V, E)$, integer K.
>
> QUESTION: Is the size of the 4-th smallest vertex cover in G at most equal to K?

> **Problem 1.16**: MAXIMUM SATISFIABILITY
>
> INSTANCE: Set C of disjunctive clauses on a set of variables V.
>
> SOLUTION: A truth assignment $f : V \mapsto \{\text{TRUE}, \text{FALSE}\}$.
>
> MEASURE: The number of clauses in C which are satisfied by f.

Exercise 1.18 Given an oracle for the evaluation version of Problem 1.16, show how it can be exploited to solve the constructive version of the same problem.

1.6 Bibliographical notes

GENERAL INTRODUCTIONS to the theory of algorithms and techniques for their analysis are given in [Knuth, 1969, Knuth, 1971, Knuth, 1973]. Other fundamental references to techniques for the design and analysis of algorithms are [Aho, Hopcroft, and Ullman, 1974] and [Cormen, Leiserson, and Rivest, 1990]. A recent text entirely devoted to formal methods for the analysis of combinatorial algorithms is [Sedgewick and Flajolet, 1996].

A detailed exposition of the theory of computational complexity of decision problems and of structural properties of related complexity classes is provided in several textbooks such as [Balcàzar, Diaz, and Gabarrò, 1988, Bovet and Crescenzi, 1994, Papadimitriou, 1994]. An excellent textbook on the theory of computability is [Rogers, 1987], where a detailed exposition of several types of reducibilities can be found.

The theory of NP-completeness is one of the bases for the topics developed in this textbook: the original concept of NP-complete problems has been introduced independently in [Cook, 1971], where SATISFIABILITY was proved to be NP-complete, and in [Levin, 1973]. The theory was further refined in [Karp, 1972], where a first set of problems were shown to be NP-complete by using reductions.

[Garey and Johnson, 1979] represents a landmark in the literature on NP-completeness and provides a detailed survey of the theory and an extensive list of NP-complete problems. Still twenty years later, this book is a fundamental reference for people interested in the assessing of the computational tractability or intractability of decision problems. The compendium of optimization problems contained in this book and the notation therein adopted are directly inspired by the list of NP-complete problems given in [Garey and Johnson, 1979]. This book is also a source of information on PSPACE-completeness and on the first examples of PSPACE-complete problems, such as QUANTIFIED BOOLEAN FORMULAS.

The characterization of the complexity of optimization problems has been first addressed in [Johnson, 1974a, Ausiello, Marchetti-Spaccamela, and Protasi, 1980, Paz and Moran, 1981], by establishing connections between combinatorial properties and complexity of decision and optimization problems. In particular, the concept of strong NP-completeness introduced in [Garey and Johnson, 1978] proved to be very fruitful in the theory of approximation of optimization problems (see Chap. 3). In [Krentel, 1988] the first structural approach to the characterization of the complexity of optimization problems is provided, by introducing a suitable computation model and the corresponding complexity classes (see Chap. 6). Further studies on the relationship between decision, evaluation, and optimization problems, including Theorem 1.5, are discussed in [Crescenzi and Silvestri, 1990].

Design Techniques for Approximation Algorithms

IN THE preceding chapter we observed that many relevant optimization problems are NP-hard, and that it is unlikely that we will ever be able to find efficient (i.e., polynomial-time) algorithms for their solution. In such cases it is worth looking for algorithms that always return a feasible solution whose measure is not too far from the optimum.

According to the general formal framework set up in Chap. 1, given an input instance x of an optimization problem, we say that a feasible solution $y \in \text{SOL}(x)$ is an *approximate solution* of the given problem and that any algorithm that always returns a feasible solution is an *approximation algorithm*.

Observe that, in most cases, it is not difficult to design a polynomial-time approximation algorithm that returns a (possibly) trivial feasible solution (e.g., in the case of MINIMUM GRAPH COLORING it is sufficient to assign each node a different color). However, we are mainly interested in approximate solutions that satisfy a specific quality criterion. The quality of an approximate solution can be defined in terms of the "distance" of its measure from the optimum, which we would like to be as small as possible. A precise characterization of how to define such distance is deferred to Chap. 3; in this chapter we will use as quality measure the *performance ratio*, that is, the ratio between the value of the approximate solution returned by the algorithm and the optimum.

In the following we present fundamental techniques for the design of approximation algorithms. In the case of problems in PO, these techniques are commonly used to design polynomial-time algorithms that always re-

Chapter 2

Design Techniques for Approximation Algorithms

turn optimal solutions: in this chapter, instead, we will use them to design efficient algorithms for the approximate solution of NP-hard optimization problems.

Each section of the chapter considers a specific algorithmic strategy that has been fruitfully used in the design of approximation algorithms. As we will see, each strategy can be applied to a variety of problems. For some problems, it is possible to show that the measure of the approximate solution is, in all cases, close enough to the optimum. For other problems, we will see that, for some instances, the measure of the obtained solution is arbitrarily far from the optimum (in such cases, a positive result will be proved for some special classes of the input instances).

We will also see that each technique allows us in many cases to define an algorithmic scheme that can be applied to obtain several algorithms for the same problem with possibly different approximation properties.

Finally, at the end of the chapter, we will briefly describe several possible approaches to the analysis of algorithms yielding approximate solutions that will be thoroughly presented in the remainder of the book.

2.1 The greedy method

THE FIRST algorithmic strategy that we consider is the *greedy method* that can be applied to optimization problems such that an instance of the problem specifies a set of items and the goal is to determine a subset of these items that satisfies the problem constraints and maximizes or minimizes the measure function. The greedy method can be applied to a variety of problems: we first consider examples of the technique applied to maximization problems whose set of feasible solutions forms an independent set. Namely, for each instance x, any feasible solution y of x is a subset of the items defined in x and any subset of y is a feasible solution as well. Note that this implies that if x has at least one feasible solution, then the empty set is a feasible solution.

The greedy method initially sorts the items according to some criterion and then incrementally builds the solution starting from the empty set. Namely, it considers items one at a time, always maintaining a set of "selected" items: when a new item is considered, it is either added to the set or it is eliminated from any further consideration. The decision whether adding an item to the set is only based on the items that have been already selected (and on the item under consideration). Clearly, different solutions might be obtained depending on the order in which items are initially sorted.

Assuming that the comparison of two items can be done in constant time,

> **Problem 2.1: Maximum Knapsack**
>
> INSTANCE: Finite set X of items, for each $x_i \in X$, value $p_i \in Z^+$ and size $a_i \in Z^+$, positive integer b.
>
> SOLUTION: A set of items $Y \subseteq X$ such that $\sum_{x_i \in Y} a_i \leq b$.
>
> MEASURE: Total value of the chosen items, i.e., $\sum_{x_i \in Y} p_i$.

Section 2.1

THE GREEDY METHOD

the running time required by the greedy algorithm is $O(n \log n)$ for sorting the n items plus the cost for n feasibility tests. Clearly this latter cost depends on the considered problem. Moreover, the quality of the approximate solution obtained depends on the initial ordering in which objects are considered; clearly, for each instance of the problem there is always an optimal ordering (i.e., an ordering that allows the greedy algorithm to find an optimal solution) but we do not expect to be able to find such an ordering in polynomial time for all instances of a computationally hard problem. However, as we will see in this section, in some cases we can find simple orderings that allow the greedy method to find good approximate solutions.

The greedy technique also applies to minimization problems. For instance, the well-known Kruskal's algorithm for the minimum spanning tree problem is based on sorting edges according to their weights and then greedily selecting edges until a spanning tree is obtained. In Sect. 2.1.3 we will see how the greedy approach can be used to design an approximation algorithm for MINIMUM TRAVELING SALESPERSON.

2.1.1 Greedy algorithm for the knapsack problem

MAXIMUM KNAPSACK (see Problem 2.1) models the problem of finding the optimal set of items to be put in a knapsack of limited capacity: to each item we associate a *profit* p_i (that represents the advantage of taking it) and an *occupancy* a_i. In general, we cannot take all items because the total occupancy of the chosen items cannot exceed the knapsack capacity b (in the sequel, without loss of generality, we assume that $a_i \leq b$, for $i = 1, 2, \ldots, n$).

Program 2.1 is a greedy algorithm for the MAXIMUM KNAPSACK problem that considers items in non-increasing order with respect to the profit/occupancy ratio (i.e., p_i/a_i). Since, after the items have been sorted, the complexity of the algorithm is linear in their number, the total running time is $O(n \log n)$. As far as the performance ratio is concerned, let us denote by $m_{Gr}(x)$ the value of the solution found by the algorithm when

Chapter 2

DESIGN TECHNIQUES FOR APPROXIMATION ALGORITHMS

Program 2.1: Greedy Knapsack

input Set X of n items, for each $x_i \in X$, values p_i, a_i, positive integer b;
output Subset $Y \subseteq X$ such that $\sum_{x_i \in Y} a_i \leq b$;
begin
    Sort X in non-increasing order with respect to the ratio p_i/a_i;
    (* Let $(x_1, x_2, \ldots, x_n)$ be the sorted sequence *)
    $Y := \emptyset$;
    for $i := 1$ **to** n **do**
      if $b \geq a_i$ **then**
      begin
        $Y := Y \cup \{x_i\}$;
        $b := b - a_i$
      end
    return Y
end.

applied to instance x. The following example shows that $m_{Gr}(x)$ can be arbitrarily far from the optimal value.

Example 2.1 ▶ Let us consider the following instance x of MAXIMUM KNAPSACK defined over n items: $p_i = a_i = 1$ for $i = 1, \ldots, n-1$, $p_n = b-1$, and $a_n = b = kn$ where k is an arbitrarily large number. In this case, $m^*(x) = b - 1$ while the greedy algorithm finds a solution whose value is $m_{Gr}(x) = n - 1$: hence, $m^*(x)/m_{Gr}(x) > k$.

An analogous result can be easily shown if the items are sorted in non-decreasing order with respect to their profit or in non-increasing order with respect to their occupancy.

A closer look at the previous example shows that the poor behavior of Program 2.1 is due to the fact that the algorithm does not include the element with highest profit in the solution while the optimal solution contains only this element. This suggests a simple modification of the greedy procedure that has a better performance, as shown in the following theorem.

Theorem 2.1 ▶ *Given an instance x of the MAXIMUM KNAPSACK problem, let $m_H(x) = \max(p_{max}, m_{Gr}(x))$, where p_{max} is the maximum profit of an item in x. Then $m_H(x)$ satisfies the following inequality:*

$$m^*(x)/m_H(x) < 2.$$

PROOF Let j be the index of the first item not inserted in the knapsack by the greedy algorithm with input x. The profit achieved by the algorithm at that point is

$$\bar{p}_j = \sum_{i=1}^{j-1} p_i \leq m_{Gr}(x)$$

and the overall occupancy is

$$\bar{a}_j = \sum_{i=1}^{j-1} a_i \leq b.$$

We first show that any optimal solution of the given instance must satisfy the following inequality: $m^*(x) < \bar{p}_j + p_j$. Since items are ordered in non-increasing order with respect to the ratio p_i/a_i, it follows that the exchange of any subset of the chosen items $x_1, x_2, \ldots, x_{j-1}$ with any subset of the unchosen items $x_j, \ldots, x_n$, that does not increase the sum of the occupancies $\bar{a}_j$, does not increase the overall profit. Therefore, the optimal value is bounded by $\bar{p}_j$ plus the maximum profit obtainable by filling the remaining available space of the knapsack (that is, $b - \bar{a}_j$) with items whose profit/weight ratio is at most p_j/a_j. Since $\bar{a}_j + a_j > b$, we obtain $m^*(x) \leq \bar{p}_j + (b - \bar{a}_j)p_j/a_j < \bar{p}_j + p_j$. To complete the proof we consider two possible cases. If $p_j \leq \bar{p}_j$ then $m^*(x) < 2\bar{p}_j \leq 2m_{Gr}(x) \leq 2m_H(x)$. On the other hand, if $p_j > \bar{p}_j$ then $p_{max} > \bar{p}_j$; in this case we have that

$$m^*(x) \leq \bar{p}_j + p_j \leq \bar{p}_j + p_{max} < 2p_{max} \leq 2m_H(x).$$

Thus, in both cases the theorem follows.   QED

As a consequence of the above theorem a simple modification of Program 2.1 allows us to obtain a provably better algorithm. The modification consists of adding one more step, in which the solution is chosen to be either the greedy solution or the item with largest profit. In Sect. 2.5 we will see that much better approximation algorithms can be obtained by applying a different approach.

2.1.2 Greedy algorithm for the independent set problem

In this section we consider the behavior of the greedy approach applied to MAXIMUM INDEPENDENT SET (see Problem 2.2). If we use a greedy algorithm for this problem, then it is reasonable to assume that vertices with smaller degree should be preferred to vertices of higher degree. In fact, whenever we add a vertex to the current solution, we also have to eliminate all its neighbors from any further consideration. Hence, first choosing vertices with smaller degrees might allow us to obtain a largest independent set: Program 2.2 is based on this criterion. Note that this procedure differs from the previously discussed greedy algorithm: indeed, at each iteration of Program 2.2 the chosen vertex is the one with minimum degree in the subgraph induced by vertices that might be added to the current independent set.

Chapter 2

DESIGN TECHNIQUES FOR APPROXIMATION ALGORITHMS

Problem 2.2: Maximum Independent Set

INSTANCE: Graph $G = (V, E)$.

SOLUTION: An independent set on G, i.e., a subset $V' \subseteq V$ such that, for any $(u, v) \in E$ either $u \notin V'$ or $v \notin V'$.

MEASURE: Cardinality of the independent set, i.e., $|V'|$.

Program 2.2: Greedy Independent Set

input Graph $G = (V, E)$;
output Independent set V' in G;
begin
   $V' := \emptyset$;
   $U := V$;
   while U is not empty **do**
   begin
     $x :=$ vertex of minimum degree in the graph induced by U;
     $V' := V' \cup \{x\}$;
     Eliminate x and all its neighbors from U
   end;
   return V'
end.

It is possible to see that there exists a sequence of graphs with increasing number of vertices for which this algorithm achieves solutions whose measures are arbitrarily far from the optimal values.

Example 2.2 ▶ Let us consider the graph of Fig. 2.1 where K_4 is a clique of four nodes and I_4 is an independent set of four nodes. In this case, the rightmost node is the first to be chosen by Program 2.2 and the resulting solution contains this node and exactly one node of K_4. The optimal solution, instead, contains the four nodes of I_4. This example can be easily generalized by substituting I_4 and K_4 with I_k and K_k, for any $k \geq 2$.

Note that, until now, no variant of the greedy approach has allowed us to obtain a result similar to Theorem 2.1 in the case of MAXIMUM INDEPENDENT SET. This limitation is not due to the used method: in fact, it can be proved that, unless $P = NP$, no polynomial-time algorithm exists that finds a good approximate solution for all graphs. As we will see in Chap. 6, this is due to the intrinsic difficulty of the problem.

We now show that the performance of the greedy approach can be bounded by a function of the "density" of the graph.

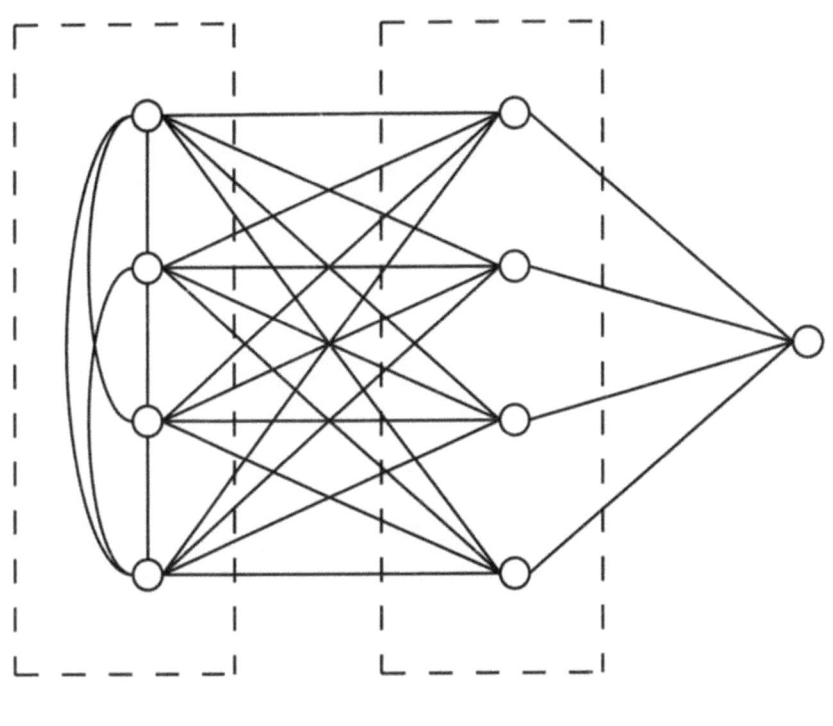

K_4 I_4

Figure 2.1
A bad example for
Program 2.2

Section 2.1

THE GREEDY METHOD

Given a graph G with n vertices and m edges, let $\delta = m/n$ be the density of G. The value $m_{Gr}(G)$ of the solution found by Program 2.2 is at least $n/(2\delta + 1)$. ◀ Theorem 2.2

Let x_i be the vertex chosen at the i-th iteration of the **while** loop of Program 2.2 and let d_i be the degree of x_i. The algorithm then deletes x_i and all its d_i neighbors from G. Hence, the number of eliminated edges is at least $d_i(d_i+1)/2$ (since x_i is the minimum degree vertex in the graph currently induced by U).

PROOF

Summing up over all iterations, we have that

$$\sum_{i=1}^{m_{Gr}(G)} \frac{d_i(d_i+1)}{2} \leq m = \delta n. \tag{2.1}$$

Since the algorithm stops when all vertices are eliminated, the following equality holds:

$$\sum_{i=1}^{m_{Gr}(G)} (d_i + 1) = n. \tag{2.2}$$

Chapter 2

DESIGN TECHNIQUES FOR APPROXIMATION ALGORITHMS

By adding Eq. (2.2) and twice Eq. (2.1) we obtain that

$$\sum_{i=1}^{m_{Gr}(G)} (d_i+1)^2 \leq n(2\delta+1).$$

The left-hand side of the above inequality is minimized when $d_i + 1 = n/m_{Gr}(G)$, for all i (this is an application of the Cauchy-Schwarz inequality; see Appendix A). It follows that

$$n(2\delta+1) \geq \sum_{i=1}^{m_{Gr}(G)} (d_i+1)^2 \geq \frac{n^2}{m_{Gr}(G)}.$$

QED Hence, $m_{Gr}(G) \geq n/(2\delta+1)$ and the theorem is proved.

The following theorem provides a relationship between the measure of the solution found by the greedy algorithm and the optimal measure.

Theorem 2.3 ▶ *Given a graph G with n vertices and m edges, let $\delta = m/n$. Program 2.2 finds an independent set of value $m_{Gr}(G)$ such that*

$$m^*(G)/m_{Gr}(G) \leq (\delta+1).$$

PROOF The proof is similar to that of the preceding theorem: in this case, we additionally count in Eq. (2.1) the number of edges that are incident to vertices of some optimal solution.

Namely, fix a maximum independent set V^* and let k_i be the number of vertices in V^* that are among the $d_i + 1$ vertices deleted in the i-th iteration of the **while** loop of Program 2.2.

Clearly, we have that

$$\sum_{i=1}^{m_{Gr}(G)} k_i = |V^*| = m^*(G). \qquad (2.3)$$

Since the greedy algorithm selects the vertex with minimum degree, the sum of the degree of the deleted vertices is at least $d_i(d_i+1)$. Since an edge cannot have both its endpoints in V^*, it follows that the number of deleted edges is at least $(d_i(d_i+1) + k_i(k_i-1))/2$.

Hence we can improve Eq. (2.1) to obtain

$$\sum_{i=1}^{m_{Gr}(G)} \frac{d_i(d_i+1) + k_i(k_i-1)}{2} \leq \delta n. \qquad (2.4)$$

Adding Eqs. (2.2), (2.3) and twice (2.4), we obtain the following bound:

$$\sum_{i=1}^{m_{Gr}(G)} ((d_i+1)^2 + k_i^2) \leq n(2\delta+1) + m^*(G).$$

By applying the Cauchy-Schwarz inequality, it is possible to show that the left-hand side of the above inequality is minimized when $d_i + 1 = n/m_{Gr}(G)$ and $k_i = m^*(G)/m_{Gr}(G)$, for $i = 1, 2, \ldots, m_{Gr}(G)$. Hence,

$$n(2\delta+1) + m^*(G) \geq \sum_{i=1}^{m_{Gr}(G)} ((d_i+1)^2 + k_i^2) \geq \frac{n^2 + m^*(G)^2}{m_{Gr}(G)};$$

that is,

$$m_{Gr}(G) \geq m^*(G) \frac{(n/m^*(G)) + (m^*(G)/n)}{2\delta + 1 + (m^*(G)/n)}.$$

To complete the proof it is sufficient to observe that the fractional term on the right-hand side of the above inequality is minimized when $m^*(G) = n$. By substituting this term, the theorem follows. QED

2.1.3 Greedy algorithm for the salesperson problem

We now show how the idea of a "greedy" selection can be applied to MINIMUM TRAVELING SALESPERSON (see Problem 1.8). Recall that an instance of this problem can be represented by a complete graph $G = (V, E)$ with positive weights on the edges. Feasible solutions of the problem are subsets I of edges such that the graph (V, I) is a cycle.

The idea of the greedy algorithm is first to find a Hamiltonian path and then to form a tour by adding the edge that connects the first and the last vertex of the path.

The Hamiltonian path is found incrementally: at the beginning, the algorithm arbitrarily chooses the first city of the path, say c_{i_1}, and then executes a loop $(n-1)$ times. In the first iteration the algorithm selects the vertex that follows c_{i_1} by choosing vertex c_{i_2} such that edge (c_{i_1}, c_{i_2}) has minimum weight among all edges with an endpoint in c_{i_1}; edge (c_{i_1}, c_{i_2}) is then added to the path. In the r-th iteration of the loop, after a path through vertices $c_{i_1}, c_{i_2}, \ldots, c_{i_r}$ has been obtained, the algorithm chooses vertex $c_{i_{r+1}}$ that follows c_{i_r} in the path as the vertex such that $(c_{i_r}, c_{i_{r+1}})$ is the minimum weight edge from c_{i_r} to a vertex that is not in the path; edge $(c_{i_r}, c_{i_{r+1}})$ is added to the path. After $(n-1)$ iterations a Hamiltonian path $c_{i_1}, c_{i_2}, \ldots, c_{i_n}$

Chapter 2

DESIGN TECHNIQUES FOR APPROXIMATION ALGORITHMS

is obtained and the last step of the algorithm completes the tour by adding edge (c_{i_n}, c_{i_1}). Since the underlying graph is complete, the algorithm is guaranteed to find a feasible solution.

The "greediness" of this algorithm lies in the fact that during the construction of the Hamiltonian path the algorithm selects the edge that minimizes the cost of extending the current path with a new vertex. For this reason, we call the algorithm *Nearest Neighbor*.

No precise estimation is known on the quality of the solutions returned by *Nearest Neighbor* in the general case (see Exercise 2.3 for a constant lower bound). However, if we restrict the set of problem instances, then it is possible to obtain bounds on the quality of the solution found by the algorithm. We do so by considering instances of MINIMUM TRAVELING SALESPERSON in which the distance matrix is symmetric (i.e., $D(i,j) = D(j,i)$, for any pair of cities c_i and c_j) and the triangular inequality is satisfied (i.e., for all triples of cities c_i, c_j, and c_k, $D(i,j) \leq D(i,k) + D(k,j)$). In the following, we will denote this problem as MINIMUM METRIC TRAVELING SALESPERSON.

In order to provide a bound on the performance of *Nearest Neighbor* we need the following technical lemma.

Lemma 2.4 ▶ *For any instance x of* MINIMUM METRIC TRAVELING SALESPERSON *with n cities, assume that there exists a mapping $l : \{c_1, \ldots, c_n\} \mapsto \mathbb{Q}$ such that:*

1. *for any two distinct cities c_i and c_j, $D(i,j) \geq \min(l(c_i), l(c_j))$;*

2. *for any city c_i, $l(c_i) \leq \frac{1}{2} m^*(x)$.*

Then such a function satisfies the following inequality:

$$\sum_{i=1}^{n} l(c_i) \leq \frac{1}{2}(\lceil \log n \rceil + 1) m^*(x).$$

PROOF Without loss of generality, we assume that cities are sorted in nonincreasing order with respect to their l-values. Let us first prove that, for all k with $1 \leq k \leq n$,

$$m^*(x) \geq 2 \sum_{i=k+1}^{\min(2k,n)} l(c_i). \tag{2.5}$$

Let I^* be an optimal tour of length $m^*(x)$ and consider the subset of cities $C_k = \{c_i \mid 1 \leq i \leq \min(2k,n)\}$. Let I_k be a tour (of length m_k^*) that traverses the cities in C_k in the same order as I^*. For each pair c_r and c_s of adjacent cities in I_k, either c_r and c_s were adjacent also in I^* or, by the triangle inequality, there exists a path in I^* starting from c_r and ending at c_s of length at least $D(r,s)$. As a consequence, $m^*(x) \geq m_k^*$, for each k.

Since $D(i,j) \geq \min(l(c_i), l(c_j))$ for all pairs of cities c_i and c_j, by summing over all edges (c_i, c_j) that belong to I_k we obtain

$$m_k^* \geq \sum_{(c_i,c_j) \in I_k} \min(l(c_i), l(c_j)) = \sum_{c_i \in C_k} \alpha_i l(c_i),$$

where α_i is the number of cities $c_j \in C_k$ adjacent to c_i in I_k and such that $i > j$ (hence, $l(c_i) \leq l(c_j)$). Clearly, $\alpha_i \leq 2$ and $\sum_{c_i \in C_k} \alpha_i$ equals the number of cities in I_k. Since the number of cities in I_k is at most $2k$, we may derive a lower bound on the quantity $\sum_{c_i \in C_k} \alpha_i l(c_i)$ by assuming that $\alpha_i = 0$ for the k cities $c_1, \ldots, c_k$ with largest values $l(c_i)$ and that $\alpha_i = 2$ for all the other $|C_k| - k$ cities. Hence, we obtain Eq. (2.5) since

$$m^*(x) \geq m_k^* \geq \sum_{c_i \in C_k} \alpha_i l(c_i) \geq 2 \sum_{i=k+1}^{\min(2k,n)} l(c_i).$$

Summing all Eqs. (2.5) with $k = 2^j$, for $j = 0, 1, \ldots, \lceil \log n \rceil - 1$, we obtain that

$$\sum_{j=0}^{\lceil \log n \rceil - 1} m^*(x) \geq \sum_{j=0}^{\lceil \log n \rceil - 1} 2 \sum_{i=2^j+1}^{\min(2^{j+1},n)} l(c_i),$$

which results in

$$\lceil \log n \rceil m^*(x) \geq 2 \sum_{i=2}^{n} l(c_i).$$

Since, by hypothesis, $m^*(x) \geq 2l(c_1)$, the lemma is then proved. QED

For any instance x of MINIMUM METRIC TRAVELING SALESPERSON with n cities, let $m_{NN}(x)$ be the length of the tour returned by Nearest Neighbor with input x. Then $m_{NN}(x)$ satisfies the following inequality: ◀ Theorem 2.5

$$m_{NN}(x)/m^*(x) \leq \frac{1}{2}(\lceil \log n \rceil + 1).$$

Let $c_{k_1}, c_{k_2}, \ldots, c_{k_n}$ be a tour resulting from the application of Nearest Neighbor to x. The proof consists in showing that, if we associate to each city c_{k_r} ($r = 1, \ldots, n$) the value $l(c_{k_r})$ corresponding to the length of edge $(c_{k_r}, c_{k_{r+1}})$ (if $r < n$) or of edge (c_{k_n}, c_{k_1}) (if $r = n$), then we obtain a mapping l that satisfies the hypothesis of Lemma 2.4. Since $\sum_{i=1}^{n} l(c_{k_i}) = m_{NN}(x)$, the theorem follows immediately by applying the lemma. PROOF

Section 2.1
THE GREEDY METHOD

Chapter 2

DESIGN TECHNIQUES FOR APPROXIMATION ALGORITHMS

Program 2.3: Partitioning Sequential Scheme

input Set I of items;
output Partition P of I;
begin
   Sort I (according to some criterion);
   (* Let $(x_1, x_2 \ldots, x_n)$ be the obtained sequence *)
   $P := \{\{x_1\}\}$;
   for $i := 2$ **to** n **do**
     if x_i can be added to a set p in P **then**
       add x_i to p
     else
       $P := P \cup \{\{x_i\}\}$;
   return P
end.

The first hypothesis of Lemma 2.4 can be proved as follows. Let c_{k_r} and c_{k_s} be any pair of cities: if $r < s$ (that is, c_{k_r} has been inserted in the tour before c_{k_s}), then $l(c_{k_r}) \leq D(k_r, k_s)$, since c_{k_s} was a possible choice as city $c_{k_{r+1}}$. Analogously, if $s < r$, then $l(c_{k_s}) \leq D(k_r, k_s)$; hence $D(k_r, k_s) \geq \min(l(c_{k_r}), l(c_{k_s}))$ and the first hypothesis of Lemma 2.4 holds.

To show the second hypothesis, assume that $l(c_{k_r}) = D(k_r, k_s)$. Observe that any optimal tour I^* is composed of two disjoint paths connecting c_{k_r} and c_{k_s}: by the triangle inequality, each such path must have length at least $D(k_r, k_s)$. The inequality $m^*(x) \geq 2D(k_r, k_s) = 2l(c_{k_r})$ hence follows and the proof of the theorem is completed.

QED

2.2 Sequential algorithms for partitioning problems

IN THIS section we consider *partitioning problems*, that is, problems in which feasible solutions are suitably defined partitions (not necessarily bipartitions) of a set of items $I = \{x_1, x_2 \ldots, x_n\}$ defined in the input instance and that satisfy the constraints specified by the particular problem.

A sequential algorithm for a partitioning problem initially sorts the items according to a given criterion and then builds the output partition P sequentially. The general scheme of such sequential procedure is given in Program 2.3.

Notice that, in the above scheme, when the algorithm considers item x_i it is not allowed to modify the partition of items x_j, for $j < i$. Therefore, if two objects are assigned to the same set of the partition then they will be in the same set of the final partition as well.

Section 2.2

SEQUENTIAL ALGORITHMS FOR PARTITIONING PROBLEMS

Problem 2.3: Minimum Scheduling on Identical Machines

INSTANCE: Set of jobs T, number p of machines, length l_j for executing job $t_j \in T$.

SOLUTION: A p-machine schedule for T, i.e., a function $f : T \mapsto [1..p]$.

MEASURE: The schedule's makespan, i.e.,

$$\max_{i \in [1..p]} \sum_{t_j \in T : f(t_j)=i} l_j.$$

To design a sequential algorithm for a particular problem, we need to specify the criteria used for (a) finding the order in which items are processed, and (b) including items in the partition. For each particular problem there are several possible criteria that can be used to sort the items. As in the case of the greedy approach, there exists at least one ordering that will allow us to obtain an optimal solution; however, we do not expect to find such an ordering in polynomial time for NP-hard combinatorial problems and we look for orderings of the items that allow us to find good approximate solutions.

2.2.1 Scheduling jobs on identical machines

In this section we apply the sequential algorithm scheme to the problem of scheduling a set of jobs on p identical machines with the goal of minimizing the time necessary to complete the execution of the jobs (see Problem 2.3). This problem is NP-hard even in the case of $p = 2$ (see Bibliographical notes).

Assume that we are given the order in which jobs are processed. In order to obtain a scheduling algorithm from the sequential scheme it is necessary to determine a rule for assigning jobs to machines. The obvious rule is to assign each job to the machine with smallest load. Namely, suppose that the first $j-1$ jobs have already been assigned and let $A_i(j-1)$ be the finish time necessary for the execution of the subset of jobs assigned to the i-th machine (i.e., $A_i(j-1) = \sum_{1 \leq k \leq j-1 : f(t_k)=i} l_k$). The j-th job is then assigned to the machine with minimum finish time (ties are arbitrarily broken). In this way the algorithm, called *List Scheduling*, minimizes the increase in the finish time required for the execution of a new job.

◀ Theorem 2.6

Given an instance x of MINIMUM SCHEDULING ON IDENTICAL MACHINES with p machines, for any order of the jobs, the List Scheduling

Chapter 2

DESIGN TECHNIQUES FOR APPROXIMATION ALGORITHMS

algorithm finds an approximate solution of measure $m_{LS}(x)$ such that

$$m_{LS}(x)/m^*(x) \leq (2 - \frac{1}{p}).$$

PROOF Let W be the sum of the processing time of all jobs, i.e., $W = \sum_{k=1}^{|T|} l_k$; the value of the optimal solution clearly satisfies the bound $m^*(x) \geq W/p$.

To bound the value of the approximate solution found by the algorithm assume that the h-th machine is the one with the highest finish time (i.e., $A_h(|T|) = m_{LS}(x)$) and let j be the index of the last job assigned to this machine. Since job t_j was assigned to the least loaded machine, then the finish time of any other machine is at least $A_h(|T|) - l_j$. This implies that $W \geq p(A_h(|T|) - l_j) + l_j$ and that the following bound on $m_{LS}(x)$ holds:

$$m_{LS}(x) = A_h(|T|) \leq \frac{W}{p} + \frac{(p-1)l_j}{p}.$$

Since $m^*(x) \geq W/p$ and $m^*(x) \geq l_j$ we have that

$$m_{LS}(x) \leq \frac{W}{p} + \frac{(p-1)l_j}{p} \leq m^*(x) + \frac{p-1}{p}m^*(x) = \left(2 - \frac{1}{p}\right)m^*(x)$$

QED and the theorem follows.

The following example shows that no better bound can be proved to hold for any ordering of the jobs: indeed, it shows that, for any number p of machines, there exists an instance of MINIMUM SCHEDULING ON IDENTICAL MACHINES on p machines with $p(p-1)+1$ jobs and an ordering of the jobs for which the bound of Theorem 2.6 is tight.

Example 2.3 ▶ Given p, let us consider the instance of MINIMUM SCHEDULING ON IDENTICAL MACHINES with p machines, $p(p-1)$ jobs of length 1 and one job of length p. Clearly, the optimal measure is p. On the other side, if the job with largest processing time is the last in the order, then *List Scheduling* will find a solution with value $2p - 1$ (see Fig. 2.2).

The bad behavior of the previous example is due to the fact that the job with largest processing time is scheduled last. A simple ordering that allows us to find a better solution is based on the *LPT* (Largest Processing Time) rule that considers jobs in non-increasing order with respect to their processing time (i.e., $l_1 \geq l_2 \geq \cdots \geq l_{|T|}$).

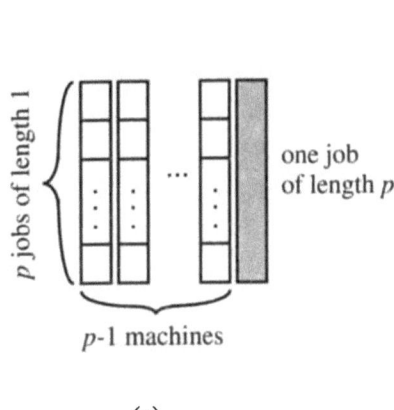
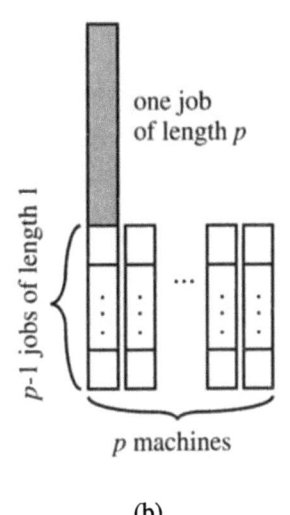

(a)   (b)

Section 2.2
SEQUENTIAL ALGORITHMS FOR PARTITIONING PROBLEMS

Figure 2.2
Example of (a) an optimal scheduling and (b) a scheduling computed by the sequential algorithm

◀ Theorem 2.7

Given an instance x of MINIMUM SCHEDULING ON IDENTICAL MACHINES *with p machines, the LPT rule provides a schedule of measure $m_{LPT}(x)$ such that $m_{LPT}(x)/m^*(x) \leq (4/3 - 1/3p)$.*

PROOF

Assume that the theorem does not hold and consider the instance x that violates the claim having the minimum number of jobs. Let j be the job of x that is last considered by the *LPT* rule and let l_{min} be its length (which is among the shortest ones). Two cases may arise. Either $l_{min} > m^*(x)/3$ or $l_{min} \leq m^*(x)/3$. In the first case, it is simple to show that at most two jobs may have been assigned to any machine and that the *LPT* rule provides the optimal solution (see Exercise 2.5). In the second case (i.e., $l_{min} \leq m^*(x)/3$), since x is a minimum counter-example it follows that the instance x' obtained from x by removing job j satisfies the claim. This implies that $m_{LPT}(x) > m_{LPT}(x')$ and, therefore, that the *LPT* rule assigns job j to a machine that will have the largest processing time at the end of the algorithm. By reasoning as in the proof of the previous theorem, we obtain that

$$m_{LPT}(x) \leq \frac{W}{p} + \frac{p-1}{p} l_{min}$$

where W is the sum of the processing times of all jobs. Since $m^*(x) \geq W/p$ and $l_{min} \leq m^*(x)/3$, we obtain

$$m_{LPT}(x) \leq m^*(x) + \frac{p-1}{3p} m^*(x) = \left(\frac{4}{3} - \frac{1}{3p}\right) m^*(x).$$

The theorem thus follows.   QED

Chapter 2

DESIGN
TECHNIQUES FOR
APPROXIMATION
ALGORITHMS

> **Problem 2.4: Minimum Bin Packing**
>
> INSTANCE: Finite multiset I of rational numbers $\{a_1, a_2, \ldots, a_n\}$ with $a_i \in (0, 1]$ for $i = 1, \ldots, n$ (in a multiset the same number may appear more than once).
>
> SOLUTION: A partition $\{B_1, B_2, \ldots, B_k\}$ of I such that $\sum_{a_i \in B_j} a_i \leq 1$ for $j = 1, \ldots, k$.
>
> MEASURE: The cardinality of the partition, i.e., k.

2.2.2 Sequential algorithms for bin packing

MINIMUM BIN PACKING (see Problem 2.4) looks for a packing of a set of weighted items using the minimum number of bins of unit capacity. The total weight of the items assigned to a bin cannot exceed its capacity.

A simple sequential algorithm for MINIMUM BIN PACKING, called *Next Fit*, processes the items one at a time in the same order as they are given in input. The first item a_1 is placed into bin B_1. Let B_j be the last used bin, when the algorithm considers item a_i: *Next Fit* assigns a_i to B_j if it has enough room, otherwise a_i is assigned to a new bin B_{j+1}.

Theorem 2.8 ▶ *Given an instance x of* MINIMUM BIN PACKING, *Next Fit returns a solution with value $m_{NF}(x)$ such that $m_{NF}(x)/m^*(x) < 2$.*

PROOF The proof estimates the value of the optimal and approximate solutions as functions of the sum of the item sizes, denoted by A (i.e., $A = \sum_{i=1}^{n} a_i$).

Observe that the number of bins used by *Next Fit* is less than $2\lceil A \rceil$: this is due to the fact that, for each pair of consecutive bins, the sum of the sizes of the items included in these two bins is greater than 1. On the other hand, since the number of bins used in each feasible solution is at least the total size of the items, we have that $m^*(x) \geq \lceil A \rceil$. It follows that
QED $m_{NF}(x) < 2m^*(x)$.

Example 2.4 ▶ The bound stated in Theorem 2.8 is asymptotically tight. In fact, for each integer n, there exists an instance of $4n$ items and an ordering of these items such that $m^*(x) = n+1$ and $m_{NF}(x) = 2n$. The instance and the order of the items are as follows: $I = \{1/2, 1/(2n), 1/2, 1/(2n), \ldots, 1/2, 1/(2n)\}$ (each pair is repeated $2n$ times). Figure 2.3 shows both the optimal and the approximate solution found by *Next Fit*.

An obvious weakness of *Next Fit* is that it tries to assign an item only to the last used bin. This suggests a new algorithm, called *First Fit*, that processes items in the input order according to the following rule: item a_i

is assigned to the first used bin that has enough available space to include it; if no bin can contain a_i, a new bin is opened.

Section 2.2

SEQUENTIAL ALGORITHMS FOR PARTITIONING PROBLEMS

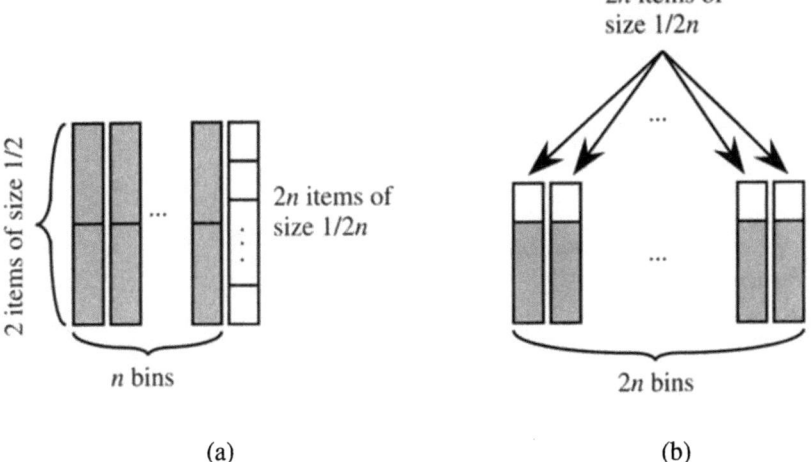

Figure 2.3
Example of (a) an optimal packing and (b) a packing found with *Next Fit*

The *First Fit* algorithm has a better performance than *Next Fit*: indeed, it finds a solution that is at most 70% far away from the optimal solution. Namely, it can be shown that *First Fit* finds a solution with value $m_{FF}(x)$ such that $m_{FF}(x) \leq 1.7m^*(x) + 2$ (see Bibliographical notes).

An even better algorithm for MINIMUM BIN PACKING is *First Fit Decreasing*: this algorithm first sorts items in non-increasing order with respect to their size and then processes items as *First Fit* (see Program 2.4).

◀ Theorem 2.9

Given an instance x of MINIMUM BIN PACKING, *the* First Fit Decreasing *algorithm finds a solution with measure* $m_{FFD}(x)$ *such that*

$$m_{FFD}(x) \leq 1.5m^*(x) + 1.$$

PROOF

Let us partition the ordered list of items $\{a_1, a_2, \ldots, a_n\}$, according to their value, into the following sets:

$$\begin{aligned} A &= \{a_i \mid a_i > 2/3\}, \\ B &= \{a_i \mid 2/3 \geq a_i > 1/2\}, \\ C &= \{a_i \mid 1/2 \geq a_i > 1/3\}, \\ D &= \{a_i \mid 1/3 \geq a_i\}. \end{aligned}$$

Consider the solution obtained by *First Fit Decreasing*. If there is at least one bin that contains only items belonging to D, then there is at most one

55

Program 2.4: First Fit Decreasing

input Set I of n positive rationals less than or equal to 1;
output Partition of I in subsets of unitary weight;
begin
    Sort elements of I in non-increasing order;
    (* Let $(a_1, a_2, \ldots, a_n)$ be the obtained sequence *)
    for $i := 1$ **to** n **do**
        if there is a bin that can contain a_i **then**
            Insert a_i into the first such bin
        else
            Insert a_i into a new bin ;
    return the partition
end.

bin (the last opened one) with total occupancy less than $2/3$, and the bound follows.

If there is no bin containing only items belonging to D, then *First Fit Decreasing* finds the optimal solution. In fact, let x' be the instance obtained by eliminating all items belonging to D. Since the value of the solution found by *First Fit Decreasing* for x and x' is the same, it is sufficient to prove the optimality of *First Fit Decreasing* for x'. To this aim, we first observe that, in each feasible solution of x', items from A cannot share a bin with any other item and that every bin contains at most two items (and only one of these two items can belong to B). The thesis follows by observing that *First Fit Decreasing* processes items in non-increasing order with respect to their weight. Therefore, it packs each item belonging to C with the largest possible item in B that might fit with it and that does not already share a bin with another item. This implies that the number of bins in the optimal solution and in the solution found by *First Fit Decreasing* are the same.

QED

By means of a detailed case analysis, the bound given in the previous theorem can be substituted by $11m^*(x)/9 + 7/9$ (see Bibliographical notes). The following example shows that no better multiplicative factor can be obtained, for sufficiently large instances.

Example 2.5 ▶ For any $n > 0$, let us consider the following instance x_n. The instance includes $5n$ items: n items of size $1/2 + \varepsilon$, n items of size $1/4 + 2\varepsilon$, n items of size $1/4 + \varepsilon$ and $2n$ items of size $1/4 - 2\varepsilon$. As can be seen in Fig. 2.4, $m_{FFD}(x_n) = \frac{11}{6}n$, while $m^*(x_n) = \frac{3}{2}n$.

Note that if there is more than one bin that can contain an item without violating the feasibility constraint, the *First Fit Decreasing* algorithm chooses the first opened bin and does not try to optimize the choice. This might give rise to bins filled only partly and, therefore, to an increased number of bins (the division of the available space into many small fragments is called *fragmentation*).

Section 2.2

SEQUENTIAL ALGORITHMS FOR PARTITIONING PROBLEMS

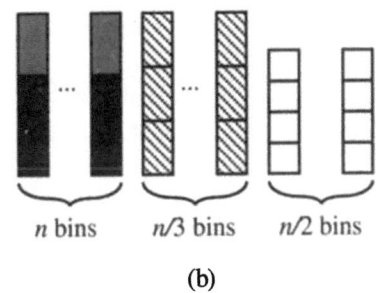

Figure 2.4
An example of (a) an optimal packing that uses $3n/2$ bins and (b) a packing produced by FFD that uses $11n/6$ bins

An apparently better algorithm for MINIMUM BIN PACKING is the *Best Fit Decreasing* algorithm. Like *First Fit Decreasing*, *Best Fit Decreasing* initially sorts the items in non-increasing order with respect to their weight and then considers them sequentially. The difference between the two algorithms is the rule used for choosing the bin in which the new item a_i is inserted: while trying to pack a_i, *Best Fit Decreasing* chooses one bin whose empty space is minimum. In this way, the algorithm tries to reduce the fragmentation by maximizing the number of bins with large available capacity.

It is possible to see that the quality of the solution found by *Best Fit Decreasing* may be worse than the quality of the solution found by *First Fit Decreasing* (see Exercise 2.9). On the other hand, as shown in the following example, it is possible to define instances for which *Best Fit Decreasing* finds an optimal solution while *First Fit Decreasing* returns a non-optimal solution.

◀ Example 2.6

For any $n > 0$, let us us consider the following instance x_n. The instance includes $6n$ items: n items of size $7/10$, $2n$ items of size $2/5$, n items of size $1/5$, and $2n$ items of size $3/20$. As can be easily seen, $m_{BFD}(x_n) = 2n$, while $m_{FFD}(x_n) = 2n + \lceil n/6 \rceil$.

Chapter 2

DESIGN TECHNIQUES FOR APPROXIMATION ALGORITHMS

However, the 11/9-bound holds for *Best Fit Decreasing* as well and the sequence of instances given in Example 2.5 shows that the bound is tight (i.e., 11/9 is the smallest multiplicative factor).

As a final remark we observe that both *First Fit Decreasing* and *Best Fit Decreasing* are *off-line* algorithms while *Next Fit* and *First Fit* are *on-line* algorithms. Indeed, these latter two algorithms assign item a_i to a bin without knowing the size of items a_j, for any $j > i$. On the other side, in an off-line algorithm the packing of the items starts when all sizes are known (as in the case of *First Fit Decreasing* and *Best Fit Decreasing* in which the first step to be performed is sorting the items).

2.2.3 Sequential algorithms for the graph coloring problem

In this section we consider the behavior of sequential algorithms for MINIMUM GRAPH COLORING. In order to apply the sequential scheme to design a coloring algorithm it is necessary to specify an order of the vertices. Assuming $(v_1, v_2, \ldots, v_n)$ is the sequence obtained, it is straightforward to obtain a sequential algorithm. Vertex v_1 is colored with color 1 and the remaining vertices are colored as follows: when vertex v_i is considered and colors $1, 2, \ldots, k$ have been used the algorithm attempts to color v_i using one of the colors $1, 2, \ldots, k$ (if there are several possibilities it chooses the minimum color); otherwise (that is, v_i has at least k adjacent vertices which have been assigned different colors) a new color $k+1$ is used to color v_i. The algorithm proceeds in this way until all vertices have been colored.

In the sequel we will see that the quality of the solution obtained using this algorithm is not as good as in the case of MINIMUM SCHEDULING ON IDENTICAL MACHINES and MINIMUM BIN PACKING. In fact, unless $P = NP$, there is no ordering of the vertices that can be found in polynomial time and that allows us to obtain a constant bound on the performance ratio that holds for all graphs.

Nevertheless, we will consider two different criteria to order the vertices of the graph: the *decreasing degree* order and the *smallest last* order. To better motivate these criteria we first analyze the number of colors used by a sequential algorithm as a function of the degree of the vertices. Namely, given an ordering $(v_1, v_2, \ldots, v_n)$ of the vertex set of G, let G_i be the graph induced by vertices $\{v_1, \ldots, v_i\}$ (clearly, $G_n = G$). Let k_i be the number of colors used by the sequential algorithm to color G_i (hence, k_n denotes the number of colors used by the algorithm with input G), and let $d_i(v)$ be the degree of v in G_i (hence, $d_n(v)$ denotes the degree of v in G).

Theorem 2.10 ▶ *Let k_n be the number of colors used by the sequential coloring algorithm* when applied to an input graph G whose vertices are considered in the order $(v_1, v_2, \ldots, v_n)$. Then, the following inequality holds:

$$k_n \leq 1 + \max_{1 \leq i \leq n} \min(d_n(v_i), i-1).$$

If the algorithm does not introduce a new color to color vertex v_i, then $k_i = k_{i-1}$. Otherwise, $k_i = k_{i-1} + 1$ and the degree of v_i in G_i must satisfy the inequality

$$d_i(v_i) \geq k_{i-1}.$$

By induction, it thus follows that

$$k_n \leq 1 + \max_{1 \leq i \leq n}(d_i(v_i)). \qquad (2.6)$$

Since $d_i(v_i)$ is clearly bounded both by $d_n(v_i)$ and by $i-1$ (that is, the number of other nodes in G_i), the theorem follows.

Section 2.2
SEQUENTIAL ALGORITHMS FOR PARTITIONING PROBLEMS

PROOF

QED

An immediate consequence of the above theorem is the following result.

For any ordering of the vertices, the sequential coloring algorithm uses at most $\Delta + 1$ colors to color a graph G, where Δ denotes the highest degree of the vertices of G. ◀ Corollary 2.11

Observe that a non-increasing ordering of the vertices with respect to their degree minimizes the upper bound in Theorem 2.10. Hence, we use this ordering to obtain the sequential algorithm called *Decreasing Degree*. Unfortunately, the number of colors used by this algorithm can be much larger than the number of colors used by the optimal solution. In fact, the following example shows that there exist 2-colorable graphs with $2n$ vertices and maximum degree $n-1$ for which *Decreasing Degree* could require n colors (as a side effect, the example also shows that the bound of Corollary 2.11 is tight).

Let n be an integer and $G(V,E)$ be a graph with $V = \{x_1, \ldots, x_n, y_1, \ldots, y_n\}$ and $E = \{(x_i, y_j) \mid i \neq j\}$ (see Fig. 2.5 where $n = 4$). Note that all vertices have the same degree $n-1$ and that G can be easily colored with 2 colors. However, if the initial ordering of the vertices is ◀ Example 2.7

$$(x_1, y_1, x_2, y_2, \ldots, x_n, y_n),$$

then *Decreasing Degree* uses n colors.

Chapter 2

DESIGN
TECHNIQUES FOR
APPROXIMATION
ALGORITHMS

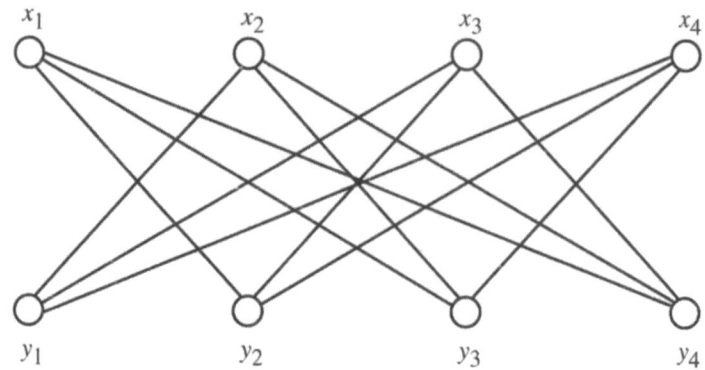

Figure 2.5
A graph for which
Decreasing Degree behaves
poorly

Many other heuristic criteria for finding an initial ordering of the vertices have been proposed for improving the bad behavior of *Decreasing Degree*. In the following we consider the *smallest last* order defined as follows: v_n, the last vertex to be considered, is the one with minimum degree in G (breaking ties arbitrarily). The order of the remaining vertices is defined backward: after vertices $v_{i+1}, \ldots, v_n$ have been inserted in the ordered sequence, v_i is the vertex with minimum degree in the subgraph induced by $V - \{v_{i+1}, \ldots, v_n\}$ (breaking ties arbitrarily). Using this ordering we obtain the *Smallest Last* algorithm).

Recall that, for any ordering $(v_1, \ldots, v_n)$ of the vertices, Eq. (2.6) implies the following bound on the number of colors used by the sequential algorithm:

$$k_n \leq 1 + \max_{1 \leq i \leq n} (d_i(v_i)).$$

The smallest last ordering of the vertices minimizes the above bound since we have that, for each i with $1 \leq i \leq n$, $d_i(v_i) = \min_{v_j \in G_i} d_i(v_j)$. Unfortunately, it is possible to show that *Smallest Last* fails to find good colorings for all graphs (see Exercise 2.14). However, its performance is satisfactory when applied to planar graphs.

Theorem 2.12 ▶ *The* Smallest Last *algorithm colors a planar graph with at most six colors.*

PROOF   It is sufficient to prove that $\max_{1 \leq i \leq n}(d_i(v_i)) \leq 5$. Indeed, this is due to the fact that, for any planar graph G, Euler's theorem implies that the vertex of smallest degree has at most 5 neighbors and that deleting a vertex from a planar graph yields a planar graph.

QED

From the above theorem, the next result follows.

Corollary 2.13 ▶ *There exists a polynomial-time coloring algorithm $\mathcal{A}$ that, when applied to a planar graph G, finds a solution with measure $m_{\mathcal{A}}(G)$ such that $m_{\mathcal{A}}(G)/m^*(G) \leq 2$.*

Program 2.5: Local Search Scheme

input Instance x;
output Locally optimal solution y;
begin
    $y :=$ initial feasible solution;
    while there exists a neighbor solution z of y better than y **do**
        $y := z$;
    return y
end.

PROOF

Since there exists a polynomial-time algorithm for optimally coloring a 2-colorable graph (see Exercise 1.15), it follows that the joint use of this algorithm and of *Smallest Last* allows us to obtain an algorithm $\mathcal{A}$ such that:

- if $m^*(G) \leq 2$, then $m_{\mathcal{A}}(G) = m^*(G)$;
- if $m^*(G) \geq 3$, then $m_{\mathcal{A}}(G) \leq 6$.

The corollary thus follows.

QED

2.3 Local search

LOCAL SEARCH algorithms start from an initial solution (found using some other algorithm) and iteratively improve the current solution by moving to a better "neighbor" solution (see Program 2.5).

Roughly speaking, if y is a feasible solution, then z is a neighbor solution if it does not differ substantially from y. Given a feasible solution y, the local search algorithm looks for a neighbor solution with an improved value of the measure function. When no improvement is possible (i.e., when the algorithm reaches a solution y such that all neighbor solutions are no better than y) then the algorithm stops in a *local optimum* (with respect to the chosen neighborhood).

To obtain a local search algorithm for a specific problem we thus need an algorithm for finding the initial feasible solution (in many cases this can be a trivial task) and a neighborhood structure of any feasible solution y (notice that it is not necessary to find all neighbors of y but it is sufficient to determine whether there exists one neighbor solution that has a better measure).

Observe that if the neighborhood structure allows us to move from one solution to a better neighbor solution in polynomial time and to find an

Chapter 2

DESIGN TECHNIQUES FOR APPROXIMATION ALGORITHMS

> **Problem 2.5: Maximum Cut**
>
> INSTANCE: Graph $G = (V, E)$.
>
> SOLUTION: Partition of V into disjoint sets V_1 and V_2.
>
> MEASURE: The cardinality of the cut, i.e., the number of edges with one endpoint in V_1 and one endpoint in V_2.

optimal solution (i.e., a global optimum) after a polynomial number of solutions have been considered, then the local search algorithm solves the problem optimally in polynomial time. Clearly, in the case of NP-hard problems, we do not expect to find a neighborhood structure that allows us to find an optimal solution in polynomial time (unless P = NP). In these cases, we look for a local search algorithm that finds an approximate solution corresponding to a local optimum with respect to the neighborhood structure.

2.3.1 Local search algorithms for the cut problem

In this section, we will describe a local search algorithm for MAXIMUM CUT (see Problem 2.5) that achieves a solution that is guaranteed to be at most a constant factor away from the optimal solution. To this aim, we need to define both the procedure for obtaining an initial feasible solution and the neighborhood structure. In the case of the MAXIMUM CUT the first task is trivial, since the partition $V_1 = \emptyset$ and $V_2 = V$ is a feasible solution.

Furthermore, we define the following neighborhood structure $\mathcal{N}$: the neighborhood of a solution (V_1, V_2) consists of all those partitions (V_{1k}, V_{2k}), for $k = 1, \ldots, |V|$, such that:

1. if vertex $v_k \in V_1$, then

$$V_{1k} = V_1 - \{v_k\} \quad \text{and} \quad V_{2k} = V_2 \cup \{v_k\};$$

2. if vertex $v_k \notin V_1$, then

$$V_{1k} = V_1 \cup \{v_k\} \quad \text{and} \quad V_{2k} = V_2 - \{v_k\}.$$

The important property of the above neighborhood structure is that each local optimum has a measure that is at least half of the optimum.

Theorem 2.14 ▶ *Given an instance x of* MAXIMUM CUT*, let (V_1, V_2) be a local optimum*

with respect to the neighborhood structure $\mathcal{N}$ and let $m_{\mathcal{N}}(x)$ be its measure. Then
$$m^*(x)/m_{\mathcal{N}}(x) \leq 2.$$

Let m be the number of edges of the graph. Since $m^*(x) \leq m$, it is sufficient to prove that $m_{\mathcal{N}}(x) \geq m/2$.

We denote by m_1 and m_2 the number of edges connecting vertices inside V_1 and V_2 respectively. We have that

$$m = m_1 + m_2 + m_{\mathcal{N}}(x). \qquad (2.7)$$

Given any vertex v_i, we define

$$m_{1i} = \{v \mid v \in V_1 \text{ and } (v, v_i) \in E\}$$

and

$$m_{2i} = \{v \mid v \in V_2 \text{ and } (v, v_i) \in E\}.$$

If (V_1, V_2) is a local optimum then, for any vertex v_k, the solution provided by (V_{1k}, V_{2k}) has a value at most $m_{\mathcal{N}}(x)$. This implies that, for every node $v_i \in V_1$,

$$|m_{1i}| - |m_{2i}| \leq 0$$

and, for every node $v_j \in V_2$,

$$|m_{2j}| - |m_{1j}| \leq 0.$$

By summing over all vertices in V_1 and V_2, we obtain

$$\sum_{v_i \in V_1} (|m_{1i}| - |m_{2i}|) = 2m_1 - m_{\mathcal{N}}(x) \leq 0$$

and

$$\sum_{v_j \in V_2} (|m_{2j}| - |m_{1j}|) = 2m_2 - m_{\mathcal{N}}(x) \leq 0.$$

Hence $m_1 + m_2 - m_{\mathcal{N}}(x) \leq 0$. From this inequality and from Eq. 2.7 it follows that $m_{\mathcal{N}}(x) \geq m/2$ and the theorem is proved.

Section 2.3
LOCAL SEARCH

PROOF

QED

For a given optimization problem, there are many possible definitions of neighborhood structure that may determine both the quality of the approximate solution and the running time of the algorithm. The main issues involved in the definition of the neighborhood structure are:

- the quality of the solution obtained (that is, how close is the value of the local optimum to the global optimal value);

Chapter 2

DESIGN TECHNIQUES FOR APPROXIMATION ALGORITHMS

- the order in which the neighborhood is searched;

- the complexity of verifying that the neighborhood does not contain any better solution;

- the number of solutions generated before a local optimum is found.

The above four issues are strongly related. In fact, if the neighborhood structure is "large" then it is likely that the value of the solution obtained is close to the optimal value. However, in this case we expect the task of checking that a better neighbor does not exist to become more complex.

As an extreme case, we can assume that any solution is neighbor of any other solution: in this case, the algorithm will find an optimal solution, but if the problem is computationally hard, then either the complexity of looking for a better neighbor is computationally hard or there is an exponential number of solutions to be considered before the global optimum is found.

2.3.2 Local search algorithms for the salesperson problem

The interesting property of the neighborhood structure defined for MAXIMUM CUT is that all local optima have a measure that is at least half of the optimum. Unfortunately, this nice property is not shared by many other combinatorial optimization problems. However, local search approximation algorithms are applied successfully due to their good practical behavior and their simple structure. For example, experimental work has shown that, in the case of MINIMUM TRAVELING SALESPERSON, there exist local search algorithms that find approximate good solutions in a reasonable amount of time for very large instances (even though there are instances in which the performance of the algorithm is quite bad).

Obtaining an initial feasible solution for an instance of MINIMUM TRAVELING SALESPERSON is easy since any permutation of the n cities is a solution: therefore, the identity permutation of the n cities $(c_1, c_2, \ldots, c_n)$ is a feasible tour. Alternatively, we can consider the solution found by any other algorithm (e.g, *Nearest Neighbor*).

A simple neighborhood structure for the symmetric case of MINIMUM TRAVELING SALESPERSON is the *2-opt* structure which is based on the following observation: given a tour I, replacing two edges (x,y) and (v,z) in I by the two edges (x,v) and (y,z) yields another tour I'. This swap operation is called a *2-move*. The 2-opt neighborhood structure of a solution I consists of those solutions that can be obtained from I using a 2-move. Observe that, according to this neighborhood structure, any solution has

$O(n^2)$ neighbors: hence, searching in a neighborhood for a better solution requires at most $O(n^2)$ operations.

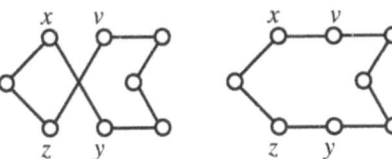

Section 2.4

LINEAR PROGRAMMING BASED ALGORITHMS

Figure 2.6
Improving a tour by performing a 2-move

Let us consider the Euclidean version of MINIMUM TRAVELING SALESPERSON in which it is assumed that cities correspond to points in the plane and that the distance between cities x and y is given by the Euclidean distance between the corresponding points in the plane. Figure 2.6 represents a tour: since edges (x,y) and (v,z) cross each other, the triangle inequality implies that replacing (x,y) and (v,z) by (x,v) and (y,z) yields a better solution.

◀ Example 2.8

Note that given a tour I the absence of crossing edges does not guarantee that I is a local optimum with respect to the 2-opt neighborhood structure (see, for instance, Fig. 2.7).

Experimental work has shown that the quality of the solution obtained by using the 2-opt neighborhood structure depends on the solution initially chosen and that if the local search algorithm is executed several times using different initial solutions, then the overall algorithm can be very effective. For example, in the case of randomly distributed points in the plane, such an algorithm finds a solution with expected value within 5.5% of the optimal value.

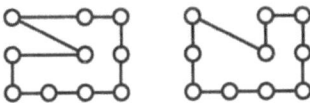

Figure 2.7
A tour with no crossing edges and its improvement

A more detailed presentation of local search algorithms for MINIMUM TRAVELING SALESPERSON is given in Chap. 10.

2.4 Linear programming based algorithms

LINEAR PROGRAMMING is one of the most successful areas of operations research and has been applied in many application contexts (we refer to Appendix A for related definitions).

Observe that, since a linear program can be solved in polynomial time, given a hard combinatorial optimization problem $\mathcal{P}$, we do not expect to

Chapter 2

DESIGN TECHNIQUES FOR APPROXIMATION ALGORITHMS

find a linear programming formulation of $\mathcal{P}$ such that, for any instance x of $\mathcal{P}$, the number of constraints of the corresponding linear program is polynomial in the size of x: in fact, this would imply that P= NP.[1]

Nevertheless, linear programming can be used as a computational step in the design of approximation algorithms. Several approaches are possible that are based on the fact that most combinatorial optimization problems can be formulated as *integer* linear programming problems. This equivalent formulation does not simplify the problem but allows us to design algorithms that, in some cases, give good approximate solutions.

2.4.1 Rounding the solution of a linear program

The simplest approach to obtain approximation algorithms based on linear programming can be described as follows. Given an integer linear program, by relaxing the integrality constraints we obtain a new linear program, whose optimal solution can be found in polynomial time. This solution, in some cases, can be used to obtain a feasible solution for the original integer linear program, by "rounding" the values of the variables that do not satisfy the integrality constraints.

As an example, let us consider the weighted version of MINIMUM VERTEX COVER (which we will denote as MINIMUM WEIGHTED VERTEX COVER) in which a non-negative weight c_i is associated with each vertex v_i and we look for a vertex cover having minimum total weight. Given a weighted graph $G = (V, E)$, MINIMUM WEIGHTED VERTEX COVER can be formulated as the following integer linear program $ILP_{VC}(G)$:

$$\begin{aligned}
\text{minimize} \quad & \sum_{v_i \in V} c_i x_i \\
\text{subject to} \quad & x_i + x_j \geq 1 \quad \forall (v_i, v_j) \in E \\
& x_i \in \{0, 1\} \quad \forall v_i \in V.
\end{aligned}$$

Let LP_{VC} be the linear program obtained by relaxing the integrality constraints to simple non-negativeness constraints (i.e., $x_i \geq 0$ for each $v_i \in V$). Let $x^*(G)$ be an optimal solution of LP_{VC}. Program 2.6 obtains a feasible solution V' for the MINIMUM WEIGHTED VERTEX COVER problem by rounding up all components of $x^*(G)$ with a sufficiently large value. Namely, it includes in the vertex cover all vertices corresponding to components whose value is at least 0.5.

Theorem 2.15 ▶ *Given a graph G with non-negative vertex weights, Program 2.6 finds a*

---

[1] Note that we might be able to find a linear programming formulation such that the number of constraints is exponential in the number of variables.

Program 2.6: Rounding Weighted Vertex Cover

input Graph $G = (V, E)$ with non-negative vertex weights;
output Vertex cover V' of G;
begin
    Let ILP_{VC} be the linear integer programming formulation of the problem;
    Let LP_{VC} be the problem obtained from ILP_{VC} by relaxing
        the integrality constraints;
    Let $x^*(G)$ be the optimal solution for LP_{VC};
    $V' := \{v_i \mid x_i^*(G) \geq 0.5\}$;
    return V'
end.

Section 2.4
LINEAR PROGRAMMING BASED ALGORITHMS

feasible solution of MINIMUM WEIGHTED VERTEX COVER *with value* $m_{LP}(G)$ *such that* $m_{LP}(G)/m^*(G) \leq 2$.

Let V' be the solution returned by the algorithm. The feasibility of V' can be easily proved by contradiction. In fact, assume that V' does not cover edge (v_i, v_j). This implies that both $x_i^*(G)$ and $x_j^*(G)$ are less than 0.5, thus contradicting the fact that $x^*(G)$ is a feasible solution of the relaxed linear program.

In order to prove that V' is a solution whose value is at most twice the optimal value, we first observe that the value $m^*(G)$ of an optimal solution satisfies the inequality:

$$m^*(G) \geq m_{LP}^*(G)$$

where $m_{LP}^*(G)$ denotes the optimal measure of the relaxed linear program. Since

$$\sum_{v_i \in V'} c_i \leq 2 \sum_{v_i \in V} c_i x_i^*(G) = 2m_{LP}^*(G) \leq 2m^*(G),$$

the theorem then follows.

PROOF

QED

2.4.2 Primal-dual algorithms

The implementation of Program 2.6 requires the solution of a linear program with a possibly large number of constraints (in fact, the number of constraints is equal to the number of edges of the graph) and, therefore, it is computationally expensive. A different approach (still based on linear programming) allows us to obtain an approximate solution more efficiently.

The method is known as *primal-dual* and uses the dual of the linear program obtained by relaxing the integrality constraints. Recall that, given a minimization linear program LP, its dual DLP is a maximization linear

Chapter 2

DESIGN TECHNIQUES FOR APPROXIMATION ALGORITHMS

program whose optimal value coincides with the optimal value of *LP* (see Appendix A). Therefore, if we consider a minimization *integer* linear program *ILP* whose relaxation provides *LP*, any feasible solution of *DLP* has a measure no greater than the optimal measure of *ILP* (which, in turn, is no greater than the value of any feasible solution of *ILP*) and can, thus, be used as a lower bound when estimating the quality of an approximate solution of *ILP* (see Fig. 2.8).

A primal-dual algorithm exploits this property to find approximate solutions of an integer linear program *ILP*: in particular, it simultaneously maintains a (possibly non-feasible) integer solution x of *ILP* and a (not necessarily optimal) feasible solution of *DLP*. At each step, x and y are examined and modified to derive a new pair of solutions x' and y' where x' is "more feasible" than x and y' has a better measure than y. The algorithm ends when the integer solution becomes feasible: the quality of this solution is evaluated by comparing it with the final dual solution. This approach allows us to obtain faster algorithms because it is not necessary to optimally solve either *ILP* or *DLP*. Moreover, as we will see in the rest of this section, there are cases in which the method allows us to obtain solutions with a good performance ratio.

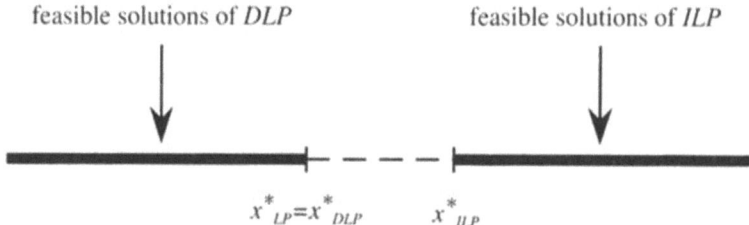

Figure 2.8
The space of values of feasible solutions of *ILP* and *DLP*

In particular, let us formulate a primal-dual algorithm for MINIMUM WEIGHTED VERTEX COVER. First observe that, given a weighted graph $G = (V,E)$, the dual of the previously defined relaxation LP_{VC} is the following linear program DLP_{VC}:

$$\begin{aligned}
\text{maximize} \quad & \sum_{(v_i,v_j) \in E} y_{ij} \\
\text{subject to} \quad & \sum_{j:(v_i,v_j) \in E} y_{ij} \leq c_i \quad \forall v_i \in V \\
& y_{ij} \geq 0 \quad \forall (v_i, v_j) \in E.
\end{aligned}$$

Note that the empty set is an unfeasible integer solution of MINIMUM WEIGHTED VERTEX COVER (that is, of the initial integer linear program) while the solution in which all y_{ij} are zero is a feasible solution with value 0 of DLP_{VC}. The primal-dual algorithm starts from this pair of solutions

Program 2.7: Primal-Dual Weighted Vertex Cover

input Graph $G = (V, E)$ with non-negative vertex weights;
output Vertex cover V' of G;
begin
    Let ILP_{VC} be the integer linear programming formulation of the problem;
    Let DLP_{VC} be the dual of the linear programming relaxation of ILP_{VC};
    for each dual variable y_{ij} of DLP_{VC} **do** $y_{ij} := 0$;
    $V' := \emptyset$;
    while V' is not a vertex cover **do**
    begin
        Let (v_i, v_j) be an edge not covered by V';
        Increase y_{ij} until a constraint of DLP_{VC} becomes tight for either i or j;
        if $\sum_{k:(v_i,v_k)\in E} y_{ik} = c_i$ **then**
            $V' := V' \cup \{v_i\}$ (* the i-th dual constraint is tight *)
        else
            $V' := V' \cup \{v_j\}$ (* the j-th dual constraint is tight *)
    end;
    return V'
end.

and constructs a vertex cover (i.e., a feasible primal solution) by looking for a better dual solution (see Program 2.7).

◂ **Theorem 2.16** Given a graph G with non-negative weights, Program 2.7 finds a feasible solution of MINIMUM WEIGHTED VERTEX COVER with value $m_{PD}(G)$ such that $m_{PD}(G)/m^*(G) \leq 2$.

PROOF Let V' be the solution obtained by the algorithm. By construction V' is a feasible solution. For the analysis of its quality, first observe that for every $v_i \in V'$ we have $\sum_{j:(v_i,v_j)\in E} y_{ij} = c_i$. Therefore,

$$m_{PD}(G) = \sum_{v_i \in V'} c_i = \sum_{v_i \in V'} \sum_{j:(v_i,v_j)\in E} y_{ij} \leq \sum_{v_i \in V} \sum_{j:(v_i,v_j)\in E} y_{ij} = 2 \sum_{(v_i,v_j)\in E} y_{ij}.$$

Since $\sum_{(v_i,v_j)\in E} y_{ij} \leq m^*(G)$ (that is, the value of the feasible dual solution obtained by the algorithm is always at most the value of the primal optimal solution), the theorem hence follows. **QED**

2.5 Dynamic programming

DYNAMIC PROGRAMMING is an algorithmic technique that, in some cases, allows us to reduce the size of the search space while looking

Chapter 2

DESIGN TECHNIQUES FOR APPROXIMATION ALGORITHMS

for an optimal solution and that, for this reason, has been applied to many combinatorial problems.

Roughly speaking, dynamic programming can be applied to any problem for which an optimal solution of the problem can be derived by composing optimal solutions of a limited set of "subproblems", regardless of how these solutions have been obtained (this is generally called the *principle of optimality*). Due to efficiency reasons, this top-down description of the technique is usually translated into a bottom-up programming implementation in which subproblems are defined with just a few indices and "subsolutions" are optimally extended by means of iterations over these indices.

In this section we will present the dynamic programming technique by first giving an (exponential-time) exact algorithm for MAXIMUM KNAPSACK and by subsequently using this algorithm to design a family of polynomial-time approximation algorithms.

In order to apply the dynamic programming technique to MAXIMUM KNAPSACK, we need to specify the notion of subproblem so that the principle of optimality is satisfied. Recall that an instance of the problem consists of a positive integer knapsack capacity b and of a finite set X of n items such that, for each $x_i \in X$, a profit $p_i \in Z^+$ and a size $a_i \in Z^+$ are specified.

For any k with $1 \leq k \leq n$ and for any p with $0 \leq p \leq \sum_{i=1}^{n} p_i$, we then consider the problem of finding a subset of $\{x_1, \ldots, x_k\}$ which minimizes the total size among all those subsets having total profit equal to p and total size at most b: we denote with $M^*(k, p)$ an optimal solution of this problem and with $S^*(k, p)$ the corresponding optimal size (we assume that, whenever $M^*(k, p)$ is not defined, $S^*(k, p) = 1 + \sum_{i=1}^{n} a_i$).

Clearly, $M^*(1, 0) = \emptyset$, $M^*(1, p_1) = \{x_1\}$, and $M^*(1, p)$ is not defined for any positive integer $p \neq p_1$. Moreover, for any k with $2 \leq k \leq n$ and for any p with $0 \leq p \leq \sum_{i=1}^{n} p_i$, the following relationship holds (this is the formal statement of the principle of optimality in the case of MAXIMUM KNAPSACK):

$$M^*(k, p) = \begin{cases} M^*(k-1, p-p_k) \cup \{x_k\} & \text{if } p_k \leq p, M^*(k-1, p-p_k) \\ & \text{is defined, } S^*(k-1, p) \text{ is at} \\ & \text{least } S^*(k-1, p-p_k) + a_k, \\ & \text{and } S^*(k-1, p-p_k) + a_k \leq b, \\ M^*(k-1, p) & \text{otherwise.} \end{cases}$$

That is, the best subset of $\{x_1, \ldots, x_k\}$ that has total profit p is either the best subset of $\{x_1, \ldots, x_{k-1}\}$ that has total profit $p - p_k$ plus item x_k or the best subset of $\{x_1, \ldots, x_{k-1}\}$ that has total profit p. Since the best subset

Section 2.5
DYNAMIC PROGRAMMING

Program 2.8: Dynamic Programming Maximum Knapsack

input Set X of n items, for each $x_i \in X$, values p_i, a_i, positive integer b;
output Subset $Y \subseteq X$ such that $\sum_{x_i \in Y} a_i \leq b$;
begin
  for $p := 0$ **to** $\sum_{i=1}^{n} p_i$ **do**
  begin
    $M^*(1,p) := $ undefined;
    $S^*(1,p) := 1 + \sum_{i=1}^{n} a_i$;
  end;
  $M^*(1,0) := \emptyset; S^*(1,0) := 0$;
  $M^*(1,p_1) := \{x_1\}; S^*(1,p_1) := a_1$;
  for $k := 2$ **to** n **do**
    for $p := 0$ **to** $\sum_{i=1}^{n} p_i$ **do**
    begin
      if $(p_k \leq p)$ **and** $(M^*(k-1, p-p_k) \neq$ undefined$)$
        and $(S^*(k-1, p-p_k) + a_k \leq S^*(k-1, p))$
        and $(S^*(k-1, p-p_k) + a_k \leq b)$ **then**
      begin
        $M^*(k,p) := M^*(k-1, p-p_k) \cup \{x_k\}$;
        $S^*(k,p) := S^*(k-1, p-p_k) + a_k$
      end
      else
      begin
        $M^*(k,p) := M^*(k-1,p)$;
        $S^*(k,p) := S^*(k-1,p)$
      end
    end;
  $p^* := $ maximum p such that $M^*(n,p) \neq$ undefined;
  return $M^*(n, p^*)$
end.

of $\{x_1, \ldots, x_k\}$ that has total profit p must either contain x_k or not, one of these two choices must be the right one.

From the above relationship, it is now possible to derive an algorithm that, for any instance of MAXIMUM KNAPSACK, computes an optimal solution: this algorithm is shown in Program 2.8.

◀ **Theorem 2.17**

Given an instance x of MAXIMUM KNAPSACK *with n items, Program 2.8 finds an optimal solution of x in time* $O(n \sum_{i=1}^{n} p_i)$ *where p_i denotes the profit of the i-th item.*

PROOF

The correctness of the algorithm is implied by the principle of optimality in the case of MAXIMUM KNAPSACK. In order to bound the running time

Chapter 2

DESIGN TECHNIQUES FOR APPROXIMATION ALGORITHMS

Program 2.9: Knapsack Approximation Scheme

input Set X of n items, for each $x_i \in X$, values p_i, a_i, positive integer b, rational number $r > 1$;
output Subset $Y \subseteq X$ such that $\sum_{x_i \in Y} a_i \leq b$;
begin
  $p_{max} :=$ maximum among values p_i;
  $t := \lfloor \log(\frac{r-1}{r} \frac{p_{max}}{n}) \rfloor$;
  $x' :=$ instance with profits $p'_i = \lfloor p_i/2^t \rfloor$;
  remove from x' items with zero profit;
  $Y :=$ solution returned by Program 2.8 with input x';
  return Y
end.

it is sufficient to observe that the execution of the body of both the first and the third **for** loop of Program 2.8 requires a constant number of steps.

QED

The running time of Program 2.8 is polynomial in the values of the profits associated with the items of the instance of the problem. These values are exponential in the length of the input if we use any reasonable encoding scheme (in fact $\log p_i$ bits are sufficient to encode the value p_i) and for this reason the algorithm is not a polynomial-time one. However, in order to stress the fact that the running time is polynomial in the value of the profits, we will say that the running time of the algorithm is *pseudo-polynomial*.

Besides being interesting by itself, Program 2.8 can be used to obtain a polynomial-time algorithm that, given an instance x of MAXIMUM KNAPSACK and a bound on the desired performance ratio, returns a feasible solution of x whose quality is within the specified bound. The algorithm works in the following way: instead of directly solving the given instance x, it solves an instance x' which is obtained by scaling down all profits by a power of 2 (depending on the desired degree of approximation). Instance x' is then solved by Program 2.8 and from its optimal solution the approximate solution of the original instance x is finally derived. The algorithm is shown in Program 2.9: since the algorithm's behavior depends on the required performance ratio, it is called an *approximation scheme* for MAXIMUM KNAPSACK.

Theorem 2.18 ▶ *Given an instance x of* MAXIMUM KNAPSACK *with n items and a rational number $r > 1$, Program 2.9 returns a solution in time $O(rn^3/(r-1))$ whose measure $m_{AS}(x,r)$ satisfies the inequality $m^*(x)/m_{AS}(x,r) \leq r$.*

PROOF   Let $Y(x,r)$ be the approximate solution computed by Program 2.9 with input x and r and let $Y^*(x)$ be an optimal solution of x, with measure $m^*(x)$.

It is easy to see that, for any item inserted in $Y(x,r)$, the largest additive error introduced by pruning the profit is at most 2^t; therefore, we have that $m^*(x) - m_{AS}(x,r) \leq n2^t$. Moreover, if p_{max} is the largest profit of an item, we then have that $np_{max} \geq m^*(x) \geq p_{max}$. Hence,

$$\frac{m^*(x) - m_{AS}(x,r)}{m^*(x)} \leq \frac{n2^t}{p_{max}},$$

that is,

$$m^*(x) \leq \frac{p_{max}}{p_{max} - n2^t} m_{AS}(x,r).$$

If we take into account that $t = \lfloor \log(\frac{r-1}{r} \frac{p_{max}}{n}) \rfloor$, then we obtain $m^*(x) \leq r \cdot m_{AS}(x,r)$.

As far as the running time is concerned, we have that it corresponds to the running time of Program 2.8 with input the scaled instance, that is, $O(n \sum_{i=1}^{n} p'_i) = O(n \sum_{i=1}^{n} p_i/2^t)$. Since p_{max} is the maximum profit of an item and because of the definition of t, we have that the running time of Program 2.9 is $O(rn^3/(r-1))$ and the theorem thus follows.

QED

The preceding result shows that Program 2.9 is an approximation scheme whose running time is polynomial both in the length of the instance and in the required quality of the solution: in the next chapter, we will see that this kind of algorithms are called *fully polynomial-time approximation schemes*.

Moreover, observe that in the preceding proof the behavior of the algorithm heavily depends (both from the point of view of the performance ratio and from that of the running time) on the upper and lower bounds to the estimated value of the optimal value $m^*(x)$, defined by the relationship $np_{max} \geq m^*(x) \geq p_{max}$. Actually, by using a tighter bound on $m^*(x)$, a better algorithm can be derived with a different definition of t whose running time is $O(rn^2/(r-1))$ (see Exercise 2.18).

The scaling technique that we have just seen is strongly related to another technique referred to as *fixed partitioning*, which we now sketch briefly. Suppose we are given an instance x of MAXIMUM KNAPSACK and a rational number $r > 1$. We divide the range of possible values of the measure function, which is bounded by np_{max}, into $\lceil np_{max}/\delta \rceil$ equal intervals of size δ, where δ is chosen according to the desired bound r on the performance ratio. Then the dynamic programming algorithm is applied by considering the problem of finding, for any k with $1 \leq k \leq n$ and for any i with $1 \leq i \leq \lceil np_{max}/\delta \rceil$, a subset of $\{x_1,\ldots,x_k\}$ which minimizes the total size among all those subsets whose total profit belongs to interval $(\delta(i-1), \delta i]$ and whose total size is at most b.

Section 2.5

DYNAMIC PROGRAMMING

It is then possible to modify Program 2.8 so that a solution of the problem relative to k and i can be computed by means of the solutions of a finite set of problems relative to $k-1$ and $j < i$ (see Exercise 2.21). At the end, the solution achieved by this procedure has a performance ratio at most r if $\delta \leq \frac{r-1}{r} p_{max}/n$. Indeed, each time we consider a new item (that is, k is increased), the absolute error of the current best solution may increase by at most δ. Thus, the absolute error of the returned solution is at most $n\delta$. Hence,

$$\frac{m^*(x) - m(x,Y)}{m^*(x)} \leq \frac{n\delta}{p_{max}} \leq \frac{r-1}{r},$$

that is,

$$1 - \frac{m(x,Y)}{m^*(x)} \leq \frac{r-1}{r}$$

which implies that the performance ratio is at most r. Notice that, since the number of intervals is $O(n^2 r/(r-1))$, the above procedure is a fully polynomial-time approximation scheme.

Clearly, the result is the same as we would have achieved by adopting a scaling factor δ: we are indeed dealing with two ways to look at the same technique rather than with two different techniques.

The scaling (or fixed partitioning) technique can be applied to construct fully polynomial-time approximation schemes for many other optimization problems (see Exercise 2.22).

2.6 Randomized algorithms

IN RECENT years there has been increased interest in the area of randomized algorithms, which have found widespread use in many areas of computer science. The main reason for this interest is the fact that, for many applications, a randomized algorithm is either simpler or faster (or both) than a deterministic algorithm. In this section we will consider randomized techniques for combinatorial optimization problems, and we will see an example of how they can be used to design simple approximation algorithms.

Roughly speaking, a randomized algorithm is an algorithm that during some of its steps performs random choices. Note that the random steps performed by the algorithm imply that by executing the algorithm several times with the same input we are not guaranteed to find the same solution. Actually, in the case of optimization problems we can find solutions whose measure might differ significantly. More precisely, given an instance of a combinatorial optimization problem, the value of the solution found by a

Program 2.10: Random Satisfiability

input Set C of disjunctive clauses on a set of variables V;
output Truth assignment $f : V \mapsto \{\text{TRUE}, \text{FALSE}\}$;
begin
    For each $v \in V$, independently set $f(v) = \text{TRUE}$ with probability $1/2$;
    return f
end.

Section 2.6
RANDOMIZED ALGORITHMS

randomized approximation algorithm is a random variable: when estimating the expected value of this random variable, we thus have to consider the behavior of the algorithm averaging over all possible executions.

In this section, we present a very simple randomized algorithm for MAXIMUM SATISFIABILITY (see Problem 1.16 and Program 2.10) which has a good average performance. Let $m_{RS}(x)$ be the random variable denoting the value of the solution found by the algorithm with input x. The following result provides a bound on its expected value, assuming that all clauses in C have at least k literals.

◀ **Theorem 2.19** Given an instance x of MAXIMUM SATISFIABILITY in which all c clauses have at least k literals, the expected measure $m_{RS}(x)$ of the solution found by Program 2.10 satisfies the following inequality:

$$E[m_{RS}(x)] \geq \left(1 - \frac{1}{2^k}\right) c.$$

PROOF The probability that any clause with k literals is not satisfied by the truth assignment found by the algorithm is 2^{-k} (which is the probability that all literals in the clause have been assigned the value FALSE). Therefore the probability that a clause with at least k literals is satisfied is at least $1 - 2^{-k}$. It follows that the expected contribution of a clause to $m_{RS}(x)$ is at least $1 - 2^{-k}$. By summing over all clauses we obtain the desired inequality and the theorem follows. **QED**

◀ **Corollary 2.20** Given an instance x of MAXIMUM SATISFIABILITY, the expected measure $m_{RS}(x)$ of the solution found by Program 2.10 satisfies the following inequality: $m^*(x)/E[m_{RS}(x)] \leq 2$.

PROOF The corollary derives immediately from the previous theorem and from the fact that the optimal solution has always measure at most c. **QED**

Chapter 2
DESIGN TECHNIQUES FOR APPROXIMATION ALGORITHMS

Note that the above corollary holds for all instances of the problem and that the average is taken over different executions of the algorithm. However, it does not guarantee that the algorithm *always* finds a good approximate solution.

2.7 Approaches to the approximate solution of problems

IN THE previous sections, given a computationally hard optimization problem, we have evaluated the performance of an approximation algorithm, that is, an algorithm that provides a non-optimal solution. In particular, given an instance of the problem, we have shown how far the solution found by the algorithm was from the optimal solution. Clearly, we have only been able to derive upper bounds on this distance since an exact evaluation of the performance ratio would require running both the approximation algorithm and an exact algorithm.

Several possible approaches to the analysis of approximation algorithms will be thoroughly presented in the remainder of the book. In this section we describe these approaches briefly.

2.7.1 Performance guarantee: chapters 3 and 4

In the performance guarantee approach, we require that, for all instances of the problem, the performance ratio of the solution found by the algorithm is bounded by a suitable function. In particular, we are mainly interested in the cases in which this function is a constant. As an example, Theorem 2.15 shows that, for any graph with positive weights on the vertices, Program 2.6 finds a feasible solution of MINIMUM WEIGHTED VERTEX COVER whose measure is bounded by twice the optimum (i.e., $m_{LP}(G) \leq 2m^*(G)$).

According to the performance guarantee approach we are interested in determining the algorithm with the minimum performance ratio. Note that if algorithm $\mathcal{A}_1$ has a better performance ratio than algorithm $\mathcal{A}_2$ this does not imply that, for all instances of the problem, the solution found by $\mathcal{A}_1$ is always closer to the optimal solution than the solution found by $\mathcal{A}_2$. In fact the performance guarantee approach uses the worst possible case to measure the performance ratio of the algorithm.

We have also seen examples in which we cannot prove similar results: in the case of MINIMUM GRAPH COLORING, all the approximate algorithms we have presented have "bad" instances in which the number of colors used cannot be bounded by a linear function of the minimum number of colors. In fact, for this problem it is possible to prove a strong negative result: if there exists an approximation algorithm $\mathcal{A}$ that, for each instance x of the problem with n vertices, finds a solution with measure $m_\mathcal{A}(x)$ such that $m_\mathcal{A}(x) \leq n^c m^*(x)$, where c is any constant with $0 < c < 1/7$, then $\text{P} = \text{NP}$.

Therefore we do not expect to design a polynomial-time algorithm for MINIMUM GRAPH COLORING that always finds a good approximate solution with respect to the performance guarantee approach. Again, this does not imply that approximation algorithms for the problem with good practical behavior cannot be designed. As we will see, similar considerations apply to MAXIMUM INDEPENDENT SET as well.

For such hard problems, it is sometimes possible to analyze the behavior of approximation algorithms by properly restricting attention to specific classes of input instances. As an example, we have seen that we can bound the performance ratio of approximation algorithms for MINIMUM GRAPH COLORING when we restrict ourselves to planar graphs.

2.7.2 Randomized algorithms: chapter 5

A randomized algorithm is an algorithm that might perform random choices among different possibilities (note that no probabilistic assumption is made on the set of instances). Clearly, repeated executions of the algorithm with the same input might give different solutions with different values depending on the random choices. More precisely, for each given instance of the problem, the value of the solution found by a randomized algorithm is a random variable defined over the set of possible choices of the algorithm. As in Theorem 2.19, we are interested in studying the expected value of this random variable by obtaining bounds on the expected quality of the solution found by the algorithm. A randomized algorithm with good expected behavior is not guaranteed to find a good solution for all executions and for all problem instances: however, in most cases, randomized algorithms are simple to implement and very efficient.

2.7.3 Probabilistic analysis: chapter 9

The performance guarantee approach analyzes the quality of an algorithm on the basis of its worst possible behavior. This might not be realistic if "bad" instances do not occur often and, for all remaining cases, the algorithm is able to find a solution that is very close to the optimum. Often we are interested in knowing the behavior of the algorithm for the set of

Chapter 2

DESIGN TECHNIQUES FOR APPROXIMATION ALGORITHMS

instances that might actually arise in practice. Since, in most cases, it is impossible to define such a class of instances, we might be interested in analyzing the behavior of the algorithm with respect to the "average input" of the problem.

Before performing the analysis, we have to choose a probability distribution on the set of the instances of the problem. In some cases, an equiprobability assumption can be acceptable, and therefore the average instance is simply determined by evaluating the statistical mean over all instances of a fixed size. However, there are problems for which it is not reasonable to assume equiprobability, and it might not be clear which input distribution is realistic for practical applications.

A final remark concerning the probabilistic analysis of approximation algorithms is that it requires the use of sophisticated mathematical tools. In fact, at the present state of the art, only simple algorithms for simple input distributions have been analyzed: in most cases, an analysis of sophisticated algorithms with respect to practical applications has not been made.

2.7.4 Heuristics: chapter 10

The previously described approaches have considerably enlarged our ability to cope with NP-hard optimization problems. However, the theoretical analysis performed using these approaches is not completely satisfactory for three main reasons.

The first reason is that there exist problems that are not efficiently solvable with any of the above approaches. The second reason is that, in the solution of large instances of a problem, it is not sufficient to devise approximation algorithms that are polynomial-time bounded because we need algorithms with a low level of complexity: in these cases, even a quadratic-time algorithm might be too expensive. Finally, there exist approximation algorithms for which we are not able to prove accurate bounds using the previous approaches.

In the last chapter of the book, we will then consider *heuristics*, that is, approximation algorithms that are characterized by a good practical behavior even though the value of the solution obtained is not provably good (either from a worst case or from a probabilistic point of view). Heuristics are examples of an approach that is different from all the others already considered and that, at the same time, has shown its usefulness for important problems.

The evaluation of a heuristic algorithm is based on executing the algorithm on a large set of instances and averaging the behavior on this set:

the set of instances can be either randomly generated or based on instances arising in real problems. The 2-opt algorithm for MINIMUM TRAVELING SALESPERSON is an example. We have seen examples in which this algorithm finds a solution that is far from the optimum and we are currently unable to perform a precise probabilistic analysis of its behavior. Moreover, no current implementation of the algorithm guarantees that the running time of the algorithm is polynomial in the number of cities. However, practical experiments with this algorithm are very satisfactory: the execution of the algorithm on an extremely large set of instances shows that the observed running time is almost linear in the number of cities and that the solution found is, on average, a few percent away from the optimum.

2.7.5 Final remarks

Before concluding this chapter, we observe that there exist approaches to the approximate solution of combinatorial problems that are not considered in this book.

For example, we are not interested in "on-line" problems in which the information concerning the input is not complete before the execution of the algorithm. Indeed, in an on-line problem the input instance is disclosed to the algorithm one piece at a time and the algorithm must take decisions concerning some input variables before knowing the next piece of information.

Furthermore, since we assume that the input instance can be represented in an exact way, we are not interested in a notion of approximation such as that of real numbers with rational numbers. It is well known that round-off errors may cause serious mistakes, and a theory of errors was created with the aim of deriving reliable computations in this setting.

Finally, we will not consider stochastic optimization that deals with the solution of problems for which the information concerning an instance is given by random variables or by a stochastic process.

2.8 Exercises

Exercise 2.1 Prove that the bound of Theorem 2.1 is tight, namely, that, for any $\varepsilon > 0$, there exists an instance x of MAXIMUM KNAPSACK such that $m^*(x)/m_H(x) > (2-\varepsilon)$.

Exercise 2.2 (*) Prove that the bounds provided by Theorems 2.2 and 2.3 are tight.

Chapter 2

DESIGN TECHNIQUES FOR APPROXIMATION ALGORITHMS

Exercise 2.3 Prove that for any constant $c > 1$ and for any $n > 3$, there exists an instance $x_{c,n}$ of MINIMUM TRAVELING SALESPERSON with n cities such that *Nearest Neighbor* achieves a solution of measure $m_{NN}(x_{c,n})$ such that $m_{NN}(x_{c,n})/m^*(x_{c,n}) > c$.

Exercise 2.4 (*) Prove that for any constant $c > 1$, there are infinitely many instances $x_{c,n}$ of MINIMUM METRIC TRAVELING SALESPERSON with n cities such that *Nearest Neighbor* achieves a solution of measure $m_{NN}(x_{c,n})$ such that $m_{NN}(x_{c,n})/m^*(x_{c,n}) > c$. (Hint: for any $i > 0$, derive an instance for which *Nearest Neighbor* finds a tour whose length is at least $(i+2)/6$ times the optimal length.)

Exercise 2.5 Consider the MINIMUM SCHEDULING ON IDENTICAL MACHINES problem. Show that the LPT rule provides an optimal solution in the case in which $l_{min} > m^*(x)/3$.

Exercise 2.6 Prove that the bound of Theorem 2.7 is tight if $p = 2$.

Exercise 2.7 Consider the variation of the LPT rule that assigns the two longest jobs to the first machine and, subsequently, applies the LPT rule to the remaining jobs. Prove that if $p = 2$ then the performance ratio of the solution provided by choosing the best solution between the one given by this variation and the one returned by the original LPT rule is strictly better than the performance ratio of the solution provided by the LPT rule.

Exercise 2.8 Prove that the bound of Theorem 2.9 is tight in the sense that no better multiplicative factor can be obtained if the additive factor is 1.

Exercise 2.9 Show that there exists an instance of MINIMUM BIN PACKING for which the number of bins used by the solution computed by *Best Fit Decreasing* is greater than the number of bins in the solution computed by *First Fit Decreasing*.

Exercise 2.10 Consider the following variant of MINIMUM BIN PACKING: the input instance is defined by a set of n items $\{x_1, x_2, \ldots, x_n\}$ whose sum is at most m. The goal is to maximize the number of items that are packed in m bins of unitary capacity. A sequential algorithm for this problems that is similar to *First Fit* considers items in the given order and tries to pack each item in the first available bin that can include it. If none of the m bins can accomodate item x_i then x_i is not packed. Prove that the above algorithm achieves a solution that packs at least $n/2$ items.

Exercise 2.11 (*) Let us consider the generalization of MINIMUM BIN PACKING to higher dimensions, known as vector packing. In this problem the size of each item x is not a single number but a d-dimensional vector

> **Problem 2.6: Minimum Multicover**
> INSTANCE: Set $S = \{s_1, \ldots, s_n\}$, collection C of multisubsets $S_1, \ldots, S_m$ of S, where each item s_i appears with multiplicity a_{ij} in S_j, weight function $w : C \mapsto \mathbf{N}$, multiplicity function $b : S \mapsto \mathbf{N}$.
> SOLUTION: A subcollection $C' \subseteq C$ such that each item s_i appears at least $b(s_i)$ times in C'.
> MEASURE: The overall weight of C'.

Section 2.8

EXERCISES

$(x_1, x_2, \ldots, x_d)$; the bin capacity is also a d-dimensional vector $(1, 1, \ldots, 1)$ and the goal is to pack all items in the minimum number of bins given that the content of any given bin must have a vector sum less than or equal to the bin capacity. Show that the approximation algorithm that generalizes *First Fit* to higher dimensions achieves a solution whose measure is not greater than $(d+1)$ times the optimal value.

Exercise 2.12 Prove that, for any graph G,

$$2\sqrt{n} \leq \chi(G) + \chi(G^c) \leq n+1$$

where $\chi(\cdot)$ denotes the chromatic number of a graph, that is, the minimum number of colors needed to color the graph, and G^c denotes the complement graph of G. (Hint: for the second inequality, use Theorem 2.10.)

Exercise 2.13 (**) Show that Δ colors are sufficient to color a graph G with maximum degree Δ if G is neither complete nor a cycle with an odd number of vertices.

Exercise 2.14 Show that, for any sufficiently large n, there exists a 3-colorable graph with n vertices such that *Smallest Last* makes use of $\Omega(n)$ colors.

Exercise 2.15 Let us consider the special case of MAXIMUM CUT in which the required partition of the node set must have the same cardinality. Define a polynomial-time local search algorithm for this problem and evaluate its performance ratio.

Exercise 2.16 Consider Problem 2.6. Define an algorithm, based on rounding the linear programming relaxation, that finds a solution whose measure is at most p times the optimal measure, where $p = \max_i(\sum_j a_{ij})$.

Chapter 2
Design Techniques for Approximation Algorithms

Problem 2.7: Minimum Hitting Set

INSTANCE: Collection C of subsets of a finite set S.

SOLUTION: A hitting set for C, i.e., a subset $S' \subseteq S$ such that S' contains at least one element from each subset in C.

MEASURE: Cardinality of the hitting set, i.e., $|S'|$.

Problem 2.8: Maximum Integer d-dimensional Knapsack

INSTANCE: Non-negative integer $d \times n$ matrix A, vector $b \in \mathbf{N}^d$, vector $c \in \mathbf{N}^n$.

SOLUTION: A vector $x \in \mathbf{N}^n$ such that $Ax \leq b$.

MEASURE: The scalar product of c and x, i.e., $\sum_{i=1}^{n} c_i x_i$.

Exercise 2.17 (*) Consider Problem 2.7. Design a primal-dual algorithm that achieves a solution whose measure is at most k times the optimal measure, where k is the maximum cardinality of an element of C. (Hint: observe that MINIMUM HITTING SET is a generalization of MINIMUM VERTEX COVER.)

Exercise 2.18 Define a modified version of Program 2.9 in order to provide an approximation scheme for MAXIMUM KNAPSACK whose running time is $O(rn^2/(r-1))$. (Hint: make use of the greedy solution found by Program 2.1 in order to define the value of t.)

Exercise 2.19 Design a dynamic programming algorithm for finding the optimal solution of Problem 2.8.

Exercise 2.20 Design a dynamic programming algorithm for finding the optimal solution of Problem 2.9.

Exercise 2.21 Define a modified version of Program 2.8 which makes use of the fixed partitioning technique.

Exercise 2.22 Construct a polynomial-time approximation scheme for MINIMUM SCHEDULING ON IDENTICAL MACHINES in the case in which we have a fixed number of machines (that is, p is not part of the instance), by making use of the fixed partitioning technique.

> **Problem 2.9:** Maximum Integer k-choice Knapsack
>
> INSTANCE: Non-negative integer $n \times k$ matrices A and C, non-negative integer b.
>
> SOLUTION: A vector $x \in \mathbf{N}^n$, function $f : \{1,\ldots,n\} \mapsto \{1,\ldots,k\}$ such that $\sum_{i=1}^{n} a_{i,f(i)} x_i \leq b$.
>
> MEASURE: $\sum_{i=1}^{n} c_{i,f(i)} x_i$.

2.9 Bibliographical notes

MANY BOOKS deal with the design and the analysis of algorithms for combinatorial optimization problems: we refer to [Papadimitriou and Steiglitz, 1982] for general reference on the first five sections. Moreover, we refer to the list of decision problems in [Garey and Johnson, 1979] for all NP-hardness results we have cited in this chapter and in the following ones.

It is well known that the greedy method finds an optimal solution when the set of feasible solutions defines a *matroid*, that is an independent system that satisfies the additional property that all maximal independent sets have the same cardinality (see for example [Edmonds, 1971]). [Graham and Hell, 1985] is a useful source of information on the history of the greedy algorithm.

A generalization of matroids, called greedoids, is studied in [Korte, Lovasz, and Schrader, 1991] where the authors prove that in a greedoid the greedy algorithm finds an optimal solution when the objective function is a bottleneck function.

The analysis of Program 2.2 can be found in [Halldórsson and Radhakrishnan, 1994b] where also tight bounds are shown. The analysis of *Nearest Neighbor* in the symmetric case of MINIMUM METRIC TRAVELING SALESPERSON is due to [Rosenkrantz, Stearns, and Lewis, 1977], that study the behavior of several other heuristics for the problem.

The early work of Graham on multiprocessor scheduling inaugurated the study of performance guarantee approximation algorithms: [Graham, 1966, Graham, 1969] give bounds on the performance of the *List Scheduling* algorithm and of other heuristics.

In [Matula, Marble, and Isaacson, 1972, Johnson, 1974c, Matula and Beck, 1983] the behavior of sequential algorithms for MINIMUM GRAPH COLORING is analysed. In particular the authors analyzed the algorithms

Chapter 2

DESIGN TECHNIQUES FOR APPROXIMATION ALGORITHMS

presented in Sect. 2.2 together with some other ones. Also, experimental results for some heuristics are given. The statement of Exercise 2.13 is due to [Brooks, 1941], and a constructive proof can be found in [Lovász, 1975b].

The worst case performance of *First Fit* for MINIMUM BIN PACKING is analyzed in [Johnson, 1972], while [Johnson, 1972, Baker, 1983] study *First Fit Decreasing*: the currently tightest analysis of *First Fit Decreasing* is presented in [Li and Yue, 2000]. Many more algorithms have been proposed and analyzed from the worst case point of view: we refer to [Johnson, 1974b, Johnson et al., 1974, Johnson and Garey, 1985, Karmarkar and Karp, 1982]. In particular, in [Johnson et al., 1974] the authors show that *Best Fit Decreasing* is at least as good as *First Fit Decreasing* if item sizes are at least 1/6, and that the two algorithms produce identical results if item sizes are ate least 1/5: moreover, they give examples that show that sometimes *Best Fit Decreasing* and sometimes *First Fit Decreasing* produces better results. [Garey and Johnson, 1981] gave a useful survey presenting many results concerning approximation algorithms for MINIMUM BIN PACKING and scheduling problems: the survey has since then been updated (see [Coffman, Garey, and Johnson, 1997]). The survey also presents complexity results for MINIMUM VECTOR PACKING (see Exercise 2.11 for the definition of the problem). [Garey et al., 1976] analyzed appropriate variations of *First Fit* and *First Fit Decreasing* and showed that an algorithm that generalizes *First Fit* achieves a solution that is at most $(d+7/10)m^*(x)+c$, where d is the number of dimensions and c is an absolute constant (in the one-dimensional case this reduces to the known 17/10 bound); in the case of *First Fit Decreasing* the measure of the solution is at most $(d+1/3)m^* + c$.

An early description of local search can be found in [Dunham et al., 1961]. However, the first proposal of a local search heuristic originates from the idea of using edge exchange procedures for improving the length of a TSP tour [Croes, 1958, Lin, 1965]. For a general introduction to MINIMUM TRAVELING SALESPERSON we refer to [Lawler et al., 1985, Junger, Reinelt, and Rinaldi, 1994]. Computational results on these and other heuristics are reported in [Golden and Stewart, 1985, Johnson, 1990, Reinelt, 1994b] (see also Chap. 10).

Linear programming has had a strong influence on the design of algorithms for combinatorial optimization problems (in Appendix A an outline of the main concepts and methods of linear programming are provided). We refer to [Papadimitriou and Steiglitz, 1982] for general references and for several examples of how to use linear programming to obtain exact polynomial-time algorithms for combinatorial optimization problems. The use of linear programming for the analysis of approximation algorithms

Section 2.9

BIBLIOGRAPHICAL NOTES

dates back to the 1970s (see [Lovász, 1975a]). [Wolsey, 1980] shows that previously known approximation algorithms can be analyzed using linear programming. The use of linear programming in the design of approximation algorithms originates in [Bar-Yehuda and Even, 1981, Hochbaum, 1982a]: Program 2.6 was proposed in [Hochbaum, 1982a], while [Bar-Yehuda and Even, 1981] proposed Program 2.7, which is the first example of a primal-dual algorithm for obtaining an approximate solution. The above two papers analysed also approximation algorithms for the MINIMUM HITTING SET problem (see Exercise 2.17), while MINIMUM MULTICOVER was considered in [Hall and Hochbaum, 1986] (see Exercise 2.16). Recently, primal-dual algorithms have been fruitfully applied to the design of approximation algorithms for several other problems (see the survey [Goemans and Williamson, 1997]).

The algorithmic method known as dynamic programming was originally proposed by Bellman: we refer to [Dreyfus and Law, 1977] for a survey of the early applications of dynamic programming.

A reference book for the analysis and the design of randomized algorithms is [Motwani and Raghavan, 1995]. Further examples of randomized algorithms for the design of approximation algorithms will be given in Chap. 5.

Chapter 3

Approximation Classes

IN THE first chapter we have seen that, due to their inherent complexity, NP-hard optimization problems cannot be efficiently solved in an exact way, unless P = NP. Therefore, if we want to solve an NP-hard optimization problem by means of an efficient (polynomial-time) algorithm, we have to accept the fact that the algorithm does not always return an optimal solution but rather an approximate one. In Chap. 2, we have seen that, in some cases, standard algorithm design techniques, such as local search or greedy techniques, although inadequate to obtain the optimal solution of NP-hard optimization problems, are powerful enough to reach good approximate solutions in polynomial time.

In this chapter and in the following one, we will formally introduce an important type of approximation algorithms. Given an instance of an NP-hard optimization problem, such algorithms provide a solution whose performance ratio is guaranteed to be bounded by a suitable function of the input size. This type of approximation is usually called *performance guarantee approximation*. In particular, while in the next chapter we will deal with slowly increasing bounding functions, we will here consider the case in which the function is a constant. An example of this kind is the greedy algorithm for MAXIMUM KNAPSACK, which we have already met in Sect. 2.1, that efficiently finds a solution whose value is always at least one half of the optimal one.

After giving the basic definitions related to the notion of performance guarantee approximation, we will state both positive results, showing that several computationally hard problems allow efficient, arbitrarily good ap-

Chapter 3
APPROXIMATION CLASSES

proximation algorithms, and negative results, showing that for other problems, by contrast, there are intrinsic limitations to approximability.

The different behavior of NP-hard optimization problems with respect to their approximability properties will be captured by means of the definition of *approximation classes*, that is, classes of optimization problems sharing similar approximability properties. We will see that, if $P \neq NP$, then these classes form a strict hierarchy whose levels correspond to different degrees of approximation.

We will, finally, discuss some conditions under which optimization problems allow approximation algorithms with arbitrarily good guaranteed performance.

3.1 Approximate solutions with guaranteed performance

BEFORE WE formally introduce algorithms that provide approximate solutions of optimization problems with guaranteed quality, let us first observe that according to the general framework set up in Chap. 1, given an input instance x of an optimization problem, any feasible solution $y \in \text{SOL}(x)$ is indeed an approximate solution of the given problem. On such a ground an approximation algorithm may be defined as follows.

Definition 3.1 ▶ *Given an optimization problem $\mathcal{P} = (I, \text{SOL}, m, \text{goal})$, an algorithm $\mathcal{A}$ is*
Approximation algorithm   *an* approximation algorithm *for $\mathcal{P}$ if, for any given instance $x \in I$, it returns an approximate solution, that is a feasible solution $\mathcal{A}(x) \in \text{SOL}(x)$.*

3.1.1 Absolute approximation

Clearly, for all practical purposes, Def. 3.1 is unsatisfactory. What we are ready to accept as an approximate solution is a feasible solution whose value is 'not too far' from the optimum. Therefore we are interested in determining how far the value of the achieved solution is from the value of an optimal one.

Definition 3.2 ▶ *Given an optimization problem $\mathcal{P}$, for any instance x and for any feasible*
Absolute error   *solution y of x, the* absolute error *of y with respect to x is defined as*

$$D(x,y) = |m^*(x) - m(x,y)|$$

where $m^*(x)$ denotes the measure of an optimal solution of instance x and $m(x,y)$ denotes the measure of solution y.

Section 3.1

APPROXIMATE SOLUTIONS WITH GUARANTEED PERFORMANCE

Indeed, given an NP-hard optimization problem, it would be very satisfactory to have a polynomial-time approximation algorithm that, for every instance x, is capable of providing a solution with a bounded absolute error, that is a solution whose measure is only a constant away from the measure of an optimal one.

Given an optimization problem $\mathcal{P}$ and an approximation algorithm $\mathcal{A}$ for $\mathcal{P}$, we say that $\mathcal{A}$ is an absolute approximation algorithm *if there exists a constant k such that, for every instance x of $\mathcal{P}$, $D(x, \mathcal{A}(x)) \leq k$.*

◀ Definition 3.3
Absolute approximation algorithm

Let us consider the problem of determining the minimum number of colors needed to color a planar graph. We have seen in Theorem 2.12 that a 6-coloring of a planar graph can be found in polynomial time. It is also known that establishing whether the graph is 1-colorable (that is, the set of edges is empty) or 2-colorable (that is, the graph is bipartite) is decidable in polynomial time whereas the problem of deciding whether three colors are enough is NP-complete. Clearly, if we provide an algorithm that returns either a 1- or a 2-coloring of the nodes if the graph is 1- or 2-colorable, respectively, and returns a 6-coloring in all other cases, we obtain an approximate solution with absolute error bounded by 3.

◀ Example 3.1

A second related (but less trivial) example, concerning the edge coloring problem, will be considered in Chap. 4.

Unfortunately, there are few cases in which we can build absolute approximation algorithms and, in general, we cannot expect such a good performance from an approximation algorithm. The knapsack problem is an example of an NP-hard problem that does not allow a polynomial-time approximation algorithm with bounded absolute error.

Unless $P = NP$, no polynomial-time absolute approximation algorithm exists for MAXIMUM KNAPSACK.

◀ Theorem 3.1

Let X be a set of n items with profits $p_1, \ldots, p_n$ and sizes $a_1, \ldots, a_n$, and let b be the knapsack capacity. If the problem would allow a polynomial-time approximation algorithm with absolute error k, then we could solve the given instance exactly in the following way. Let us create a new instance by multiplying all profits p_i by $k+1$. Clearly, the set of feasible solutions of the new instance is the same as that of the original instance. On the other side, since the measure of any solution is now a multiple of $k+1$, the only solution with absolute error bounded by k is the optimal solution. Hence, if we knew how to find such a solution in polynomial time, we would also be able to exactly solve the original instance in polynomial time.

PROOF

QED

Similar arguments apply to show that most other problems, such as MINIMUM TRAVELING SALESPERSON and MAXIMUM INDEPENDENT SET, do not allow polynomial-time absolute approximation algorithms.

Chapter 3
APPROXIMATION CLASSES

3.1.2 Relative approximation

In order to express the quality of an approximate solution, more interesting and more commonly used notions are the relative error and the performance ratio.

Definition 3.4 ▶ *Given an optimization problem $\mathcal{P}$, for any instance x of $\mathcal{P}$ and for any feasible solution y of x, the* relative error *of y with respect to x is defined as*
Relative error

$$E(x,y) = \frac{|m^*(x) - m(x,y)|}{\max\{m^*(x), m(x,y)\}}.$$

Both in the case of maximization problems and of minimization problems, the relative error is equal to 0 when the solution obtained is optimal, and becomes close to 1 when the approximate solution is very poor.

Definition 3.5 ▶ *Given an optimization problem $\mathcal{P}$ and an approximation algorithm $\mathcal{A}$ for*
ε-approximate algorithm *$\mathcal{P}$, we say that $\mathcal{A}$ is an ε-approximate algorithm for $\mathcal{P}$ if, given any input instance x of $\mathcal{P}$, the relative error of the approximate solution $\mathcal{A}(x)$ provided by algorithm $\mathcal{A}$ is bounded by ε, that is:*

$$E(x, \mathcal{A}(x)) \leq \varepsilon.$$

The greedy algorithm we analyzed in Sect. 2.1 for MAXIMUM KNAPSACK is an example of a polynomial-time 1/2-approximate algorithm. In fact, such an algorithm always provides a solution whose relative error is at most 1/2.

As an alternative to the relative error, we can express the quality of the approximation by means of a different, but related, measure.

Definition 3.6 ▶ *Given an optimization problem $\mathcal{P}$, for any instance x of $\mathcal{P}$ and for any*
Performance ratio *feasible solution y of x, the* performance ratio *of y with respect to x defined as*

$$R(x,y) = \max\left(\frac{m(x,y)}{m^*(x)}, \frac{m^*(x)}{m(x,y)}\right).$$

Both in the case of minimization problems and of maximization problems, the value of the performance ratio is equal to 1 in the case of an optimal solution, and can assume arbitrarily large values in the case of poor approximate solutions. The fact of expressing the quality of approximate solutions by a number larger than 1 in both cases is slightly counterintuitive, but it yields the undoubted advantage of allowing a uniform treatment of minimization and maximization problems.

Clearly the performance ratio and the relative error are related. In fact, in any case, the relative error of solution y on input x is equal to $E(x,y) = 1 - 1/R(x,y)$.

As in the case of the relative error, also for the performance ratio it is interesting to consider situations in which such a quality measure is bounded by a constant for all input instances.

Section 3.1
APPROXIMATE SOLUTIONS WITH GUARANTEED PERFORMANCE

Given an optimization problem $\mathcal{P}$ and an approximation algorithm $\mathcal{A}$ for $\mathcal{P}$, we say that $\mathcal{A}$ is an r-approximate algorithm for $\mathcal{P}$ if, given any input instance x of $\mathcal{P}$, the performance ratio of the approximate solution $\mathcal{A}(x)$ is bounded by r, that is:

$$R(x, \mathcal{A}(x)) \leq r.$$

◀ Definition 3.7
r-approximate algorithm

For example, the greedy algorithm for MAXIMUM KNAPSACK is an example of a 2-approximate algorithm since it always provides a solution whose value is at least one half of the optimal value.

Notice that, according to our definition, if, for a given optimization problem $\mathcal{P}$ and a given algorithm $\mathcal{A}$ for $\mathcal{P}$, we have that, for all instances x of $\mathcal{P}$, $m_{\mathcal{A}}(x,y) \leq rm^*(x) + k$ we do not say that algorithm $\mathcal{A}$ is r-approximate, but that it is at most $r+k$-approximate. In the literature it is widely accepted that, in such case, the algorithm is called r-approximate, under an asymptotic point of view. For example, Theorem 2.9 states that, given an instance x of MINIMUM BIN PACKING, *First Fit Decreasing* returns a solution such that $m_{FFD}(x) \leq \frac{3}{2}m^*(x) + 1$. Such an algorithm is indeed usually known as a $\frac{3}{2}$-approximate algorithm. The asymptotic point of view in the evaluation of the performance ratio of algorithms will be taken into consideration in Chap. 4.

As we have seen, in order to qualify the degree of approximation achieved by an approximation algorithm we may refer either to the bound ε on its relative error (and speak of an ε-approximate algorithm) or to the bound r on its performance ratio (and speak of an r-approximate algorithm). In the following, we will mainly refer to the performance ratio in order to estimate the quality of the computed solutions: however, since the relative error is always smaller than 1 and the performance ratio is always larger than or equal to 1, it will be always clear which approximation quality measure we are referring to.

The existence of polynomial-time approximation algorithms qualifies different kinds of NP-hard optimization problems.

An NP-hard optimization problem $\mathcal{P}$ is ε-approximable (respectively, r-approximable) if there exists a polynomial-time ε-approximate (respectively, r-approximate) algorithm for $\mathcal{P}$.

◀ Definition 3.8
ε-approximable problem

Chapter 3

APPROXIMATION
CLASSES

Program 3.1: Greedy Sat

input Set C of disjunctive clauses on a set of variables V;
output Truth assignment $f : V \mapsto \{\text{TRUE}, \text{FALSE}\}$;
begin
  for all $v \in V$ **do** $f(v) := \text{TRUE}$;
  repeat
    Let l be the literal that occurs in the maximum number of
        clauses in C (solve ties arbitrarily);
    Let C_l be the set of clauses in which l occurs;
    Let $C_{\bar{l}}$ be the set of clauses in which $\bar{l}$ occurs;
    Let v_l be the variable corresponding to l;
    if l is positive **then**
    begin
      $C := C - C_l$;
      Delete $\bar{l}$ from all clauses in $C_{\bar{l}}$; Delete all empty clauses in C
    end
    else
    begin
      $f(v_l) := \text{FALSE}; C := C - C_l$;
      Delete $\bar{l}$ from all clauses in $C_{\bar{l}}$; Delete all empty clauses in C
    end
  until $C = \emptyset$;
  return f
end.

MAXIMUM KNAPSACK is an example of a 2-approximable problem. In the previous chapter, we have seen several other examples of approximable problems, and more examples will be shown in this chapter and in the following ones.

Let us now consider MAXIMUM SATISFIABILITY. For this problem, we have already shown in Sect. 2.6 a randomized algorithm whose expected performance ratio is bounded by 2. We now show a deterministic 2-approximate algorithm for MAXIMUM SATISFIABILITY which runs in polynomial time (namely, Program 3.1), that is a simple example of applying the greedy technique and can be considered as a "derandomized" version of the algorithm shown in the previous chapter (a different approach based on the local search technique can also be followed in order to obtain a similar result, as stated in Exercise 3.1).

Theorem 3.2 ▶ *Program 3.1 is a polynomial-time 2-approximate algorithm for* MAXIMUM SATISFIABILITY.

PROOF   Given an instance with c clauses, we prove, by induction on the number of variables, that Program 3.1 always satisfies at least $c/2$ clauses. Since no optimal solution can have value larger than c, the theorem will follow.

The result is trivially true in the case of one variable. Let us assume that it is true in the case of $n-1$ variables ($n>1$) and let us consider the case in which we have n variables. Let v be the variable corresponding to the literal which appears in the maximum number of clauses. Let c_1 be the number of clauses in which v appears positive and c_2 be the number of clauses in which v appears negative. Without loss of generality suppose that $c_1 \geq c_2$, so that the algorithm assigns the value TRUE to v. After this assignment, at least $c - c_1 - c_2$ clauses, on $n-1$ variables, must be still considered. By inductive hypothesis, Program 3.1 satisfies at least $(c-c_1-c_2)/2$ such clauses. Hence, the overall number of satisfied clauses is at least $c_1 + (c-c_1-c_2)/2 \geq c/2$.

QED

Within the class NPO, the class of problems that allow polynomial-time r-approximate algorithms (or, equivalently, ε-approximate algorithms) plays a very important role. In fact, the existence of a polynomial-time r-approximate algorithm for an NP-hard optimization problem shows that, despite the inherent complexity of finding the exact solution of the problem, such a solution can somehow be approached.

APX *is the class of all* NPO *problems P such that, for some $r \geq 1$, there exists a polynomial-time r-approximate algorithm for P.*

◄ Definition 3.9
Class APX

As shown above and in the previous chapter, MAXIMUM SATISFIABILITY, MAXIMUM KNAPSACK, MAXIMUM CUT, MINIMUM BIN PACKING, MINIMUM GRAPH COLORING restricted to planar graphs, MINIMUM SCHEDULING ON IDENTICAL MACHINES, and MINIMUM VERTEX COVER are all in APX.

The definition of the class APX provides a first important notion for characterizing NPO problems with respect to their degree of approximability. For several important NPO problems, in fact, it can be shown that they do not allow any r-approximate algorithm, unless P = NP. In other words, for these problems, approximate solutions with guaranteed performance are as hard to determine as the optimal solutions. This means that, under the hypothesis that P ≠ NP, the class APX is strictly contained in the class NPO.

In the next subsection we will show that MINIMUM TRAVELING SALESPERSON is a problem for which determining approximate solutions with constant bounded performance ratio is computationally intractable. Other NPO problems that do not belong to APX, such as MAXIMUM CLIQUE and MAXIMUM INDEPENDENT SET, will be seen in Chap. 6, where some techniques needed to prove such results will also be provided.

Section 3.1

APPROXIMATE SOLUTIONS WITH GUARANTEED PERFORMANCE

Chapter 3
APPROXIMATION CLASSES

3.1.3 Approximability and non-approximability of TSP

MINIMUM TRAVELING SALESPERSON is an important example of an optimization problem that cannot be r-approximated, no matter how large is the performance ratio r.

Theorem 3.3 ▶ *If* MINIMUM TRAVELING SALESPERSON *belongs to* APX, *then* P = NP.

PROOF The result is proved by reduction from the HAMILTONIAN CIRCUIT decision problem. HAMILTONIAN CIRCUIT is the problem of deciding whether a directed graph admits a circuit which passes exactly once through every node: this problem is known to be NP-complete. Let $G = (V, E)$ be an instance of HAMILTONIAN CIRCUIT with $|V| = n$. For any $r \geq 1$, we construct an instance of MINIMUM TRAVELING SALESPERSON such that if we had a polynomial-time r-approximate algorithm for MINIMUM TRAVELING SALESPERSON, then we could decide whether the graph G has a Hamiltonian circuit in polynomial time. From this construction the theorem will then follow.

Given $G = (V, E)$, the instance of MINIMUM TRAVELING SALESPERSON is defined on the same set of nodes V and with distances

$$d(v_i, v_j) = \begin{cases} 1 & \text{if } (v_i, v_j) \in E, \\ 1 + nr & \text{otherwise.} \end{cases}$$

This instance of MINIMUM TRAVELING SALESPERSON has a solution of measure n if and only if the graph G has a Hamiltonian circuit: in that case, the next smallest approximate solution has measure at least $n(1+r)$ and its performance ratio is hence greater than r. Besides, if G has no Hamiltonian circuit, then the optimal solution has measure at least $n(1+r)$. Therefore, if we had a polynomial-time r-approximate algorithm for MINIMUM TRAVELING SALESPERSON, we could use it to decide whether G has a Hamiltonian circuit in the following way: we apply the approximation algorithm to the instance of MINIMUM TRAVELING SALESPERSON corresponding to G and we answer YES if and only if it returns a solution of measure n.

QED

Corollary 3.4 ▶ *If* P $\neq$ NP, *then* APX $\subset$ NPO.

The hardness of approximating MINIMUM TRAVELING SALESPERSON significantly reduces if we make suitable assumptions on the problem instances. In particular, recall that MINIMUM METRIC TRAVELING SALESPERSON is defined as MINIMUM TRAVELING SALESPERSON restricted to instances that satisfy the triangular inequality (see Sect. 2.1.3). Note also that, as a particular case, all instances of MINIMUM TRAVELING SALESPERSON which make use of the Euclidean distance satisfy the triangular inequality.

From a complexity point of view, MINIMUM TRAVELING SALESPERSON does not really become easier when the triangular inequality is satisfied: indeed, it remains NP-hard. From an approximation point of view the situation is quite different and MINIMUM METRIC TRAVELING SALESPERSON can be shown to be, in a sense, "easier" than MINIMUM TRAVELING SALESPERSON. In fact, we now define a 3/2-approximate polynomial-time algorithm for MINIMUM METRIC TRAVELING SALESPERSON, called *Christofides' algorithm*.

To this aim, let us first introduce some definitions and preliminary results that are needed for understanding and analyzing the algorithm.

A *multigraph* is a pair $M = (V, F)$ where V is a finite set of nodes and F is a multiset of edges. A *weighted multigraph* is a multigraph where a weight $c(e)$ is associated to each edge $e \in F$. A *walk* on a multigraph is a sequence of nodes $(v_1, \ldots, v_m)$, where each node may appear more than once and such that, for every i with $1 \leq i < m$, there is an edge connecting v_i and v_{i+1}. A walk $(v_1, \ldots, v_m)$ is said to be *closed* if $v_1 = v_m$. An *Eulerian walk* is a closed walk in which each node is visited at least once and each edge is traversed exactly once. A multigraph is *Eulerian* if it admits an Eulerian walk. For example, the multigraph of Fig. 3.1 is an Eulerian graph since the walk

$$(v_1, v_2, v_3, v_5, v_6, v_4, v_3, v_2, v_1, v_6, v_5, v_4, v_1)$$

is Eulerian.

Section 3.1

APPROXIMATE SOLUTIONS WITH GUARANTEED PERFORMANCE

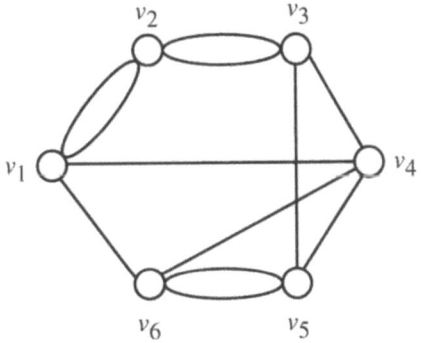

Figure 3.1
An Eulerian multigraph

It is well-known that a polynomial-time algorithm exists that, given in input a multigraph M, decides whether M is Eulerian and, eventually, returns an Eulerian walk on M (see Bibliographical notes). Christofides'

Chapter 3

APPROXIMATION
CLASSES

algorithm is based on this fact and on the possibility of extracting a tour from an Eulerian walk, as stated in the following lemma.

Lemma 3.5 ▶ *Let $G = (V,E)$ be a complete weighted graph satisfying the triangular inequality and let $M = (V,F)$ be any Eulerian multigraph over the same set of nodes such that all edges between two nodes u and v in M have the same weight as the edges (u,v) in G. Then, we can find in polynomial time a tour I in G whose measure is at most equal to the sum of the weights of the edges in M.*

PROOF

Let w be any Eulerian walk over M. Since all nodes $v_1, \ldots, v_n$ appear at least once in the walk, there must exist at least one permutation $\pi(1), \ldots, \pi(n)$ of the integers in $\{1, \ldots, n\}$ such that w can be written as $(v_{\pi(1)}, \alpha_1, v_{\pi(2)}, \alpha_2, \ldots, v_{\pi(n)}, \alpha_n, v_{\pi(1)})$ where the symbols $\alpha_1, \ldots, \alpha_n$ denote (possibly empty) sequences of nodes (for example, such a permutation can be obtained by considering the first occurrences of all nodes). Due to the triangular inequality, the weight of any edge $(v_{\pi(j)}, v_{\pi(j+1)})$, with $1 \leq j < n$, is no greater than the total weight of the path $(v_{\pi(j)}, \alpha_j, v_{\pi(j+1)})$ and the weight of the edge $(v_{\pi(n)}, v_{\pi(1)})$ is no greater than the total weight of the path $(v_{\pi(n)}, \alpha_n, v_{\pi(1)})$. Hence, if we consider the tour I corresponding to permutation π, the measure of I is at most equal to the total weight of the Eulerian walk w, that is, the sum of the weights of the edges in M.

QED

Given an instance G of MINIMUM METRIC TRAVELING SALESPERSON, a general approach to approximately solve this instance could then consists of the following steps: (1) compute a spanning tree T of G in order to visit all nodes, (2) starting from T, derive a multigraph M satisfying the hypothesis of the previous lemma, (3) compute an Eulerian walk w on M, and (4) extract from w a tour according to the proof of the lemma.

The only unspecified step of this approach is how the multigraph M is derived from T. A simple way to perform this step consists of just doubling all edges in T: that is, for each edge (u,v) in T, M contains two copies of this edge with the same weight. It can be easily shown that such an algorithm returns a tour whose performance ratio is bounded by 2 (see Exercise 3.5).

A more sophisticated application of this approach is shown in Program 3.2 where the multigraph M is obtained by adding to T the edges of a minimum weight perfect matching among the vertices in T of odd degree. The following result shows that this new algorithm provides a sensibly better performance ratio.

Theorem 3.6 ▶ *Given an instance G of MINIMUM METRIC TRAVELING SALESPERSON,*

Section 3.1

APPROXIMATE SOLUTIONS WITH GUARANTEED PERFORMANCE

Program 3.2: Christofides

input Weighted complete graph $G = (V, E)$;
output Tour I;
begin
   Find a minimum spanning tree $T = (V, E_T)$ in G;
   Let C be the set of nodes in T with odd degree;
   Find a minimum weight perfect matching H in the subgraph of G induced by C;
   Create a multigraph $M = (V, E_T \cup H)$;
   Find a Eulerian walk w in M;
   Extract a tour I from w (according to the proof of Lemma 3.5);
   return I
end.

Christofides' algorithm returns, in polynomial time, a solution of G whose performance ratio is at most 3/2.

PROOF

Let us consider the multigraph $M = (V, E_T \cup H)$ and an Eulerian walk w on M. Let us denote by $c(T)$ and $c(H)$ the sums of the weights of the edges belonging to T and H, respectively. Since M clearly satisfies the hypothesis of Lemma 3.5, we can find in polynomial time a tour I such that
$$m(G, I) \leq c(T) + c(H). \tag{3.1}$$

We now prove the following two facts.

Fact 1: $m^*(G) \geq 2c(H)$. Let $(v_{i_1}, \ldots, v_{i_{2|H|}})$ be the sequence of odd-degree vertices of T in the order in which they appear in an optimal tour I^*. Let H_1 and H_2 be the following two matchings:

$$H_1 = \{(v_{i_1}, v_{i_2}), (v_{i_3}, v_{i_4}), \ldots, (v_{i_{2|H|-1}}, v_{i_{2|H|}})\}$$

and

$$H_2 = \{(v_{i_2}, v_{i_3}), \ldots, (v_{i_{2|H|}}, v_{i_1})\}.$$

By making use of the triangular inequality, we have that $m^*(G) \geq c(H_1) + c(H_2)$, where $c(H_i)$ denotes the sum of the weights of the edges in H_i, for $i = 1, 2$. Since H is a minimum weight perfect matching, we have that both $c(H_1)$ and $c(H_2)$ are greater than or equal to $c(H)$ and, hence, $m^*(G) \geq 2c(H)$.

Fact 2: $c(T) \leq m^*(G)$. In order to prove this fact, it is enough to observe that a tour is also a spanning tree (plus one additional edge): since T is a minimum spanning tree, we have $c(T) \leq m^*(G)$.

Chapter 3

APPROXIMATION CLASSES

From Eq. 3.1 and from the above two facts, we then have that $m(G,I) \leq m^*(G) + m^*(G)/2 = 3m^*(G)/2$.

As far as the running time is concerned, it is easy to see that the most expensive step is the computation of the minimum weight perfect matching. This task can, indeed, be performed in polynomial time by means of a primal-dual algorithm (see Bibliographical notes).

QED

Example 3.2 ▶ Let us consider the instance G of MINIMUM METRIC TRAVELING SALESPERSON consisting of eight cities (i.e., Amsterdam, Berlin, Geneva, Milan, Prague, Rome, Warsaw, and Wien) with distances shown in Fig. 3.2 (these distances are the real road-distances between the specified cities).

AMS	685	925	1180	960	1755	1235	1180
	BER	1160	1105	340	1530	585	630
		GEN	325	950	880	1575	1025
			MIL	870	575	1495	830
				PRA	1290	625	290
					ROM	1915	1130
						WAR	795
							WIE

Figure 3.2
An instance of MINIMUM METRIC TRAVELING SALESPERSON

In Fig. 3.3(a) the minimum spanning tree T of G computed in the first step of Christofides' algorithm is shown: note that, in this tree, there are six odd-degree nodes (that is, Amsterdam, Berlin, Geneva, Milan, Rome, and Warsaw). A minimum weight perfect matching H among these six nodes consists of edges Amsterdam-Geneva, Berlin-Warsaw, and Milan-Rome. In Fig. 3.3(b) the multigraph M obtained by joining T and H is given: clearly, this is an Eulerian graph satisfying the hypothesis of Lemma 3.5. An Eulerian walk on M starting from Amsterdam is

AMS-BER-WAR-BER-PRA-WIE-MIL-ROM-MIL-GEN-AMS.

By considering only the first occurrence of each city in the walk, we obtain the approximate tour shown in Fig 3.3(c) whose measure is 5395. In Fig 3.3(d) the optimal tour is given: the corresponding optimal measure is 5140, that is, about 5% better than the measure of the approximate tour.

As a consequence of Theorem 3.6, we have that MINIMUM METRIC TRAVELING SALESPERSON belongs to the class APX. The next example shows that the bound of the theorem is tight: that is, there exists a family of weighted complete graphs such that Christofides' algorithm returns a solution whose measure is asymptotically 50% greater than the measure of an optimal tour.

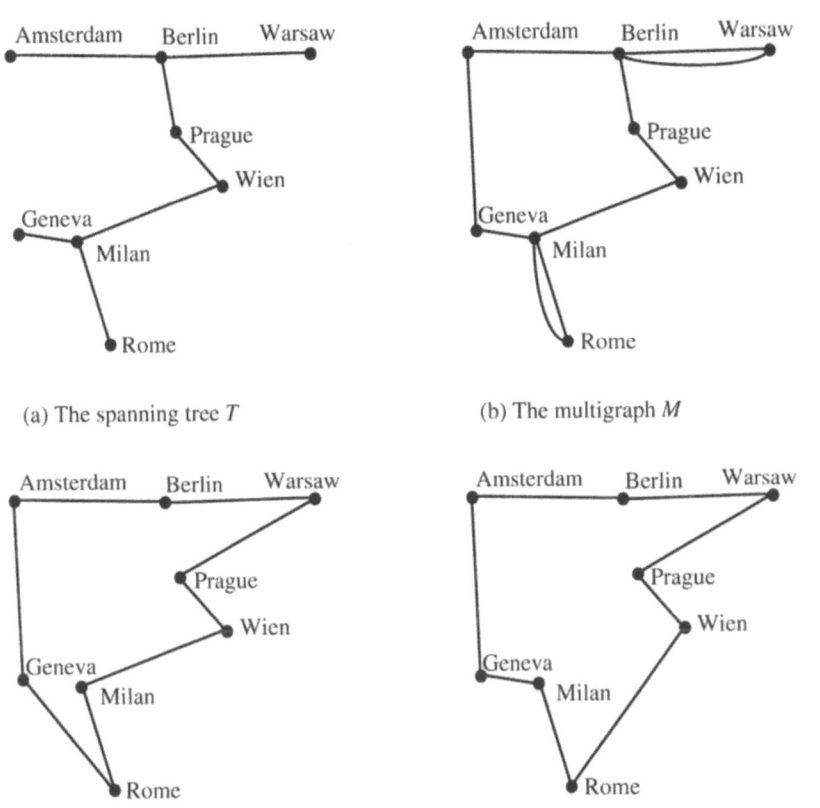

Figure 3.3
A sample application of Christofides' algorithm

Section 3.1
APPROXIMATE SOLUTIONS WITH GUARANTEED PERFORMANCE

◀ Example 3.3

For any positive integer n, let us consider the instance G_n of MINIMUM METRIC TRAVELING SALESPERSON shown in Fig. 3.4(a), where all distances that are not explicitly specified must be assumed to be computed according to the Euclidean distance. It is easy to see that one possible minimum spanning tree is the tour shown in the figure without the edge (a_1, a_{n+1}). If Christofides' algorithm chooses this spanning tree, then the approximate tour shown in Fig. 3.4(a) results: note that this tour has measure $3n + 2\varepsilon$. On the contrary, the optimal tour is shown in Fig. 3.4(b) and has measure $2n + 1 + 4\varepsilon$. As n grows to infinity, the ratio between these two measures approaches the bound $3/2$.

Until now Christofides' algorithm is the best known approximation algorithm for MINIMUM METRIC TRAVELING SALESPERSON. While in the Euclidean case arbitrarily good polynomial-time approximation algorithms can be found (see Bibliographical notes), no polynomial-time approximation algorithm with a better guaranteed performance ratio is known for MINIMUM METRIC TRAVELING SALESPERSON, neither it is known whether the existence of any such algorithm would imply $P = NP$.

Chapter 3
APPROXIMATION CLASSES

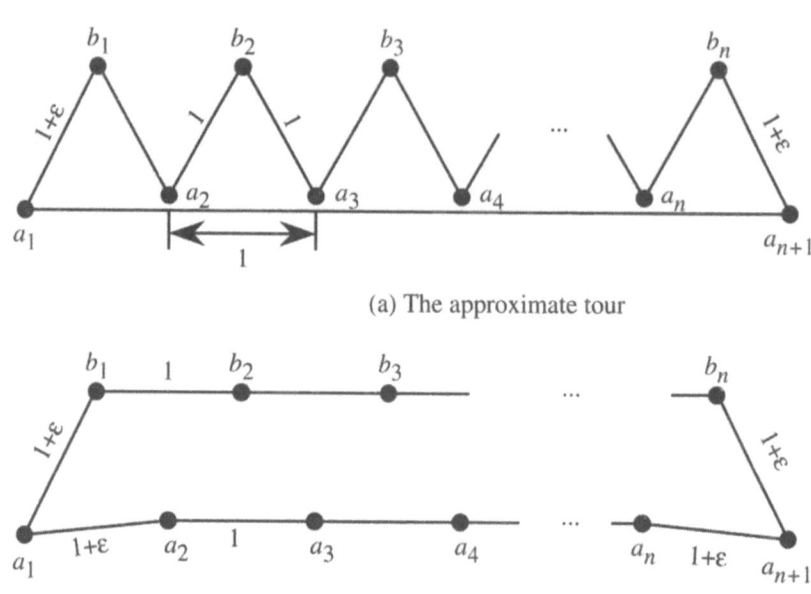

Figure 3.4
Worst case of Christofides' algorithm

3.1.4 Limits to approximability: The gap technique

From the point of view of practical applications, it may be observed that knowing that a problem belongs to APX, while interesting in itself, is only partially satisfactory. In fact, in some cases, the existence of guaranteed approximate solutions with large relative errors (e.g., a 50% error as in the case of MAXIMUM SATISFIABILITY or MINIMUM VERTEX COVER) may not be enough for practical purposes. We may be interested in finding stronger approximations, with much smaller relative errors (say 1%). With respect to this need, the optimization problems that belong to APX may behave quite differently. For some problems, not only we can find tight approximate solutions but we can even find arbitrarily good approximate solutions. For such problems we can construct particular polynomial-time algorithms, called polynomial-time approximation schemes (see Sects. 2.5 and 3.2), that, given an instance x of the problem and a constant $r > 1$, produce an r-approximate solution for x. For other problems (unfortunately, the vast majority of problems in APX) the performance ratio can only be reduced up to a certain point: sometimes the approximation techniques can even lead to very tight approximate solutions, but then a threshold t exists such that r-approximability, with $r < t$, becomes computationally intractable.

In order to prove this latter type of result, a simple but powerful technique is frequently used. Such technique is known as *gap technique* and is strictly related to the technique that we have used for proving the non-approximability of MINIMUM TRAVELING SALESPERSON. The technique will now be described in the case of minimization problems but it can also be applied to maximization problems by performing simple modifications to our exposition.

Section 3.1
APPROXIMATE SOLUTIONS WITH GUARANTEED PERFORMANCE

◀ Theorem 3.7

Let $\mathcal{P}'$ be an NP-complete decision problem and let $\mathcal{P}$ be an NPO minimization problem. Let us suppose that there exist two polynomial-time computable functions $f : I_{\mathcal{P}'} \mapsto I_{\mathcal{P}}$ and $c : I_{\mathcal{P}'} \mapsto \mathbf{N}$ and a constant $\text{gap} > 0$, such that, for any instance x of $\mathcal{P}'$,

$$m^*(f(x)) = \begin{cases} c(x) & \text{if } x \text{ is a positive instance,} \\ c(x)(1+\text{gap}) & \text{otherwise.} \end{cases}$$

Then no polynomial-time r-approximate algorithm for $\mathcal{P}$ with $r < 1+\text{gap}$ can exist, unless $P = NP$.

PROOF

Suppose we have a polynomial-time r-approximate algorithm $\mathcal{A}$ with $r < 1+\text{gap}$ for problem $\mathcal{P}$. We can make use of this algorithm for solving problem $\mathcal{P}'$ in polynomial time in the following way. Given an instance x of $\mathcal{P}'$, we compute $f(x)$ and then we apply the approximation algorithm $\mathcal{A}$ to $f(x)$. Let us distinguish the following two cases.

1. x is a negative instance. By hypothesis, in this case $m^*(f(x)) \geq c(x)(1+\text{gap})$ and, *a fortiori*, $m(f(x), \mathcal{A}(x)) \geq c(x)(1+\text{gap})$.

2. x is a positive instance. In this case, since $\mathcal{A}$ is an r-approximate algorithm, we have that

$$\frac{m(f(x), \mathcal{A}(x))}{m^*(f(x))} \leq r < 1+\text{gap}.$$

By hypothesis, $m^*(f(x)) = c(x)$. Hence, $m(f(x), \mathcal{A}(x)) < c(x)(1+\text{gap})$.

Therefore, x is a positive instance of $\mathcal{P}'$ if and only if $m(f(x), \mathcal{A}(x)) < c(x)(1+\text{gap})$, and we would be able to solve problem $\mathcal{P}'$ in polynomial time. Since $\mathcal{P}'$ is NP-complete, this would imply $P = NP$.

QED

Let us consider MINIMUM GRAPH COLORING. For this problem, the gap technique can be applied by reduction from the coloring problem for planar graphs. In fact, while a well-known result states that any planar graph is colorable with

◀ Example 3.4

Chapter 3

APPROXIMATION CLASSES

at most four colors (see Bibliographical notes), the problem of deciding whether a planar graph is colorable with at most three colors is NP-complete. Hence, in this case, the gap is $1/3$ and we can hence conclude that no polynomial-time r-approximate algorithm for MINIMUM GRAPH COLORING can exist with $r < 4/3$, unless $P = NP$. Actually, much stronger results hold for the graph coloring problem: it has been proved that, if $P \neq NP$, then no polynomial-time algorithm can provide an approximate solution whatsoever (that is, MINIMUM GRAPH COLORING belongs to NPO $-$ APX).

The considerations of the previous example can be extended to show that, for any NPO minimization problem $\mathcal{P}$, if there exists a constant k such that it is NP-hard to decide whether, given an instance x, $m^*(x) \leq k$, then no polynomial-time r-approximate algorithm for $\mathcal{P}$ with $r < (k+1)/k$ can exist, unless $P = NP$ (see Exercise 3.8). Another application of the gap technique has been shown in the proof of Theorem 3.3: in that case, actually, we have seen that the constant **gap** can assume any value greater than 0. Other results which either derive bounds on the performance ratio that can be achieved for particular optimization problems or prove that a problem does not allow a polynomial-time approximation scheme can be obtained by means of a sophisticated use of the gap technique. Such results will be discussed in Chap. 6.

3.2 Polynomial-time approximation schemes

AS WE noticed before, for most practical applications, we need to approach the optimal solution of an optimization problem in a stronger sense than it is allowed by an r-approximate algorithm. Clearly, if the problem is intractable, we have to restrict ourselves to approximate solutions, but we may wish to find better and better approximation algorithms that bring us as close as possible to the optimal solution. Then, in order to obtain r-approximate algorithms with better performances, we may be also ready to pay the cost of a larger computation time, a cost that, as we may expect, will increase with the inverse of the performance ratio.

Definition 3.10 ▶
Polynomial-time approximation scheme

Let $\mathcal{P}$ be an NPO *problem. An algorithm $\mathcal{A}$ is said to be a* polynomial-time approximation scheme (PTAS) *for $\mathcal{P}$ if, for any instance x of $\mathcal{P}$ and any rational value $r > 1$, $\mathcal{A}$ when applied to input (x, r) returns an r-approximate solution of x in time polynomial in $|x|$.*

While always being polynomial in $|x|$, the running time of a PTAS may also depend on $1/(r-1)$: the better is the approximation, the larger may be the running time. In most cases, we can indeed approach the optimal solution of a problem arbitrarily well, but at the price of a dramatic increase

Problem 3.1: Minimum Partition

INSTANCE: Finite set X of items, for each $x_i \in X$ a weight $a_i \in Z^+$.

SOLUTION: A partition of the items into two sets Y_1 and Y_2.

MEASURE: $\max\{\sum_{x_i \in Y_1} a_i, \sum_{x_i \in Y_2} a_i\}$.

Section 3.2

POLYNOMIAL-TIME APPROXIMATION SCHEMES

Program 3.3: Partition PTAS

input Set of items X with integer weights a_i, rational $r > 1$;
output Partition of X into two sets Y_1 and Y_2;
begin
  if $r \geq 2$ **then return** $X, \emptyset$
  else
  begin
    Sort items in non-increasing order with respect to their weight;
    (*Let $(x_1, \ldots, x_n)$ be the obtained sequence*)
    $k(r) := \lceil (2-r)/(r-1) \rceil$;
    (* First phase *)
    Find an optimal partition Y_1, Y_2 of $x_1, \ldots, x_{k(r)}$;
    (* Second phase *)
    for $j := k(r) + 1$ **to** n **do**
      if $\sum_{x_i \in Y_1} a_i \leq \sum_{x_i \in Y_2} a_i$ **then**
        $Y_1 := Y_1 \cup \{x_j\}$
      else
        $Y_2 := Y_2 \cup \{x_j\}$;
    return Y_1, Y_2
  end;
end.

in the computation cost. In other cases, we may construct approximation schemes whose running time is polynomial both in the size of the instance and in the inverse of the required degree of approximation. In such cases the possibility of approaching the optimal solution in practice with arbitrarily small error is definitely more concrete. Problems that allow this stronger form of approximation are very important from the practical point of view and will be discussed in the Sect. 3.3.

Let us now consider MINIMUM PARTITION, that is, Problem 3.1: a simple approach to obtain an approximate solution for this problem is based on the greedy technique. Such an approach consists in sorting items in non-increasing order and then inserting them into set Y_1 or into set Y_2 according to the following rule: always insert the next item in the set of smaller overall weight (breaking ties arbitrarily). It is possible to show

Chapter 3
APPROXIMATION CLASSES

that this procedure always returns a solution whose performance ratio is bounded by 7/6 and that this bound is indeed tight (see Theorem 2.7, in the case $p = 2$, and Exercise 2.6).

Program 3.3 basically consists in deriving an optimal solution of the subinstance including the k heaviest items and, subsequently, extending this solution by applying the greedy approach previously described. The next result shows that this algorithm provides a stronger approximation for MINIMUM PARTITION.

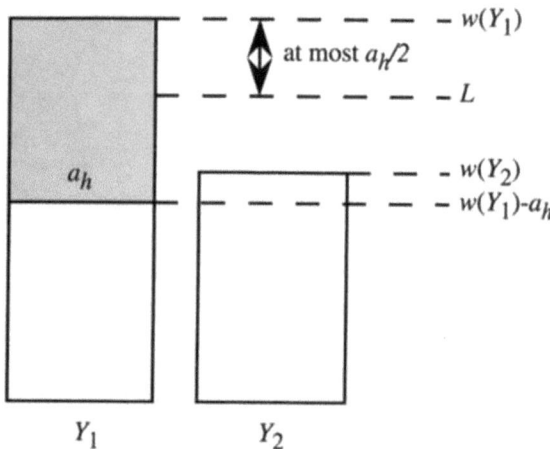

Figure 3.5
The analysis of Program 3.3

Theorem 3.8 ▶ *Program 3.3 is a polynomial-time approximation scheme for* MINIMUM PARTITION.

PROOF Let us first prove that, given an instance x of MINIMUM PARTITION and a rational $r > 1$, the algorithm provides an approximate solution (Y_1, Y_2) whose performance ratio is bounded by r. If $r \geq 2$, then the solution $(X, \emptyset)$ is clearly an r-approximate solution since any feasible solution has measure at least equal to half of the total weight $w(X) = \sum_{x_i \in X} a_i$. Let us then assume that $r < 2$ and let $w(Y_i) = \sum_{x_j \in Y_i} a_j$, for $i = 1, 2$, and $L = w(X)/2$. Without loss of generality, we may assume that $w(Y_1) \geq w(Y_2)$ and that x_h is the last item that has been added to Y_1 (see Fig. 3.5). This implies that $w(Y_1) - a_h \leq w(Y_2)$. By adding $w(Y_1)$ to both sides and dividing by 2 we obtain that

$$w(Y_1) - L \leq \frac{a_h}{2}.$$

If x_h has been inserted in Y_1 during the first phase of the algorithm, then it is easy to see that the obtained solution is indeed an optimal solution. Otherwise (that is, x_h has been inserted during the second phase), we have that $a_h \leq a_j$, for any j with $1 \leq j \leq k(r)$, and that $2L \geq a_h(k(r)+1)$. Since $w(Y_1) \geq L \geq w(Y_2)$ and $m^*(x) \geq L$, the performance ratio of the computed solution is

$$\frac{w(Y_1)}{m^*(x)} \leq \frac{w(Y_1)}{L} \leq 1 + \frac{a_h}{2L} \leq 1 + \frac{1}{k(r)+1} \leq 1 + \frac{1}{\frac{2-r}{r-1}+1} = r.$$

Finally, we prove that the algorithm works in time $O(n\log n + n^{k(r)})$. In fact, we need time $O(n\log n)$ to sort the n items. Subsequently, the first phase of the algorithm requires time exponential in $k(r)$ in order to perform an exhaustive search for the optimal solution over the $k(r)$ heaviest items $x_1, \ldots, x_{k(r)}$ and all other steps have a smaller cost. Since $k(r)$ is $O(1/(r-1))$, the theorem follows.

QED

3.2.1 The class PTAS

Let us now define the class of those NPO problems for which we can obtain an arbitrarily good approximate solution in polynomial time with respect to the size of the problem instance.

PTAS *is the class of* NPO *problems that admit a polynomial-time approximation scheme.*

◀ Definition 3.11
Class PTAS

The preceding result shows that MINIMUM PARTITION belongs to PTAS. Let us now see other examples of problems in PTAS. The first example will also show another application of the algorithmic technique of Program 3.3, which essentially consists of optimally solving a "subinstance" and, then, extending the obtained solution by applying a polynomial-time procedure.

Problem 3.2 models a variant of MAXIMUM KNAPSACK in which there is a set of types of items and we can take as many copies as we like of an item of a given type, provided the capacity constraint is not violated (observe that this problem is equivalent to Problem 2.8 with $d = 1$).

MAXIMUM INTEGER KNAPSACK *belongs to the class* PTAS.

◀ Theorem 3.9

Given an instance I of MAXIMUM INTEGER KNAPSACK, we first notice that if we relax the integrality constraint on the values of function c (that is, if we allow fractions of items to be included in the knapsack), then the optimal solution can be easily computed as follows. Let x_{md} be a type of item with maximal value/size ratio: then, the optimal assignment for the relaxed problem is given by $c_r(x_{\text{md}}) = b/a_{\text{md}}$ and $c_r(x_i) = 0$ for all

PROOF

Chapter 3

APPROXIMATION
CLASSES

Problem 3.2: Maximum Integer Knapsack

INSTANCE: Finite set X of types of items, for each $x_i \in X$, value $p_i \in Z^+$ and size $a_i \in Z^+$, positive integer b.

SOLUTION: An assignment $c : X \mapsto \mathbf{N}$ such that $\sum_{x_i \in X} a_i c(x_i) \leq b$.

MEASURE: Total value of the assignment, i.e., $\sum_{x_i \in X} p_i c(x_i)$.

Program 3.4: Integer Knapsack Scheme

input Set X of n types of items, values p_i, sizes a_i, $b \in \mathbf{N}$, rational $r > 1$;
output Assignment $c : X \mapsto \mathbf{N}$ such that $\sum_{x_i \in X} a_i c(x_i) \leq b$;
begin
$\quad \delta := \lceil \frac{1}{r-1} \rceil$;
$\quad$ Sort types of items in non-increasing order with respect to their values;
$\quad$ (* Let $(x_1, x_2, \ldots, x_n)$ be the sorted sequence *)
$\quad$ **for** $i := 1$ **to** n **do** $c(x_i) := 0$;
$\quad F := \{f \mid f : X \mapsto \mathbf{N} \wedge \sum_{i=1}^{n} f(x_i) \leq \delta \wedge \sum_{i=1}^{n} a_i f(x_i) \leq b\}$;
$\quad$ **for all** f **in** F **do**
$\quad$ **begin**
$\quad\quad k :=$ maximum i such that $f(x_i) \neq 0$;
$\quad\quad b_k := b - \sum_{i=1}^{k} a_i f(x_i)$;
$\quad\quad$ Let x_{md} be the type of items with maximal value/size ratio in $\{x_k, \ldots, x_n\}$;
$\quad\quad f(x_{\mathrm{md}}) := f(x_{\mathrm{md}}) + \lfloor b_k / a_{\mathrm{md}} \rfloor$;
$\quad\quad$ **if** $\sum_{x_i \in X} p_i f(x_i) \geq \sum_{x_i \in X} p_i c(x_i)$ **then**
$\quad\quad\quad$ **for** $i := 1$ **to** n **do** $c(x_i) := f(x_i)$;
$\quad$ **end**;
$\quad$ **return** c
end.

other types of items. For any type of item x, let $c(x) = \lfloor c_r(x) \rfloor$: since $m^*(I) \leq b p_{\mathrm{md}} / a_{\mathrm{md}}$, we have that function c is a feasible solution of I such that the following holds:

$$m^*(I) - m(I, c) \leq b p_{\mathrm{md}} / a_{\mathrm{md}} - p_{\mathrm{md}} \lfloor b / a_{\mathrm{md}} \rfloor \leq p_{\mathrm{md}} \leq p_{\max}$$

where $p_{\max}$ is the maximal value.

The approximation scheme described in Program 3.4 makes use of the above observation in order to extend a partial solution. Let us first show that, for any instance I of MAXIMUM INTEGER KNAPSACK and for any $r > 1$, the algorithm indeed returns an r-approximate solution of I. Let c^* be an optimal solution of I. If $\sum_{x \in X} c^*(x) \leq \delta$, then $m(I, c) = m^*(I)$ where c is the solution returned by the algorithm.

Otherwise (that is, $\sum_{x \in X} c^*(x) > \delta$), let $(x_1, x_2, \ldots, x_n)$ be the sequence of types of items sorted in non-increasing order with respect to their values and let c_δ^* be an assignment defined as follows:

$$c_\delta^*(x_i) = \begin{cases} c^*(x_i) & \text{if } \sum_{j=1}^{i} c^*(x_j) \leq \delta, \\ \delta - \sum_{j=1}^{i-1} c^*(x_j) & \text{if } \sum_{j=1}^{i-1} c^*(x_j) \leq \delta \text{ and } \sum_{j=1}^{i} c^*(x_j) > \delta, \\ 0 & \text{otherwise.} \end{cases}$$

Let k be the maximum index i such that $c_\delta^*(x_i) \neq 0$: observe that $c_\delta^*(x_k) \leq c^*(x_k)$. Clearly, $c_\delta^* \in F$ since $\sum_{j=1}^{n} c_\delta^*(x_j) = \delta$. Let $b_k = b - \sum_{i=1}^{k} a_i c_\delta^*(x_i)$, and let x_{md} be the type of items with maximal value/size ratio among $\{x_k, \ldots, x_n\}$. Then, $m(I, c) \geq m(I, c_\delta^*) + p_{md} \lfloor b_k/a_{md} \rfloor$ and $m^*(I) \leq m(I, c_\delta^*) + b_k p_{md}/a_{md}$. Therefore,

$$\begin{aligned}\frac{m^*(I)}{m(I,c)} &\leq \frac{m(I, c_\delta^*) + b_k p_{md}/a_{md}}{m(I, c_\delta^*) + p_{md} \lfloor b_k/a_{md} \rfloor} \\ &\leq \frac{m(I, c_\delta^*) + (b_k/a_{md} - \lfloor b_k/a_{md} \rfloor) p_{md}}{m(I, c_\delta^*)} \\ &= 1 + \frac{(b_k/a_{md} - \lfloor b_k/a_{md} \rfloor) p_{md}}{m(I, c_\delta^*)} \\ &\leq 1 + \frac{p_k}{\delta p_k} = \frac{\delta + 1}{\delta},\end{aligned}$$

where the last inequality is due to the fact that $(b_k/a_{md} - \lfloor b_k/a_{md} \rfloor) p_{md} \leq p_{md} \leq p_k$ and that $m(I, c_\delta^*) \geq \delta p_k$. From the definition of δ, it follows that the performance ratio is at most r.

Finally, let us estimate the running time of Program 3.4. Since $|F| = O(n^\delta)$ (see Exercise 3.9), the overall time is clearly $O(n^{1+\delta})$, that is, $O(n^{1+\frac{1}{r-1}})$.
QED

The last example of a polynomial-time approximation scheme that we consider is a scheme for computing approximate solutions of MAXIMUM INDEPENDENT SET restricted to planar graphs. The algorithm is based on the fact that MAXIMUM INDEPENDENT SET is polynomial-time solvable when restricted to a special class of graphs, called k-outerplanar and defined below.

Given a planar embedding of a planar graph G, the *level* of a node is inductively defined as follows:

1. All nodes that lie on the border of the exterior face are at level 1.

2. For any $i > 1$, if we remove from the embedding all nodes of level j with $1 \leq j < i$, then all nodes (if any) that lie on the border of the exterior face of the remaining embedding are at level i.

Chapter 3

APPROXIMATION CLASSES

A planar graph is said to be *k-outerplanar* if it admits a *k-outerplanar embedding*, that is, a planar embedding with maximal node level at most k.

Example 3.5 ▶ The embedding shown in Fig. 3.6 is a 6-outerplanar embedding. Indeed, for $i = 1, \ldots, 18$, the level of node v_i is $\lceil i/3 \rceil$.

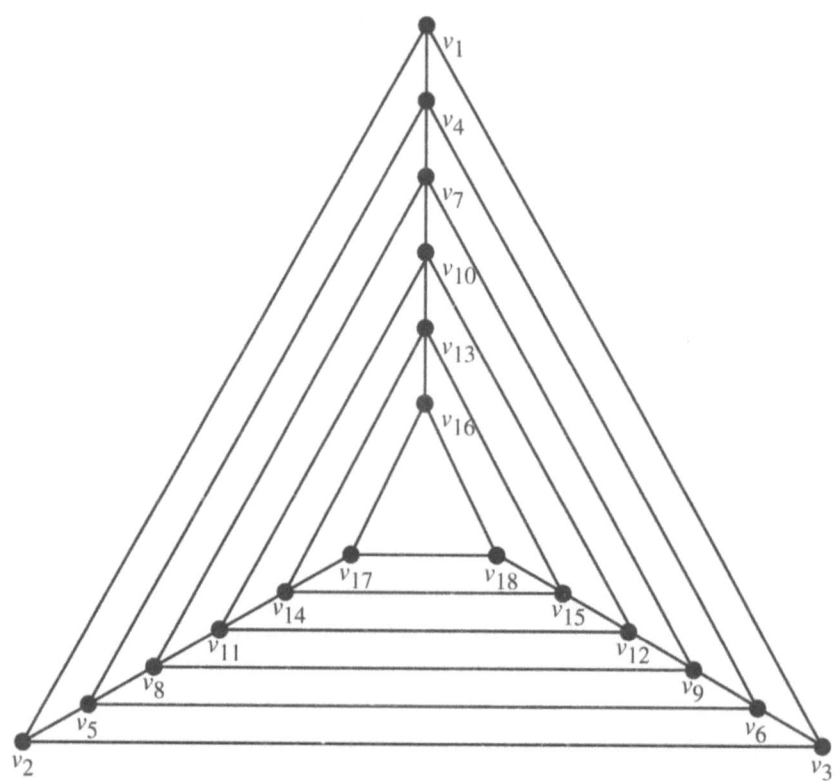

Figure 3.6
An example of a 6-outerplanar embedding

The following result, whose proof is here omitted (see Bibliographical notes), will be used by the approximation scheme.

Theorem 3.10 ▶ *For any k,* MAXIMUM INDEPENDENT SET *restricted to k-outerplanar graphs can be solved optimally in time $O(8^k n)$ where n is the number of nodes.*

The approximation scheme for MAXIMUM INDEPENDENT SET restricted to planar graphs exploits this result by considering a cover of the original graph formed by *k*-outerplanar graphs, where the parameter *k* depends on the required approximation ratio. In particular, the approximation scheme works in the following way (see Program 3.5).

Program 3.5: Independent Set Scheme

input Planar graph $G = (V,E)$, rational $r > 1$;
output Independent set $I \subseteq V$;
begin
   $k := \lceil 1/(r-1) \rceil$;
   Compute a planar embedding of G (see Bibliographical notes);
   Compute node levels;
   (* Let V_i be the set of nodes of level i *)
   for $i := 0$ **to** k **do**
     begin
       Let $\overline{V}_i$ be the union of all sets V_j with $j \equiv i \pmod{k+1}$;
       Let $\overline{G}_i$ be the subgraph of G induced by $V - \overline{V}_i$;
       (* $\overline{G}_i$ is k-outerplanar *)
       Compute a maximal independent set I_i on $\overline{G}_i$;
    end
   $I := I_m$ such that $|I_m| = \max_{0 \leq i \leq k}(|I_i|)$;
   return I
end.

Section 3.2
POLYNOMIAL-TIME APPROXIMATION SCHEMES

Let r be the required approximation ratio and $k = \lceil 1/(r-1) \rceil$. Given a planar embedding of a planar graph G, for all i with $0 \leq i \leq k$, let $\overline{V}_i$ be the set of nodes whose level is congruent to i modulo $k+1$ and let $\overline{G}_i$ be the subgraph of G induced by all nodes not in $\overline{V}_i$. Since we have deleted at least one level every $k+1$ levels, it is easy to verify that $\overline{G}_i$ is a k-outerplanar graph: indeed, $\overline{G}_i$ is a collection of connected components, each with a k-outerplanar embedding. Therefore, we can compute the maximal independent set I_i of $\overline{G}_i$ in time $O(8^k n)$. Let I_m be the maximal cardinality independent set among $\{I_0, \ldots, I_k\}$.

If I^* denotes a maximal independent set of G, then there must exist an integer r with $0 \leq r \leq k$ such that $|\overline{V}_r \cap I^*| \leq |I^*|/(k+1)$. Hence, the maximal independent set I_r of $\overline{G}_r$ contains at least $k|I^*|/(k+1)$ nodes. Since $|I_m| \geq |I_r|$, we have that the performance ratio of I_m is at most $(k+1)/k \leq r$.

The running time of the algorithm is $O(8^k k n)$, since we apply $k+1$ times the exact algorithm for k-outerplanar graphs implied by Theorem 3.10. Hence, we have proved the following result.

MAXIMUM INDEPENDENT SET *restricted to planar graphs belongs to the class* PTAS. ◀ Theorem 3.11

Let us consider the graph G in Fig. 3.6 and assume we wish to find an independent set on G with performance ratio $3/2$. By applying the algorithm described in Program 3.5, we obtain $k = 2$, which implies that we obtain ◀ Example 3.6

Chapter 3

APPROXIMATION
CLASSES

three sets $\overline{V}_0 = \{v_7, v_8, v_9, v_{16}, v_{17}, v_{18}\}$, $\overline{V}_1 = \{v_1, v_2, v_3, v_{10}, v_{11}, v_{12}\}$, and $\overline{V}_2 = \{v_4, v_5, v_6, v_{13}, v_{14}, v_{15}\}$. Three possible corresponding maximal independent sets of $\overline{G}_i$ are $I_0 = \{v_1, v_5, v_{10}, v_{14}\}$, $I_1 = \{v_4, v_8, v_{13}, v_{18}\}$, and $I_2 = \{v_1, v_7, v_{11}, v_{16}\}$. Hence, we obtain that the solution returned by the algorithm contains 4 nodes. Clearly, $m^*(G) = 6$: indeed, we can choose exactly one node for every triangle formed by nodes of level i, for $i = 1, \ldots, 6$. As a consequence, we have that $m^*(G)/m(G, I) = 6/4 = r$.

3.2.2 APX versus PTAS

By definition, it is clear that the class PTAS is contained in the class APX. Henceforth, those problems that do not belong to APX, such as MINIMUM TRAVELING SALESPERSON, *a fortiori* cannot have a polynomial-time approximation scheme. A natural question hence arises at this point: Does there exist any NP optimization problem that for some value of r can be r-approximated in polynomial time, but which does not allow a polynomial-time approximation scheme? In other words the question is whether PTAS is *strictly* contained in APX. The answer is yes, provided that $P \neq NP$. Again the main technique for proving the non-existence of a PTAS for an NPO problem is the gap technique.

Theorem 3.12 ▶ *If $P \neq NP$, then* MINIMUM BIN PACKING *does not belong to the class* PTAS.

PROOF

QED

We show that, if $P \neq NP$, then no r-approximate algorithm for MINIMUM BIN PACKING can be found with $r \leq 3/2 - \varepsilon$, for any $\varepsilon > 0$. To this aim, let us consider the PARTITION decision problem which consists in deciding whether a given set of weighted items can be partitioned into two sets of equal weight: this problem is NP-complete (see Bibliographical notes of Chap. 2). Given an instance x of PARTITION, let B be the sum of the weights of all items. We then define the corresponding instance x' of MINIMUM BIN PACKING as follows: for each item of x of weight w, x' has an item of size $2w/B$ (observe that we can always assume $w \leq B/2$ since, otherwise, x is a negative instance). If x is a positive instance, then $m^*(x') = 2$. Otherwise, $m^*(x') = 3$. From Theorem 3.7, it then follows the $3/2$ lower bound on the approximability of MINIMUM BIN PACKING. Therefore, unless $P = NP$, MINIMUM BIN PACKING does not admit a PTAS.

Since in Sect. 2.2.2 we have shown that MINIMUM BIN PACKING belongs to the class APX, the next corollary immediately follows.

Corollary 3.13 ▶ *If $P \neq NP$, then* PTAS $\subset$ APX.

Other examples of problems that belong to APX but do not admit a PTAS are MAXIMUM SATISFIABILITY, MAXIMUM CUT, and MINIMUM METRIC TRAVELING SALESPERSON. Actually, the negative results on the existence of a PTAS for these problems make use of the gap technique in a more sophisticated way, as we will see in later chapters.

Section 3.3

FULLY POLYNOMIAL-TIME APPROXIMATION SCHEMES

3.3 Fully polynomial-time approximation schemes

THE RUNNING time of the approximation schemes we have described in Chap. 2 and in this chapter depends on both the size of the input x and the inverse of the desired degree of approximation $r-1$: the better the approximation, the greater the running time. While, by definition of a PTAS, the running time must be polynomial in the size of the input x, the dependency on the quality of approximation may be very heavy. For example, in the case of MINIMUM PARTITION and MAXIMUM INDEPENDENT SET restricted to planar graphs, the dependency of the running time on the performance ratio has been shown to be exponential in $1/(r-1)$.

In some cases, such bad behavior strongly hampers the advantages of having a polynomial-time approximation scheme. In fact, the increase in the running time of the approximation scheme with the degree of approximation may prevent any practical use of the scheme.

3.3.1 The class FPTAS

A much better situation arises when the running time is polynomial *both* in the size of the input *and* in the inverse of the performance ratio.

Let $\mathcal{P}$ be an NPO problem. An algorithm $\mathcal{A}$ is said to be a fully polynomial-time approximation scheme (FPTAS) for $\mathcal{P}$ if, for any instance x of $\mathcal{P}$ and for any rational value $r > 1$, $\mathcal{A}$ with input (x, r) returns an r-approximate solution of x in time polynomial both in $|x|$ and in $1/(r-1)$.

◀ Definition 3.12
Fully polynomial-time approximation scheme

MAXIMUM KNAPSACK is an example of an optimization problem that admits an FPTAS: Program 2.9, in fact, is a fully polynomial-time approximation scheme for this problem, since its running time is $O(r|x|^3/(r-1))$.

FPTAS *is the class of* NPO *problems that admit a fully polynomial-time approximation scheme.*

◀ Definition 3.13
Class FPTAS

Chapter 3

APPROXIMATION CLASSES

Clearly, the class FPTAS is contained in the class PTAS. Henceforth, those problems that do not belong to PTAS (such as MINIMUM TRAVELING SALESPERSON, which is not even in APX, or MINIMUM BIN PACKING, which is in APX but not in PTAS) *a fortiori* cannot have a fully polynomial-time approximation scheme.

The existence of a FPTAS for an NP-hard combinatorial optimization problems provides evidence that, for such a problem, despite the difficulty of finding an optimal solution, for most practical purposes the solution can be arbitrarily and efficiently approached.

3.3.2 The variable partitioning technique

In Sect. 2.5 we have shown how to derive a FPTAS for MAXIMUM KNAPSACK. Let us now consider the related MAXIMUM PRODUCT KNAPSACK problem in which the measure function is the product of the values of the chosen items. We will prove that also MAXIMUM PRODUCT KNAPSACK admits a FPTAS. In order to prove the result, we introduce a new technique, called *variable partitioning*, which is a variant of the fixed partitioning technique that we described in Sect. 2.5.

Observe that, in the case of MAXIMUM PRODUCT KNAPSACK, the value of the objective function can be as large as p_{max}^n, where p_{max} is the maximum value and n is the number of items. Therefore, if we make use of the fixed partitioning technique, it is possible to see that, when we divide the range into a polynomial number of equal size intervals (in order to obtain a polynomial-time algorithm), the relative error of the obtained solution cannot be bounded by any constant smaller than 1. The idea of the variable partitioning technique then consists in dividing the range of possible measures into a suitable collection of polynomially many *variable size* intervals.

More precisely, given an arbitrary instance x and given a rational $r > 1$, we divide the range of possible measures into the following intervals:

$$\{(0, \delta_1], (\delta_1, \delta_2], \ldots, (\delta_{t-1}, \delta_t]\},$$

where $\varepsilon = (r-1)/r$ and $\delta_j = (1 + \frac{\varepsilon}{2n})^j$ for $j = 1, \ldots, t$. Notice that, since the range of the possible measures is $(0, p_{max}^n]$, t is the smallest integer such that $(1 + \varepsilon/(2n))^t \geq p_{max}^n$. Hence, t is $O(\frac{n^2}{\varepsilon} \log p_{max})$ and the algorithm has time complexity $O(\frac{n^3}{\varepsilon} \log p_{max})$.

Concerning the approximation, let Y be the solution returned by the algorithm and assume that $m(x,Y) \in (\delta_{i-1}, \delta_i]$: hence, the optimal measure must be contained in $(\delta_{i-1}, \delta_{i+n-1}]$. It is then possible to show that Y satisfies the following inequality:

$$\frac{m^*(x) - m(x,Y)}{m^*(I)} \leq \varepsilon = \frac{r-1}{r}$$

(see Exercise 3.11), which implies that the performance ratio of Y is at most r. Thus the following theorem derives.

MAXIMUM PRODUCT KNAPSACK *belongs to the class* FPTAS.   ◀ Theorem 3.14

3.3.3 Negative results for the class FPTAS

We now present some negative results which show that, unfortunately, many problems are not in FPTAS.

Actually, a first general result drastically reduces the class of combinatorial problems in PTAS which admit a FPTAS since it excludes the existence of a fully polynomial-time approximation scheme for all those optimization problems for which the value of the optimal measure is polynomially bounded with respect to the length of the instance. This result, in turn, will allow us to show that the containment of FPTAS in PTAS is proper.

An optimization problem is polynomially bounded *if there exists a polynomial p such that, for any instance x and for any $y \in \mathrm{SOL}(x)$, $m(x,y) \leq p(|x|)$.*   ◀ Definition 3.14
Polynomially bounded optimization problem

No NP-*hard polynomially bounded optimization problem belongs to the class* FPTAS *unless* P $=$ NP.   ◀ Theorem 3.15

Let $\mathcal{P}$ be an NP-hard polynomially bounded maximization problem (the minimization case would lead to a similar proof). Suppose we had a FPTAS $\mathcal{A}$ for $\mathcal{P}$ which, for any instance x and for any rational $r > 1$, runs in time bounded by $q(|x|, 1/(r-1))$ for a suitable polynomial q. Since $\mathcal{P}$ is polynomially bounded, there exists a polynomial p such that, for any instance x, $m^*(x) \leq p(|x|)$. If we choose $r = 1 + 1/p(|x|)$, then $\mathcal{A}(x,r)$ provides an optimal solution of x. Indeed, since $\mathcal{A}$ is a FPTAS, we have that

PROOF

$$\frac{m^*(x)}{m(x,\mathcal{A}(x,r))} \leq r = \frac{p(|x|)+1}{p(|x|)},$$

that is,

$$m(x,\mathcal{A}(x,r)) \geq m^*(x)\frac{p(|x|)}{p(|x|)+1} = m^*(x) - \frac{m^*(x)}{p(|x|)+1} > m^*(x) - 1,$$

where the last inequality is due to the fact that $m^*(x) \leq p(|x|)$. From the integrality constraint on the measure function m, it follows that $m(x, \mathcal{A}(x,r)) = m^*(x)$, that is, $\mathcal{A}(x,r)$ is an optimal solution.

Since the running time of $\mathcal{A}(x,r)$ is bounded by $q(|x|, p(|x|))$, we have that $\mathcal{P}$ is solvable in polynomial time. From the NP-hardness of $\mathcal{P}$, the theorem thus follows.

QED

Corollary 3.16 ▶ *If* $P \neq NP$, *then* FPTAS $\subset$ PTAS.

PROOF   As we have seen in Sect. 3.2.1, MAXIMUM INDEPENDENT SET restricted to planar graphs belongs to class PTAS. On the other side, this problem is clearly polynomially bounded and by the previous theorem it does not belong to the class FPTAS (unless $P = NP$).

QED

3.3.4 Strong NP-completeness and pseudo-polynomiality

We conclude this section by giving another general condition that assures that a problem is not in FPTAS. Let us first introduce some definitions which are intrinsically interesting because they allow us to relate the approximability properties of a problem to its combinatorial structure. In particular, we are going to study the different ways in which numbers play a role in an NPO problem.

Let us consider, for example, MAXIMUM CUT. This problem is NP-hard: since no number appears in its instances (but the vertex indices), we may conclude that it is the combinatorial structure of MAXIMUM CUT, i.e. the property that the graph has to satisfy, that makes the problem hard.

When we consider other problems the situation is somewhat different. Let us consider MAXIMUM KNAPSACK. As we already noted in Sect. 2.5, by using a dynamic programming algorithm, we can solve this problem in time $O(n^2 p_{max})$: moreover, since p_{max} is an integer contained in the instance whose encoding requires $\lceil \log p_{max} \rceil$ bits, this algorithm is not a polynomial-time one. However, if we restrict ourselves to instances in which the numbers p_i have values bounded by a polynomial in the length of the instance, we obtain a polynomial-time algorithm. This means that the complexity of MAXIMUM KNAPSACK is essentially related to the size of the integers that appear in the input.

For any NPO problem $\mathcal{P}$ and for any instance x of $\mathcal{P}$, let $\max(x)$ denote the value of the largest number occurring in x. We note that, from a formal point of view, the definition of the function max depends on the encoding of the instance. However, we can repeat for the function max the same kind of considerations that are usually made when considering the computational complexity of a problem assuming the length of the instance as the main parameter. In fact, if we choose two different functions max and max' for the same problem, the results we are going to present do not change in the case that these two functions are polynomially related, that is, two polynomials p and q exist such that, for any instance x, both $\max(x) \leq p(\max'(x))$ and $\max'(x) \leq q(\max(x))$ hold.

For the NPO problems we are interested in, all the intuitive max functions we can think of are polynomially related. Thus, the concept of max is sufficiently flexible to be used in practice without any limitation.

Section 3.3
FULLY POLYNOMIAL-TIME APPROXIMATION SCHEMES

◀ **Definition 3.15**
Pseudo-polynomial problem

An NPO problem $\mathcal{P}$ is pseudo-polynomial if it can be solved by an algorithm that, on any instance x, runs in time bounded by a polynomial in $|x|$ and in $\max(x)$.

◀ **Example 3.7**

In the case of MAXIMUM KNAPSACK, the max function can be defined as

$$\max(x) = \max(a_1,\ldots,a_n, p_1,\ldots,p_n, b).$$

The dynamic programming algorithm for MAXIMUM KNAPSACK thus shows that this problem is pseudo-polynomial. Indeed, for any instance x, the running time of this algorithm is $O(n^2 p_{max})$ and, hence, $O(n^2 \max(x))$.

The following result shows an interesting relationship between the concepts of pseudo-polynomiality and full approximability.

◀ **Theorem 3.17**

Let $\mathcal{P}$ be an NPO problem in FPTAS. If a polynomial p exists such that, for every input x, $m^*(x) \leq p(|x|, \max(x))$, then $\mathcal{P}$ is a pseudo-polynomial problem.

PROOF

Let $\mathcal{A}$ be a fully polynomial-time approximation scheme for $\mathcal{P}$. We will now exhibit an algorithm $\mathcal{A}'$ that solves any instance x of $\mathcal{P}$ in time polynomial both in $|x|$ and in $\max(x)$. This algorithm is simply defined as:

$$\mathcal{A}'(x) = \mathcal{A}\left(x, 1 + \frac{1}{p(|x|, \max(x)) + 1}\right).$$

Since the optimal measure is bounded by $p(|x|, \max(x))$, $\mathcal{A}'(x)$ must be an optimal solution. Regarding the running time of $\mathcal{A}'$, recall that $\mathcal{A}$ operates within time $q(|x|, 1/(r-1))$, for some polynomial q. Therefore, $\mathcal{A}'$ operates in time $q(|x|, p(|x|, \max(x)) + 1)$, that is, a polynomial in both $|x|$ and $\max(x)$.

QED

Chapter 3
APPROXIMATION CLASSES

Let $\mathcal{P}$ be an NPO problem and let p be a polynomial. We denote by $\mathcal{P}^{\max,p}$ the problem obtained by restricting $\mathcal{P}$ to only those instances x for which $\max(x) \leq p(|x|)$. The following definition formally introduces the notion of an optimization problem whose computational hardness does not depend on the values of the numbers included in its instances.

Definition 3.16 ▶ An NPO *problem $\mathcal{P}$ is said to be* strongly NP-hard *if a polynomial p exists such that $\mathcal{P}^{\max,p}$ is NP-hard.*
Strongly NP-hard problem

Theorem 3.18 ▶ *If* $P \neq NP$, *then no strongly NP-hard problem can be pseudo-polynomial.*

PROOF Let us assume that $\mathcal{P}$ is a strongly NP-hard problem, which is also pseudo-polynomial. This means that an algorithm exists that solves $\mathcal{P}$ in time $q(|x|, \max(x))$ for a suitable polynomial q. Then, for any polynomial p, $\mathcal{P}^{\max,p}$ can be solved in time $q(|x|, p(|x|))$. From the strong NP-hardness of $\mathcal{P}$, it also follows that a polynomial p exists such that $\mathcal{P}^{\max,p}$ is NP-hard. Hence, $P = NP$ and the theorem follows.

QED

Example 3.8 ▶ MAXIMUM CUT is an example of strongly NP-hard problem. Indeed, it is sufficient to consider the polynomial $p(n) = n$. Therefore, unless $P = NP$, MAXIMUM CUT is not pseudo-polynomial.

From Theorems 3.17 and 3.18, the following result can be immediately derived.

Corollary 3.19 ▶ *Let $\mathcal{P}$ be a strongly NP-hard problem that admits a polynomial p such that $m^*(x) \leq p(|x|, \max(x))$, for every input x. If $P \neq NP$, then $\mathcal{P}$ does not belong to the class FPTAS.*

The concepts of pseudo-polynomiality and strong NP-hardness allow us to classify NPO problems in different classes. Once we have shown that an NPO problem is pseudo-polynomial, we can think that it is computationally easier than a problem that is strongly NP-hard (see Theorem 3.18). On the other hand, we can capture some connections of these concepts with the approximability properties. Also from this point of view, even if we only have partial relationships, it is clear that pseudo-polynomiality is linked to well-approximable problems (see Theorem 3.17) while strong NP-hardness seems to be one of the characteristics of problems that behave badly with respect to approximability (see Corollary 3.19).

3.4 Exercises

Exercise 3.1 Define a polynomial-time local search algorithm for MAXIMUM SATISFIABILITY. Prove that the algorithm finds a solution with measure at least one half of the optimum.

Section 3.4

EXERCISES

Program 3.6: Gavril

input Graph $G = (V, E)$;
output Vertex cover V';
begin
   $V' := \emptyset$;
   repeat
      Choose any edge $e = (v_i, v_j) \in E$;
      $V' := V' \cup \{v_i, v_j\}$;
      $E := E - \{e' \mid e' \in E \text{ incident to } v_i \text{ or } v_j\}$;
   until $E = \emptyset$;
   return V'
end.

Problem 3.3: Maximum k-Dimensional Matching

INSTANCE: Set $M \subseteq X_1 \times X_2 \times \ldots \times X_k$ where $X_1, X_2, \ldots, X_k$ are disjoint sets having the same number q of elements.

SOLUTION: Subset $M' \subseteq M$ such that no two elements of M' agree in any coordinate.

MEASURE: Cardinality of M'.

Exercise 3.2 Prove that Program 3.1 can be extended to the case in which each clause has an associated weight preserving the performance ratio.

Exercise 3.3 Show that Program 3.6, also known as *Gavril's algorithm*, is a 2-approximate algorithm for MINIMUM VERTEX COVER. (Hint: observe that the algorithm computes a maximal matching of the input graph.)

Exercise 3.4 Consider Problem 3.3. Show that, for any $k \geq 3$, MAXIMUM k-DIMENSIONAL MATCHING is k-approximable. (Hint: consider maximal matchings, that is, matchings that cannot be extended without violating the feasibility constraints.)

Exercise 3.5 Consider the following variant of Christofides' algorithm (known as the tree algorithm for MINIMUM TRAVELING SALESPERSON): after finding the minimum spanning tree T, create the multigraph M by using two copies of each edge of T. Show that this algorithm is 2-approximate and that the bound 2 is tight.

Exercise 3.6 Consider Problem 3.4. Prove that a minimum spanning tree on S is a 2-approximate solution for this problem.

Chapter 3
APPROXIMATION CLASSES

> **Problem 3.4: Minimum Metric Steiner Tree**
>
> INSTANCE: Complete graph $G = (V, E)$, edge weight function $w : E \mapsto \mathbf{N}$ satisfying the triangle inequality, and subset $S \subseteq V$ of required vertices.
>
> SOLUTION: A Steiner tree T, i.e., a subgraph of G that is a tree and includes all the vertices in S.
>
> MEASURE: The sum of the weights of the edges in T.

> **Problem 3.5: Minimum Knapsack**
>
> INSTANCE: Finite set X of items, for each $x_i \in X$, value $p_i \in Z^+$ and size $a_i \in Z^+$, positive integer b.
>
> SOLUTION: A set of items $Y \subseteq X$ such that $\sum_{x_i \in Y} p_i \geq b$.
>
> MEASURE: Total size of the chosen items, i.e., $\sum_{x_i \in Y} a_i$.

Exercise 3.7 Show that the greedy heuristic for MINIMUM VERTEX COVER based on repeatedly choosing a vertex with highest degree does not provide a constant approximation ratio.

Exercise 3.8 Prove that, for any NPO minimization problem $\mathcal{P}$, if there exists a constant k such that it is NP-hard to decide whether, given an instance x, $m^*(x) \leq k$, then no polynomial-time r-approximate algorithm for $\mathcal{P}$ with $r < (k+1)/k$ can exist, unless P = NP.

Exercise 3.9 Prove that, for any integer c and for any rational $\delta > 0$, the number of ways of choosing c positive integers whose sum is less than δ is equal to $\binom{c + \lfloor \delta \rfloor}{\lfloor \delta \rfloor}$.

Exercise 3.10 By making use of a technique similar to the one used for MINIMUM PARTITION, show that, for any integer k, there is a $k/(k+1)$-approximate algorithm for MAXIMUM KNAPSACK.

Exercise 3.11 Fill in all the details of the proof of Theorem 3.14.

Exercise 3.12 Construct a FPTAS for MINIMUM KNAPSACK (see Problem 3.5) by making use of the variable partitioning technique.

Exercise 3.13 An NPO problem $\mathcal{P}$ is *simple* if, for every positive integer k, the problem of deciding whether an instance x of $\mathcal{P}$ has optimal measure at most k is in P. Prove that MAXIMUM CLIQUE is simple and that MINIMUM GRAPH COLORING is not simple (unless P = NP).

Exercise 3.14 Prove that a problem is simple if it belongs to the class PTAS.

Exercise 3.15 An NPO maximization problem $\mathcal{P}$ satisfies the *boundedness condition* if an algorithm $\mathcal{A}$ and a positive integer constant b exist such that the following hold: (a) for every instance x of $\mathcal{P}$ and for every positive integer c, $\mathcal{A}(x,c)$ is a feasible solution y of x such that $m^*(x) \leq m(x,y) + cb$, and (b) for every instance x of $\mathcal{P}$ and for every positive integer c, the time complexity of $\mathcal{A}(x,c)$ is a polynomial in $|x|$ whose degree depends only on the value $m(x,\mathcal{A}(x,c))/c$. Prove that MAXIMUM KNAPSACK verifies the boundedness condition.

Exercise 3.16 (*) Prove that an NPO maximization problem $\mathcal{P}$ admits a PTAS if and only if it is simple and satisfies the boundedness condition.

Exercise 3.17 An NPO problem $\mathcal{P}$ is *p-simple* if, for every positive integer k, the problem of deciding whether an instance x of $\mathcal{P}$ has optimal measure at most k is solvable in time bounded by a polynomial in $|x|$ and k. Prove that MAXIMUM KNAPSACK is p-simple.

Exercise 3.18 An NPO maximization problem $\mathcal{P}$ satisfies the *polynomial boundedness condition* if an algorithm $\mathcal{A}$ and a univariate polynomial p exist such that the following hold: (a) for every instance x of $\mathcal{P}$ and for every positive integer c, $\mathcal{A}(x,c)$ is a feasible solution y of x such that $m^*(x) \leq m(x,y) + cp(|x|)$, and (b) for every instance x of $\mathcal{P}$ and for every positive integer c, the time complexity of $\mathcal{A}(x,c)$ is a polynomial in $|x|$ whose degree depends only on the value $m(x,\mathcal{A}(x,c))/c$. Prove that MAXIMUM KNAPSACK verifies the polynomial boundedness condition.

Exercise 3.19 (*) Prove that an NPO maximization problem $\mathcal{P}$ admits a FPTAS if and only if it is p-simple and satisfies the polynomial boundedness condition.

3.5 Bibliographical notes

THE CONCEPT of approximation algorithm with "guaranteed performance" was introduced in the 1970s in the context of the first attempts to provide a formal analysis of computer heuristics for the solution of difficult optimization problems. The problem of designing efficient algorithms capable of achieving "good" feasible solutions of optimization problems in those cases in which the exact solution could not be achieved unless running exponential-time algorithms, has, of course, been addressed since the beginning of computational studies in operations research. But

Chapter 3

APPROXIMATION
CLASSES

it was not until the late 1960s that people started to perceive the need to go beyond computational experiments and to provide the formal analysis of the performance of an approximation algorithm in terms of quality of approximation. One of the first papers in which the performance af an approximation algorithm was analyzed is [Graham, 1966] which deals with multiprocessor scheduling.

In the subsequent years the study of approximation algorithms started to become more systematic. Among the papers that are considered to have laid down the first basic concepts relative to approximation algorithms, we may refer the reader to [Garey, Graham, and Ullman, 1972, Sahni, 1973, Johnson, 1974a, Nigmatullin, 1976]. In particular, in Johnson's paper the first examples of $f(n)$-approximate algorithms (that is with non-constant approximation ratio), and the first examples of approximation schemes are shown. In [Garey and Johnson, 1976b] the first survey with a complete view of the earliest results concerning approximation algorithms is provided.

In the late 1970s a large body of literature on approximation algorithms and on classes of approximability already existed. In [Horowitz and Sahni, 1978, Garey and Johnson, 1979] the authors provide the basic concepts and the related terminology (approximate algorithm, polynomial-time approximation scheme, fully polynomial-time approximation scheme). Those books have been the common ground for all work in the field.

The greedy algorithm for MAXIMUM SATISFIABILITY is due to [Johnson, 1974a]: a careful analysis of this algorithm is given in [Chen, Friesen, and Zheng, 1997] where it is shown that its performance ratio is at most 3/2. The approximation algorithm for MINIMUM VERTEX COVER based on the matching construction (see Exercise 3.3) is due to Gavril.

The proof that MINIMUM TRAVELING SALESPERSON is not approximable unless P = NP appears in [Sahni and Gonzalez, 1976] where the first examples of NP-hardness of approximation are presented. Note that such problems are called P-complete, a terminology never used afterwards with this meaning. The 2-approximate algorithm for the MINIMUM METRIC TRAVELING SALESPERSON based on the spanning tree construction (see Exercise 3.5) appears for the first time in [Korobkov and Krichevskii, 1966]. In [Rosenkrantz, Stearns, and Lewis, 1977] several heuristics for MINIMUM METRIC TRAVELING SALESPERSON are analyzed and various 2-approximate algorithms are shown. The 1.5-approximate algorithm (that is, Christofides' algorithm) appears in [Christofides, 1976]. It is worth noting that in the metric case this is still the best result known in terms of approximation ratio.

The primal-dual algorithm for the minimum weight perfect matching which is used in the proof of Theorem 3.6 is due to [Gabow, 1990]. Currently, the best known algorithm for finding a minimum weight maximal matching in a graph satisfying the triangular inequality is due to [Vaidya, 1990] and takes time $O(n^{2.5}(\log n)^4)$. Since in the case of Christofides' algorithm we have exactly this type of instances, we may also apply Vaidya's algorithm and the overall time of Christofides' algorithm becomes $O(n^{2.5}(\log n)^4)$.

The gap technique has been implicitly used for some time (for example, in the cited results on NP-hardness of approximation of MINIMUM TRAVELING SALESPERSON). The first proof of hardness of approximability (up to ratio 2) for the MINIMUM GRAPH COLORING problem, based on the gap technique, appeared in [Garey and Johnson, 1976a].

For a long time only a few problems admitting a PTAS were known. Among them there was MAXIMUM INDEPENDENT SET restricted to planar graphs (several drawing algorithms for planar graphs are described in [Di Battista et al., 1999]). A PTAS for this problem was due to [Lipton and Tarjan, 1980] and, independently, to [Chiba, Nishizeki, and Saito, 1982]. In [Baker, 1994], by means of a new technique, it is proved a more general result for such problem. The author proved that the MAXIMUM INDEPENDENT SET can be solved in polynomial time for k-outerplanar graphs and, as a consequence, showed that several problems restricted to planar graphs admit a PTAS: beside MAXIMUM INDEPENDENT SET, such problems include MINIMUM VERTEX COVER and MINIMUM DOMINATING SET.

More recently, by means of new techniques, polynomial-time approximation schemes have been designed for a large group of geometric problems in the Euclidean plane by [Arora, 1997]. Among them, the most relevant are MINIMUM TRAVELING SALESPERSON and MINIMUM STEINER TREE. Notice that the result is remarkable, because in [Papadimitriou and Yannakakis, 1993] it is proved that in the general metric case MINIMUM TRAVELING SALESPERSON does not allow a PTAS.

One of the first examples of a fully polynomial-time approximation scheme, namely the FPTAS for the knapsack problem, was given by [Ibarra and Kim, 1975]. Both notions are extensively discussed in [Horowitz and Sahni, 1978] where more examples of problems in PTAS and FPTAS are shown.

The technique of variable partitioning used in the FPTAS for MAXIMUM PRODUCT KNAPSACK has been introduced in [Marchetti-Spaccamela and Romano, 1985].

The strict inclusion of FPTAS in the class PTAS was proved in [Korte, and Schrader, 1981] by showing that MAXIMUM INTEGER m-DIMENSIONAL KNAPSACK (previously shown to be in PTAS by [Chandra, Hirschberg, and Wong, 1976]) is not in FPTAS. In the same paper, necessary and suf-

Section 3.5

BIBLIOGRAPHICAL NOTES

Chapter 3

APPROXIMATION
CLASSES

ficient conditions for showing the existence of PTAS and FPTAS (based on generalizations of the dominance rules introduced for knapsack-type problems) were presented. Other necessary and sufficient conditions, of more general applicability, were provided by [Paz and Moran, 1981]. In [Ausiello, Marchetti-Spaccamela, and Protasi, 1980] it is shown that these conditions are strictly related to the notions of strong NP-completeness and pseudo-polynomiality introduced by [Garey and Johnson, 1978].

Finally it is worth noting that even though the standard approach to the study of approximation algorithms was based on the notions of relative error and performance ratio, as defined in Sect. 3.1.2, in [Hsu and Nemhauser, 1979, Ausiello, Marchetti-Spaccamela, and Protasi, 1980, Demange and Pascos, 1996] it is suggested to consider the so-called "differential performance measure" which relates the error made by an algorithm to the range of possible values of the measure.

Input-Dependent and Asymptotic Approximation

IN THE previous two chapters we have seen examples of NPO problems that can be approximated either within a specific constant factor or within any constant factor. We also saw examples of NPO problems for which no approximation algorithm exists (unless P=NP) and examples of NPO problems for which an approximation algorithm but no approximation scheme exists (unless P=NP). To deal with these two latter kinds of problem, in this chapter we will relax the constraint on the performance ratio in two ways.

We will first allow the performance ratio to be dependent on the input size. Clearly, any NPO problem for which a solution can be computed in polynomial time is approximable with respect to this weaker notion of performance ratio. Indeed, since the measure function can be computed in polynomial time, the ratio between the optimal measure and the measure of *any* feasible solution is always bounded by a function exponential in the length of the input. However, our goal is to find algorithms that produce solutions whose ratio is bounded by a more slowly increasing function. In particular, we will provide an $O(\log n)$-approximate algorithm for MINIMUM SET COVER, where n denotes the cardinality of the set to be covered, an $O(n/\log n)$-approximate algorithm for MINIMUM GRAPH COLORING, where n denotes the number of nodes of the graph, and an $O(\log k)$-approximate algorithm for MINIMUM MULTI-CUT, where k denotes the number of pairs of vertices that have to be disconnected.

We will then consider *asymptotic approximation schemes*, that is, schemes whose performance ratio is bounded by a constant, for any in-

Chapter 4

INPUT-DEPENDENT AND ASYMPTOTIC APPROXIMATION

> **Problem 4.1: Minimum Set Cover**
> INSTANCE: Collection C of subsets of a finite set S.
> SOLUTION: A set cover for S, i.e., a subset $C' \subseteq C$ such that every element in S belongs to at least one member of C'.
> MEASURE: $|C'|$.

stance whose optimal measure is large enough. In this case, we will show that MINIMUM EDGE COLORING and MINIMUM BIN PACKING admit an asymptotic approximation scheme. We have already seen in Sect. 2.2.2 that the latter problem belongs to the class APX and in Sect. 3.2.2 that it does not belong to PTAS (unless P=NP). We will see in this chapter that the same holds for the former problem. In a certain sense, the existence of an asymptotic approximation scheme shows that these two problems are easier than other problems in APX − PTAS.

4.1 Between APX and NPO

As we already saw in the preceding chapters, for several problems (such as MINIMUM TRAVELING SALESPERSON) it is possible to prove that a constant performance ratio is not achievable unless $P = NP$. In these cases we can relax the constant requirement on the ratio by looking for algorithms whose performance depends on the length of the instance. In this section we will give three examples of these algorithms.

4.1.1 Approximating the set cover problem

The first problem we consider is MINIMUM SET COVER (see Problem 4.1). For this problem, let us consider Program 4.1, which is a simple polynomial-time greedy procedure to cover set S. At each iteration, the algorithm chooses the set that covers the maximum number of uncovered elements (breaking ties arbitrarily) and updates the remaining sets.

Theorem 4.1 ▶ *Program 4.1 is a $(\lfloor \ln n \rfloor + 1)$-approximate algorithm for* MINIMUM SET COVER, *where n denotes the number of elements of the universe S.*

PROOF Given an instance of MINIMUM SET COVER, let k denote the largest cardinality of the sets in the collection. We will now prove that, for any optimal

Section 4.1
BETWEEN APX AND NPO

Program 4.1: Greedy Set Cover

input Collection C of subsets of a finite set S;
output Set cover C';
begin
   $U := S$;
   for each set c_i in C **do** $c'_i := c_i$;
   $C' := \emptyset$;
   while $U \neq \emptyset$ **do**
   begin
     Let c'_j be a set with maximum cardinality;
     $C' := C' \cup \{c_j\}$;
     $U := U - c'_j$;
     for each c'_i **do** $c'_i := c'_i - c'_j$
   end;
   return C'
end.

solution C^*,
$$\sum_{c_i \in C^*} \mathcal{H}(|c_i|) \geq |C'|, \quad (4.1)$$

where, for any integer $r > 0$, $\mathcal{H}(r)$ denotes the r-th harmonic number (that is, $\mathcal{H}(r) = \sum_{i=1}^{r} 1/i$) and C' is the set cover obtained by Program 4.1. Since $|c_i| \leq k$, for any i with $1 \leq i \leq m$, $\mathcal{H}(k) \leq \lfloor \ln k \rfloor + 1$, and $k \leq n$, the above inequality implies that

$$|C'| \leq \sum_{c_i \in C^*} \mathcal{H}(k) \leq \mathcal{H}(n)|C^*| \leq (\lfloor \ln n \rfloor + 1)|C^*|.$$

That is, the performance ratio of Program 4.1 is at most $\lfloor \ln n \rfloor + 1$.

For any instance $x = (S, \{c_1, \ldots, c_m\})$ of MINIMUM SET COVER, let us denote by $a_1, \ldots, a_{|C'|}$ the sequence of indices of the subsets that belong to the set cover C'. Moreover, for any $j \in \{1, \ldots, |C'|\}$ and for any $i \in \{1, \ldots, m\}$, let c_i^j be the *surviving* part of c_i before index a_j has been chosen (that is, the subset of c_i that does not intersect any chosen set). Clearly, $c_i^1 = c_i$, for any i. Moreover, for the sake of simplicity, we will adopt the convention that $c_i^{|C'|+1} = \emptyset$, for any $i \in \{1, \ldots, m\}$. The set of elements of c_i that are covered for the first time by c_{a_j} is given by

$$c_i \cap c_{a_j}^j = c_i^j \cap c_{a_j}^j = c_i^j - c_i^{j+1}. \quad (4.2)$$

For any $i \in \{1, \ldots, m\}$, let l_i denote the largest index such that $|c_i^{l_i}| > 0$: that is, after $c_{a_{l_i}}$ has been included into C' the subset c_i has been covered.

The proof of Eq. (4.1) consists of the following two steps:

Chapter 4

INPUT-DEPENDENT AND ASYMPTOTIC APPROXIMATION

1. Prove that, for any $i \in \{1,\ldots,m\}$,

$$\mathcal{H}(|c_i|) \geq \sum_{j=1}^{|C'|} \frac{|c_i \cap c_{a_j}^j|}{|c_{a_j}^j|}.$$

2. Prove that, for any set cover C'' (and, therefore, for any optimal solution),

$$\sum_{c_i \in C''} \sum_{j=1}^{|C'|} \frac{|c_i \cap c_{a_j}^j|}{|c_{a_j}^j|} \geq |C'|.$$

Both steps consist of simple algebraic calculations. For what regards the first step, notice that, for any i with $1 \leq i \leq m$,

$$\sum_{j=1}^{|C'|} \frac{|c_i \cap c_{a_j}^j|}{|c_{a_j}^j|} = \sum_{j=1}^{|C'|} \frac{|c_i^j| - |c_i^{j+1}|}{|c_{a_j}^j|} \leq \sum_{j=1}^{l_i} \frac{|c_i^j| - |c_i^{j+1}|}{|c_i^j|}$$

$$= \sum_{j=1}^{l_i} \sum_{k=|c_i^{j+1}|+1}^{|c_i^j|} \frac{1}{|c_i^j|} \leq \sum_{j=1}^{l_i} \sum_{k=1}^{|c_i^j|-|c_i^{j+1}|} \frac{1}{k+|c_i^{j+1}|}$$

$$\leq \sum_{j=1}^{l_i} \left(\mathcal{H}(|c_i^j|) - \mathcal{H}(|c_i^{j+1}|) \right) = \mathcal{H}(|c_i^1|) = \mathcal{H}(|c_i|)$$

where the first equality follows from Eq. (4.2) and the first inequality is due to the fact that, for any j with $1 \leq j \leq |C'|$, $|c_i^j| \leq |c_{a_j}^j|$ (since the algorithm always chooses the biggest surviving set).

The second step follows as easily as the first one. Namely, for any set cover C'',

$$\sum_{c_i \in C''} \sum_{j=1}^{|C'|} \frac{|c_i \cap c_{a_j}^j|}{|c_{a_j}^j|} = \sum_{j=1}^{|C'|} \frac{1}{|c_{a_j}^j|} \sum_{c_i \in C''} |c_i \cap c_{a_j}^j| \geq \sum_{j=1}^{|C'|} \frac{|c_{a_j}^j|}{|c_{a_j}^j|} = |C'|$$

where the inequality is due to the fact that C'' is a set cover. In conclusion, Eq. (4.1) is satisfied and the theorem follows. **QED**

Example 4.1 ▶ Let us consider the instance of MINIMUM SET COVER with $S = \{1,\ldots,24\}$ and with C given by the following seven subsets of S: $c_1 = \{1,\ldots,8\}$, $c_2 = \{9,\ldots,16\}$, $c_3 = \{17,\ldots,24\}$, $c_4 = \{5,6,7,8,13,14,15,16,21,22,23,24\}$, $c_5 = \{3,4,11,12,19,20\}$, $c_6 = \{2,10,18\}$, $c_7 = \{1,9,17\}$. The first subset chosen by Program 4.1 is c_4. As a consequence, the first three subsets become $c_1^2 = \{1,\ldots,4\}$, $c_2^2 = \{9,\ldots,12\}$, and $c_3^2 = \{17,\ldots,20\}$, respectively. The next subset chosen is c_5 and, once again, the algorithm modifies the first three subsets that become $c_1^3 = \{1,2\}$, $c_2^3 = \{9,10\}$, and $c_3^3 = \{17,18\}$, respectively. At the next two iterations of the loop, the subsets chosen are c_6 and c_7, respectively, so that the solution computed by the algorithm uses four subsets. On the other hand, it is easy to verify that the first three subsets form an optimal solution.

Program 4.2: Greedy Graph Coloring

```
input Graph G = (V, E);
output Node coloring of G;
begin
   i := 0;
   U := V;
   while U ≠ ∅ do
   begin
      i := i + 1;
      W := U;
      while W ≠ ∅ do
      begin
         H := graph induced by W;
         v := node of minimum degree in H;
         f(v) := i;
         W := W − {v} − {z|z is a neighbor of v in H}
      end;
      U := U − f⁻¹(i);
   end;
   return f
end.
```

Section 4.1

BETWEEN APX
AND NPO

The analysis of Theorem 4.1 is tight: indeed, it is possible to generalize the previous example in order to prove such statement (see Exercise 4.1).

4.1.2 Approximating the graph coloring problem

In this section we provide an $O(n/\log n)$-approximate algorithm for MINIMUM GRAPH COLORING based on repeatedly finding an independent set (i.e., a set of vertices that can be colored with one color) and coloring it with a new color, until all vertices have been colored. This algorithm is given as Program 4.2 and, clearly, runs in polynomial time. The next result gives an upper bound on its performance ratio.

◀ **Lemma 4.2**

Program 4.2 colors any k-colorable graph $G = (V, E)$ with at most $3|V|/\log_k |V|$ colors.

PROOF

Let H be the graph considered by the algorithm at the beginning of a generic iteration of the inner loop and let W be the corresponding set of vertices. Since H is a subgraph of a k-colorable graph, H is a k-colorable graph and, therefore, it must contain an independent set of size at least $|W|/k$. Each vertex of this set has degree at most $|W| - |W|/k =$

Chapter 4
INPUT-DEPENDENT AND ASYMPTOTIC APPROXIMATION

$|W|(k-1)/k$. This implies that the minimum degree of H is at most $|W|(k-1)/k$ so that at least $|W| - |W|(k-1)/k = |W|/k$ nodes will still be in W at the beginning of the next iteration. Since the inner loop ends when W becomes empty, we have that at least $\lceil \log_k |W| \rceil$ iterations have to be performed, which in turn implies that, at the end of the inner loop,

$$|\{v \mid v \in W \wedge f(v) = i\}| \geq \lceil \log_k |W| \rceil.$$

Hence, for each used color i, the number of vertices colored with i is at least $\lceil \log_k |U| \rceil$, where U denotes the set of nodes uncolored just before color i is considered.

Let us now analyze the size of U at the beginning of an iteration of the outer loop. If $|U| \geq |V|/\log_k |V|$, then we have that

$$\lceil \log_k |U| \rceil \geq \log_k |U| \geq \log_k (|V|/\log_k |V|) > \log_k \sqrt{|V|} = \frac{1}{2} \log_k |V|.$$

This implies that the size of U decreases by at least $\frac{1}{2}\log_k |V|$ at each iteration: as a consequence, the first time $|U|$ becomes smaller than $|V|/\log_k |V|$, the algorithm has used no more than $2|V|/\log_k |V|$ colors.

If $|U| < |V|/\log_k |V|$, it is clear that the algorithm colors all remaining nodes in U with at most $|U| < |V|/\log_k |V|$ colors. It follows that the algorithm uses at most $3|V|/\log_k |V|$ colors.

QED

Theorem 4.3 ▶ MINIMUM GRAPH COLORING *admits an $O(n/\log n)$-approximate algorithm, where n denotes the number of nodes.*

PROOF   The previous lemma implies that, for any input graph G with n nodes, the solution returned by Program 4.2 uses at most $3n/\log_{m^*(G)} n = 3n \log(m^*(G))/\log n$ colors, where $m^*(G)$ denotes the minimum number of colors necessary to color G. This implies that the performance ratio of this algorithm is at most

$$\frac{3n \log(m^*(G))/\log n}{m^*(G)} = O(n/\log n)$$

QED   and the theorem follows.

Example 4.2 ▶ Let us consider the graph of Fig. 4.1(a). In this case $m^*(G) = 3$: indeed, we can assign to nodes a, d, and h the first color, to nodes b, e, and g the second color, and to nodes c and f the third color. Program 4.2, instead, could first choose node a and assign to it the first color. The resulting graph H is shown in Fig. 4.1(b): at this step, node e is chosen and the first color is assigned to it. The new graph H is shown in Fig. 4.1(c): in this case, the algorithm could choose node h and assign to it the first color. In conclusion, after exiting the inner loop, nodes a, e, and h have been assigned the first color.

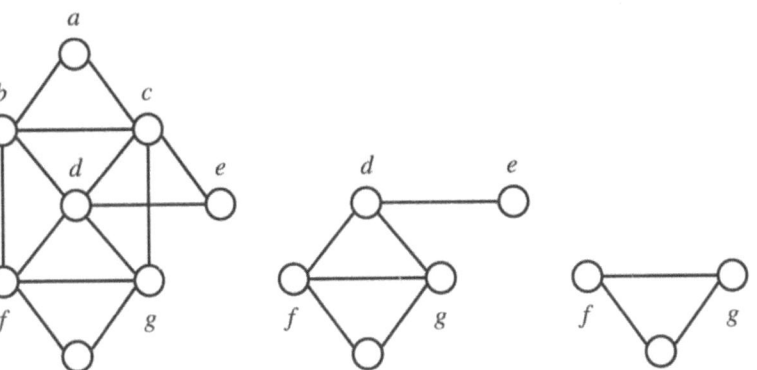

Figure 4.1
The first iteration of the outer loop of Program 4.2

Since uncolored nodes exist (see Fig. 4.2(a)), the algorithm will continue executing the outer loop and a possible run of the algorithm could assign the second color to nodes b and g. Successively, the third color could be assigned to nodes c and f and, finally, the fourth color is assigned to node d (see Figs. 4.2(a) and (b)). Hence, in this case the algorithm produces a 4-coloring.

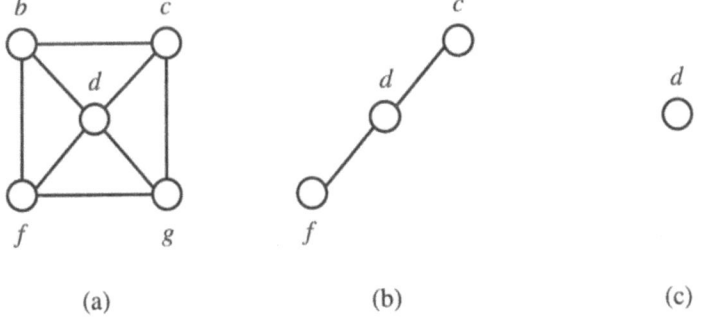

Figure 4.2
The next steps of Program 4.2

4.1.3 Approximating the minimum multi-cut problem

In this section, we give another example of how duality relationships between integer linear programming problems (see Appendix A) can be exploited to derive efficient approximation algorithms (a first example of an approximation algorithm based on the duality relationship has been given in Sect. 2.4.2).

First of all, let us recall that MAXIMUM FLOW (see Problem 4.2) consists in finding the maximum flow that can be routed in a network from a

Chapter 4

INPUT-DEPENDENT
AND ASYMPTOTIC
APPROXIMATION

> **Problem 4.2: Maximum Flow**
>
> INSTANCE: Graph $G = (V, E)$, edge capacity function $c : E \mapsto \mathbb{Q}^+$, pair $(s, t) \in V \times V$.
>
> SOLUTION: A flow from s to t, that is, a function $f : V \times V \mapsto \mathbb{Q}^+$ such that:
>
> 1. $f(u, v)$ is defined if and only if $(u, v) \in E$;
>
> 2. $f(u, v) + f(v, u) \leq c(u, v)$, for each $(u, v) \in E$, that is, the total flow in an edge cannot exceed its capacity;
>
> 3. $\sum_{u \in V - \{v\}} f(u, v) = \sum_{u \in V - \{v\}} f(v, u)$ for each $v \in V - \{s, t\}$, that is, the flow entering in a node must equate the flow exiting the same node, for all nodes except the source and the destination.
>
> MEASURE: $\sum_{u \in V - \{t\}} f(u, t) - \sum_{u \in V - \{t\}} f(t, u)$, that is, the total flow entering the destination.

source s to a destination t without violating some specified edge capacity constraints. Moreover, MINIMUM CUT (see Problem 4.3) requires to find the set of arcs of minimum total capacity whose removal disconnects the source from the destination in a network. These two problems are related through the maxflow-mincut theorem (see Bibliographical notes), a central result in flow theory, that lies at the basis of the solution of many relevant optimization problems. Such theorem states that, since a duality relationship holds between MAXIMUM FLOW and MINIMUM CUT, their optimal solutions have the same value.

Let us now consider the generalization of MAXIMUM FLOW to the case in which more than one commodity has to flow in the network (see Problem 4.4). This problem has the following linear programming formulation LP_{MMF}:

$$
\begin{aligned}
\text{maximize} \quad & \sum_{i=1}^{k} \sum_{v \in V - \{t_i\}} (f^i_{v,t_i} - f^i_{t_i,v}) \\
\text{subject to} \quad & \sum_{(v,u) \in E} f^i_{vu} - \sum_{(u,v) \in E} f^i_{uv} = 0 \quad \forall v \in V - \{s_i, t_i\}, \, i = 1, \ldots, k \\
& \sum_{i=1}^{k} f^i_{uv} + \sum_{i=1}^{k} f^i_{vu} \leq c_{uv} \quad \forall (u, v) \in E \\
& f^i_{uv} \geq 0 \quad \forall (u, v) \in E,
\end{aligned}
$$

where f^i_{uv} is the flow of commodity i along edge (u, v) from u to v and c_{uv} denotes the capacity of edge (u, v). The first set of constraints represents

> **Problem 4.3: Minimum Cut**
> INSTANCE: Graph $G = (V, E)$, edge capacity function $c : E \mapsto \mathbb{Q}^+$, pair $(s, t) \in V \times V$.
> SOLUTION: A set $E' \subseteq E$ of edges whose removal disconnects s and t.
> MEASURE: The overall capacity of E', i.e., $\sum_{e \in E'} c(e)$.

flow conservation at vertices, while the second set of inequalities states that the total flow value assigned to each edge does not exceed its capacity.

Let us also consider the corresponding generalization of MINIMUM CUT, shown in Problem 4.5. This problem has the following integer linear programming formulation:

$$\begin{aligned}
\text{minimize} \quad & \sum_{(u,v) \in E} d_{uv} c_{uv} \\
\text{subject to} \quad & \sum_{(u,v) \in E} d_{uv} \pi_i^j(u,v) \geq 1 \quad \forall \pi_i^j \in \mathcal{P}_i, i = 1, \ldots, k \\
& d_{uv} \in \mathbb{N} \quad \forall (u,v) \in E,
\end{aligned}$$

where $\mathcal{P}_i$ denotes the set of all paths connecting s_i to t_i, π_i^j is the j-th of such paths, $\pi_i^j(u,v)$ is 1 if (u,v) is an edge in π_i^j, and $\pi_i^j(u,v)$ is 0 otherwise. Notice that, indeed, any solution of the above problem corresponds to a multi-cut, defined as the set of edges (u,v) such that $d_{uv} > 0$. Notice also that the above formulation has an exponential number of constraints: however, an equivalent, less natural, formulation ILP_{MC} with polynomially many constraints is the following one (see Exercise 4.8):

$$\begin{aligned}
\text{minimize} \quad & \sum_{(u,v) \in E} d_{uv} c_{uv} \\
\text{subject to} \quad & d_{uv} \geq p_u^i - p_v^i \quad \forall (u,v) \in E, i = 1, \ldots, k \\
& p_s^i - p_t^i \geq 1 \quad i = 1, \ldots, k \\
& p_v^i \in \mathbb{N} \quad \forall v \in V, i = 1, \ldots, k \\
& d_{uv} \in \mathbb{N} \quad \forall (u,v) \in E,
\end{aligned}$$

where d_{uv} is a variable that, by the minimization requirement, will assume value at most 1 in the optimal solution and with such a value will denote that edge (u,v) is in the multi-cut. Intuitively, variables p_v^i denote a *potential* associated to flow i at vertex v. In this way, the constraints impose that, for each pair (s_i, t_i), they must be sufficiently "far apart" to always have at least one edge in the multi-cut between them.

It is possible to see that, by applying the well-known primal-dual transformation (see Appendix A) to LP_{MMF}, we obtain the same linear program

Chapter 4

INPUT-DEPENDENT AND ASYMPTOTIC APPROXIMATION

Problem 4.4: Maximum Multi-commodity Flow

INSTANCE: Graph $G = (V, E)$, edge capacity function $c : E \mapsto \mathbb{Q}^+$, k pairs $(s_i, t_i) \in V \times V$.

SOLUTION: A set of k flows f^i, each from s_i to t_i, for $i = 1, \ldots, k$, such that the total flow value $\sum_{\forall i} f^i(u, v)$ that passes through edge (u, v) does not exceed its capacity.

MEASURE: The sum of the total flows entering all destinations.

Problem 4.5: Minimum Multi-Cut

INSTANCE: Graph $G = (V, E)$, edge capacity function $c : E \mapsto \mathbb{Q}^+$, k pairs $(s_i, t_i) \in V \times V$.

SOLUTION: A subset $E' \subseteq E$ of edges whose removal disconnects s_i and t_i, $i = 1, \ldots, k$.

MEASURE: The overall capacity $\sum_{e \in E'} c(e)$ of E'.

---

as by relaxing the integrality constraints of ILP_{MC}. Such a program is the linear programming formulation of MINIMUM FRACTIONAL MULTI-CUT. This situation is similar to what happens in the case of MAXIMUM FLOW, whose dual is MINIMUM FRACTIONAL CUT, i.e., the relaxation of MINIMUM CUT to include also non integral solutions. However, the combinatorial structure of MINIMUM FRACTIONAL CUT guarantees that any optimal solution is an integral one and, as a consequence, an optimal solution also for MINIMUM CUT. Unfortunately, the equality does not hold for MINIMUM FRACTIONAL MULTI-CUT and all we can say in this case is that the capacity of the minimum multi-cut in a graph is *at least* the value of the maximum multi-commodity flow in the same graph.

Actually, a stronger relationship has been shown between MINIMUM MULTI-CUT and MAXIMUM MULTI-COMMODITY FLOW: given a graph $G = (V, E)$ with k commodities, the capacity of the minimum multi-cut is at most $O(\log k)$ times the maximum flow in G.

In the following, we present an approximation algorithm that, given an instance of MINIMUM MULTI-CUT, returns a solution whose capacity is bounded by $O(\log k)$ times the maximum multi-commodity flow in the graph. This guarantees that the value of such a solution is at most $O(\log k)$ times the capacity of the minimum multi-cut. The algorithm exploits the knowledge of an optimal fractional multi-cut, which can be efficiently computed by solving in polynomial time either MINIMUM FRACTIONAL

MULTI-CUT or its dual MAXIMUM MULTI-COMMODITY FLOW.

Let $d: E \mapsto \mathbb{Q}^+$ be a function that associates to each edge (u,v) the value of variable d_{uv} in the optimal solution of MINIMUM FRACTIONAL MULTI-CUT. If we consider the graph $G = (V,E)$ with functions $c: E \mapsto \mathbb{Q}^+$ and $d: E \mapsto \mathbb{Q}^+$ and k pairs $\{(s_1,t_1), \ldots, (s_k,t_k)\}$, we may see it as a set of pipes (edges) connected at nodes, where pipe (u,v) has length $d(u,v)$ and cross section $c(u,v)$ (and thus volume $c(u,v)d(u,v)$). In particular, this pipe system is the one of minimum volume such that, for each pair (s_i,t_i), the distance between s_i and t_i is at least 1. Moreover, let $\Psi = \sum_{(u,v) \in E} c(u,v)d(u,v)$ be the overall volume of G: clearly, since MINIMUM FRACTIONAL MULTI-CUT is a relaxation of MINIMUM MULTI-CUT, Ψ is a lower bound on the optimal measure of the corresponding MINIMUM MULTI-CUT problem.

The algorithm (see Program 4.3) works by iteratively producing a sequence $\mathcal{V} = (V_1, V_2, \ldots, V_q)$ of disjoint subsets of V such that, for each $r = 1, \ldots, q$:

1. at most one between s_i and t_i is contained in V_r, for each $i = 1, \ldots, k$;

2. there exists $l \in \{1, \ldots, k\}$ such that either s_l or t_l is contained in V_r.

Note that, for any r with $1 \leq r \leq q$, the subsequence $(V_1, \ldots, V_r)$ *separates* at least r pairs (s_i, t_i).

The algorithm works in phases, where at each phase a new set V_r is added to $\mathcal{V}$, until each vertex is included in one subset of $\mathcal{V}$. In particular, let $\overline{G}_r = (\overline{V}_r, \overline{E}_r)$ be the graph induced by $\overline{V}_r = V - (\cup_{j=1}^{r-1} V_j)$: if there exists no pair (s_i, t_i) such that $s_i \in \overline{V}_r$ and $t_i \in \overline{V}_r$, then the algorithm sets $V_r = \overline{V}_r$ and returns the multi-cut $\{(u,v) \in E \mid u \in V_p \wedge v \in V_q \wedge p \neq q\}$.

Otherwise, let (s_i, t_i) be an arbitrarily chosen pair such that both $s_i \in \overline{V}_r$ and $t_i \in \overline{V}_r$. The algorithm then computes a suitable set $V_r \subseteq \overline{V}_r$ such that exactly one between s_i and t_i is contained in V_r. This is performed by selecting a set of vertices that are within a certain distance ρ from s_i. This set of vertices is a *ball* centered at s_i and of radius ρ. The algorithm starts with $\rho = 0$ and iteratively increases ρ until a suitable condition is verified.

In order to describe more in detail how this is done, let us first introduce some definitions. For any $u, v \in V$, let $\delta(u,v)$ be the length of a shortest path from u to v, with respect to edge length $d: E \mapsto \mathbb{Q}^+$. For any $v \in \overline{V}_r$ and for any $\rho \in \mathbb{Q}$, the *ball* of radius ρ around v (with respect to δ) is defined as

$$\mathcal{B}_\delta(v, \rho) = \{u \in \overline{V}_r \mid \delta(u,v) \leq \rho\}.$$

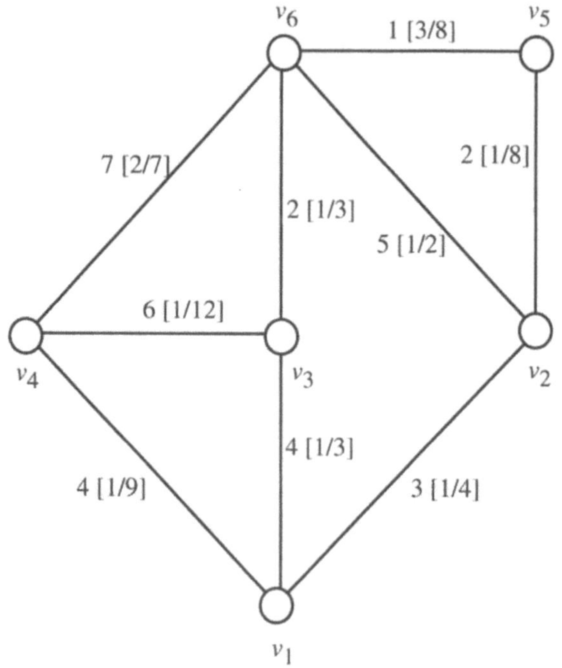

Chapter 4

INPUT-DEPENDENT
AND ASYMPTOTIC
APPROXIMATION

Figure 4.3
Sample graph G: square brackets contain the values of function d

Moreover, let

$$\mathcal{E}_\delta(v,\rho) = \{(w,z) \in \overline{E}_r \mid w \in B_\delta(v,\rho) \wedge z \in B_\delta(v,\rho)\}$$

be the set of edges completely contained in $\mathcal{B}_\delta(v,\rho)$ and let

$$\overline{\mathcal{E}}_\delta(v,\rho) = \{(w,z) \in \overline{E}_r - \mathcal{E}_\delta(v,\rho) \mid w \in B_\delta(v,\rho) \vee z \in B_\delta(v,\rho)\}$$

be the set of edges partially contained in $\mathcal{B}_\delta(v,\rho)$. Finally, the *volume* of $\mathcal{B}_\delta(v,\rho)$ is defined as

$$V_\delta(v,\rho) = \frac{\Psi}{k} + \sum_{(w,z)\in \mathcal{E}_\delta(v,\rho)} c(w,z)d(w,z) + \sum_{(w,z)\in \overline{\mathcal{E}}_\delta(v,\rho)} c(w,z)(\rho - \min(\delta(v,w), \delta(v,z))).$$

while the *cost* of $\mathcal{B}_\delta(v,\rho)$ is defined as

$$C_\delta(v,\rho) = \sum_{(w,z)\in \overline{\mathcal{E}}_\delta(v,\rho)} c(w,z).$$

Note that, apart from a fixed amount of volume Ψ/k (that we assume located on v), each edge contributes to the volume of $\mathcal{B}_\delta(v,\rho)$ for its fraction contained in $\mathcal{B}_\delta(v,\rho)$. On the other side, an edge contributes to the cost of $\mathcal{B}_\delta(v,\rho)$ for its capacity, if it is in the cut separating $\mathcal{B}_\delta(v,\rho)$ from $V - \mathcal{B}_\delta(v,\rho)$.

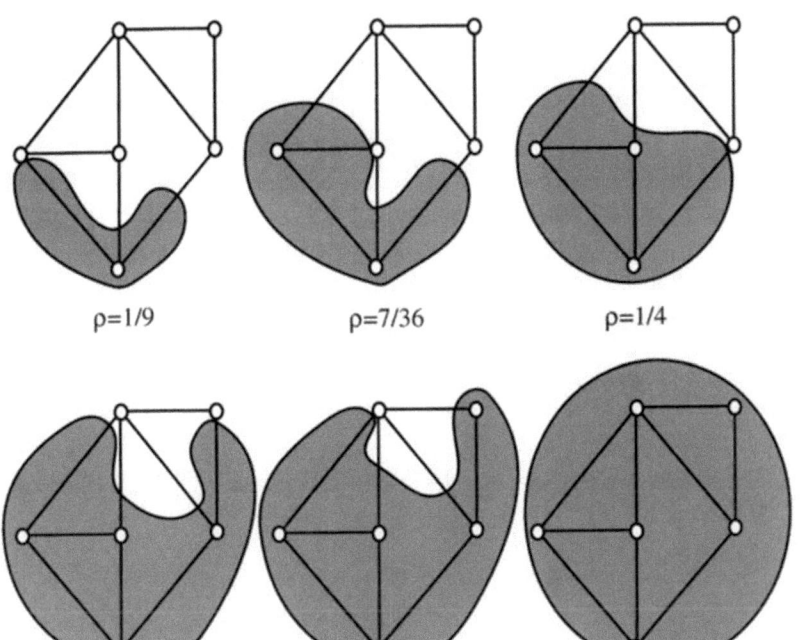

Figure 4.4
The balls $B_\delta(v_1, \rho)$ for some values of ρ

Section 4.1

BETWEEN APX
AND NPO

◀ Example 4.3

Let us consider the graph shown in Fig. 4.3, where the square brackets contain the values of function d, and assume that $k = 2$. Hence, the amount of volume that the algorithm assumes to be located at each node is

$$\frac{\Psi}{k} = \frac{213/24}{2} = \frac{213}{48} \approx 4.4.$$

Moreover, if we consider v_1 as the center of the balls, we have that $\delta(v_1, v_2) = 1/4$, $\delta(v_1, v_3) = 7/36$, $\delta(v_1, v_4) = 1/9$, $\delta(v_1, v_5) = 3/8$, $\delta(v_1, v_6) = 25/63$. In Fig. 4.4 we show the balls centered in v_1 corresponding to $\rho = \delta(v_1, v_4) - \varepsilon$, $\rho = \delta(v_1, v_3) - \varepsilon$, $\rho = \delta(v_1, v_2) - \varepsilon$, $\rho = \delta(v_1, v_5) - \varepsilon$, $\rho = \delta(v_1, v_6) - \varepsilon$, and $\rho = 1 - \varepsilon$, where ε is abitrarily small. Let us now consider, for example, $\mathcal{B}_\delta(v_1, 1/4 - \varepsilon)$. This ball contains vertices v_1, v_3, and v_4 and we have that $\mathcal{E}_\delta(v_1, 1/4 - \varepsilon) = \{(v_1, v_3), (v_1, v_4), (v_3, v_4)\}$ and that $\overline{\mathcal{E}}_\delta(v_1, 1/4 - \varepsilon) = \{(v_1, v_2), (v_3, v_6), (v_4, v_6)\}$. Hence, it results

$$\begin{aligned} V_\delta(v_1, \frac{1}{4} - \varepsilon) &= \frac{\Psi}{k} + \frac{4}{3} + \frac{4}{9} + \frac{1}{2} + 3(\frac{1}{4} - \varepsilon) + 2(\frac{1}{4} - \varepsilon - \frac{7}{36}) + 7(\frac{1}{4} - \varepsilon - \frac{1}{9}) \\ &= \frac{213}{48} + 16.8 + \frac{10}{9} - 3\varepsilon \approx 22.31 \end{aligned}$$

and

$$C_\delta(v_1, 1/4 - \varepsilon) = 3 + 2 + 7 = 12.$$

Chapter 4

INPUT-DEPENDENT AND ASYMPTOTIC APPROXIMATION

Observe that we can consider the ratio $C_\delta(v,\rho)/V_\delta(v,\rho)$ as a function of ρ. Let $\{u_1, u_2, \ldots, u_{|\overline{V}_r|-1}\}$ be the set of nodes in $\overline{V}_r - \{v\}$ ordered by non decreasing distance from v (that is, $i < j \Rightarrow \delta(v,u_i) \leq \delta(v,u_j)$) and let $\{r_1, r_2, \ldots, r_{|\overline{V}_r|-1}\}$ be the corresponding distances from v. It is easy to see that the function $C_\delta(v,\rho)/V_\delta(v,\rho)$ is continuous and non-increasing, for $0 < \rho < r_1$ and for $r_i < \rho < r_{i+1}$ ($i = 1, \ldots, |\overline{V}_r| - 2$). This implies that it tends to its minimum as it approaches some r_j from below. Notice, moreover, that the derivative of $C_\delta(v,\rho)/V_\delta(v,\rho)$ with respect to ρ is well defined for all values $\rho \notin \{r_1, \ldots, r_{|\overline{V}_r|-1}\}$ (see Fig. 4.3).

Example 4.4 ▶ Let us consider again the graph shown in Fig. 4.3 and assume that v_1 is the center of the balls. In this case, $u_1 = v_4$, $u_2 = v_3$, $u_3 = v_2$, $u_4 = v_5$, and $u_5 = v_6$. Moreover, $r_1 = 1/9$, $r_2 = 7/36$, $r_3 = 1/4$, $r_4 = 3/8$, and $r_5 = 25/63$. In Fig. 4.5 the behavior of function $C_\delta(v,\rho)/V_\delta(v,\rho)$ is shown in the interval of interest (that is, $(0, 25/63)$).

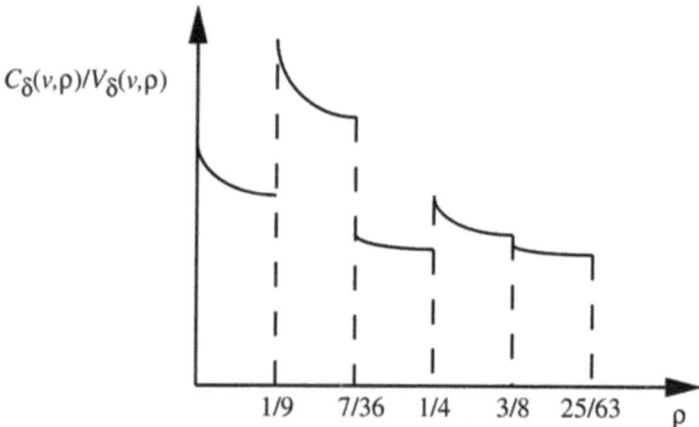

Figure 4.5
The function $C_\delta(v,\rho)/V_\delta(v,\rho)$

Let us now prove that there exists a value $\rho_r < 1/2$ for which the corresponding value $C_\delta(v,\rho_r)/V_\delta(v,\rho_r)$ is sufficiently small.

Lemma 4.4 ▶ *There exists a value $\rho_r < 1/2$ such that*

$$\frac{C_\delta(v,\rho_r)}{V_\delta(v,\rho_r)} \leq 2\ln 2k.$$

PROOF   Let us first notice that $C_\delta(v,\rho)$ is the derivative of $V_\delta(v,\rho)$ with respect to ρ. As a consequence, whenever such a derivative is defined, we have

$$\frac{C_\delta(v,\rho)}{V_\delta(v,\rho)} = \frac{\partial}{\partial \rho}(\ln V_\delta(v,\rho)).$$

Assume now, by contradiction, that, for every $\rho \in (0, 1/2)$, we have $C_\delta(v,\rho)/V_\delta(v,\rho) > 2\ln 2k$. Two cases are possible:

Section 4.1

BETWEEN APX AND NPO

1. $r_1 \geq 1/2$. This implies that $C_\delta(v,\rho)/V_\delta(v,\rho)$ is differentiable in $(0,1/2)$. As a consequence, we have that

$$\frac{\partial}{\partial \rho}(\ln V_\delta(v,\rho)) > 2\ln 2k$$

and, by integrating both sides in $(0,1/2)$,

$$\ln \frac{V_\delta(v,1/2)}{V_\delta(v,0)} > \ln 2k,$$

which implies $V_\delta(v,1/2) > 2kV_\delta(v,0) = 2\Psi$. This contradicts the fact that $\Psi + \Psi/k$ is the maximum possible value of $V_\delta(v,\rho)$, which is obtained when $\rho \geq r_{|\bar{V}_r|-1}$, i.e., when all the graph is included.

2. There exists t, with $1 \leq t \leq |\bar{V}_r| - 1$, such that $r_1, \ldots, r_t \in (0,1/2)$ (therefore, $C_\delta(v,\rho)/V_\delta(v,\rho)$ might not be differentiable in $(0,1/2)$). Let $r_0 = 0$ and $r_{t+1} = 1/2$. If we apply the same considerations above to each interval (r_i, r_{i+1}), for $i = 0, \ldots, t$, we obtain that

$$\ln V_\delta(v, r_{i+1}) - \ln V_\delta(v, r_i) > (r_{i+1} - r_i)2\ln 2k.$$

Since the volume does not decrease in correspondence to each vertex, by summing over all intervals we get

$$\ln \frac{V_\delta(v,1/2)}{V_\delta(v,0)} = \ln V_\delta(v,1/2) - \ln V_\delta(v,0) > \ln 2k,$$

thus reducing to the previous case.

The lemma is thus proved. QED

We are now ready to prove the main result of this section.

Given an instance x of MINIMUM MULTI-CUT, *the solution returned by Program 4.3 with input x is a feasible solution whose cost $m_{MC}(x)$ satisfies the inequality*

$$m_{MC}(x) < 4\ln(2k)m^*(x).$$

◀ Theorem 4.5

PROOF

Let us first observe that the solution returned by the algorithm is feasible. Indeed, since distances $d(u,v)$ are derived by solving MINIMUM FRACTIONAL MULTI-CUT, we have that $d(s_i,t_i) \geq 1$, for $i = 1,\ldots,k$. As we may observe, for all $i = 1,\ldots,k$, Program 4.3 separates s_i from all vertices at distance at least $1/2$: in particular, it separates s_i from t_i. Hence, the solution returned by the algorithm is a multi-cut.

Program 4.3: Multi-cut

input Graph $G = (V,E)$, pairs (s_i, t_i), functions c, d;
output Multi-cut E';
begin
    $r := 1$;
    $\overline{V}_r := V$;
    while $\overline{V}_r \neq \emptyset$ **do begin**
        if there exists i such that $s_i \in \overline{V}_r \wedge t_i \in \overline{V}_r$ **then**
        begin
            Let ρ_r be such that $C_\delta(s_i, \rho_r)/V_\delta(s_i, \rho_r) \leq C_\delta(s_i, \rho)/V_\delta(s_i, \rho)$
            for all $\rho = \delta(s_i, u) - \varepsilon, u \in \overline{V}$ and $\rho < 1/2$;
            $V_r := \mathcal{B}_\delta(s_i, \rho_r)$;
            $r := r + 1$
            $\overline{V}_r := \overline{V}_{r-1} - V_{r-1}$;
        end
        else
            $\overline{V}_r := \emptyset$
    end;
    $E' :=$ set of edges between vertices in V_i and vertices in V_j, for $i \neq j$;
    return E'
end.

For what concerns the performance ratio, let us now prove that $m_{MC}(x) < \ln(2k)\Psi$. The theorem will then follow since Ψ is a lower bound on the optimal measure of MINIMUM MULTI-CUT.

Let $V_1, V_2, \ldots, V_h$ ($h \leq k$) be the subsets of V produced by Program 4.3. Let $V_\delta(s_{i_1}, \rho_1), V_\delta(s_{i_2}, \rho_2), \ldots, V_\delta(s_{i_h}, \rho_h)$ be the volumes of the corresponding balls and let $C_\delta(s_{i_1}, \rho_1), C_\delta(s_{i_2}, \rho_2), \ldots, C_\delta(s_{i_h}, \rho_h)$ be the costs of the corresponding cuts.

By Lemma 4.4, the cost of the multi-cut returned by the algorithm is

$$m_{MC}(x) = \sum_{j=1}^{h} C_\delta(s_{i_j}, \rho_j) \leq 2\ln(2k) \sum_{j=1}^{h} V_\delta(s_{i_j}, \rho_j).$$

Notice now that $\sum_{j=1}^{h} V_\delta(s_{i_j}, \rho_j)$ is equal to the overall volume Ψ of the system, plus $h \leq k$ contributions of size Ψ/k. As a consequence,

$$m_{MC}(x) \leq 2\ln(2k) \sum_{j=1}^{h} V_\delta(s_{i_j}, \rho_j) \leq 4\ln(2k)\Psi,$$

QED and the theorem is proved.

> **Problem 4.6: Minimum Edge Coloring**
>
> INSTANCE: Graph $G = (V, E)$.
>
> SOLUTION: A coloring of E, i.e., a partition of E into disjoint sets $E_1, E_2, \ldots, E_k$ such that, for $1 \leq i \leq k$, no two edges in E_i share a common endpoint in G.
>
> MEASURE: The number of colors, i.e., k.

Section 4.2
BETWEEN APX AND PTAS

4.2 Between APX and PTAS

AN ASYMPTOTIC approximation scheme is a weaker form of approximation scheme based on the idea that the performance ratio of the returned solution may improve as the optimal measure increases. As we will see in this section, NPO problems exist for which no PTAS can be developed (unless P = NP) but that admit an asymptotic approximation scheme. Let us start with the formal definition of this latter notion.

◀ **Definition 4.1**
Asymptotic approximation scheme

An NPO *problem* $\mathcal{P}$ *admits an* asymptotic approximation scheme *if there exist an algorithm* $\mathcal{A}$ *and a constant* k *such that, for any instance* x *of* $\mathcal{P}$ *and for any rational* $r \geq 1$, $\mathcal{A}(x, r)$ *returns a solution whose performance ratio is at most* $r + k/m^*(x)$. *Moreover, for any fixed* r, *the running time of* $\mathcal{A}$ *is polynomial.*

The class PTAS$^\infty$ is the set of all NPO problems that admit an asymptotic approximation scheme. Clearly, PTAS $\subseteq$ PTAS$^\infty$ $\subseteq$ APX. Moreover, it is possible to prove that these inclusions are strict if and only if P $\neq$ NP (see Exercise 4.2).

4.2.1 Approximating the edge coloring problem

In this section we consider MINIMUM EDGE COLORING, that is, the problem of coloring the edges of a graph with the minimum number of colors (see Problem 4.6).

Given a graph G whose maximum degree is Δ, it is easy to see that the optimal value of MINIMUM EDGE COLORING is greater than or equal to Δ. The next result will show that, on the other hand, this value is never greater than $\Delta + 1$. Deciding which one of the two cases holds is an NP-complete problem (see Bibliographical notes) thus implying that MINIMUM EDGE COLORING is not in PTAS (see Exercise 4.9).

Chapter 4

INPUT-DEPENDENT
AND ASYMPTOTIC
APPROXIMATION

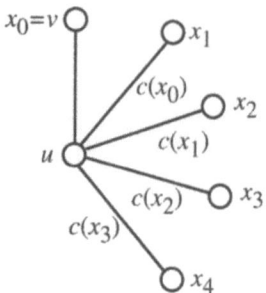

Figure 4.6
The star starting from (u,v)

Theorem 4.6 ▶ *There exists a polynomial-time algorithm $\mathcal{A}$ such that, for any graph G of maximum degree Δ, $\mathcal{A}(G)$ returns an edge-coloring with at most $\Delta+1$ colors.*

PROOF Let $G=(V,E)$ be an instance of MINIMUM EDGE COLORING and let Δ denote the maximum degree of G. The algorithm starts from a new graph $G'=(V,E'=\emptyset)$ (which can be trivially edge-colored with 0 colors) and repeatedly performs the following operations until all edges of E have been moved to G':

1. Consider an uncolored edge $(u,v) \notin E'$.

2. Extend the edge-coloring of $G'=(V,E')$ into an edge-coloring of $G''=(V,E'\cup\{(u,v)\})$ with at most $\Delta+1$ colors.

3. Delete (u,v) from E and add it to E'.

We now present step 2 above in detail. Let us assume that a (partial) edge-coloring of G' with at most $\Delta+1$ colors is available such that all edges are colored but edge (u,v). For any node v, in the following we denote by $\mu(v)$ the set of colors that are not used to color edges incident to v. Clearly, if the coloring uses $\Delta+1$ colors, then, for any v, $\mu(v) \neq \emptyset$. In this case, we denote by $c(v)$ one of the colors in $\mu(v)$ (for instance, the color with minimum index).

We can easily find in linear time a sequence $(u,x_0),\ldots,(u,x_s)$ of distinct edges of G' incident to u such that (see Fig. 4.6 where $s=4$):

1. $x_0 = v$.

2. For any i with $1 \leq i \leq s$, the color of edge (u,x_i) is $c(x_{i-1})$, that is, edge (u,x_i) is colored with a color not used for any edge incident to x_{i-1}.

3. The sequence is maximal, i.e., there is no other edge (u,w) which does not belong to the sequence such that its color is $c(x_s)$.

Note that if $s = 1$, that is, the sequence contains only (u,v), then $\mu(u) \cap \mu(v) \neq \emptyset$: in this case, coloring (u,v) is a trivial task.
We distinguish the following two cases:

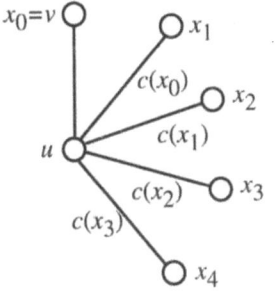
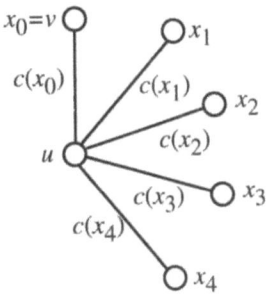

Figure 4.7
The case in which $c(x_4) \in \mu(u)$

1. $c(x_s) \in \mu(u)$: in this case, we can extend the coloring of G' in order to include (u,v) by *shifting* the colors of the sequence. More formally, we start coloring (u, x_s) with $c(x_s)$; then, $c(x_{s-1})$ (that is, the previous color of (u, x_s)) is now in $\mu(u)$, that is, $c(x_{s-1}) \in \mu(u)$. By repeatedly applying this reasoning, we modify the colors of all edges in the sequence but the first one. At the end of this process, we have that $c(v) \in \mu(u)$ so that (u,v) can be colored with $c(v)$ (see Fig. 4.7).

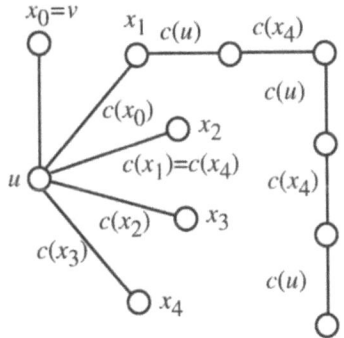
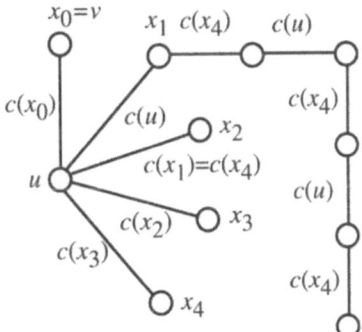

Figure 4.8
The case in which $c(x_4) \notin \mu(u)$ and P_1 does not end in u

2. $c(x_s) \notin \mu(u)$: in this case, one edge (u, x_i) must have been colored with $c(x_s)$ (otherwise, the sequence is not maximal). This implies that $c(x_{i-1}) = c(x_s)$. Let P_{i-1} be the maximal path starting from x_{i-1}

Chapter 4

INPUT-DEPENDENT
AND ASYMPTOTIC
APPROXIMATION

formed by edges whose colors are, alternatively, $c(u)$ and $c(x_s)$ and let w be the last node of this path. Notice that P_{i-1} may have length 0: in such a case, clearly, $w = x_{i-1}$. Two cases may arise.

(a) $w \neq u$: in this case we can interchange colors $c(u)$ and $c(x_s)$ in P_{i-1} and assign color $c(u)$ to (u, x_{i-1}). Note that in this way $c(x_{i-1}) \in \mu(u)$ and, thus, the subsequence preceeding x_{i-1} can be dealt as in Case 1 above (see Fig. 4.8).

(b) $w = u$: in this case, we derive in linear time the path P_s starting from x_s formed by edges whose colors are, alternatively, $c(u)$ and $c(x_s)$. This path cannot intersect P_{i-1}. On the contrary, let z be the first node in P_s that belongs to P_{i-1}. If $z = u$ then there are two edges incident in u with color $c(x_s)$ contradicting the property of a feasible edge coloring of G'. If $z \neq u$, then there must exist three edges incident to z colored with $c(u)$ or $c(x_s)$: once again this contradicts the property of a coloring (see Fig. 4.9).

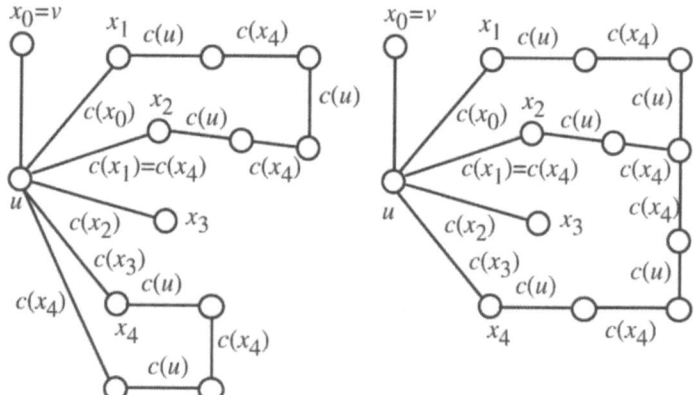

Figure 4.9
P_1 cannot intersect P_4

Since P_s does not intersect P_{i-1}, it cannot end in u: thus, analogously to Case 2.1 above, we can interchange colors $c(u)$ and $c(x_s)$ in P_s and assign color $c(u)$ to (u, x_s). Finally, the subsequence preceeding x_s can be dealt as in Case 1 above (see Fig. 4.10).

In both cases, we have obtained a valid edge-coloring of G' with at most $\Delta + 1$ colors. Since the updating of the coloring of G' has to be done $|E|$ times, it also follows that the running time of the algorithm is bounded by a polynomial.

QED

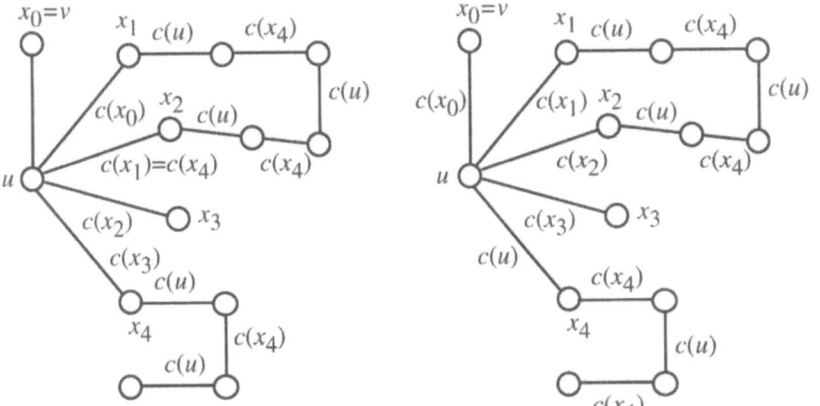

Figure 4.10
The interchange of colors in P_4 and the shift in the remaining sequence

The above theorem implies that a polynomial-time algorithm exists such that, for any graph G, the performance ratio of the solution returned by the algorithm is at most

$$\frac{m^*(G)+1}{m^*(G)} = 1 + \frac{1}{m^*(G)};$$

that is, MINIMUM EDGE COLORING admits an asymptotic approximation scheme with $k=1$ (observe that a similar argument applies to any NPO problem that admits a polynomial-time algorithm whose absolute error is bounded by a constant). In conclusion, we have the following result.

MINIMUM EDGE COLORING *belongs to the class* PTAS$^\infty$. ◀ Theorem 4.7

4.2.2 Approximating the bin packing problem

In this section we will prove that MINIMUM BIN PACKING belongs to PTAS$^\infty$: note that, in this case, no polynomial-time algorithm whose absolute error is bounded by a constant exists (unless P=NP). We will provide an asymptotic approximation scheme for MINIMUM BIN PACKING, which is based on a combination of the First Fit algorithm and a partitioning technique and is structured into the following five steps.

1. Eliminate *small* items.

2. Group the remaining items into a constant number of size values.

3. Find an optimal solution of the resulting instance.

Chapter 4

INPUT-DEPENDENT AND ASYMPTOTIC APPROXIMATION

4. Ungroup the items.

5. Reinsert *small* items.

We will first describe step 3, then steps 1 and 2 along with their "inverse" steps 5 and 4, respectively.

Solving instances of restricted bin packing

For any integer constant $c > 0$ and for any rational constant $\delta \leq 1$, let us consider a restricted version of MINIMUM BIN PACKING in which, for any instance, there are at most c different sizes for the items (i.e., the cardinality of the range of the size function s is bounded by c) and the size of each item is at least $B \cdot \delta$, where B denotes the size of the bins. For the sake of clarity, we will denote this restriction as MINIMUM (c, δ)-RESTRICTED BIN PACKING. Observe that an instance of this problem can be described by the size B of the bins and a multiset $I = \{s_1 : n_1, s_2 : n_2, \ldots, s_c : n_c\}$ in which, the pair $s_i : n_i$, with $1 \leq i \leq c$, denotes the number n_i of items having size s_i.

Example 4.5 ▶ Let us consider the following instance of MINIMUM $(3, 3/8)$-RESTRICTED BIN PACKING:

$$I = \{3 : 4, 5 : 2, 7 : 1\} \quad \text{and} \quad B = 8.$$

In this case we have four items with size 3, two with size 5, and one with size 7.

We now show how an instance of MINIMUM (c, δ)-RESTRICTED BIN PACKING can be solved in time $O(n^q)$ where n denotes the number of items and q depends on c and δ only. Equivalently, we show that, for any two constants c and δ, MINIMUM (c, δ)-RESTRICTED BIN PACKING is solvable in polynomial-time.

Let (I, B) be an instance of MINIMUM (c, δ)-RESTRICTED BIN PACKING. The *type* of a bin is a c-vector $\vec{t} = (t_1, \ldots, t_c)$ of integers with $0 \leq t_i \leq n_i$ such that $\sum_{i=1}^{c} t_i s_i \leq B$. In other words, the type specifies for each element s_i of the range of s the number of items having size s_i that are contained in the bin. Observe that for each type $\vec{t}$,

$$\sum_{i=1}^{c} t_i \leq \frac{1}{\delta} \sum_{i=1}^{c} t_i \frac{s_i}{B} \leq \frac{1}{\delta}.$$

This implies that the number of distinct bin types is bounded by the number of ways of choosing c integers whose sum is less than $\lfloor 1/\delta \rfloor$. It is easy to prove that such a number is equal to

$$q = \binom{c + \lfloor \frac{1}{\delta} \rfloor}{\lfloor \frac{1}{\delta} \rfloor}$$

(see Exercise 3.9). Therefore, in any instance of MINIMUM (c,δ)-RESTRICTED BIN PACKING, the number of possible bin types depends on c and δ and does not depend on n.

Section 4.2
BETWEEN APX AND PTAS

Let us consider the instance of Example 4.5. In this case we have that $c = 3$ and $\lfloor 1/\delta \rfloor = 2$. Then

◀ Example 4.6

$$q = \binom{5}{2} = 10.$$

Indeed, the possible bin types are only six (i.e., $(0,0,0)$, $(0,0,1)$, $(0,1,0)$, $(1,0,0)$, $(1,1,0)$, and $(2,0,0)$) since the other four bin types (i.e., $(0,0,2)$, $(0,1,1)$, $(0,2,0)$, and $(1,0,1)$) are not feasible because they violate the bound given by the size of the bin.

Since there are at most q different bin types, a feasible solution can now be described by a q-vector $\vec{y} = (y_1, \ldots, y_q)$ where y_i for $i = 1, \ldots, n$ specifies, for each bin type, the number of bins of that type (observe that $0 \leq y_i \leq n$).

Let us consider again the instance of Example 4.5. A feasible solution can be the one using two bins of type $(2,0,0)$, two bins of type $(0,1,0)$, and one bin of type $(0,0,1)$. One optimal solution, instead, uses two bins of type $(1,1,0)$, one bin of type $(2,0,0)$, and one bin of type $(0,0,1)$.

◀ Example 4.7

It is clear that the number of feasible solutions is bounded by $O(n^q)$, which implies that the instance can be solved in time $O(n^q p(n))$ where p is a polynomial by exaustively generating all these feasible solutions (see Exercise 4.10).

Grouping and ungrouping items
Given an instance x of MINIMUM BIN PACKING, let us assume that the n items are ordered according to their size values so that

$$s(u_1) \geq s(u_2) \geq \cdots \geq s(u_n).$$

For any integer $k \leq n$, let $m = \lfloor n/k \rfloor$ and partition the n items into $m+1$ groups G_i with $G_i = \{u_{(i-1)k+1}, \ldots, u_{ik}\}$ for $i = 1, \ldots, m$ and $G_{m+1} = \{u_{mk+1}, \ldots, u_n\}$.

We then define a new instance x_g of MINIMUM BIN PACKING with the same bin size B that, for $i = 2, 3, \ldots, m+1$, contains an item of size $s(u_{(i-1)k+1})$ for each item of instance x that belongs to G_i (that is, the size of all items in the ith group are made equal to that of the largest item in G_i). Note that there at most mk items in x_g.

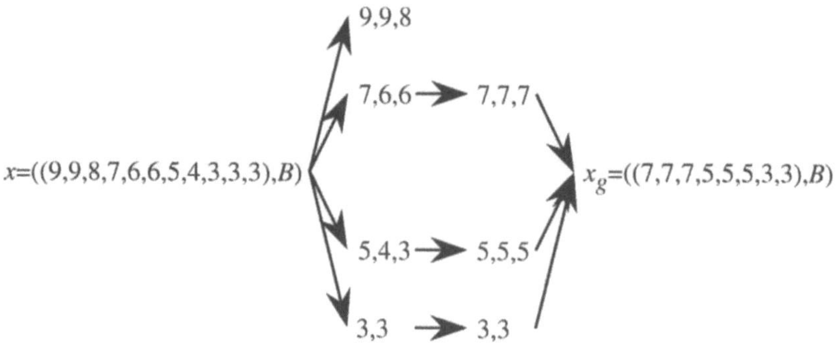

Chapter 4

INPUT-DEPENDENT
AND ASYMPTOTIC
APPROXIMATION

Figure 4.11
An example of grouping items

Example 4.8 ▶ Let us consider the instance x formed by 11 items whose size values are 9, 9, 8, 7, 6, 6, 5, 4, 3, 3, and 3, respectively, and let $k = 3$. We have four groups: $G_1 = \{u_1, u_2, u_3\}$, $G_2 = \{u_4, u_5, u_6\}$, $G_3 = \{u_7, u_8, u_9\}$, and $G_4 = \{u_{10}, u_{11}\}$. The instance x_g contains eight items (see Fig. 4.11): three items of size 7 (corresponding to the items in G_2), three items of size 5 (corresponding to the items in G_3), and two items of size 3 (corresponding to the items in G_4).

Clearly, any packing for the items in x can be transformed into a packing for the items in x_g with the same number of bins by eliminating items in the last group and then substituting each item u in G_i with an item of x_g whose size is equal to $s(u_{ik+1})$ (that is, less than or equal to $s(u)$). Moreover, given a packing of the items in x_g we can obtain a packing of the items in x by simply adding k bins in which we can put the first k items. This implies that

$$m^*(x_g) \leq m^*(x) \leq m^*(x_g) + k, \tag{4.3}$$

that is, if we are able to optimally solve x_g, then we can find a solution for x whose absolute error is at most k.

Dealing with small items

Let x be an instance of bin packing and, for any rational constant $\delta \in (0, 1/2]$, let x_δ be the instance obtained by eliminating all items whose size is less than δB. Given a packing of x_δ with M bins, we can use the First Fit approach to reinsert small items. That is, for each of these items, we insert it into the first bin that can contain it: if it does not fit into any of the currently available bins, then a new bin is created.

At the end of the above procedure two cases are possible.

1. No new bin has been created and the packing uses M bins.

2. $M' \geq 1$ new bins have been created. In this case, similarly to the analysis made in Sect. 2.2.2, we can show that all bins except at most one have an empty space that is at most δB. This implies that

$$(1-\delta)(M+M'-1) \leq \frac{\sum_{i=1}^n s(u_i)}{B} \leq m^*(x),$$

that is,

$$M+M' \leq \frac{1}{1-\delta}m^*(x)+1 \leq (1+2\delta)m^*(x)+1.$$

In conclusion, given a packing of x_δ with M bins, we can find in polynomial time a solution for x whose measure is at most

$$\max(M, (1+2\delta)m^*(x)+1). \tag{4.4}$$

We have thus completed the description of the five steps of the proposed algorithm for MINIMUM BIN PACKING which is summarized in Program 4.4. Observe that if $r \geq 2$, the First Fit algorithm achieves the desired performance ratio: hence, the input of the algorithm, without loss of generality, is restricted to values of r smaller than 2. We can finally state the main result of this section.

Program 4.4 is an asymptotic polynomial-time approximation scheme for ◀ Theorem 4.8
MINIMUM BIN PACKING.

To show that, for any fixed $r < 2$, Program 4.4 runs in polynomial time, PROOF
it suffices to prove that an optimal solution for $x_{\delta,g}$ can be found in polynomial time. Indeed, $x_{\delta,g}$ is an instance of MINIMUM $(\lfloor n'/k \rfloor, \delta)$-RESTRICTED BIN PACKING: since, for any fixed r, both $\lfloor n'/k \rfloor$ and δ are constant, it follows that an optimal solution for $x_{\delta,g}$ can be computed in time $O(n^q p(n))$ where q depends on r and p is a polynomial.

Let us now compute the performance ratio of the packing obtained by the algorithm. To this aim, first observe that the measure of the solution computed by Program 4.4 is $m^*(x_{\delta,g})+k$. Since all items in x_δ have size at least δB, it follows that $\delta n' \leq m^*(x_\delta)$ and, therefore,

$$k \leq \frac{(r-1)^2}{2}n'+1 = (r-1)\delta n' + 1 \leq (r-1)m^*(x_\delta)+1.$$

From Eq. 4.3, it follows that the algorithm packs the items of x_δ into at most

$$m^*(x_{\delta,g}) + k \leq m^*(x_\delta) + (r-1)m^*(x_\delta) + 1 = rm^*(x_\delta)+1.$$

Chapter 4

INPUT-DEPENDENT AND ASYMPTOTIC APPROXIMATION

Program 4.4: Asymptotic Bin Packing

input Instance x of MINIMUM BIN PACKING, rational r with $1 < r < 2$;
output Solution whose measure is at most $rm^*(x) + 1$;
begin
  $\delta := (r-1)/2$;
  Let x_δ be the instance obtained by removing items whose size is less than δB;
  $k := \lceil (r-1)^2 n'/2 \rceil$ where n' is the number of items of x_δ;
  Let $x_{\delta,g}$ be the instance obtained by grouping x_δ with respect to k;
  Find an optimal solution for $x_{\delta,g}$ with measure $m^*(x_{\delta,g})$;
  Insert the first k items of x_δ into k new bins;
  Apply First Fit approach to reinsert the small items;
  return packing obtained
end

Finally, by plugging $r = (1 + 2\delta)$ into Eq. 4.4, we obtain that the total number of used bins is at most

$$\max(rm^*(x_\delta) + 1, rm^*(x) + 1) \leq rm^*(x) + 1,$$

QED which provides the desired performance ratio.

The above theorem states that MINIMUM BIN PACKING belongs to PTAS$^\infty$. It is indeed possible to prove a stronger result that states the existence of an asymptotic approximation scheme whose running time is polynomial both in the length of the input and in $1/(r-1)$ (see Bibliographical notes).

4.3 Exercises

Exercise 4.1 Prove that the analysis of Theorem 4.1 is tight. (Hint: generalize the instance of Example 4.1.)

Exercise 4.2 (*) Prove that if P $\neq$ NP then PTAS $\neq$ PTAS$^\infty$ and PTAS$^\infty \neq$ APX.

Exercise 4.3 Prove that Program 4.5 colors any graph G with n vertices using $O(m^*(G) \log n)$ colors.

Exercise 4.4 Prove that any 2-colorable graph G can be colored in polynomial time using at most 2 colors.

Exercise 4.5 Prove that any graph G of maximum degree $\Delta(G)$ can be colored in polynomial time using at most $\Delta(G) + 1$ colors.

Program 4.5: OptIS Graph Coloring

input A graph $G = (V, E)$;
output A coloring of G with i colors;
begin
   $i := 1$;
   while G is not empty **do**
   begin
      find a maximum independent set S_i in G;
      color vertices of S_i with color i;
      delete from G all vertices in S_i;
      $i := i + 1$
   end
end.

Program 4.6: 3-Coloring

input 3-colorable graph $G = (V, E)$;
output Coloring of vertices of G;
begin
   $H := G; n := |V|$;
   $i := 1$;
   while the maximum degree in H is at least $\lceil \sqrt{n} \rceil$ **do**
   begin
      Let v be the vertex of maximum degree in H;
      Let $H_N(v)$ be the graph induced on H by the neighbors of v;
      Color $H_N(v)$ with colors $i, i+1$;
      Color v with color $i+2$;
      $i := i + 2$;
      $H :=$ the subgraph of H obtained by deleting v and its neighbors
   end;
   Color all nodes in H with $\Delta(H) + 1$ colors
end.

Section 4.3

EXERCISES

Exercise 4.6 Prove that, for any 3-colorable graph G, Program 4.6 colors G in polynomial time with at most $3\lceil \sqrt{n} \rceil$ colors. (Hint: use the previous two exercises in order to evaluate the running time.)

Exercise 4.7 (**) Prove that any k-colorable graph G, can be colored in polynomial time with at most $2k\lceil n^{1-1/(k-1)} \rceil$ colors. (Hint: extend the ideas contained in Program 4.6.)

Exercise 4.8 Prove that the two integer linear programming formulations of MINIMUM MULTI-CUT are, indeed, equivalent.

Exercise 4.9 Use the gap technique to show that MINIMUM EDGE COLORING is not in PTAS.

Exercise 4.10 For any fixed c and δ, give a polynomial-time algorithm to solve MINIMUM (c,δ)-RESTRICTED BIN PACKING.

4.4 Bibliographical notes

THE NOTION of input-dependent approximation algorithm is as old as that of constant approximation algorithm: they both appeared in [Johnson, 1974a] which is widely considered as one of the starting points of the systematic study of the complexity of approximation.

The approximation algorithm for MINIMUM SET COVER, which is described in Sect. 4.1.1, was proposed in [Johnson, 1974a] and in [Chvátal, 1979]. This algorithm is optimal: in [Feige, 1996, Raz and Safra, 1997], in fact, it is proved that, for any $\varepsilon > 0$, there is no $(\ln n - \varepsilon)$-approximation algorithm for MINIMUM SET COVER, unless some likely complexity theoretic conjectures fail. It is worth pointing out that it took more than twenty years to close this gap and that MINIMUM SET COVER is one of the very few problems for which optimal approximation algorithms have been proved.

The $O(n/\log n)$-approximate algorithm for MINIMUM GRAPH COLORING is due to [Johnson, 1974a] (from this reference, Exercise 4.3 is also taken): this algorithm is not optimal. Indeed, the best known approximation algorithm for MINIMUM GRAPH COLORING is due to [Halldórsson, 1993a] and has a performance ratio $O(n(\log \log n)^2/(\log n)^3)$. On the other hand, in [Bellare, Goldreich, and Sudan, 1998] it is shown that this problem is not approximable within $n^{1/7-\varepsilon}$ for any $\varepsilon > 0$ unless P = NP. Exercises 4.6 and 4.7 are due to [Wigderson, 1983].

The max-flow min-cut theorem in single-commodity networks was introduced in [Ford and Fulkerson, 1956]. The first corresponding approximation result for the multi-commodity case has been presented in [Leighton and Rao, 1988]. In this paper, the authors consider the related SPARSEST CUT problem, where a cut E' is required on a weighted graph $G = (V,E,c)$ on which k pairs $\{s_i,t_i\} \in V^2$ are defined, such that the ratio $\rho(E') = \sum_{e \in E'} c(e)/\sum_{i \in I(E')} d_i$ is minimum, where $i \in I(E')$ if and only if s_i and t_i are disconnected by the removal of E'. In the same paper an $O(\log n)$ approximation algorithm for this problem is given and a relationship between this problem and the BALANCED CUT is shown, where BALANCED CUT requires, given a value $\alpha \in (0,1)$, to find a cut of minimum cost which disconnects the graph into two subgraphs each of size at least αn. Such a

Section 4.4

BIBLIOGRAPHICAL NOTES

problem presents particular relevance as a building block for the design of divide and conquer algorithms on graphs.

The algorithm presented in Sect. 4.1.3 for MINIMUM MULTI-CUT is from [Garg, Vazirani, and Yannakakis, 1996] and has been proved to be applicable to the approximate solution of SPARSEST CUT (and of BALANCED CUT) in [Kahale, 1993]. Other approximation results for SPARSEST CUT has been derived in [Klein, Rao, Agrawal, Ravi, 1995, Plotkin, Tardos, 1995, Linial, London, Rabinovich, 1995, Aumann and Rabani, 1995].

The definition of an asymptotic approximation scheme is taken from [Motwani, 1992] and is based on the notion of asymptotic performance ratio as given in [Garey and Johnson, 1979].

Theorem 4.6 is a fundamental result of graph theory and appeared in [Vizing, 1964]: observe that this theorem also implies that MINIMUM EDGE COLORING is 4/3-approximable. Deciding whether a graph is edge-colorable with Δ colors was shown to be NP-complete in [Holyer, 1981].

The asymptotic approximation scheme for MINIMUM BIN PACKING appeared in [Fernandez de la Vega and Lueker, 1981]: our presentation follows that of [Motwani, 1992]. Indeed, this scheme has been improved in [Karmarkar and Karp, 1982]: in this paper, an asymptotic fully polynomial-time approximation scheme is obtained by means of mathematical programming relaxation techniques and of the ellipsoid method introduced in [Khachian, 1979].

Approximation through Randomization

RANDOMIZATION IS one of the most interesting and useful tools in designing efficient algorithms. Randomized algorithms, indeed, have been proposed for many problems arising in different areas: taking into account the scope of this book, however, we will limit ourselves to considering randomized approximation algorithms for NP-hard optimization problems.

As it can be observed also in the case of the example given in Sect. 2.6, where a randomized algorithm for MAXIMUM SATISFIABILITY was given and analyzed, a remarkable property of randomized algorithms is their structural simplicity. For some problems, in fact, it happens that the only known efficient deterministic algorithms are quite involved, while it is possible to introduce a randomized efficient algorithm which is much easier to code. This happens also in the case of approximation algorithms, where we are interested in achieving good approximate solutions in polynomial time. For example, in this chapter we will describe a simple randomized approximation algorithm for the weighted version of MINIMUM VERTEX COVER, which achieves an expected performance ratio comparable to that of the best deterministic algorithms based on linear programming techniques.

Randomized algorithms can sometimes be even more efficient than deterministic ones in terms of the quality of the returned solution. This will be shown in the case of the weighted versions of MAXIMUM SATISFIABILITY and MAXIMUM CUT: indeed, we will present a randomized 4/3-approximation algorithm for the former problem and a randomized 1.139-

Chapter 5
APPROXIMATION THROUGH RANDOMIZATION

approximation algorithm for the latter one.

On the other hand, the main drawback of the randomized approach is that we may only derive statistical properties of the solution returned (in particular, with respect to its expected value): this means that, even if we prove that an algorithm returns solutions of expected good quality, we may, nevertheless, get poor approximate solutions in some cases. However, it is sometimes possible to overcome this drawback by *derandomizing* the algorithm, that is, by transforming the given randomized algorithm into a deterministic one, which always returns in polynomial time a solution whose performance ratio is no more than the expected performance ratio of the solution computed by the randomized algorithm. As we will see at the end of this chapter, this can be done by applying a general technique, called the method of conditional probabilities.

5.1 Randomized algorithms for weighted vertex cover

In this section we present a randomized approximation algorithm for the weighted version of MINIMUM VERTEX COVER. This algorithm achieves an approximate solution whose expected measure is at most twice the optimum measure. Deterministic algorithms for this problem that find approximate solutions whose performance ratio is at most 2 (or even slightly better, i.e., at most $2 - \frac{\log \log n}{2 \log n}$) are known (see, for example, Sect. 2.4). In spite of this fact, even though the randomized algorithm does not improve the quality of approximation, it presents a remarkable simplicity, when compared with its deterministic counterparts.

The randomized algorithm (see Program 5.1) exploits the following idea: while there are edges which are not covered, randomly choose a vertex which is an endpoint of an uncovered edge and add this vertex to the vertex cover. The selection of vertices is done by flipping a biased coin that favors the choice of vertices with small weight. Note that, while the choice of an edge is assumed to be done deterministically and how it is performed will not be significant in order to analyze the algorithm behavior, the selection of an endpoint is assumed to be done randomly under a given probability distribution. Note also that if the graph is unweighted (that is, each vertex has weight 1), then an endpoint is chosen with probability 1/2.

Clearly, Program 5.1 runs in polynomial time. To analyze the performance ratio of the solution obtained, let $m_{RWVC}(x)$ be the random variable denoting the value of the solution found by the algorithm on instance x.

Theorem 5.1 ▶ *Given an instance x of the weighted version of* MINIMUM VERTEX COVER, *the expected measure of the solution returned by Program 5.1*

Section 5.1
RANDOMIZED ALGORITHMS FOR WEIGHTED VERTEX COVER

Program 5.1: Random Weighted Vertex Cover

input Graph $G = (V, E)$, weight function $w : V \to \mathbf{N}$;
output Vertex cover U;
begin
  $U := \emptyset$;
  while $E \neq \emptyset$ **do**
  begin
    Select an edge $e = (v, t) \in E$;
    Randomly choose x from $\{v, t\}$ with $\Pr\{x = v\} = \frac{w(t)}{w(v)+w(t)}$;
    $U := U \cup \{x\}$;
    $E := E - \{e \mid x \text{ is an endpoint of } e\}$
  end;
  return U
end.

satisfies the following inequality:

$$\mathrm{E}[m_{RWVC}(x)] \leq 2m^*(x).$$

PROOF

Let U be the vertex cover found by the algorithm with input the instance x formed by the graph $G = (V, E)$ and the weight function w, and let U^* be an optimum vertex cover for the same instance. Given any $v \in V$, we define a random variable X_v as follows:

$$X_v = \begin{cases} w(v) & \text{if } v \in U, \\ 0 & \text{otherwise.} \end{cases}$$

Since

$$\mathrm{E}[m_{RWVC}(x)] = \mathrm{E}[\sum_{v \in V} X_v] = \sum_{v \in V} \mathrm{E}[X_v]$$

and

$$\sum_{v \in U^*} \mathrm{E}[X_v] = \mathrm{E}[\sum_{v \in U^*} X_v] \leq \mathrm{E}[\sum_{v \in U^*} w(v)] = m^*(x),$$

in order to prove the theorem, it suffices to show that

$$\sum_{v \in V} \mathrm{E}[X_v] \leq 2 \sum_{v \in U^*} \mathrm{E}[X_v]. \tag{5.1}$$

Given an edge (v, t) selected by the algorithm at the first step of the loop, we say that (v, t) *picks* vertex v if v is randomly chosen at the next step. We also denote as $N(v)$ the set of vertices adjacent to v, i.e., $N(v) = \{u \mid u \in V \land (v, u) \in E\}$.

Chapter 5
APPROXIMATION THROUGH RANDOMIZATION

Let us now define the random variable $X_{(v,t),v}$ as

$$X_{(v,t),v} = \begin{cases} w(v) & \text{if } (v,t) \text{ is chosen and picks } v, \\ 0 & \text{otherwise.} \end{cases}$$

Note that if $X_{(v,t),v} = w(v)$, then $X_{(v,t'),v} = 0$ for each $t' \in N(v)$ with $t' \neq t$. This implies that

$$X_v = \sum_{t \in N(v)} X_{(v,t),v}$$

and, by the linearity of expectation, that

$$E[X_v] = \sum_{t \in N(v)} E[X_{(v,t),v}].$$

Moreover, $E[X_{(v,t),v}] = E[X_{(v,t),t}]$: in fact, we have that

$$\begin{aligned} E[X_{(v,t),v}] &= w(v) \Pr\{(v,t) \text{ picks } v\} \\ &= w(v) \Pr\{(v,t) \text{ is chosen}\} \frac{w(t)}{w(v)+w(t)} \\ &= w(t) \Pr\{(v,t) \text{ is chosen}\} \frac{w(v)}{w(v)+w(t)} \\ &= w(t) \Pr\{(v,t) \text{ picks } t\} \\ &= E[X_{(v,t),t}]. \end{aligned}$$

Let us now notice that

$$\sum_{v \in U^*} E[X_v] = \sum_{v \in U^*} \sum_{t \in N(v)} E[X_{(v,t),v}] \qquad (5.2)$$

and that

$$\sum_{v \notin U^*} E[X_v] = \sum_{v \notin U^*} \sum_{t \in N(v)} E[X_{(v,t),v}] = \sum_{v \notin U^*} \sum_{t \in N(v)} E[X_{(v,t),t}]. \qquad (5.3)$$

Observe also that, since, for any $v \notin U^*$, each vertex $t \in N(v)$ must be in U^*, then, for each term $E[X_{(v,t),t}]$ in Eq. (5.3), an equal term $E[X_{(v,t),v}]$ must appear in Eq. (5.2). This implies that

$$\sum_{v \notin U^*} E[X_v] \leq \sum_{v \in U^*} E[X_v]$$

and, as an immediate consequence, that

$$\sum_{v \in V} E[X_v] = \sum_{v \in U^*} E[X_v] + \sum_{v \notin U^*} E[X_v] \leq 2 \sum_{v \in U^*} E[X_v].$$

QED   The theorem thus follows.

5.2 Randomized algorithms for weighted satisfiability

In Sect. 3.1 we presented Program 3.1, a 2-approximate greedy algorithm for MAXIMUM SATISFIABILITY. In this section, we will concentrate on MAXIMUM WEIGHTED SATISFIABILITY, whose input is given by a set of clauses C and a weight function w, which associates to each clause a positive weight, and whose goal is to find a truth assignment that maximizes the sum of the weights of the satisfied clauses.

It can be easily shown that Program 3.1 can be extended to the weighted case preserving the performance ratio (see Exercise 3.2). In this section we present two different randomized algorithms that can be combined in order to achieve an expected performance ratio equal to $4/3$. As we will see in the last section of this chapter, it is possible to derandomize these two algorithms and, hence, to obtain a deterministic algorithm with the same performance ratio for every instance.

The first of the two randomized algorithms is Program 2.10, which can, clearly, be applied also to the case in which the clauses are weighted. It is possible to modify the proof of Theorem 2.19 to show that if k is the minimum number of literals in a clause, then the expected performance ratio of the algorithm is at most $2^k/(2^k-1)$, which, in particular, is equal to 2 when $k=1$ and at most $4/3$ for $k \geq 2$ (see Exercise 5.3).

In the following, we will denote as $m_{RWS}(x)$ the random variable denoting the value of the solution found by Program 2.10 on instance x.

5.2.1 A new randomized approximation algorithm

In Program 2.10 the truth value of every variable is independently and randomly chosen with probability $1/2$. Let us now consider a generalization of that algorithm, which independently assigns the value TRUE to variable x_i, for $i=1,2,\ldots,n$, with probability p_i, where p_i is suitably chosen.

Let $m_{GRWS}(x)$ be the random variable denoting the value of the solution found by this generalization on instance x. It is then easy to see that, for any instance x, the following equality holds:

$$E[m_{GRWS}(x)] = \sum_{c \in C} w(c)(1 - \prod_{i \in V_c^+}(1-p_i) \prod_{i \in V_c^-} p_i)$$

where V_c^+ (respectively, V_c^-) denotes the set of indices of the variables appearing positive (respectively, negative) in clause c.

In the following, we show that it is possible to compute in polynomial time suitable values p_i such that $E[m_{GRWS}(x)]$ is at most $4/3$ when $k \leq 2$ and is at most $e/(e-1)$ for $k \geq 3$. In order to reach this aim,

Chapter 5

APPROXIMATION THROUGH RANDOMIZATION

Program 5.2: General Random Weighted Satisfiability

input Instance x, i.e., set C of clauses on set of variables V, function $w : C \mapsto \mathbf{N}$;
output Truth assignment $f : V \mapsto \{\text{TRUE}, \text{FALSE}\}$;
begin
    Find the optimum value (y^*, z^*) of *LP-SAT(x)*;
    for each variable v_i **do**
    begin
        $p_i := g(y_i^*)$ (for a suitable function g);
        $f(v_i) := \text{TRUE}$ with probability p_i
    end;
    return f
end.

we first represent each instance of MAXIMUM WEIGHTED SATISFIABILITY as an integer linear program. Namely, given an instance x of MAXIMUM WEIGHTED SATISFIABILITY formed by a set $C = \{c_1, c_2, \ldots, c_t\}$ of clauses defined over the Boolean variables $v_1, \ldots, v_n$ and a weight function w, we define the following integer liner program *IP-SAT(x)*:

$$\begin{aligned}
\text{maximize} \quad & \sum_{c_j \in C} w(c_j) z_j \\
\text{subject to} \quad & \sum_{i \in V_{c_j}^+} y_i + \sum_{i \in V_{c_j}^-} (1 - y_i) \geq z_j \quad \forall c_j \in C \\
& y_i \in \{0, 1\} \quad & 1 \leq i \leq n \\
& z_j \in \{0, 1\} \quad & 1 \leq j \leq t.
\end{aligned}$$

Observe that we may define a one-to-one correspondence between feasible solutions of x and feasible solutions of *IP-SAT(x)* as follows:

- $y_i = 1$ if and only if variable x_i is true;

- $z_j = 1$ if and only if clause C_j is satisfied.

Let *LP-SAT(x)* be the linear program obtained by relaxing the integrality constraints of *IP-SAT(x)* and let $(y^* = (y_1^*, \ldots, y_n^*), z^* = (z_1^*, \ldots, z_t^*))$ be an optimal solution of *LP-SAT(x)*: clearly, $m_{LP-SAT}^*(x) \geq m_{IP-SAT}^*(x)$.

Given an instance x of MAXIMUM WEIGHTED SATISFIABILITY, Program 5.2 first solves *LP-SAT(x)* obtaining an optimal solution (y^*, z^*). Then, given a function g to be specified later, it computes probabilities $p_i = g(y_i^*)$, for $i = 1, \ldots, n$ and assigns the truth values according to these probabilities. If the function g can be computed in polynomial time, then Program 5.2 clearly runs in polynomial time.

The performance ratio of the returned solution depends on the choice of the function g. Let us suppose that there exists a real number α, with $0 < \alpha < 1$, such that

$$\left(1 - \prod_{i \in V_{c_j}^+}(1-p_i) \prod_{i \in V_{c_j}^-} p_i\right) \geq \alpha z_j^*,$$

for each clause c_j. Since

$$\sum_{j=1}^{t} w(c_j) z_j^* = m_{LP-SAT}^*(x)$$

and

$$E[m_{GRWS}(x)] = \sum_{c_j \in C} w(c_j)\left(1 - \prod_{i \in V_{c_j}^+}(1-p_i) \prod_{i \in V_{c_j}^-} p_i\right),$$

the solution returned by Program 5.2 has expected performance ratio at most $1/\alpha$.

A first interesting choice of the function g consists in setting $g(y_i) = y_i^*$, for $i = 1, 2, \ldots, n$: in other words, each variable v_i is independently set to TRUE with probability y_i^*.

◀ Lemma 5.2

Given an instance x of MAXIMUM WEIGHTED SATISFIABILITY, let (y^*, z^*) be an optimal solution of LP-SAT(x). Then, for any clause c_j in x with k literals, we have

$$\left(1 - \prod_{i \in V_{c_j}^+}(1-y_i^*) \prod_{i \in V_{c_j}^-} y_i^*\right) \geq \alpha_k z_j^*$$

where

$$\alpha_k = 1 - \left(1 - \frac{1}{k}\right)^k.$$

PROOF

Without loss of generality, we assume that every variable in clause c_j is positive (i.e., $c_j = v_{j_1} \vee \ldots \vee v_{j_k}$). The lemma is proved by showing that

$$1 - \prod_{i=1}^{k}(1 - y_{j_i}^*) \geq \alpha_k z_j^*.$$

To this aim, recall that, given a set of nonnegative numbers $\{a_1, \ldots, a_k\}$, we have that

$$\frac{a_1 + \cdots + a_k}{k} \geq \sqrt[k]{a_1 \cdots a_k}.$$

By applying the above inequality to the set $\{1-y^*_{j_1},\ldots,1-y^*_{j_k}\}$ and recalling that $\sum_{i=1}^{k} y^*_{j_i} \geq z^*_j$ we obtain

$$1-\prod_{i=1}^{k}(1-y^*_{j_i}) \geq 1-\left(\frac{\sum_{i=1}^{k}(1-y^*_{j_i})}{k}\right)^k \geq 1-\left(1-\frac{\sum_{i=1}^{k} y^*_{j_i}}{k}\right)^k$$

$$\geq 1-\left(1-\frac{z^*_j}{k}\right)^k \geq \alpha_k z^*_j,$$

where the last inequality is due to the fact that

$$f(z^*_j) = 1-\left(1-\frac{z^*_j}{k}\right)^k$$

is a concave function in the interval of interest, i.e., $[0,1]$, and $f(z^*_j) \geq \alpha_k z^*_j$ at the extremal points of the interval. The lemma is thus proved. QED

Since α_k is a decreasing function with respect to k, given an instance of MAXIMUM WEIGHTED SATISFIABILITY such that each clause has at most k literals, the previous lemma implies that choosing g as the identity function in Program 5.2 yields a randomized algorithm whose expected performance ratio is at most $1/\alpha_k$.

In particular, if $k \leq 2$, then the expected performance ratio is bounded by $4/3$, while if $k \geq 3$, since $\lim_{k\to\infty}(1-(1/k))^k = 1/e$, the expected performance ratio is at most $e/(e-1) \approx 1.582$.

5.2.2 A 4/3-approximation randomized algorithm

In this section we show that an appropriate combination of the two randomized approximation algorithm described above allows to obtain a 4/3-approximation randomized algorithm.

First note that Program 2.10 has expected performance ratio bounded by 4/3 if we deal with clauses with *at least* 2 literals. On the other hand, Program 5.2 with g equal to the identity function has the same expected performance ratio if we deal with clauses with *at most* 2 literals.

We can then derive a new algorithm, which simply chooses the best truth assignment returned by the previous two algorithms. The expected performance ratio of this new algorithm is analyzed in the following lemma.

Lemma 5.3 ▶ *Given an instance x of* MAXIMUM WEIGHTED SATISFIABILITY, *let W_1 be the expected measure of the solution returned by Program 2.10 on input*

x and let W_2 be the expected measure of the solution returned by Program 5.2, with g equal to the identity function, on input x. Then the following inequality holds:

$$\max(W_1, W_2) \geq \frac{3}{4} m^*(x).$$

Since $\max(W_1, W_2) \geq (W_1 + W_2)/2$ and $m^*_{LP-SAT}(x) \geq m^*(x)$, it is sufficient to show that $(W_1 + W_2)/2 \geq 3 m^*_{LP-SAT}(x)/4$. Let us denote by C^k the set of clauses with exactly k literals. From the proof of Theorem 2.19 (see also Exercise 5.3), it follows that each clause $c_j \in C^k$ is satisfied by the truth assignment returned by Program 2.10 with probability $1 - 1/2^k$. Hence,

$$W_1 \geq \sum_{k \geq 1} \sum_{c_j \in C^k} \gamma_k w(c_j) \geq \sum_{k \geq 1} \sum_{c_j \in C^k} \gamma_k w(c_j) z_j^* \quad (5.4)$$

where

$$\gamma_k = \left(1 - \frac{1}{2^k}\right)$$

and the last inequality is due to the fact that $0 \leq z_j^* \leq 1$. Moreover, by Lemma 5.2, we have that

$$W_2 \geq \sum_{k \geq 1} \sum_{c_j \in C^k} \alpha_k w(c_j) z_j^* \quad (5.5)$$

where

$$\alpha_k = 1 - (1 - \frac{1}{k})^k.$$

By summing Eqs. (5.4) and (5.5), we obtain

$$\frac{W_1 + W_2}{2} \geq \sum_{k \geq 1} \sum_{c_j \in C^k} \frac{\gamma_k + \alpha_k}{2} w(c_j) z_j^*.$$

Notice that $\gamma_1 + \alpha_1 = \gamma_2 + \alpha_2 = 3/2$. Moreover, for $k \geq 3$, we have that

$$\gamma_k + \alpha_k \geq 7/8 + 1 - \frac{1}{e} \geq 3/2.$$

Hence, it follows that

$$\frac{W_1 + W_2}{2} \geq \sum_{k \geq 1} \sum_{c_j \in C^k} \frac{3}{4} w(c_j) z_j^* = \frac{3}{4} m^*_{LP-SAT}(x)$$

and the lemma is proved.

Section 5.2
RANDOMIZED ALGORITHMS FOR WEIGHTED SATISFIABILITY

PROOF

QED

Chapter 5
APPROXIMATION THROUGH RANDOMIZATION

Note that it is not necessary to separately apply Programs 2.10 and 5.2 and then choose the best between the two returned solutions. Indeed, it is possible to obtain the same expected performance ratio by randomly choosing one of the two algorithms with probability $1/2$.

The proof of the following theorem easily follows from the previous lemma and is, hence, omitted.

Theorem 5.4 ▶ *There exists a randomized algorithm for* MAXIMUM WEIGHTED SATIS- FIABILITY *whose expected performance ratio is at most* $4/3$.

5.3 Algorithms based on semidefinite programming

In the last section we have seen that it is possible to design good randomized approximation algorithms for MAXIMUM WEIGHTED SATISFIABILITY by first relaxing the integrality constraint of an integer program and, subsequently, probabilistically rounding the optimal solution of the linear programming relaxation. This technique can be fruitfully applied to a limited number of cases. However, the underlying idea of relaxing and rounding is extremely powerful and it can be applied to other significant problems if a suitable relaxation can be found.

In this section we present a randomized approximation algorithm for the weighted version of MAXIMUM CUT, called MAXIMUM WEIGHTED CUT: given a graph $G = (V, E)$ and a weight function $w : E \mapsto \mathbf{N}$, we want to find a partition (V_1, V_2) of V such that the total weight of the corresponding cut, i.e., the set of edges with an endpoint in V_1 and the other endpoint in V_2, is maximized. We now present a randomized algorithm based on a *semidefinite* relaxation of an integer quadratic formulation of the problem, which returns a solution whose expected performance ratio is at most 1.139.

Let us first express an instance x of MAXIMUM WEIGHTED CUT as an integer quadratic program *IQP-CUT(x)*. To this aim, let us associate to each pair $v_i, v_j \in V$ a value w_{ij} defined as $w_{ij} = w(v_i, v_j)$ if $(v_i, v_j) \in E$, $w_{ij} = 0$ otherwise. The integer quadratic program *IQP-CUT(x)* is then defined as

$$\texttt{maximize} \quad \frac{1}{2} \sum_{j=1}^{n} \sum_{i=1}^{j-1} w_{ij}(1 - y_i y_j)$$
$$\texttt{subject to} \quad y_i \in \{-1, 1\} \qquad 1 \leq i \leq n,$$

where n denotes the number of vertices of the graph. Observe that an assignment of values to variables y_i naturally corresponds to a partition

Section 5.3
ALGORITHMS BASED ON SEMIDEFINITE PROGRAMMING

Program 5.3: Random Weighted Cut

input Instance x, i.e., graph $G = (V,E)$ and weight function w;
output Partition $\{V_1, V_2\}$ of V;
begin
    Find an optimal solution $(\mathbf{y}_1^*, \ldots, \mathbf{y}_n^*)$ of QP-$CUT(x)$;
    Randomly choose a vector $\mathbf{r} \in S_2$ according to the uniform distribution;
    $V_1 := \{v_i \in V \mid \mathbf{y}_i^* \cdot \mathbf{r} \geq 0\}$;
    $V_2 := V - V_1$;
    return $\{V_1, V_2\}$
end.

(V_1, V_2) of V with cut weight equal to $\frac{1}{2} \sum_{j=1}^{n} \sum_{i=1}^{j-1} w_{ij}(1 - y_i y_j)$. Indeed, let us consider two adjacent vertices v_i and v_j: if either $v_i, v_j \in V_1$ or $v_i, v_j \in V_2$ (that is, $y_i = y_j$), then $1 - y_i y_j = 0$; on the other hand, if v_i and v_j do not belong to the same set (that is, $y_i \neq y_j$), then $\frac{1}{2}(1 - y_i y_j) = 1$.

Notice that each variable y_i can be considered as a vector of unit norm in the 1-dimensional space. Let us now relax IQP-$CUT(x)$ by substituting each y_i with a 2-dimensional vector $\mathbf{y}_i$ of unit norm. The relaxation QP-$CUT(x)$ is then defined as

$$\text{maximize} \quad \frac{1}{2} \sum_{j=1}^{n} \sum_{i=1}^{j-1} w_{ij}(1 - \mathbf{y}_i \cdot \mathbf{y}_j)$$
$$\text{subject to} \quad \mathbf{y}_i \cdot \mathbf{y}_i = 1 \qquad \mathbf{y}_i \in \mathbf{R}^2, 1 \leq i \leq n,$$

where $\mathbf{y}_i \cdot \mathbf{y}_j$ denotes the inner product of vectors $\mathbf{y}_i$ and $\mathbf{y}_j$ (that is, $\mathbf{y}_i \cdot \mathbf{y}_j = y_{i,1} y_{j,1} + y_{i,2} y_{j,2}$).

QP-$CUT(x)$ is clearly a relaxation of IQP-$CUT(x)$. Indeed, given a feasible solution $Y = (y_1, \ldots, y_n)$ of IQP-$CUT(x)$, we can obtain the following feasible solution of QP-$CUT(x)$: $\mathbf{Y} = (\mathbf{y}_1, \ldots, \mathbf{y}_n)$ where for all $\mathbf{y}_i$, $\mathbf{y}_i = (y_i, 0)$. Clearly, the measures of Y and $\mathbf{Y}$ coincide.

Let us now consider a randomized approximation algorithm for MAXIMUM WEIGHTED CUT, which, given an instance x, behaves as follows (see Program 5.3): it first finds an optimal solution $(\mathbf{y}_1^*, \ldots, \mathbf{y}_n^*)$ of QP-$CUT(x)$, and then computes an approximate solution of MAXIMUM WEIGHTED CUT by randomly choosing a 2-dimensional vector $\mathbf{r}$ of unit norm and putting each vertex v_i in V_1 or in V_2 depending on whether the corresponding vector $\mathbf{y}_i^*$ is above or below the line normal to $\mathbf{r}$. An example of how the algorithm decides in which set a vertex has to be put is shown in Fig. 5.1: in this case, we have that v_2, v_4, v_5, and v_7 are included in V_1 while v_1, v_3, and v_6 are included in V_2.

Chapter 5

APPROXIMATION THROUGH RANDOMIZATION

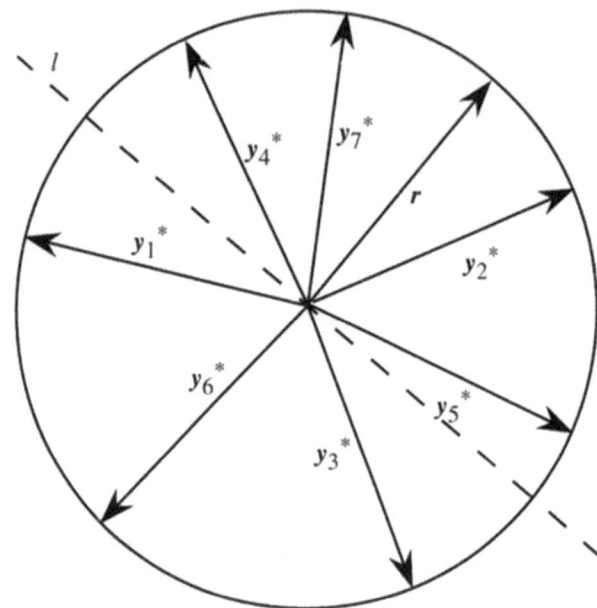

Figure 5.1
Finding a cut by separating vectors on the unit sphere

Let us now show that the expected weight of the cut returned by the algorithm is at least 0.87856 times the optimal measure, that is, the expected performance ratio of the algorithm is at most 1.139.

Lemma 5.5 ▶ *Given an instance x of* MAXIMUM WEIGHTED CUT, *let $m_{RWC}(x)$ be the measure of the solution returned by Program 5.3. Then, the following equality holds:*

$$\mathrm{E}[m_{RWC}(x)] = \frac{1}{\pi} \sum_{j=1}^{n} \sum_{i=1}^{j-1} w_{ij} \arccos(\mathbf{y}_i^* \cdot \mathbf{y}_j^*).$$

PROOF Let us first define the function *sgn* as

$$sgn(x) = \begin{cases} 1 & \text{if } x \geq 0, \\ -1 & \text{otherwise.} \end{cases}$$

Observe that the expected value $\mathrm{E}[m_{RWC}(x)]$ clearly verifies the following equality:

$$\mathrm{E}[m_{RWC}(x)] = \sum_{j=1}^{n} \sum_{i=1}^{j-1} w_{ij} \Pr\{sgn(\mathbf{y}_i^* \cdot \mathbf{r}) \neq sgn(\mathbf{y}_j^* \cdot \mathbf{r})\}$$

where $\mathbf{r}$ is a randomly and uniformly chosen vector of unit norm. Therefore, to prove the lemma it is sufficient to show that

$$\Pr\{sgn(\mathbf{y}_i^* \cdot \mathbf{r}) \neq sgn(\mathbf{y}_j^* \cdot \mathbf{r})\} = \frac{\arccos(\mathbf{y}_i^* \cdot \mathbf{y}_j^*)}{\pi}. \tag{5.6}$$

Note that $sgn(\mathbf{y}_i^* \cdot \mathbf{r}) \neq sgn(\mathbf{y}_j^* \cdot \mathbf{r})$ if and only if the random line l normal to $\mathbf{r}$ separates $\mathbf{y}_i^*$ and $\mathbf{y}_j^*$. The random choice of $\mathbf{r}$ implies that l has two opposite intersecting points s and t with the unit circle that are uniformly distributed. Moreover, $\mathbf{y}_i^*$ and $\mathbf{y}_j^*$ are separated by l if and only if either s or t lies on the shorter arc of the circle between $\mathbf{y}_i^*$ and $\mathbf{y}_j^*$ (see Fig. 5.2). The probability that either s or t lies on this arc is

$$\frac{\arccos(\mathbf{y}_i^* \cdot \mathbf{y}_j^*)}{2\pi} + \frac{\arccos(\mathbf{y}_i^* \cdot \mathbf{y}_j^*)}{2\pi} = \frac{\arccos(\mathbf{y}_i^* \cdot \mathbf{y}_j^*)}{\pi}.$$

Hence, Eq. (5.6) follows and the lemma is proved. QED

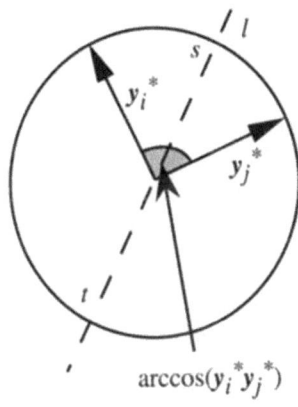
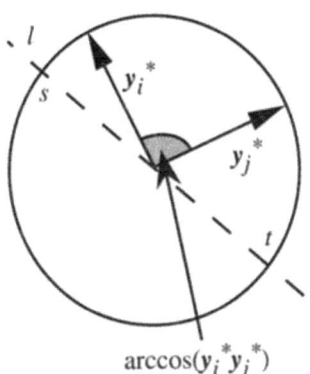

Figure 5.2
The probability of separating two vectors

◂ Theorem 5.6

For any instance of MAXIMUM WEIGHTED CUT, *Program 5.3 returns a solution whose expected measure is at least 0.8785 times the optimum measure.*

PROOF

Let us define

$$\beta = \min_{0 < \alpha \leq \pi} \frac{2\alpha}{\pi(1 - \cos\alpha)}.$$

Given an instance x of MAXIMUM WEIGHTED CUT with optimal measure $m^*(x)$, let $\mathbf{y}_1^*, \ldots, \mathbf{y}_n^*$ be an optimal solution of $QP\text{-}CUT(x)$ with measure

$$m_{QP-CUT}^*(x) = \frac{1}{2} \sum_{j=1}^{n} \sum_{i=1}^{j-1} w_{ij}(1 - \mathbf{y}_i^* \cdot \mathbf{y}_j^*).$$

If we consider the change of variables $\mathbf{y}_i^* \cdot \mathbf{y}_j^* = \cos\alpha_{ij}$, we have, by definition of β,

$$\beta \leq \frac{2\alpha_{ij}}{\pi(1 - \cos\alpha_{ij})} = \frac{2}{\pi} \frac{\arccos(\mathbf{y}_i^* \cdot \mathbf{y}_j^*)}{1 - (\mathbf{y}_i^* \cdot \mathbf{y}_j^*)}.$$

Chapter 5

APPROXIMATION THROUGH RANDOMIZATION

or, equivalently,

$$\frac{\arccos(\mathbf{y}_i^* \cdot \mathbf{y}_j^*)}{\pi} \geq \frac{\beta}{2}(1 - (\mathbf{y}_i^* \cdot \mathbf{y}_j^*)).$$

Since $QP\text{-}CUT(x)$ is a relaxation of $IQP\text{-}CUT(x)$, we have that

$$\begin{aligned}
E[m_{RWC}(x)] &\geq \frac{1}{2}\beta \sum_{j=1}^{n}\sum_{i=1}^{j-1} w_{ij}(1 - \mathbf{y}_i^* \cdot \mathbf{y}_j^*) \\
&= \beta m_{QP\text{-}CUT}^*(x) \geq \beta m_{IQP\text{-}CUT}^*(x) = \beta m^*(x),
\end{aligned}$$

where $m_{RWC}(x)$ is the measure of the solution returned by Program 5.3.

Since it is possible to show that $\beta > 0.8785$ (see Exercise 5.10), the Lemma is thus proved.

QED

Regarding the time complexity of Program 5.3, it is clear that the algorithm runs in polynomial time if and only if it is possible to solve $QP\text{-}CUT(x)$ in polynomial time. Unfortunately, it is not known whether this is possible. However, the definition of $QP\text{-}CUT(x)$ can be slightly modified in order to make it efficiently solvable: the modification simply consists in considering variables $\mathbf{y}_i$ as vectors in the n-dimensional space instead that in the 2-dimensional one. In particular, the n-dimensional version of $QP\text{-}CUT(x)$ is defined as

$$\begin{aligned}
\texttt{maximize} \quad & \frac{1}{2}\sum_{j=1}^{n}\sum_{i=1}^{j-1} w_{ij}(1 - \mathbf{y}_i \cdot \mathbf{y}_j) \\
\texttt{subject to} \quad & \mathbf{y}_i \cdot \mathbf{y}_i = 1 \qquad \mathbf{y}_i \in \mathbf{R}^n, 1 \leq i \leq n,
\end{aligned}$$

Observe that, clearly, the above analysis of the expected performance ratio of Program 5.3 can still be carried out if we refer to this new version of $QP\text{-}CUT(x)$.

In order to justify this modification, we need some definitions and results from linear algebra. First of all, we say that a $n \times n$ matrix M is *positive semidefinite* if, for every vector $x \in R^n$, $x^T M x \geq 0$. It is known that a $n \times n$ symmetric matrix M is positive semidefinite if and only if there exists a matrix P such that $M = P^T P$, where P is an $m \times n$ matrix for some $m \leq n$. Moreover, if M is positive semidefinite, then matrix P can be computed in polynomial time (see Bibliographical notes).

Observe now that, given n vectors $\mathbf{y}_1, \ldots, \mathbf{y}_n \in S_n$, the matrix M defined as $M_{i,j} = \mathbf{y}_i \cdot \mathbf{y}_j$ is positive semidefinite. On the other hand, from the above properties of positive semidefinite matrices, it follows that, given a $n \times n$ positive semidefinite matrix M such that $M_{i,i} = 1$ for $i = 1, \ldots, n$, it is possible to compute, in polynomial time, a set of n vectors $\mathbf{y}_1,\ldots,\mathbf{y}_n$ of uniti norm such that $M_{i,j} = \mathbf{y}_i \cdot \mathbf{y}_j$.

In other words, QP-CUT(x) is equivalent to the following *semidefinite program* SD-CUT(x):

$$\text{maximize} \quad \frac{1}{2}\sum_{j=1}^{n}\sum_{i=1}^{j-1} w_{ij}(1-M_{i,j})$$
$$\text{subject to} \quad M \text{ is positive semidefinite}$$
$$M_{i,i} = 1 \qquad 1 \leq i \leq n.$$

It can be proven that, for any instance x of MAXIMUM WEIGHTED CUT, if $m^*_{SD-CUT}(x)$ is the optimal value of SD-CUT(x), then, for any $\varepsilon > 0$, it is possible to find a solution with measure greater than $m^*_{SD-CUT}(x) - \varepsilon$ in time polynomial both in $|x|$ and in $\log(1/\varepsilon)$ (see Bibliographical notes). It is also possible to verify that solving SD-CUT(x) with $\varepsilon = 10^{-5}$ does not affect the previously obtained performance ratio of 0.8785. Therefore, the following theorem holds.

Program 5.3, where the optimal solution of QP-CUT(x) *is obtained by solving the equivalent program* SD-CUT(x), *runs in polynomial time.* ◀ Theorem 5.7

As a consequence of Theorems 5.6 and 5.7, it thus follows that MAXIMUM WEIGHTED CUT admits a polynomial-time randomized algorithm whose expected performance ratio is at most 1.139.

5.3.1 Improved algorithms for weighted 2-satisfiability

The approach based on semidefinite programming can be applied to other problems and, in particular, to satisfiability problems. Let us consider, for example, MAXIMUM WEIGHTED 2-SATISFIABILITY, that is, the weighted satisfiability problem in which every clause has at most two literals.

Given an instance x of MAXIMUM WEIGHTED 2-SATISFIABILITY with n variables $v_1,\ldots,v_n$, let us define the following integer quadratic program IQP-SAT(x):

$$\text{maximize} \quad \sum_{j=0}^{n}\sum_{i=0}^{j-1}[a_{ij}(1-y_iy_j) + b_{ij}(1+y_iy_j)]$$
$$\text{subject to} \quad y_i \in \{-1,1\} \qquad i = 0,1,\ldots,n,$$

where a_{ij} and b_{ij} are non-negative coefficients that will be specified later, y_i is a variable associated with v_i, for $i = 1,\ldots,n$, and y_0 denotes the boolean value TRUE, that is, v_i is TRUE if and only if $y_i = y_0$, for $i = 1,\ldots,n$.

Section 5.3

ALGORITHMS
BASED ON
SEMIDEFINITE
PROGRAMMING

In order to define the values of the coefficients a_{ij} and b_{ij}, let us define the value $t(c_j)$ of a clause c_j as follows:

$$t(c_j) = \begin{cases} 1 & \text{if } c_j \text{ is satisfied,} \\ 0 & \text{if } c_j \text{ is not satisfied.} \end{cases}$$

According to the previous definitions, it results that if c_j is a unit clause, then

$$t(c_j) = \frac{1 + y_i y_0}{2}$$

if $c_j = v_i$, and

$$t(c_j) = \frac{1 - y_i y_0}{2}$$

otherwise. If c_j contains two literals, then its value can be inductively computed: for example, if $c_j = v_i \vee v_k$, then

$$\begin{aligned} t(c_j) &= 1 - t(\bar{v}_i \wedge \bar{v}_k) = 1 - t(\bar{v}_i)t(\bar{v}_k) = 1 - \frac{1 - y_i y_0}{2} \frac{1 - y_k y_0}{2} \\ &= \frac{1}{4}(3 + y_i y_0 + y_k y_0 - y_i y_k y_0^2) \\ &= \frac{1}{4}[(1 + y_i y_0) + (1 + y_k y_0) + (1 - y_i y_k)] \end{aligned}$$

(the cost of the other possible clauses with two literals can be computed in a similar way).

Hence, it is possible, for any instance x of MAXIMUM WEIGHTED 2-SATISFIABILITY, to compute suitable values of a_{ij} and b_{ij} such that the resulting program $IQP\text{-}SAT(x)$ is an equivalent formulation of instance x.

Program $IQP\text{-}SAT(x)$ can be relaxed using the same approach used for MAXIMUM WEIGHTED CUT. By introducing unit norm $(n+1)$-dimensional vectors y_i, for $i = 0, 1, \ldots, n$, we can indeed obtain the semidefinite relaxation of $IQP\text{-}SAT(x)$ and, then, prove the following result (see Exercise 5.11).

Theorem 5.8 ▶ *There exists a randomized polynomial-time algorithm for* MAXIMUM WEIGHTED 2-SATISFIABILITY *whose expected performance ratio is at most* 1.139.

5.4 The method of the conditional probabilities

In this section we will see that a randomized approximation algorithm $\mathcal{A}$ can sometimes be *derandomized*, that is, a deterministic algorithm can be

Section 5.4
THE METHOD OF THE CONDITIONAL PROBABILITIES

derived whose running time is comparable to $\mathcal{A}$'s running time and whose performance ratio is no more than the expected performance ratio of $\mathcal{A}$. In particular, we will briefly describe a general technique known as the *method of conditional probabilities* and we will show how it can be applied to derandomize Program 2.10, when applied to MAXIMUM WEIGHTED SATISFIABILITY.

The method of conditional probabilities is based on viewing the behavior of a randomized approximation algorithm on a given input as a computation tree. To this aim, we assume, without loss of generality, that $\mathcal{A}$, on input x, independently performs $r(|x|)$ random choices each with exactly two possible outcomes, denoted by 0 and 1. According to this hypothesis, we can then define, for any input x, a complete binary tree of height $r(|x|)$ in which each node of level i is associated with the i-th random choice of $\mathcal{A}$ with input x, for $i = 1, \ldots, r(|x|)$: the left subtree of the node corresponds to outcome 0, while the right subtree corresponds to outcome 1. In this way, each path from the root to a leaf of this tree corresponds to a possible computation of $\mathcal{A}$ with input x.

Notice that, to each node u of level i, it is possible to associate a binary string $\sigma(u)$ of length $i - 1$ representing the random choices performed so far. Moreover, we can associate to each leaf l a value m_l, which is the measure of the solution returned by the corresponding computation, and to each inner node u the average measure $E(u)$ of the values of all leaves in the subtree rooted at u. Clearly, $E(u)$ is the expected measure of the solution returned by $\mathcal{A}$ with input x, assumed that the outcomes of the first $|\sigma(u)|$ random choices are consistent with $\sigma(u)$. It is easy to show that, for any inner node u, if v and w are the two children of u, then either $E(v) \geq E(u)$ or $E(w) \geq E(u)$.

The derandomization is then based on the following observation: if r is the root of the computation tree, then there must exists a path from r to a leaf l such that $m_l \geq E(r)$, that is, the measure of the solution returned by the corresponding computation is at least equal to the expected measure of the solution returned by $\mathcal{A}$ with input x. This path can be deterministically derived if, in order to choose which of the children v and w to proceed from a node u, we are able to efficiently determine which of $E(v)$ and $E(w)$ is greater.

In the following we will show how this approach can be applied to derandomize Program 2.10 in order to obtain a deterministic 2-approximation algorithm for the weighted version of MAXIMUM SATISFIABILITY.

Given an instance x of MAXIMUM WEIGHTED SATISFIABILITY, let $v_1, \ldots, v_n$ be the Boolean variables in x, which can be considered as $\{0,1\}$-variables, where the boolean values TRUE and FALSE are represented by 1 and 0, respectively. The deterministic algorithm consists of n iterations

Chapter 5
APPROXIMATION THROUGH RANDOMIZATION

corresponding to the n random choices performed by Program 2.10. At the i-th iteration, the value of variable v_i is determined as follows: let $\bar{v}_1, \ldots, \bar{v}_{i-1}$ be the values of variables $v_1, \ldots, v_{i-1}$ determined so far, and let

$$m_{RWS}(x \mid \bar{v}_1, \ldots, \bar{v}_{i-1})$$

be the random variable denoting the measure of the solution found by Program 2.10 when applied to the instance obtained from x by assuming that the value of variables $v_1, \ldots, v_{i-1}$ is $\bar{v}_1, \ldots, \bar{v}_{i-1}$ and applying Program 2.10 to determine the values of variables $v_i, \ldots, v_n$.

Given $\bar{v}_1, \ldots, \bar{v}_{i-1}$, the value of v_i is determined by computing

$$E[m_{RWS}(x \mid \bar{v}_1, \ldots, \bar{v}_{i-1}, 0)]$$

and

$$E[m_{RWS}(x \mid \bar{v}_1, \ldots, \bar{v}_{i-1}, 1)].$$

If $E[m_{RWS}(x \mid \bar{v}_1, \ldots, \bar{v}_{i-1}, 0)] \leq E[m_{RWS}(x \mid \bar{v}_1, \ldots, \bar{v}_{i-1}, 1)]$ then the value of v_i is set to 1, otherwise it is set to 0. After n iterations, a truth assignment $\bar{v}_1, \ldots, \bar{v}_n$ has been obtained with value

$$m_{\mathcal{A}}(x) = E[m_{RWS}(x \mid \bar{v}_1, \ldots, \bar{v}_n)].$$

We first show that the computation of $E[m_{RWS}(x \mid \bar{v}_1, \ldots, \bar{v}_{i-1}, 0)]$ and $E[m_{RWS}(x \mid \bar{v}_1, \ldots, \bar{v}_{i-1}, 1)]$ can be performed in deterministic polynomial time and, then, that $m_{\mathcal{A}}(x)$ is at least one half of the optimal measure. We will show how to compute $E[m_{RWS}(x \mid \bar{v}_1, \ldots, \bar{v}_{i-1}, 1)]$ in polynomial time: the computation of $E[m_{RWS}(x \mid \bar{v}_1, \ldots, \bar{v}_{i-1}, 0)]$ is analogous and it is omitted.

Assume that x contains t clauses $c_1, \ldots, c_t$. We have

$$E[m_{RWS}(x \mid \bar{v}_1, \ldots, \bar{v}_{i-1}, 1)] = \sum_{j=1}^{t} w(c_j) \Pr\{c_j \text{ is satisfied} \mid \bar{v}_1, \ldots, \bar{v}_{i-1}, 1\}$$

where

$$\Pr\{c_j \text{ is satisfied} \mid \bar{v}_1, \ldots, \bar{v}_{i-1}, 1\}$$

denotes the probability that a random truth assignment of variables $v_{i+1}, \ldots, v_n$ satisfies clause c_j given that $\bar{v}_1, \ldots, \bar{v}_{i-1}, 1$ are the truth assignments of variables $v_1, \ldots, v_{i-1}, v_i$, respectively.

Let W_i be the sum of the weights of the clauses that are satisfied by values $\bar{v}_1, \ldots, \bar{v}_{i-1}$ of variables $v_1, \ldots, v_{i-1}$ and let $C^-(i)$ be the set of clauses that are not satisfied by $v_1, \ldots, v_{i-1}$ and could be satisfied by a suitable assignment of values to variables $v_i, \ldots, v_n$.

Let c_j be a clause in $C^-(i)$. If v_i occurs positive in c_j then

$$\Pr\{c_j \text{ is satisfied} \mid v_1, \ldots, v_{i-1}, 1\} = 1.$$

If v_i occurs negative or does not occur in c_j, let d_j be the number of variables occurring in c_j that are different from $v_1, \ldots, v_i$. The probability that a random assignment of values to variables $v_{i+1}, \ldots, v_n$ satisfies clause c_j is

$$\Pr\{c_j \text{ is satisfied} \mid v_1, \ldots, v_{j-1}, 1\} = 1 - \frac{1}{2^{d_j}}.$$

Summing over all the clauses we have that

$$\mathrm{E}[m_{RWS}(x \mid \bar{v}_1, \ldots, \bar{v}_{i-1}, 1)] = W_i + \sum_{\substack{c_j \in C^-(i), \text{ s.t. } v_i \\ \text{occurs positive}}} 1 + \sum_{\substack{c_j \in C^-(i), \text{ s.t. } v_i \\ \text{occurs negative}}} \left(1 - \frac{1}{2^{d_j}}\right).$$

It is clear that the above computation can be performed in polynomial time.

In order to analyze the quality of the obtained solution observe that the chosen value $\bar{v}_i$, for $i = 1, \ldots, n$, satisfies

$$\mathrm{E}[m_{RWS}(x \mid \bar{v}_1, \ldots, \bar{v}_i)] \geq \mathrm{E}[m_{RWS}(x \mid \bar{v}_1, \ldots, \bar{v}_{i-1})].$$

Hence we have

$$\begin{aligned}
\mathrm{E}[m_{RWS}(x)] &\leq \mathrm{E}[m_{RWS}(x \mid \bar{v}_1)] \\
&\leq \mathrm{E}[m_{RWS}(x \mid \bar{v}_1, \bar{v}_2)] \\
&\leq \ldots \\
&\leq \mathrm{E}[m_{RWS}(x \mid \bar{v}_1, \ldots, \bar{v}_n)] = m_{\mathcal{A}}(x).
\end{aligned}$$

Since we have seen in Sect. 5.2 that $\mathrm{E}[m_{RWS}(x)] \geq m^*(x)/2$, it derives that $m_{\mathcal{A}}(C)$ is at least one half of the optimal measure.

The method of conditional probabilities can be successfully applied to derandomize the 4/3-approximation randomized algorithm for MAXIMUM WEIGHTED SATISFIABILITY presented in this chapter. It can also be used to derandomize the semidefinite programming based algorithm for the weighted version of MAXIMUM CUT, even though this derandomization requires a more sophisticated version of the method.

5.5 Exercises

Exercise 5.1 Consider a greedy algorithm for the weighted version of MINIMUM VERTEX COVER that at each step chooses the vertex with minimum weight among vertices that are an endpoint of an uncovered edge.

Chapter 5
APPROXIMATION THROUGH RANDOMIZATION

Show that the algorithm has an unbounded ratio in the worst case. (Hint: consider a star graph, i.e., a graph in which there exists a vertex v_1 that is connected to all the other $n-1$ vertices $v_2 \ldots v_n$ and in which no other edge exists.)

Exercise 5.2 Consider a greedy algorithm for the weighted version of MINIMUM VERTEX COVER that at each step chooses the vertex that has the least ratio weight/degree among vertices that are an endpoint of an uncovered edge. Show that the algorithm has an unbounded ratio in the worst case. (Hint: Consider an unweighted bipartite graph $G = (V \cup R, E)$, where V has n vertices and R is divided into n subsets $R_1, \ldots, R_n$. Every vertex in R_i is connected to i vertices in V and no two vertices in R_i have a common endpoint in V.)

Exercise 5.3 Modify the proof of Theorem 2.19 to show that if k is the minimum number of literals in a clause, then the expected performance ratio of the algorithm applied to weighted clauses is 2 when $k=1$ and is at most $4/3$ for $k \geq 2$.

Exercise 5.4 A function $g : [0,1] \mapsto [0,1]$ verifies the *3/4-property* if it satisfies the following inequality

$$1 - \prod_{i=1}^{l}(1 - g(y_i)) \prod_{i=l+1}^{k} g(y_i) \geq \frac{3}{4} \min(1, \sum_{i=1}^{l} y_i + \sum_{i=l+1}^{k}(1-y_i))$$

for any pair of integers k and l with $k \geq l$, and for any $y_1, \ldots, y_k \in [0,1]$. Prove that if a function g with the 3/4-property is used in Program 5.2, then the expected performance ratio of the algorithm is at most 4/3.

Exercise 5.5 Show that if a function $g : [0,1] \mapsto [0,1]$ satisfies the following conditions:

1. $g(y) \leq 1 - g(1-y)$,
2. $1 - \prod_{i=1}^{k}(1 - g(y_i)) \geq \frac{3}{4} \min(1, \sum_{i=1}^{k} y_i)$,

for any integer k, for any $y \in [0,1]$, and for any tuple $y_1, \ldots, y_n \in [0,1]$, then g verifies the 3/4-property.

Exercise 5.6 Show that the following function verifies the 3/4-property:

$$g_\alpha(y) = \alpha + (1 - 2\alpha)y,$$

where

$$2 - \frac{3}{\sqrt[3]{4}} \leq \alpha \leq \frac{1}{4}.$$

Problem 5.1: Maximum Subgraph

INSTANCE: Directed graph $G = (V, A)$.

SOLUTION: An acyclic spanning subgraph $G' = (V, A')$ of G.

MEASURE: $|A'|$.

Section 5.6
BIBLIOGRAPHICAL NOTES

Exercise 5.7 Show that the following function verifies the 3/4-property:

$$f(y) = \begin{cases} \frac{3}{4}y + \frac{1}{4} & \text{if } 0 \leq y < \frac{1}{3}, \\ \frac{1}{2} & \text{if } \frac{1}{3} \leq y < \frac{2}{3}, \\ \frac{3}{4}y & \text{if } \frac{2}{3} \leq y \leq 1. \end{cases}$$

Exercise 5.8 (*) Apply randomized rounding to MINIMUM SET COVER. Namely, consider the integer programming relaxation I of MINIMUM SET COVER and set each variable to be 1 with probability given by the value of the optimal solution of the linear programming relaxation of I. Show that the probability that a set S_i is covered is at least $1 - (1/e)$.

Exercise 5.9 Apply the result of Exercise 5.8 to show that there exists a randomized algorithm that finds an $O(\log m)$-approximate solution with probability at least 0.5, where m is the number of sets to be covered.

Exercise 5.10 Show that

$$\min_{0 < \alpha \leq \pi} \frac{2\alpha}{\pi(1 - \cos \alpha)} > 0.87856.$$

Exercise 5.11 Prove Theorem 5.8.

Exercise 5.12 Consider Problem 5.1 and consider the randomized algorithm that chooses a random ordering of the vertices and picks either the arcs that go forward or the arcs that go backward. Prove that this algorithm has expected performance ratio at most 2.

5.6 Bibliographical notes

The first algorithms in which randomization is used explicitly were introduced in the mid-seventies. A classical paper [Rabin, 1976] on primality test is considered to have started the field of randomized algorithms. In the

Chapter 5

APPROXIMATION THROUGH RANDOMIZATION

same period, [Solovay and Strassen, 1977] introduced another randomized algorithm for the same problem. Since then, a lot of problems that arise in many different areas have been studied from this point of view. Here we will limit ourselves to considering approximation algorithms for combinatorial optimization problems. A wide study of randomized algorithms and a rich bibliography can be found in [Motwani and Raghavan, 1995]. A description of the technique of the conditional probabilities can be found in [Alon and Spencer, 1992].

The randomized approximation algorithm for the weighted version of MINIMUM VERTEX COVER is presented in [Pitt, 1985].

The randomized 2-approximation algorithm for MAXIMUM WEIGHTED SATISFIABILITY follows the greedy approach used in [Johnson, 1974a], while the two 4/3-approximation algorithms were presented in [Goemans and Williamson, 1994]. Another 4/3-approximation deterministic algorithm for MAXIMUM WEIGHTED SATISFIABILITY was given in [Yannakakis, 1994]. The approach here followed is rather different and exploits techniques from the theory of maximum flows. Further improvements to the approximation of MAXIMUM WEIGHTED SATISFIABILITY based on semidefinite programming achieve a performance ratio of 1.318 [Goemans and Williamson, 1995b]. In the specific case of MAXIMUM WEIGHTED 2-SATISFIABILITY, [Feige and Goemans, 1995] have achieved a stronger result obtaining a 1.066 bound. Instead, just considering satisfiable formulas, [Karloff and Zwick, 1997] have shown that it is possible to approximate MAXIMUM WEIGHTED 3-SATISFIABILITY with approximation ratio 8/7. Further improvements are based on combining together almost all the known techniques used to approximate MAXIMUM WEIGHTED SATISFIABILITY, obtaining an approximation ratio of 1.29 (see [Ono, Hirata and Asano, 1996] and [Asano, 1997]).

The technique of randomized rounding was introduced in [Raghavan and Thompson, 1987] and [Raghavan, 1988] while studying a wire routing problem. Randomized rounding algorithms that improve the bound given in Exercise 5.9 for the MINIMUM SET COVER problem have been proposed in [Bertsimas and Vohra, 1994, Srinivasan, 1995, Srinivasan, 1996]. In particular, in [Bertsimas and Vohra, 1994] the technique is applied to a variety of covering problems.

The randomized approximation algorithm for the MAXIMUM CUT problem based on semidefinite programming is presented in [Goemans and Williamson, 1995b] while [Mahajan and Ramesh, 1995] give a derandomized version of the algorithm. A proof that semidefinite programming is solvable efficiently can be found in [Alizadeh, 1995]. [Karger, Motwani, and Sudan, 1998] applied semidefinite programming to MINIMUM GRAPH COLORING.

NP, PCP and Non-approximability Results

IN THIS chapter we first present a formal treatment of the complexity concepts introduced in Chap. 1. This treatment will provide the reader with a precise machine-based characterization that will be used in Sect. 6.3 to develop the notion of *probabilistically checkable proof* (in short, PCP). In Sect. 6.4 we will see how probabilistically checkable proofs can be used in a rather surprising way to show non-approximability results for NP-hard optimization problems.

6.1 Formal complexity theory

Historically, all basic concepts of the theory of computational complexity have been stated in terms of *Turing machines*. Let us introduce this machine model in the form of an *acceptor*, a simple form needed for discussing the complexity of recognition or decision problems.

6.1.1 Turing machines

A Turing machine $\mathcal{M}$ can be seen as a computing device (see Fig. 6.1) provided with:

1. A set Q of internal states, including a *start* state q_S and an *accepting* state q_A.

Chapter 6

NP, PCP AND NON-APPROXIMABILITY RESULTS

2. An infinite memory, represented by an infinite *tape*[1] consisting of *cells*, each of which contains either a symbol in a work alphabet Γ or the special *blank* symbol $\square$.

3. A *tape head* that spans over the tape cells and at any moment identifies the current cell.

4. A finite control (program) Δ whose elements are called *transition rules*: any such rule $((q_i, a_k), (q_j, a_l, r))$ specifies that if q_i is the current state and a_k is the symbol in the cell currently under the tape head, then a computing step can be performed that makes q_j the new current state, writes a_l in the cell, and either moves the tape head to the cell immediately to the right (if $r = \text{R}$) or to the left (if $r = \text{L}$) or leaves the tape head on the same cell (if $r = \text{N}$).

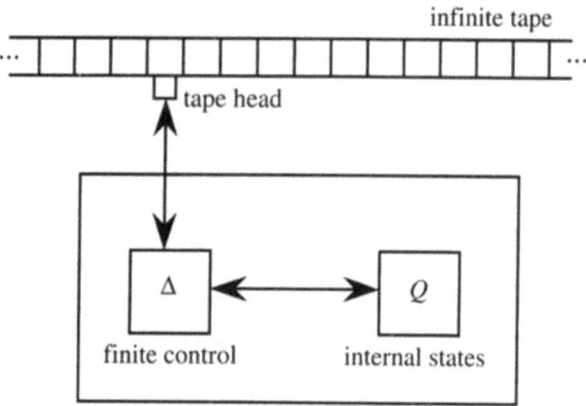

Figure 6.1
A Turing machine

We can view $\mathcal{M}$ as a computer with a fixed single program. The software is thus represented by the set of transition rules and the hardware is given by the tape, the tape head, and the mechanism that controls the head and changes states. More formally, a Turing machine can be defined as follows.

Definition 6.1 ▶ A Turing machine $\mathcal{M}$ is a 6-tuple $\mathcal{M} = (Q, \Sigma, \Gamma, \Delta, q_S, q_A)$ where:
Turing machine

1. Q is a finite set of internal states.

2. The input alphabet Σ is a finite set of symbols (not including the special symbol $\square$).

3. The work alphabet Γ is a finite set of symbols that includes all symbols in Σ and does not include $\square$.

---

[1] All the following considerations can be easily extended to the case when $\mathcal{M}$ has more than one tape.

4. The set of transition rules Δ is a subset of $(Q \times (\Gamma \cup \{\sqsubset\})) \times (Q \times (\Gamma \cup \{\sqsubset\}) \times \{L, R, N\})$.

5. $q_S, q_A \in Q$ are the starting and the accepting states, respectively.

Section 6.1
FORMAL COMPLEXITY THEORY

As a particular case, a Turing machine $\mathcal{M}$ is said to be *deterministic* (see Sect. 6.1.2) if Δ is a (partial) function $\delta : Q \times (\Gamma \cup \{\sqsubset\}) \mapsto Q \times (\Gamma \cup \{\sqsubset\}) \times \{L, R, N\}$. That is, $\mathcal{M}$ is deterministic if and only if, for any pair (q_i, a_k), there exists at most one transition that can fire if q_i is the current state and a_k is the currently read tape symbol. In this case, δ is said to be the *transition function*.

In the general case, a Turing machine is said to be *nondeterministic* (see Sect. 6.1.3). In this case, for the same pair (q_i, a_k) there might be several possible transitions.

A *configuration* of $\mathcal{M}$ is a description of the current overall status of the computation, including:

1. The current state.

2. The current content of the tape.

3. The current position of the tape head.

Formally, a configuration is an element of

$$Q \times (\Gamma \cup \{\sqsubset\})^* \{\#\} (\Gamma \cup \{\sqsubset\})^+$$

where $\# \notin \Gamma$. A configuration $(q_i, x_1 x_2 \cdots x_{k-1} \# x_k \cdots x_n)$ means that q_i is the current state, $x_1 x_2 \ldots x_{k-1} x_k \ldots x_n$ is the current tape content,[2] and the tape head is positioned on the cell containing x_k.

Given any string $\sigma = a_1 a_2 \cdots a_n \in \Sigma^*$ as input to $\mathcal{M}$, the machine starts in the initial configuration $C_0 = (q_S, \# a_1 a_2 \cdots a_n)$: that is, $\mathcal{M}$ is in the initial state, with the tape containing only σ, and the tape head on the first symbol.

Any computing step is represented by the application of one transition rule in Δ to the current configuration C, and brings $\mathcal{M}$ into a different configuration C'. We denote the occurrence of such a computing step as $C \vdash C'$. Thus, in general, we may define a (finite) *computation path* as a sequence $C_0, C_1, \ldots, C_m$ such that C_0 is an initial configuration and $C_i \vdash C_{i+1}$, for $i = 0, \ldots, m - 1$ (notice that it is immediately possible to define also infinite computation paths).

---

[2] It is assumed that all cells to the left of x_1 and to the right of x_n contain the blank symbol $\sqsubset$. Observe that if the input is a finite string then, at any step of the computation, only a finite portion of the tape contains symbols different from $\sqsubset$.

We assume that a computation halts when either the current state is q_A or no transition rule is applicable to the current configuration. More formally, for any configuration $C = (q_i, x_1 x_2 \cdots x_{k-1} \# x_k \cdots x_n)$, we say that:

1. C is *accepting* if $q_i = q_A$.

2. C is *rejecting* if $q_i \neq q_A$ and there is no pair in Δ whose first element is (q_i, x_k).

Similarly, we say that a finite computation path $C_0, C_1, \ldots, C_m$ is accepting (respectively, rejecting) if C_m is an accepting (respectively, rejecting) configuration. In an accepting path, C_m must be the first configuration that contains q_A.

If an accepting (respectively, rejecting) configuration corresponds to the value TRUE (respectively, FALSE), then we have that Turing machines can be used to compute Boolean functions. It is not hard to extend the definition of Turing machines so that they can express functions returning outputs of arbitrary types. For example, when a Turing machine accepts, the result of the computation may be defined as the content of the tape between two occurrences of a special output symbol in Σ not used for anything else (for example, the symbol $).

6.1.2 Deterministic Turing machines

As stated above, we say that a Turing machine $\mathcal{M}$ is deterministic if and only if for any possible configuration there is at most one applicable computing step.

We say that a string $\sigma \in \Sigma^*$ is *accepted* by $\mathcal{M}$ if it leads the Turing machine to halt in state q_A. That is, σ is accepted by $\mathcal{M}$ if there exists $m \geq 0$ such that the unique computation path $C_0, C_1, \ldots, C_m$ starting from the initial configuration $C_0 = (q_S, \#\sigma)$ is an accepting path.

Similarly, σ is *rejected* by $\mathcal{M}$ if there exists $m \geq 0$ such that the unique computation path $C_0, C_1, \ldots, C_m$ starting from the initial configuration $C_0 = (q_S, \#\sigma)$ is a rejecting path.

Definition 6.2 ▶ Recognized language A language $L \subseteq \Sigma^*$ is recognized by a deterministic Turing machine $\mathcal{M}$ if and only if:

1. Each string $\sigma \in L$ is accepted by $\mathcal{M}$.

2. Each string $\sigma \in \overline{L}$ is rejected by $\mathcal{M}$.

Notice that if L is recognized by $\mathcal{M}$ then $\mathcal{M}$ halts for all input strings.[3] We will refer to the language recognized by a deterministic Turing machine $\mathcal{M}$ as $L(\mathcal{M})$.

Section 6.1
FORMAL COMPLEXITY THEORY

◀ Example 6.1

Consider the (deterministic) Turing machine with $Q = \{q_S, q_A\}$, $\Sigma = \{a, b\}$, $\Gamma = \{a, b\}$, and the following transition function δ:

	q_S	q_A
a	(q_A, a, N)	-
b	(q_S, b, R)	-
$\square$	-	-

It is possible to see that such a machine recognizes the language $L \subseteq \{a,b\}^*$ consisting of all strings containing at least one a (see Exercise 6.2). The behavior of this machine with input bba is shown in Fig. 6.2.

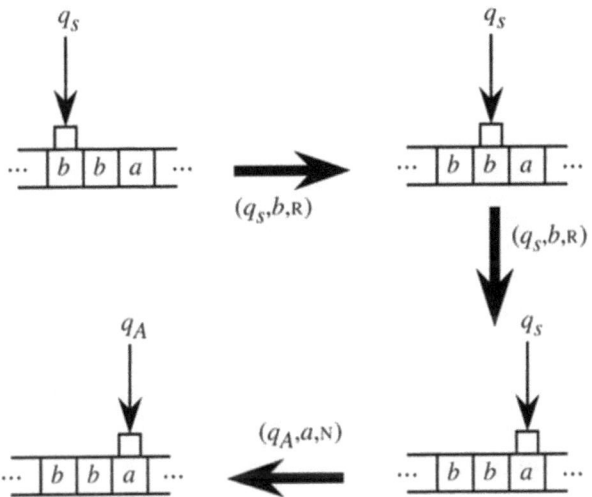

Figure 6.2
An accepting computation path of the machine of Example 6.1

Let us consider the (deterministic) Turing machine with $Q = \{q_S, q_1, q_A\}$, $\Sigma = \{a,b\}$, $\Gamma = \{a,b\}$, and the following transition function δ:

◀ Example 6.2

	q_S	q_1	q_A
a	(q_1, a, R)	(q_S, a, R)	-
b	(q_S, b, R)	(q_1, b, R)	-
$\square$	$(q_A, \square, N)$	-	-

It is possible to see that such a machine recognizes the language $L \subseteq \{a,b\}^*$ consisting of all strings with zero or an even number of a's (see Exercise 6.1). The behavior of this machine with input $abaab$ is shown in Fig. 6.3.

---

[3] That is, all languages we will deal with are decidable: for the purpose of studying the complexity of decision problems, we do not have to take into consideration semi-decidable languages.

Chapter 6

NP, PCP AND NON-APPROXIMABILITY RESULTS

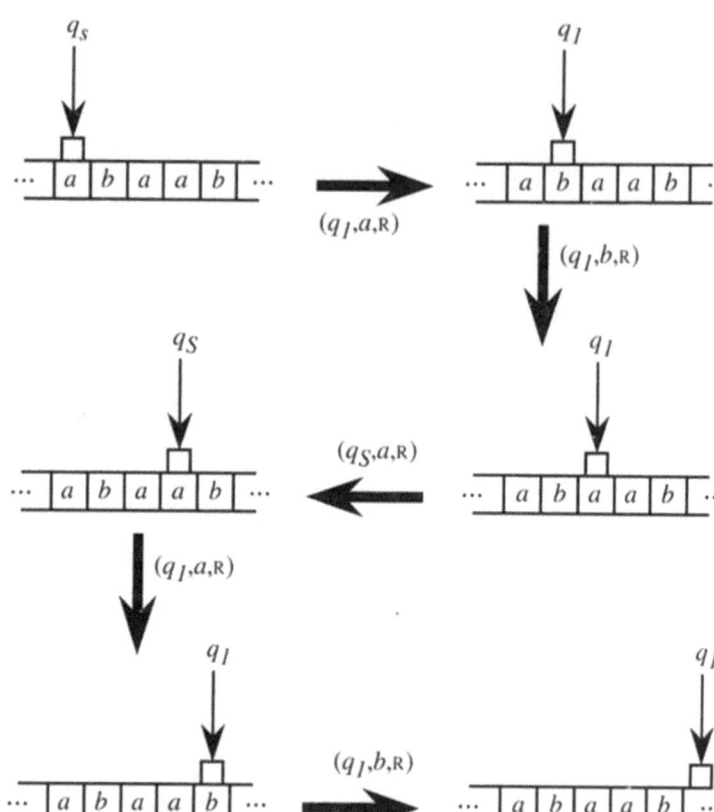

Figure 6.3
A rejecting computation path of the machine of Example 6.2

6.1.3 Nondeterministic Turing machines

Nondeterministic Turing machines correspond to the general definition of such computing devices. In general, for a nondeterministic Turing machine $\mathcal{M}$, an arbitrary finite number of computing steps can be applicable to a given configuration C, i.e., there may exist a set $\{C^1, C^2, \ldots, C^r\}$ of configurations such that $C \vdash C^i$, for $i = 1, \ldots, r$. This implies that, given a nondeterministic Turing machine, there exists a tree of computation paths starting from the same initial configuration.

We say that a string $\sigma \in \Sigma^*$ is *accepted* by $\mathcal{M}$ if at least one such path leads the Turing machine to halt in state q_A. That is, σ is accepted by $\mathcal{M}$ if there exists an accepting computation path $C_0, C_1, \ldots, C_m$ starting from the initial configuration $C_0 = (q_S, \#\sigma)$.

On the other hand, σ is *rejected* by $\mathcal{M}$ if no computation path leads the Turing machine to halt in state q_A. That is, σ is rejected by $\mathcal{M}$ if all computation paths starting from the initial configuration are rejecting (notice the asymmetry between the definitions of acceptance and rejection).

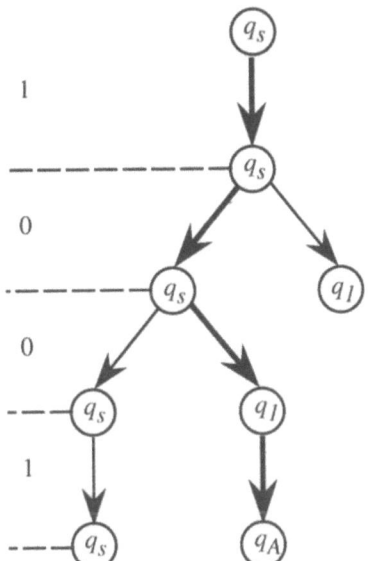

Figure 6.4
A tree of computation paths of the machine of Example 6.3

Section 6.1
FORMAL COMPLEXITY THEORY

◀ Example 6.3

Let us consider the nondeterministic Turing machine with $Q = \{q_S, q_1, q_A\}$, $\Sigma = \{0,1\}$, $\Gamma = \{0,1\}$, and the following transition rules:

	q_S	q_1	q_A
0	$\{(q_S, 0, R), (q_1, 0, R)\}$	-	-
1	$(q_S, 1, R)$	$(q_A, 1, R)$	-
☐	-	-	-

The tree of computation paths relative to the input string $\sigma = 1001$ is shown in Fig. 6.4 where an accepting computation path is also highlighted.

The definition of a language recognized by a nondeterministic Turing machine is then similar to the one given in the deterministic case. As before, we will refer to the language recognized by a nondeterministic Turing machine M as $L(M)$.

6.1.4 Time and space complexity

In order to determine the execution cost of a (deterministic or nondeterministic) Turing machine on some input we may refer to two possible measures:

1. The number of computing steps performed by the machine (time complexity).

181

Chapter 6
NP, PCP AND NON-APPROXIMABILITY RESULTS

2. The amount of different tape cells visited during the computation (space complexity).

Definitions similar to those in Chap. 1 for decision problems can then be stated.

Definition 6.3 ▶ *Complexity bounds of Turing machines*
A language L is recognized by a deterministic Turing machine $\mathcal{M}$ in time $O(t(n))$ (respectively, space $O(s(n))$) if $L = L(\mathcal{M})$ and, for each input string $\sigma \in \Sigma^*$ with $|\sigma| = n$, $\mathcal{M}$ performs $O(t(n))$ computing steps (respectively, accesses $O(s(n))$ tape cells).[4]

The Turing machine given in Example 6.2 recognizes language L in time $O(n)$ and space $O(n)$. Actually, it is worth noting that the tape is accessed by the machine only for reading the input string: the machine never writes anything, since it does not need to store any information. This is due to the simplicity of language L, that in fact could be accepted also by simpler types of acceptors. The acceptance or recognition of more complex languages would indeed require the Turing machine to use the full power of reading and writing on its tape.

Example 6.4 ▶ Consider the Turing machine with $Q = \{q_S, q_1, q_2, q_3, q_A\}$, $\Sigma = \{a, b\}$, $\Gamma = \{a, b, \alpha, \beta\}$, and the following transition function:

	q_S	q_1	q_2	q_3	q_A
a	(q_1, α, R)	(q_1, a, R)	(q_2, a, L)	-	-
b	-	(q_2, β, L)	-	-	-
α	-	-	(q_S, α, R)	-	-
β	(q_3, β, R)	(q_1, β, R)	(q_2, β, L)	(q_3, β, R)	-
$\square$	-	-	-	$(q_A, \square, N)$	-

Such a machine recognizes the language $L = \{a^n b^n \mid n \geq 1\}$ in time $O(n^2)$ and space $O(n)$. Indeed, the machine starts by marking an a (transforming it into an α) at the beginning of the input string and then goes to mark the first b (transforming it into a β); it then returns back to the first a and proceeds in the same way until all a's are marked, and the string is accepted if an equal number of b's has been marked and there are no more bs in the tape. Any other string on the alphabet $\{a, b\}$ would be rejected. The first five steps of the machine with input $aabbb$ are shown in Fig. 6.5.

The space required by the machine is $O(n)$ for storing the string and the time is $O(n^2)$ because, after marking any of the a's, the machine scans $O(n)$ symbols until it finds the first unmarked b.

---

[4]In general, we want to consider only the amount of tape used for the computation, excluding cells necessary to hold the input string. If we want to consider the case in which the space complexity is smaller than the length of the input string we then have to use a slightly different model of Turing machine with two tapes: the first one is a read-only *input tape* that contains the input string and can be scanned only from left to right, the other one is a (read/write) *work tape* used for the computation. However, we will not need this model of computation in this book.

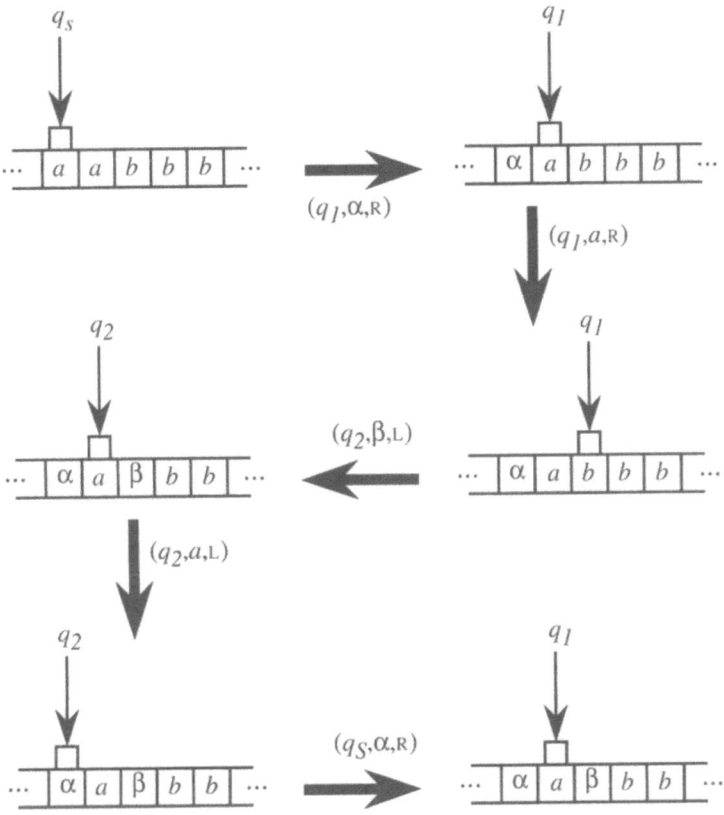

Section 6.1

FORMAL
COMPLEXITY
THEORY

Figure 6.5
The first five steps of the machine of Example 6.4 with input *aabbb*

We will now relate the complexity of a Turing machine computation to the complexity of a Pascal program, the computation model informally described in Chap. 1. Indeed, one can show that each Pascal program can be simulated by a Turing machine. Moreover, simulating one Pascal statement does not take more steps than a polynomial in the size of the input. This means that if the Pascal program runs in polynomial time, then one can construct a Turing machine that executes the same algorithm in polynomial time. Since this is true also in the opposite situation of simulating a Turing machine by a Pascal program, we say that the Pascal program and the Turing machine models are *polynomially related*.

Thus, we can define complexity classes as P, PSPACE, and NP using Turing machines instead of Pascal programs, and using analogous definitions as in Chap. 1 (recall that with any decision problem $\mathcal{P}$ we can associate a language $L_{\mathcal{P}}$).

For any function $f : \mathbf{N} \mapsto \mathbf{R}$, the complexity classes TIMETM$(f(n))$ *and* SPACETM$(f(n))$ *are the sets of decision problems $\mathcal{P}$ whose corresponding language $L_{\mathcal{P}}$ can be recognized by a deterministic Turing machine*

◀ Definition 6.4
Time and space complexity classes

183

Chapter 6

NP, PCP AND NON-APPROXIMABILITY RESULTS

in time and space $O(f(n))$, respectively. Moreover, the complexity class NTIMETM$(f(n))$ *is the set of decision problems $\mathcal{P}$ whose corresponding language $L_\mathcal{P}$ can be solved in time $O(f(n))$ by a nondeterministic Turing machine.*

According to the above definition and because the Pascal program and Turing machine models are polynomially related, we have that

$$P = \cup_{k=0}^{\infty} \text{TIMETM}(n^k),$$

$$\text{PSPACE} = \cup_{k=0}^{\infty} \text{SPACETM}(n^k),$$

and

$$\text{NP} = \cup_{k=0}^{\infty} \text{NTIMETM}(n^k).$$

Observe that, according to the above definitions, classes P, PSPACE, and NP contain decision problems. In the following, however, we will, in some cases, identify a decision problem with its corresponding language and we will view these classes as sets of languages.

Before concluding this section, we observe that Turing machines are in practice not used to design algorithms. However, defining the complexity classes using Turing machines is useful when proving hardness results. The Turing machine is such a simple computation model that it is relatively easy to express any computation on it using a simple structure like a Boolean formula. We will soon use this to show a hardness result for SATISFIABILITY.

6.1.5 NP-completeness and Cook-Levin theorem

Now that we have defined P and NP using the Turing machine computation model we will proceed to define NP-completeness in this model. Recall from Chap. 1 that a decision problem $\mathcal{P}$ is NP-complete if it is included in NP and if every problem in NP is polynomial-time Karp-reducible to $\mathcal{P}$. The Karp-reducibility defined in Sect. 1.3 is stated using a Pascal program that transforms an instance of one problem into an instance of another problem. Instead of a polynomial-time Pascal program we can use a polynomial-time Turing machine that simulates the Pascal program in polynomial time. Observe that such a Turing machine does not behave as an acceptor machine as defined in Def. 6.2, but it must return an output value and is usually called a *transducer*: the formal definition of a transducer is left to the reader (see also the end of Sect. 6.1.1).

As stated in Chap. 1, the standard way to show the NP-completeness of a decision problem is to find a polynomial-time Karp-reduction from some other problem that is already known to be NP-complete. Clearly, such a process needs some *initial* NP-complete problem, whose NP-completeness must be proved by using some other technique.

We now see an important result, known as *Cook-Levin theorem*, which shows that SATISFIABILITY is such an initial NP-complete problem.

Section 6.1

FORMAL
COMPLEXITY
THEORY

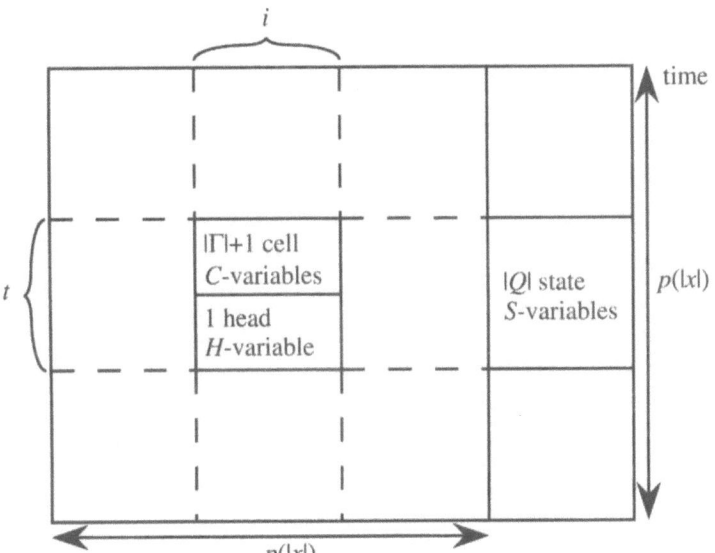

Figure 6.6
The Boolean variables of Cook-Levin theorem

SATISFIABILITY *is* NP-*complete*.   ◀ Theorem 6.1

SATISFIABILITY has already been shown to belong to NP by providing a nondeterministic polynomial-time algorithm that solves it (see Algorithm 1.3). We have then only to show that, given any decision problem $\mathcal{P} \in$ NP, we can (Karp-) reduce $\mathcal{P}$ to SATISFIABILITY in polynomial time.

PROOF

Let $\mathcal{M}$ be a nondeterministic Turing machine that recognizes $L_\mathcal{P}$ in time $p(n)$, for a suitable polynomial p. Given $x \in \Sigma^*$, we will construct a formula w in conjunctive normal form such that w is satisfiable if and only if $\mathcal{M}$ accepts x in time $p(|x|)$, that is, if and only if $x \in L$. Intuitively, w will be constructed in such a way that any satisfying truth assignment to its variables will *encode* an accepting computation path of $\mathcal{M}$ on input x.

In order to simplify the proof, and without loss of generality, we make the assumption on $\mathcal{M}$ that its tape is semi-infinite and that, for any input, there is no transition trying to overpass the leftmost tape cell: it can be shown (see Exercise 6.8) that this assumption is justified. Since $\mathcal{M}$ on input x runs in time $p(|x|)$, only the first $p(|x|)$ tape cells may be accessed during the computation. Finally, we will denote with $a \Rightarrow b$ (respectively, $a \equiv b$) the Boolean formula $\bar{a} \vee b$ (respectively, $(\bar{a} \vee b) \wedge (a \vee \bar{b})$).

The formula w we are going to derive is the conjunction of several formulas, that is, $w = w_M \wedge w_I \wedge w_A \wedge w_T$, where:

1. w_M specifies the general properties of Turing machines.

2. w_I specifies that $\mathcal{M}$ has been given string x as input on the tape.

3. w_A specifies that $\mathcal{M}$ has eventually entered an accepting configuration.

4. w_T specifies the particular transition function of $\mathcal{M}$.

The formula w contains the following variables with the corresponding interpretations (see Fig. 6.6):

1. $S(t,k)$ ($0 \le t < p(|x|)$, $1 \le k \le |Q|$) where Q is the set of states of $\mathcal{M}$: $S(t,k)$ takes value TRUE if and only if $\mathcal{M}$ is in state q_k at time t (we let q_1 be the initial state q_S and $q_{|Q|}$ be the accepting state q_A).

2. $H(t,i)$ ($0 \le t < p(|x|)$, $0 \le i < p(|x|)$): $H(t,i)$ takes value TRUE if and only if $\mathcal{M}$'s head scans cell i at time t.

3. $C(t,i,h)$ ($0 \le t < p(|x|)$, $0 \le i < p(|x|)$, $0 \le h \le |\Gamma|$) where Γ is the tape alphabet of $\mathcal{M}$: $C(t,i,h)$ takes value TRUE if and only if cell i at time t contains the symbol σ_h (we let $\sigma_0 = \square$).

Let us now see how the various subformulas of w are built:

1. w_M is the conjunction of four formulas w_{MS}, w_{MH}, w_{MC}, and w_{MT} that specify the following properties, respectively:[5]

    - At any time t, $\mathcal{M}$ must be in exactly one state, that is, w_{MS} is the formula
    $$\bigwedge_t \left(\bigvee_k S(t,k) \wedge \bigwedge_{k_1 \neq k_2} (S(t,k_1) \Rightarrow \bar{S}(t,k_2)) \right).$$

    - At any time t, the head must be on exactly one tape cell, that is, w_{MH} is the formula
    $$\bigwedge_t \left(\bigvee_i H(t,i) \wedge \bigwedge_{i_1 \neq i_2} (H(t,i_1) \Rightarrow \bar{H}(t,i_2)) \right).$$

---

[5]Note that here the quantifiers $\bigwedge$ and $\bigvee$ are used just as abbreviations: for example, $\bigwedge_t F(\ldots t \ldots)$ stands for $F(\ldots 0 \ldots) \wedge F(\ldots 1 \ldots) \wedge \ldots \wedge F(\ldots p(|x|) - 1 \ldots)$.

- At any time t, each cell must contain exactly one character, that is, w_{MC} is the formula

$$\bigwedge_{t,i} \left(\bigvee_h C(t,i,h) \wedge \bigwedge_{h \neq h'} (C(t,i,h) \Rightarrow \overline{C}(t,i,h')) \right).$$

- At any two successive times t and $t+1$, the tape contents are the same, except possibly for the cell scanned at time t. That is, w_{MT} is the formula

$$\bigwedge_{t,i,h} (\overline{H}(t,i) \Rightarrow (C(t,i,h) \equiv C(t+1,i,h))).$$

Notice that w_{MT} is not in conjunctive normal form: however, it can be easily put in conjunctive normal form with a linear increase of its length (see Exercise 6.9).

Taking into account the ranges of the values assumed by t, i, h, and k, the overall length of w_M is $O(p^3(|x|))$.

2. If $x = \sigma_{h_0}\sigma_{h_1}\ldots\sigma_{h_{|x|-1}}$, then w_I is the formula

$$S(0,1) \wedge H(0,0) \wedge C(0,0,h_0) \wedge C(0,1,h_1) \wedge \cdots$$
$$\cdots \wedge C(0,|x|-1,h_{|x|-1}) \wedge C(0,|x|,0) \wedge \cdots$$
$$\cdots \wedge C(0,p(|x|)-1,0).$$

3. $w_A = S(0,|Q|) \vee S(1,|Q|) \vee \cdots \vee S(p(|x|)-1,|Q|).$

4. w_T encodes the set of transition rules for $\mathcal{M}$. It is structured as the conjunction of $p^2(|x|)|Q|(|\Gamma|+1)$ formulas $w_{t,i,k,h}$:

$$\bigwedge_{t=0}^{p(|x|)-1} \bigwedge_{i=0}^{p(|x|)-1} \bigwedge_{k=1}^{|Q|} \bigwedge_{h=0}^{|\Gamma|} w_{t,i,k,h}.$$

For any t,i,k,h, if

$$\delta(q_k, \sigma_h) = \{(q_{k_1}, \sigma_{h_1}, r_{k_1}), \ldots, (q_{k_d}, \sigma_{h_d}, r_{k_d})\},$$

then (assuming $\mu(\text{R}) = 1$, $\mu(\text{L}) = -1$, and $\mu(\text{N}) = 0$)

$$w_{t,i,k,h} = (S(t,k) \wedge H(t,i) \wedge C(t,i,h)) \Rightarrow \bigvee_{j=1}^{d} w_{t,i,k,h}^j$$

Section 6.1

FORMAL
COMPLEXITY
THEORY

Chapter 6

NP, PCP AND NON-APPROXIMABILITY RESULTS

where

$$w^j_{t,i,k,h} = H(t+1, i+\mu(r_{k_j})) \wedge C(t+1,i,h_j) \wedge S(t+1,k_j).$$

If $\delta(q_k, \sigma_h) = \emptyset$, then $w_{t,i,k,h}$ is equal to TRUE, for any t and i. The length of formula w_T can be seen to be $O(p^2(|x|))$. As in the case of w_{MT}, w_T is not in conjunctive normal form but it can be put in conjunctive normal form with a linear increase of its length.

Taking into consideration the whole formula w, it can easily be checked that its length is $O(p^3(|x|))$ and that it can be derived from x (and from $\mathcal{M}$ and p) in time proportional to its length. Moreover, it is not difficult to verify that the formula w is satisfiable if and only if $\mathcal{M}$ accepts the string x in time $p(|x|)$.

QED

We have thus shown that SATISFIABILITY is NP-complete. Proving the NP-completeness of other problems is, usually, an easier task since we can use a decision problem that is already known to be NP-complete.

Example 6.5 ▶ Let us consider E3-SATISFIABILITY which is the restriction of SATISFIABILITY to instances with exactly three literals per clause. It is clear that such a problem belongs to NP. To prove that it is NP-complete, we will define a polynomial-time reduction from SATISFIABILITY to E3-SATISFIABILITY. The reduction will transform the clauses of the instance of SATISFIABILITY into a set of "equivalent" clauses containing exactly three (different) literals. More precisely, let C_i be any clause of the instance of SATISFIABILITY. Then C_i is tranformed into the following subformula C'_i, where the y variables are new ones:

1. If $C_i = l_{i_1}$, then $C'_i = (l_{i_1} \vee y_{i,1} \vee y_{i,2}) \wedge (l_{i_1} \vee y_{i,1} \vee \overline{y_{i,2}}) \wedge (l_{i_1} \vee \overline{y_{i,1}} \vee y_{i,2}) \wedge (l_{i_1} \vee \overline{y_{i,1}} \vee \overline{y_{i,2}})$.
2. If $C_i = l_{i_1} \vee l_{i_2}$, then $C'_i = (l_{i_1} \vee l_{i_2} \vee y_i) \wedge (l_{i_1} \vee l_{i_2} \vee \overline{y_i})$.
3. If $C_i = l_{i_1} \vee l_{i_2} \vee l_{i_3}$, then $C'_i = l_{i_1} \vee l_{i_2} \vee l_{i_3}$.
4. If $C_i = l_{i_1} \vee l_{i_2} \vee \ldots \vee l_{i_k}$ with $k > 3$, then $C'_i = (l_{i_1} \vee l_{i_2} \vee y_{i,1}) \wedge (\overline{y_{i,1}} \vee l_{i_3} \vee y_{i,2}) \wedge \ldots \wedge (\overline{y_{i,k-4}} \vee l_{i_{k-2}} \vee y_{i,k-3}) \wedge (\overline{y_{i,k-3}} \vee l_{i_{k-1}} \vee l_{i_k})$.

Clearly, such a reduction can be done in polynomial time. Moreover, it is easy to prove that the original formula is satisfiable if and only if the transformed formula is satisfiable (see Exercise 6.12). That is, E3-SATISFIABILITY is NP-complete.

6.2 Oracles

ORACLES WERE mentioned already in Chap. 1: we will here define them more formally. Oracles will later be used for defining probabilistically checkable proofs, a basic tool for proving non-approximability results (see Sect. 6.3).

6.2.1 Oracle Turing machines

An *oracle Turing machine* is a (deterministic or nondeterministic) Turing machine with the following additional features:

1. An associated *oracle language* $A \subseteq \Sigma^*$.

2. An *oracle tape* with alphabet Σ.

3. Three states q_Q, q_Y, q_N.

The Turing machine behaves as usual, except that, any time it enters state q_Q, the next step enters state q_Y if the string σ currently on the oracle tape belongs to A, otherwise the next state entered is q_N.

Thus, we assume that, when the machine is in state q_Q, it is able to check whether $\sigma \in A$ (that is, to solve the decision problem for A) in just one step.

It is immediate to see that, for an oracle Turing machine $\mathcal{M}$, changing the oracle language may imply that also the language recognized by $\mathcal{M}$ changes. For such a reason, we shall denote the language recognized by $\mathcal{M}$ with oracle language A as $L(\mathcal{M}, A)$.

For any pair $(\mathcal{M}, A)$, the usual complexity measures can be defined. So, for example, we may say that a language L is recognized in polynomial time by $\mathcal{M}$ with oracle language A if and only if $L(\mathcal{M}, A) = L$ and, for any string $x \in \Sigma^*$, $\mathcal{M}$ with oracle language A and input x halts after a number of steps polynomial with respect to $|x|$.

◀ Example 6.6

Turing-reducibility (see Sect. 1.3) can easily be defined using the oracle notation. There is a Turing-reduction from a language L_1 to a language L_2 if and only if there is an oracle Turing machine $\mathcal{M}$ such that $L(\mathcal{M}, L_2) = L_1$.

On the ground of the above considerations, it is possible to introduce complexity classes defined with respect to oracle Turing machines.

◀ Definition 6.5
Relativized complexity class

For any complexity class C and any language A, let C^A be the set of languages recognized with complexity C by an oracle Turing machine with oracle language A.

◀ Example 6.7

From Cook-Levin theorem and from the fact that Karp-reducibility is a particular case of Turing-reducibility, it easily follows that $\text{NP} \subseteq \text{P}^{L_{\text{SAT}}}$, where L_{SAT} is the language associated with SATISFIABILITY. Since any deterministic complexity class is closed under complement (see, for example, Exercise 6.6) it also follows that $\text{co-NP} \subseteq \text{P}^{L_{\text{SAT}}}$: this suggests that Turing-reducibility may be more powerful than Karp-reducibility.

Chapter 6
NP, PCP AND NON-APPROXIMABILITY RESULTS

We also extend the notation C^A by allowing, as superscript to the complexity class, another complexity class. For example, if C_1 and C_2 are complexity classes, we can write $C_1^{C_2}$: the meaning of this notation is

$$C_1^{C_2} = \bigcup_{\mathcal{P} \in C_2} C_1^{L_\mathcal{P}}.$$

6.3 The PCP model

IN THIS section, we will define the notion of probabilistically checkable proofs (in short, PCP) and we will state a new characterization of NP in terms of PCP. This characterization will then be used in deriving non-approximability results.

6.3.1 Membership proofs

Recall that a problem $\mathcal{P}$ has been defined as belonging to NP if all its positive instances are accepted by some nondeterministic polynomial-time algorithm. In particular, it is not difficult to show that $\mathcal{P}$ can be decided by some algorithm that first nondeterministically guesses a string of polynomial length and, next, performs a polynomial-time deterministic computation on the problem instance and the guess (see Exercise 6.18).

This can also be seen as a situation where the guessed string is a candidate to be a membership *proof*, that is, a proof that the instance under consideration is a positive one, and there exists an algorithm that checks the proof in deterministic polynomial time.

Example 6.8 ▶ Suppose we are given an instance of the decision problem relative to MINIMUM VERTEX COVER, that is, a graph $G = (V, E)$ and an integer K. A simple membership proof is represented by a subset $V' \subseteq V$ which is a vertex cover and has size at most K. It is immediate to see that, given $G = (V, E)$, K, and a subset V' of V, a deterministic polynomial-time algorithm can easily verify whether V' is such a membership proof.

Thus, we may define the set of problems in NP as all problems with membership proofs of polynomial length that can be verified in polynomial time. More formally, given a decision problem $\mathcal{P}$, we can associate with $\mathcal{P}$ a unary predicate $\phi_\mathcal{P}$ such that, for every input instance x, $\phi_\mathcal{P}(x) = \text{TRUE}$ if and only if x is a positive instance. We defined complexity classes such as P and NP by considering the worst case cost of computing $\phi_\mathcal{P}$.

According to the new approach of this section, given an instance x and a proposed proof π_x that $\phi_\mathcal{P}(x)$ is TRUE, we are instead interested in the cost of *verifying* that π_x is a correct proof. In the case of NP, we assume that there exists a related polynomial-time computable binary predicate $v_\mathcal{P}$ such that $v_\mathcal{P}(x,\pi_x)$ is TRUE if and only if π_x is a correct proof of $\phi_\mathcal{P}(x)$.

Thus, NP can be seen as the set of decision problems $\mathcal{P}$ such that

$$\forall x : \phi_\mathcal{P}(x) \Leftrightarrow (\exists \pi_x : |\pi_x| \leq p(|x|))[v_\mathcal{P}(x,\pi_x)]$$

where p is a polynomial.

◀ Example 6.9

Let x be an instance of SATISFIABILITY. The associated predicate $\phi(x)$ is TRUE if and only if x is satisfiable. A proof π of an instance x is, for example, a truth-assignment to the variables that occur in x. The verification predicate $v(x,\pi)$ then should be TRUE if and only if π satisfies x. It is easy to compute $v(x,\pi)$ on a deterministic Turing machine, by substituting the variables in x with the corresponding values in the truth-assignment π. This can be done in polynomial time. We have thus shown that SATISFIABILITY is in NP also with this alternative definition of NP.

6.3.2 Probabilistic Turing machines

The class NP has been defined in terms of polynomial-time *deterministic* verification of proofs. However, important developments in complexity theory are obtained if such an approach is generalized to consider *probabilistic* verification of proofs. In order to introduce this concept, we first have to define a suitable computation model, which can be seen as a Turing machine equipped with an additional source of random bits (this model also corresponds to Pascal programs which can make use of a **random** instruction that, each time is called, returns a random binary value).

A probabilistic Turing machine *(PTM) is a Turing machine with an additional read-only tape, denoted as* random tape, *and an additional state, denoted as* q_r. *The machine behaves as an ordinary Turing machine (that is, without regard to the random tape) if it is not in state* q_r. *Otherwise, it performs a* random *step of the computation by considering the current symbol on the random tape only: this step consists in changing the state according to some specified transition rules and in advancing the random tape head by one cell to the right.*

◀ Definition 6.6
Probabilistic Turing machine

Clearly, for any string x, the computation of a PTM with input x depends on the random string initially contained in the random tape. We say that a PTM is $r(n)$-restricted if, for any input x of length n, it enters at most $r(n)$

Chapter 6

NP, PCP AND NON-APPROXIMABILITY RESULTS

> **Problem 6.1: Non-Zero Polynomial**
>
> INSTANCE: A multivariate polynomial $p(x_1,\ldots,x_n)$ of degree d.
>
> QUESTION: Is p not identically zero?

times state q_r. We can then define the probability that an $r(n)$-restricted PTM accepts an input x as the probability, taken over all random strings of length $r(n)$, that, starting on input x, the PTM halts in an accepting state.

Different criteria for defining the notion of a language accepted by a PTM can be defined. For instance, the class RP is defined as the set of languages L for which there exists a polynomial-time PTM $\mathcal{M}$ such that, for any string x,

1. If $x \in L$, then $\mathcal{M}$ accepts x with probability at least $1/2$.

2. If $x \notin L$, then $\mathcal{M}$ rejects x with probability 1.

Example 6.10 ▶ Consider Problem 6.1 where the *degree* of a multivariate polynomial is defined as the largest among the degrees of its variables. Note that if p is given in standard simplified polynomial form, then it is easy to test whether p is identically zero: we simply determine whether all the coefficients of p are zero. However, if p is given as an arbitrary arithmetic expression, then no polynomial-time algorithm is known for solving this problem. It is not difficult to prove that, for any multivariate polynomial $p(x_1,\ldots,x_n)$ of degree d, if p is not identically zero, then the number of n-tuples $(a_1,\ldots,a_n)$ of integers between $-nd$ and nd such that $p(a_1,\ldots,a_n) = 0$ is, at most, $nd(2nd+1)^{n-1}$ (see Exercise 6.21 and Lemma 7.8). This result allows us to derive a simple polynomial-time PTM to solve the NON-ZERO POLYNOMIAL problem: this algorithm randomly generates an n-tuple $(a_1,\ldots,a_n)$ of integers, each integer between $-nd$ and nd, and checks whether $p(a_1,\ldots,a_n) = 0$. Since the number of n-tuples of integers between $-nd$ and nd is equal to $(2nd+1)^n$, it follows from the above result that if the polynomial is not identically zero then the probability that the polynomial will take on the value zero (and thus the algorithm gives a wrong answer) is, at most,

$$\frac{nd(2nd+1)^{n-1}}{(2nd+1)^n} = \frac{1}{2+1/nd} < 1/2.$$

Note that it is trivial to compute the degree of a polynomial. Thus, the previous algorithm is a polynomial-time one and NON-ZERO POLYNOMIAL belongs to RP.

Note that PTMs can also be defined as computing functions instead of accepting languages.

Example 6.11 ▶ All the randomized approximation algorithms in Chaps. 2 and 5 can be implemented on probabilistic Turing machines.

6.3.3 Verifiers and PCP

In this section, we will see that PTMs play a major role in the field of proof verification. Suppose that, given a language L and a string x, a proof π of the fact that $x \in L$ is given (without loss of generality, we assume that a proof is a binary string). In the case of a deterministic check, the whole proof π, in general, has to be read. For instance, if the proof that a formula is satisfiable consists in a satisfying truth-assignment, it is clear that, in general, the values of all the variables have to be read in order to check the correctness of the proof. In the case of a probabilistic proof check, instead of reading the entire proof and conducting a deterministic computation, the proof checker may randomly select a set of locations in the proof and, then, deterministically decide whether the proof is correct by only considering the content of these locations.

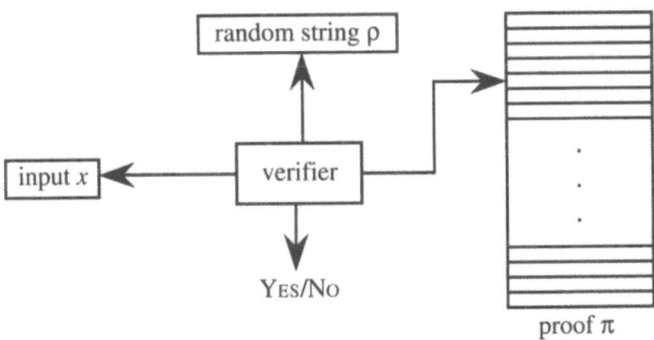

Figure 6.7
A probabilistic verifier

A verifier *is a polynomial-time oracle PTM (see Fig. 6.7), which uses the oracle to access a proof π on a random access basis: when the oracle is given a position (address) in π it returns the value of the corresponding bit (more formally, given a proof π, the corresponding oracle language X_π is the set of addresses in π corresponding to 1-bits).*[6]

◀ Definition 6.7
Verifier

The computation of the verifier is divided into two phases. In the first phase the verifier uses the random tape to determine which bits in the proof will be probed. In the second phase the verifier deterministically reads these bits and, finally, accepts (answering YES*) or rejects (answering* NO*) depending on their values.*

Note that, in the above definition, the verifier has to decide the addresses of the bits in the proof it wants to access before reading any bit of the proof:

---

[6]Observe that in this way we are collapsing the two cases in which an address points to a 0-bit and in which an address points outside of π. As it can be easily verified, this will not be significant in our context.

Chapter 6

NP, PCP AND NON-APPROXIMABILITY RESULTS

therefore, it acts in a nonadaptive fashion (i.e., it cannot read different bits depending on the values of previously read bits).

For any verifier V, let $V(x, \rho, \pi)$ be the outcome of the computation performed by V on input x, random string ρ, and proof π.

Definition 6.8 ▶ *A verifier V is $(r(n), q(n))$-restricted if V is an $r(n)$-restricted PTM and, for any input x with $|x| = n$, V accesses $q(n)$ bits of the proof (q is also called the* query complexity*).*
Restricted verifier

For any pair $(r(n), q(n))$ we may now define the complexity class $\text{PCP}[r(n), q(n)]$ as follows.

Definition 6.9 ▶ *A language L belongs to the class $\text{PCP}[r(n), q(n)]$ if and only if there exists a $(r(n), q(n))$-restricted verifier V such that:*
PCP classes

1. *For any $x \in L$, there exists a proof π such that*

$$\Pr\{V(x, \rho, \pi) = \text{YES}\} = 1.$$

2. *For any $x \notin L$ and for each proof π,*

$$\Pr\{V(x, \rho, \pi) = \text{NO}\} \geq \frac{1}{2}.$$

In both cases, the probability is taken over all random binary strings ρ of length $r(|x|)$, uniformly chosen.

In the above definition, we used the error probability 1/2. Notice that any positive constant error probability can be obtained by applying the verifier a suitable constant number of times and accepting only if the verifier accepts each time.

Observe also that, since the number of random bits is $r(n)$, the number of random strings is $2^{r(n)}$, and since for each random string the verifier uses $q(n)$ bits of the proof, the total number of bits that may actually be used in any proof is $q(n)2^{r(n)}$. Therefore, we can restrict our attention to proofs of length at most $q(n)2^{r(n)}$.

6.3.4 A different view of NP

Probabilistic verifiers can be used to introduce different characterizations of NP, in terms of PCP classes. To this aim, we first extend the notation of

PCP classes in order to allow the usage of classes of functions. In particular, given two classes of functions $\mathcal{R}$ and Q,

$$\text{PCP}[\mathcal{R}, Q] = \bigcup_{r \in \mathcal{R}, q \in Q} \text{PCP}[r(n), q(n)].$$

A first, immediate characterization is

$$\text{NP} = \text{PCP}[0, \text{poly}]$$

where $\text{poly} = \bigcup_{k \in \mathbb{N}} n^k$. Indeed, this is just a restatement of the polynomial-time deterministic verifiability of proofs for NP languages, given in Sect. 6.3.1.

A different, somewhat surprising, characterization that fully exploits the use of randomness is the following theorem, known as the *PCP theorem*.

$$\text{NP} = \text{PCP}[O(\log n), O(1)].$$

◀ Theorem 6.2

The relation $\text{PCP}[O(\log n), O(1)] \subseteq \text{NP}$ is easily derived by observing that the number of different random strings of logarithmic length is polynomial. Thus, for any input instance x, a polynomial-time nondeterministic Turing machine can guess a polynomial-length proof π, simulate the behavior $V(x, \rho, \pi)$ of the probabilistic verifier on each random string ρ, and accept if and only if all such strings lead to acceptance.

The proof of the opposite relation $\text{NP} \subseteq \text{PCP}[O(\log n), O(1)]$ is instead quite long and complex, involving sophisticated techniques from the theory of error-correcting codes and the algebra of polynomials in finite fields. This proof is contained in the next chapter. In the rest of this chapter, instead, we will show some applications of the PCP theorem to obtain limits on the approximability of specific problems.

Section 6.4

USING PCP TO PROVE NON-APPROXIMABILITY RESULTS

6.4 Using PCP to prove non-approximability results

WE SAW in Sects. 3.1.3 and 3.1.4 non-approximability results for MINIMUM TRAVELING SALESPERSON and MINIMUM GRAPH COLORING. In the proof of these results we showed that if the problems were approximable, then some NP-complete problem would be solvable in polynomial time, and therefore we would have P=NP. By making use of the PCP theorem, this method can now be applied to show non-approximability results for several other problems.

In particular, in this section we will show non-approximability results for MAXIMUM 3-SATISFIABILITY and MAXIMUM CLIQUE. For the

Chapter 6
NP, PCP AND NON-APPROXIMABILITY RESULTS

former problem we will start with the PCP characterization of NP, create an instance of the problem, and use the gap technique presented in Sect. 3.1.4 to show the non-approximability result. For the latter problem, we will instead make use of an approximation preserving reduction (see also Chap. 8) and of the *self-improvability* of the problem.

6.4.1 The maximum satisfiability problem

MAXIMUM 3-SATISFIABILITY is the restricted version of MAXIMUM SATISFIABILITY in which each clause contains at most three literals. From Example 6.5 it follows that MAXIMUM 3-SATISFIABILITY is NP-hard. Moreover, it is clearly approximable within 2 since it is a restriction of MAXIMUM SATISFIABILITY, that is approximated within 2 by Algorithm 3.1. For this problem, it is then interesting to know whether it can be approximated within *any* constant, that is, if it is included in PTAS or not. The PCP theorem allows us to show that the answer to this question is negative.

Theorem 6.3 ▶ MAXIMUM 3-SATISFIABILITY *does not admit a polynomial-time approximation scheme unless* P=NP.

PROOF Let L be an NP-complete language, for example the language corresponding to SATISFIABILITY. The PCP theorem ensures that there exists a $(r(n), q)$-restricted verifier for L, where $r(n)$ is $O(\log n)$ and q is a constant that we may assume greater than 2.

Without loss of generality, we can also assume that the $(r(n), q)$-restricted verifier for L always asks for *exactly* q bits of the proof (even though it does not need to use the value of every bit it asks for).

Given x, we will construct in polynomial time an instance (U, C) of MAXIMUM 3-SATISFIABILITY such that if $x \in L$, then (U, C) is satisfiable and if $x \notin L$ there is a constant $\varepsilon > 0$ such that at least a fraction ε of the clauses in C cannot be satisfied. By applying the gap technique described in Sect. 3.1.4, it thus follows that MAXIMUM 3-SATISFIABILITY does not belong to PTAS.

For each bit of the possible proof string π we introduce a Boolean variable in U. Since we do not need to consider proofs longer than $q2^{r(n)}$ and since $r(n) \leq c \log n$, for some constant c, the number of variables in U is bounded by $q \cdot n^c$. The i-th variable stands for the statement "*the i-th bit in π is 1*".

For each possible ρ, we let $v_{\rho[1]}, \ldots, v_{\rho[q]}$ be the q variables that correspond to the q bits that the verifier will read given the random string ρ. For some q-tuples of values of these bits, the verifier will accept and, for some tuples, it will reject. For example, Fig. 6.8 represents a computation that probes three bits (assume that the left branches correspond to 0 and the right ones to 1). Let A_ρ be the set of q-tuples for which the verifier rejects: note that $|A_\rho| \leq 2^q$ (in the case of Fig. 6.8, we have $A_\rho = \{(0,0,0), (0,1,1), (1,0,0), (1,1,1)\}$).

Section 6.4

USING PCP TO PROVE NON-APPROXIMABILITY RESULTS

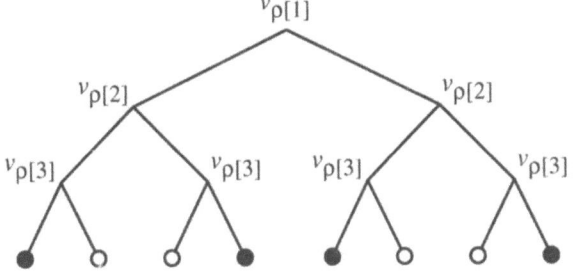

Figure 6.8
The query tree for the random string ρ

Now, for each tuple $(a_1, \ldots, a_q) \in A_\rho$, we construct a clause of q literals which is true if and only if the proof bits $v_{\rho[1]}, \ldots, v_{\rho[q]}$ do not take the values $a_1, \ldots, a_q$, respectively. For example, in the case of Fig. 6.8 we obtain the following four clauses:

$$v_{\rho[1]} \vee v_{\rho[2]} \vee v_{\rho[3]},$$

$$v_{\rho[1]} \vee \bar{v}_{\rho[2]} \vee \bar{v}_{\rho[3]},$$

$$\bar{v}_{\rho[1]} \vee v_{\rho[2]} \vee v_{\rho[3]},$$

and

$$\bar{v}_{\rho[1]} \vee \bar{v}_{\rho[2]} \vee \bar{v}_{\rho[3]}.$$

We thus have a set C' of at most $2^{q+r(n)} \leq 2^q n^c$ clauses, that is, polynomial in n. From Example 6.5 it follows that C' can be transformed into a set C of clauses with exactly 3 literals per clause such that C is satisfiable if and only if C' is satisfiable and $|C| = (q-2)|C'|$.

We will now show that if $x \in L$ there is a truth-assignment satisfying all the clauses of C. In fact, if $x \in L$, then there is a proof $\pi(x)$ such that the verifier will accept for all random strings ρ: if we set the variables as $\pi(x)$ shows, then every clause will be satisfied, since, for every ρ, the corresponding q-tuple of proof bits will never be in A_ρ.

If $x \notin L$, then we know that regardless of the proof π, there will be at least $2^{r(n)}/2$ of the random strings ρ for which the verifier will reject. For each such ρ, the q-tuple corresponding to the values of the proof bits that the verifier will read on random string ρ by definition is in A_ρ. Thus,

Chapter 6

NP, PCP AND NON-APPROXIMABILITY RESULTS

regardless of the truth-assignment, for each random string ρ for which the verifier will reject, at least one clause in the group of clauses constructed from A_ρ is not satisfied. Hence there will be at least $2^{r(n)}/2$ clauses that are not satisfied. The fraction of unsatisfiable clauses is therefore at least $(2^{r(n)}/2)/(q-2)2^{q+r(n)} \leq (2^{r(n)}/2)/2^{q+r(n)} = 2^{-(q+1)}$, that is a constant fraction.

QED

It is important to observe that, if we have a precise information on the number of bits queried by the verifier, the proof of the above theorem not only shows that MAXIMUM 3-SATISFIABILITY is not in PTAS but it also allows to explicitly derive a precise bound on the approximability of this problem (see Exercise 6.23).

6.4.2 The maximum clique problem

In this final section, we show non-approximability results for MAXIMUM CLIQUE. To this aim, we first show that this problem is at least as hard to approximate as MAXIMUM 3-SATISFIABILITY.

Lemma 6.4 ▶ MAXIMUM CLIQUE *is not in* PTAS *unless* P=NP.

PROOF

We will show that MAXIMUM 3-SATISFIABILITY is "reducible" to MAXIMUM CLIQUE in such a way that any approximate solution for the latter problem could be transformed in polynomial time into an approximate solution for the former problem having at most the same performance ratio. From this fact and from Theorem 6.3, the lemma will follow.

Given an instance (U, C) of MAXIMUM 3-SATISFIABILITY where U is the set of variables and C is the set of clauses, we define a graph $G = (V, E)$ such that:

$$V = \{(l, c) : l \in c \text{ and } c \in C\}$$

and

$$E = \{((l_1, c_1), (l_2, c_2)) : l_1 \neq \overline{l_2} \wedge c_1 \neq c_2\}$$

where l, l_1, l_2 denote literals, i.e., variables or negation of variables in U, and c, c_1, c_2 denote clauses. For any clique V' in G, let τ be a truth assignment defined as follows: for any variable u, $\tau(u)$ is TRUE if and only if a clause c exists such that $(u, c) \in V'$. This assignment is consistent since no variable receives both the value TRUE and the value FALSE. In fact, if this happens for a variable u then two clauses c_1 and c_2 exist such that

$(u,c_1), (\bar{u},c_2) \in V'$. From the definition of E, these two nodes are not connected thus contradicting the hypothesis that V' is a clique. Moreover, τ satisfies any clause c containing a literal l such that $(l,c) \in V'$: indeed, either $l = u$ and $\tau(u) = $ TRUE or $l = \bar{u}$ and, since V' is a clique, there is no clause c' such that $(u,c') \in V'$, so that $\tau(u) = $ FALSE. From the definition of E the number of these clauses is equal to $|V'|$. Since τ may satisfy some more clauses, we have that $m((U,C),\tau) \geq |V'|$.

It remains to show that the maximum number of satisfiable clauses is equal to the size of the maximum clique in G. Given a satisfiable set of clauses $C' \subseteq C$, for any truth-assignment satisfying C' and for any $c \in C'$, let l_c be any literal in c which is assigned the value TRUE. The set of nodes (l_c,c) defined in this way is clearly a clique in G.

Hence the performance ratio of τ is no worse than the performance ratio of V' and the lemma is proved. QED

By exploiting an interesting property of the MAXIMUM CLIQUE problem, known as *self-improvability*, we can now strenghten the previous non-approximability result to the following.

Section 6.4
USING PCP TO PROVE NON-APPROXIMABILITY RESULTS

◀ Theorem 6.5

MAXIMUM CLIQUE *is not in* APX *unless* P=NP.

PROOF

Let us suppose that there is an algorithm $\mathcal{A}$ approximating MAXIMUM CLIQUE within a constant $\alpha > 1$ in polynomial time. Instead of applying $\mathcal{A}$ directly on an instance $G = (V,E)$ of MAXIMUM CLIQUE, we will transform G into another, larger, instance $f(G)$ of MAXIMUM CLIQUE, and apply $\mathcal{A}$ to $f(G)$. Then we will use the approximate solution $\mathcal{A}(f(G))$ to find a better approximate solution to G than we would have got by applying $\mathcal{A}$ directly to G.

As the transformation f we will use the k-th graph product of G, defined in the following way (where k is a constant to be specified later). Let the vertex set V^k of $f(G)$ be the k-th Cartesian product of V. This means that each vertex in V^k can be denoted by $(v_1,\ldots,v_k)$ where $v_i \in V$ for every $1 \leq i \leq k$. The number of vertices in V^k is $|V|^k$. There is an edge between $(v_1,\ldots,v_k)$ and $(w_1,\ldots,w_k)$ in $f(G)$ if and only if, for every $1 \leq i \leq k$, either $v_i = w_i$ or $(v_i,w_i) \in E$ (see Fig. 6.9).

Now suppose that there is a clique $C \subseteq V$ in G. Then it is easy to verify that
$$\{(v_1,\ldots,v_k) : v_i \in C \text{ for every } 1 \leq i \leq k\}$$
is a clique in $f(G)$ (for example, from the clique $\{v_1,v_2\}$ of graph G of Fig. 6.9 we obtain the clique $\{(v_1,v_1),(v_1,v_2),(v_2,v_1),(v_2,v_2)\}$ of its second product). Since this clique is of size $|C|^k$, we have shown that $m^*(f(G)) \geq (m^*(G))^k$.

Chapter 6

NP, PCP AND NON-APPROXIMABILITY RESULTS

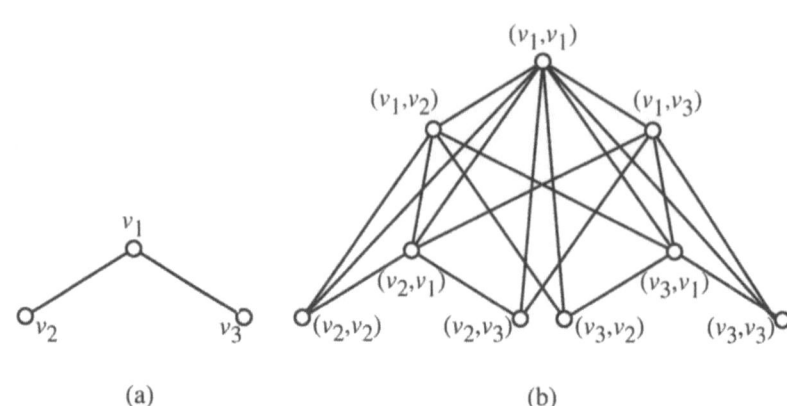

Figure 6.9
The graph product: (a) a graph G and (b) its second product

Next we suppose that there is a clique $C' \subseteq V^k$ in $f(G)$ with at least m^k vertices, for some integer m. There must then be a coordinate i between 1 and k such that, in the vertices of C' written as k-tuples, there are at least m different elements in coordinate i. Each such element corresponds to a vertex in V, and moreover these elements must form a clique in V. Thus, given a clique of size $|C'|$ in $f(G)$, we can use this procedure to find a clique C of size at least $|C'|^{1/k}$ in G. Call this procedure g.

The new approximation algorithm that we get by applying g to the clique obtained by running the approximation algorithm $\mathcal{A}$ on the constructed graph $f(G)$ has the following performance ratio:

$$\frac{m^*(G)}{m(G, g(\mathcal{A}(f(G))))} \leq \left(\frac{m^*(f(G))}{m(f(G), \mathcal{A}(f(G)))} \right)^{1/k} \leq \alpha^{1/k}.$$

That is, for any $r > 1$, if we choose $k \geq \log \alpha / \log r$, then we have a polynomial-time r-approximate algorithm for MAXIMUM CLIQUE. But Theorem 6.4 said that MAXIMUM CLIQUE is not in PTAS unless P=NP. Thus, unless P=NP, we have a contradiction.

QED

Theorems 6.3 and 6.5 are two examples of non-approximability results that can be derived from the PCP theorem. The following chapter will be devoted to the proof of this theorem while in Chap. 8 we will see other non-approximability results for large classes of NP optimization problems that do not require the explicit use of the PCP theorem but are an indirect consequence of it.

6.5 Exercises

Exercise 6.1 Prove that the language recognized by the Turing machine of Example 6.1 consists of all strings with at least one a.

Exercise 6.2 Prove that the language recognized by the Turing machine of Example 6.2 consists of all strings over the alphabet $\{a,b\}$ with an even number of a's.

Exercise 6.3 Define Turing machines that recognize the following languages over the alphabet $\{0,1\}$:

1. The set of strings that contain an equal number of 0s and 1s.
2. The set of strings that contain twice as many 0s as 1s.
3. The set of strings that do not contain twice as many 0s as 1s.

Exercise 6.4 Which language is recognized by the Turing machine of Example 6.3?

Exercise 6.5 Prove that every nondeterministic Turing machine has an equivalent deterministic Turing machine. (Hint: use a breadth-first search technique.)

Exercise 6.6 Show that P is closed under union and complement.

Exercise 6.7 Show that NP is closed under union. Is NP also closed under complement?

Exercise 6.8 In the proof of Theorem 6.1 it is said that without loss of generality one can suppose that the tape of the Turing machine is semi-infinite. Show that this is true, that is, show that any deterministic Turing machine with an infinite tape can be simulated by a Turing machine with a semi-infinite tape, and that this simulation does not blow up the time (or space) complexity. (Hint: imagine the semi-infinite tape divided into two tracks, one corresponding to the left side of the infinite tape, the other corresponding to the right side.)

Exercise 6.9 Show that any Boolean formula f can be transformed in polynomial-time into a Boolean formula g in conjunctive normal form such that f is satisfiable if and only if g is satisfiable.

Exercise 6.10 Show that if P = NP, then there exists a polynomial-time algorithm that, given a Boolean formula f in conjunctive normal form, actually produces a satisfying truth-assignment for f (if f is satisfiable). (Hint: use techniques similar to those of Example 1.22.)

Exercise 6.11 Let EXPTIME $= \bigcup_{k=0}^{\infty}$ TIMETM(2^{n^k}). Prove that PSPACE $\subseteq$ EXPTIME. (Hint: note that any Turing machine which terminates without using more than M tape cells during its computation may pass through no more than $O(2^M)$ distinct configurations.)

Section 6.5

EXERCISES

Chapter 6

NP, PCP AND NON-APPROXIMABILITY RESULTS

Problem 6.2: 3-Colorability

INSTANCE: Graph $G = (V, E)$.

QUESTION: Is G 3-colorable, i.e., does it exist a function $f : V \mapsto \{1, 2, 3\}$ such that $\forall (u, v) \in E$, $f(u) \neq f(v)$.

Exercise 6.12 Prove that the reduction of Example 6.5 is indeed a Karp-reduction.

Exercise 6.13 Consider 3-SATISFIABILITY(B), that is, the restriction of 3-SATISFIABILITY to instances in which each variable appears at most B times. Prove that 3-SATISFIABILITY(3) is NP-complete. (Hint: for each occurence of a variable, create a new variable; impose the equivalence of all new variables corresponding to the same original variable.)

Exercise 6.14 Prove that VERTEX COVER (see Problem 1.7) is NP-complete. (Hint: give a reduction from 3-SATISFIABILITY.)

Exercise 6.15 Prove that Problem 6.2 is NP-complete. (Hint: give a reduction from 3-SATISFIABILITY.)

Exercise 6.16 Show that the problem to decide whether the maximum clique in a graph is unique is in P^{NP}.

Exercise 6.17 Prove that $P^{L}SAT = P^{NP}$.

Exercise 6.18 Show that any $\mathcal{P} \in NP$ can be decided by a nondeterministic algorithm that first guesses a string of polynomial length in the size of the instance and, next, performs a polynomial-time deterministic computation on the problem instance and the guessed string. (Hint: the guessed string should be interpreted as a path in the computation tree of a nondeterministic machine.)

Exercise 6.19 We have defined class RP in Sect. 6.3.2. In the same way the class co-RP can be defined. A co-RP problem thus admits a polynomial-time PTM that with probability 1 says YES for YES instances, and with probability at least 1/2 says NO for NO instances. Show that $P \subseteq RP \subseteq NP$ and $P \subseteq$ co-RP $\subseteq$ co–NP.

Exercise 6.20 Show that, for any $Q \in RP$, there exists a polynomial-time PTM that with probability 1 says NO for NO instances, and with probability at most 1% says NO for YES instances. (Hint: exploit the fact that the probability of two independent events is the product of the probabilities of each event.)

Exercise 6.21 (*) Show that, for any n-variate polynomial p of degree at most d in each variable, if p is not identically zero, then the number of n-tuples of integers between $-nd$ and nd which are roots of p is, at most, $nd(2nd+1)^{n-1}$. (Hint: by induction on the number n of variables. In case of need, see Lemma 7.8.)

Exercise 6.22 (*) Show that if a verifier is only allowed to look at one bit in the proof and to use a logarithmic number of random bits, then it cannot do anything more than a deterministic Turing machine can do in polynomial time. (Hint: for every bit in the proof, set it so that the probability of acceptance is the greatest possible.)

It is in fact possible to prove that this result holds even if the verifier can read two bits of the proof. Try to show this!

Exercise 6.23 In Def. 6.9 the class $\text{PCP}[r(n), q(n)]$ is defined. We can define the generalization $\text{PCP}_{c,s}[r(n), q(n)]$ consisting of every language L for which there exists an $(r(n), q(n))$-restricted verifier V such that:

1. For any $x \in L$, there exists a proof π such that

$$\Pr_\rho\{V(x, \rho, \pi) = \text{YES}\} \geq c.$$

2. For any $x \notin L$ and for each proof π,

$$\Pr_\rho\{V(x, \rho, \pi) = \text{YES}\} < s.$$

The parameter c is called the *completeness probability* and tells how good the verifier is to accept true instances. If $c = 1$ (as in Def. 6.9) we say that the verifier has *perfect completeness*. The parameter s is called the *soundness probability* and gives an upper bound on the probability that the verifier accepts false instances. The greater the gap c/s between completeness and soundness, the better is the verifier at telling the truth.

We can now see that $\text{PCP}[r(n), q(n)] = \text{PCP}_{1,1/2}[r(n), q(n)]$. The ordinary deterministic proof system for NP (in Sect. 6.3.4) says that $\text{NP} = \text{PCP}_{1,0}[0, \text{poly}]$, where $\text{poly} = \cup_{k \in \mathbb{N}} n^k$.

Using the parameters c and s it is possible to prove some other characterizations of NP:

$$\begin{aligned} \text{NP} &= \text{PCP}_{1, 0.32}[O(\log n), 9] \\ &= \text{PCP}_{1, 0.76}[O(\log n), 3] \\ &= \text{PCP}_{0.99, 0.51}[O(\log n), 3]. \end{aligned}$$

Chapter 6

NP, PCP AND NON-APPROXIMABILITY RESULTS

In what way do these alternative characterizations of NP influence the non-approximability property of MAXIMUM 3-SATISFIABILITY shown in Sect. 6.4.1?

Exercise 6.24 (**) Prove Lemma 6.4 without using Theorem 6.3. (Hint: use the PCP theorem to construct a graph whose nodes are "transcripts" of the verifier computation and whose edges denote the consistency between two transcripts.)

6.6 Bibliographical notes

REFERENCES ON formal complexity theory have already been provided at the end of Chap. 1. The same references contain definitions and basic properties of oracle and probabilistic Turing machines. The proof of Cook-Levin's theorem has been given in [Cook, 1971], and independently in [Levin, 1973].

The result presented in Example 6.10 and in Exercise 6.21 is originally due to [Schwartz, 1980] where probabilistic algorithms for testing several properties of polynomials are described (see also Lemma 7.8).

The bibliographical notes of Chap. 7 give a detailed account of the development of the concept of probabilisticaly checkable proof together with extensive reference to the literature.

The first connection between probabilistically checkable proofs and approximability of optimization problems has been shown in [Feige et al., 1991] where the non-approximability of MAXIMUM CLIQUE was proved modulo a complexity theoretic conjecture stronger than P ≠ NP: Exercise 6.24 actually requires the reader to reconstruct the proof of this result starting from the PCP theorem.

The application of probabilistically checkable proofs to proving non-approximability results has then been improved in [Arora and Safra, 1998] and in [Arora et al., 1992]. The former paper proves that MAXIMUM CLIQUE is not in APX (unless P = NP), while the latter one contains a proof that MAXIMUM 3-SATISFIABILITY is not in PTAS (unless P = NP).

After these two papers, three basic approaches have been followed in order to obtain new non-approximability results. The first one consists of deriving stronger versions of the PCP theorem with respect to several parameters in order to obtain tight lower bounds on the approximability of specific problems. This approach initiated with [Bellare, Goldwasser, Lund, and Russell, 1993] and culminated with the analysis of the usage of the so-called free bits developed in [Bellare, Goldreich, and Sudan, 1998]. As a consequence of this approach a tight bound on the approximability of

MAXIMUM k-SATISFIABILITY (something which is not known for most optimization problems) has been shown in [Håstad, 1997]. In this latter paper, besides various other non-approximability results for important NPO problems, the PCP-characterizations of NP shown in Exercise 6.23 are also proved.

The second approach, instead, consists of considering different interactive proof systems and obtaining results similar to the PCP theorem to be then applied for proving non-approximability results. For example, this approach has been followed in [Lund and Yannakakis, 1994] and in [Raz and Safra, 1997] to show tight negative results for MINIMUM SET COVER.

The last approach makes use of approximation preserving reductions and will be extensively discussed in Chap. 8. We used this approach in the proof of Theorem 6.4. Note that the reduction described in that proof is exactly the same as the one usually applied to prove the NP-completeness of the corresponding decision problem.

The above three approaches have rapidly led to hundreds of non-approximability results: most of these results are contained in the compendium at the end of the book.

Finally, we remind that the self-improvability of MAXIMUM CLIQUE is mentioned in [Garey and Johnson, 1976a] where it is shown that this property is also shared by other problems, such as the graph coloring problem.

Section 6.6

BIBLIOGRAPHICAL NOTES

The PCP theorem

IN THIS chapter we will give a complete proof of the PCP theorem, that is, Theorem 6.2. In particular, after a brief presentation of the main ingredients of a PCP result, we first show that any NP problem admits long (that is, exponential) proofs which can be verified with a constant number of queries. To this aim we introduce the arithmetization of a Boolean formula and we make use of some properties of linear functions.

We then show that any NP language admits short (that is, polynomial) proofs which are verifiable with a polylogarithmic number of queries. This step is mainly based on the very well-known property of polynomials that two distinct multi-variate polynomials of low degree (with respect to the size of their domain) can agree on a small fraction of points. This property has been already partially stated in Exercise 6.21: in this chapter we will give a proof of it in its generality.

The presentation of the above two PCP results is intentionally very similar. Indeed, both presentations are structured into three parts. The first part proves the basic properties of the algebraic tools (that is, linear functions and low-degree polynomials): these properties are the verifiability and the possibility of reconstructing correct values. The second part shows how a Boolean formula can be arithmetized so that checking its satisfiability can be reduced to asking an algebraic question. Finally, the third part puts together the previous pieces in order to prove the corresponding PCP result.

Subsequently, we describe the overall structure of the proof of the PCP theorem and we prove its final main ingredient, that is, the composition lemma. To prove this lemma, we introduce a restricted version of proof

Chapter 7
THE PCP THEOREM

verifiers (normal form) which, intuitively, have access to a *split encoding* of the proof (instead of an encoding of the entire proof).

The proof of the PCP theorem (along with some exercises) is almost completely self-contained: indeed, the only missing part is the proof of the correctness of the low-degree test, which is used in the second PCP result. This proof is quite long and has mostly a technical flavor, without adding new insights on the basic ideas of the overall PCP theorem. Moreover, alternative analyses of the low-degree test, which are based on combinatorial rather than algebraic arguments, are currently under examination in order to make this part of the proof of the PCP theorem simpler. The reader can hence try to prove the correctness of the low-degree test by him/herself or look at the references contained in the bibliographical notes.

Throughout this chapter, we will focus our attention on the SATISFIABILITY problem (in particular, the restriction E3-SATISFIABILITY of this problem to instances with clauses containing exactly three literals). Since this problem is NP-complete (see Example 6.5), the PCP theorem will follow (see Exercise 7.1).

7.1 Transparent long proofs

IN THIS section and in the next one we will construct transparent proofs of membership for E3-SATISFIABILITY (informally, a transparent proof either proves a correct statement or contains mistakes almost everywhere). Such membership proofs will be suitable encodings of satisfying truth assignments for the input Boolean formulas.

Given a Boolean formula φ and a candidate proof of the satisfiability of φ (recall that, according to Def. 6.7, the proof is accessed by means of an oracle), the verifier has to perform the following two steps:

1. It has to check that the oracle is a proof, that is, a *codeword*. Actually, since the verifier will not be allowed to access the entire proof, we will only ask that it checks that the oracle is *close* to a codeword c, where "close" here means that the oracle agrees with c on most bits. The possibility of efficiently performing such a test will be guaranteed by a property of the used encoding, called *verifiability*.

2. It has to obtain the value of some bits of c in order to verify that c is a membership proof, that is, it encodes a satisfying truth assignment. Since, however, the verifier has access to something which is only close to c, each time it needs a bit of c it has to perform a *correcting* procedure in order to obtain the right value of the bit (this procedure

Section 7.1
TRANSPARENT LONG PROOFS

might need to access several bits of the oracle). The possibility of efficiently performing such a procedure will be guaranteed by another property of the used encoding, called *error correction*.

Observe that, since the verifier is a probabilistic Turing machine, the above two steps may be correctly performed with high probability (that is, we allow a non-zero error probability).

In this section and in the following one, we will describe two encodings that satisfy both the verifiability and the error correction property. While the first encoding, which is based on the use of linear functions, will not require anything else but the codeword in order to satisfy the two properties, the second encoding, which is based on the use of low-degree polynomials, will need some additional information to be given along with the codeword. On the other hand, making use of linear functions will result in the need of many variables, which, in turn, will cause an explosion of the length of the proof, while making use of polynomials of higher degree will make it possible to reduce this length.

Both encodings make use of algebraic functions (either linear functions or low-degree polynomials) over a finite field. For this reason, we need to translate a combinatorial problem such as E3-SATISFIABILITY into an algebraic problem: this is achieved by arithmetizing Boolean formulas (see Sects. 7.1.2 and 7.2.2).

The following definition formally introduces, in this framework, the notion of closeness referenced in the first step of the verifier described above.

◀ **Definition 7.1**
δ-*close functions*

Let $0 < \delta < 1$ and let D and R be two finite sets. Two functions $f, g : D \mapsto R$ are said to be δ-close if

$$\frac{|\{x \in D \mid f(x) \neq g(x)\}|}{|D|} \leq \delta.$$

In the following, given a finite set D and a Boolean predicate $P : D^k \mapsto \{\text{TRUE}, \text{FALSE}\}$, we will denote with

$$\Pr_{x_1,\ldots,x_k \in D} \{P(x_1,\ldots,x_k)\}$$

the following ratio:

$$\frac{|\{(x_1,\ldots,x_k) \mid x_1,\ldots,x_k \in D \wedge P(x_1,\ldots,x_k) = \text{TRUE}\}|}{|D|^k}$$

(that is, the probability that P is satisfied by a k-tuple of elements of D randomly chosen with respect to the uniform distribution). According to this notation, the condition of δ-closeness used in the previous definition can be rewritten as

$$\Pr_{x \in D} \{f(x) \neq g(x)\} \leq \delta.$$

Chapter 7
THE PCP THEOREM

7.1.1 Linear functions

The two key ingredients of the first encoding will be the self-testing and the self-correcting properties with respect to multi-variate linear functions. While the self-testing property allows us to verify whether a given function is close to a linear function, the self-correcting property states that if a small fraction of all values of a linear function are erroneous, it is not difficult to reconstruct the correct values. In this section we prove these properties of linear functions.

A linear function over the field Z_2 is a polynomial of total degree one. Equivalently, it can be defined as follows.

Definition 7.2 ▶ *A function $f : Z_2^m \mapsto Z_2$ is linear if, for any $\mathbf{x}, \mathbf{y} \in Z_2^m$,*
Linear function

$$f(\mathbf{x}+\mathbf{y}) = f(\mathbf{x}) + f(\mathbf{y}).$$

The linearity test

The next result formally states that it is easy to verify whether a function g is close to a linear function f.

Lemma 7.1 ▶ *Let $\delta < 1/3$ be a constant and let $g : Z_2^m \mapsto Z_2$ be a function such that*

$$\Pr_{\mathbf{x},\mathbf{y} \in Z_2^m} \{g(\mathbf{x}+\mathbf{y}) \neq g(\mathbf{x}) + g(\mathbf{y})\} \leq \delta/2.$$

Then there exists a linear function $f : Z_2^m \mapsto Z_2$ such that f and g are δ-close.

PROOF We will show that the function f defined as

$$f(\mathbf{x}) = b \in Z_2 \text{ such that } \frac{|\{\mathbf{y} \in Z_2^m \mid g(\mathbf{x}+\mathbf{y}) - g(\mathbf{y}) = b\}|}{2^m} \geq \frac{1}{2}$$

has the desired property. First, we show that f and g are δ-close. Indeed, if

$$\Pr_{\mathbf{x} \in Z_2^m} \{f(\mathbf{x}) \neq g(\mathbf{x})\} > \delta$$

and since, by definition of f, for all $\mathbf{x} \in Z_2^m$,

$$\Pr_{\mathbf{y} \in Z_2^m} \{f(\mathbf{x}) = g(\mathbf{x}+\mathbf{y}) - g(\mathbf{y})\} \geq 1/2,$$

we have that

$$\Pr_{\mathbf{x},\mathbf{y} \in Z_2^m} \{g(\mathbf{x}+\mathbf{y}) - g(\mathbf{y}) \neq g(\mathbf{x})\} > \delta/2,$$

thus contradicting the hypothesis of the lemma.

Section 7.1

TRANSPARENT
LONG PROOFS

Program 7.1: Linearity Test

input Value $\delta < 1/3$;
oracle Function $g : Z_2^m \mapsto Z_2$;
output YES if g is δ-close to a linear function;
begin
  $k := \lceil 2/\delta \rceil$;
  for $i := 1$ **to** k **do**
  begin
    Randomly pick $\mathbf{x}, \mathbf{y} \in Z_2^m$;
    if $g(\mathbf{x}) + g(\mathbf{y}) \neq g(\mathbf{x} + \mathbf{y})$ **then return** NO
  end;
  return YES
end.

To show that f is linear, we first show that, for any $\mathbf{a} \in Z_2^m$,

$$p_\mathbf{a} = \Pr_{\mathbf{x} \in Z_2^m} \{f(\mathbf{a}) = g(\mathbf{a}+\mathbf{x}) - g(\mathbf{x})\} \geq 1 - \delta \qquad (7.1)$$

(observe that from the definition of f it follows $p_\mathbf{a} \geq 1/2$). Let $\mathbf{a} \in Z_2^m$. From the hypothesis of the lemma, it follows that

$$\Pr_{\mathbf{x},\mathbf{y} \in Z_2^m} \{g(\mathbf{x}+\mathbf{a}+\mathbf{y}) \neq g(\mathbf{x}+\mathbf{a}) + g(\mathbf{y})\} \leq \delta/2;$$

and that

$$\Pr_{\mathbf{x},\mathbf{y} \in Z_2^m} \{g(\mathbf{x}+\mathbf{y}+\mathbf{a}) \neq g(\mathbf{x}) + g(\mathbf{y}+\mathbf{a})\} \leq \delta/2.$$

Hence,

$$1 - \delta \leq \Pr_{\mathbf{x},\mathbf{y} \in Z_2^m} \{g(\mathbf{x}+\mathbf{a}) + g(\mathbf{y}) = g(\mathbf{x}) + g(\mathbf{y}+\mathbf{a})\}.$$

Now observe that

$$\Pr_{\mathbf{x},\mathbf{y} \in Z_2^m} \{g(\mathbf{x}+\mathbf{a}) + g(\mathbf{y}) = g(\mathbf{x}) + g(\mathbf{y}+\mathbf{a})\}$$

$$= \sum_{z \in Z_2} \Pr_{\mathbf{x} \in Z_2^m} \{g(\mathbf{x}+\mathbf{a}) - g(\mathbf{x}) = z\} \Pr_{\mathbf{y} \in Z_2^m} \{g(\mathbf{y}+\mathbf{a}) - g(\mathbf{y}) = z\}$$

$$= \sum_{z \in Z_2} \left(\Pr_{\mathbf{x} \in Z_2^m} \{g(\mathbf{x}+\mathbf{a}) - g(\mathbf{x}) = z\} \right)^2.$$

Assume $f(\mathbf{a}) = 0$ (the other case is similar). We then have

$$1 - \delta \leq p_\mathbf{a}^2 + (1 - p_\mathbf{a})^2 \leq p_\mathbf{a}(p_\mathbf{a} + (1 - p_\mathbf{a})) = p_\mathbf{a}$$

Chapter 7
THE PCP THEOREM

where the second inequality is due to the fact that $p_\mathbf{a} \geq 1/2$. In order to complete the proof of the lemma, fix $\mathbf{a}, \mathbf{b} \in Z_2^m$. If we apply Eq. 7.1 three times, we obtain

$$\Pr_{\mathbf{x} \in Z_2^m} \{f(\mathbf{a}) + f(\mathbf{b}) + g(\mathbf{x}) \neq g(\mathbf{a}+\mathbf{x}) + f(\mathbf{b})\} \leq \delta,$$

$$\Pr_{\mathbf{x} \in Z_2^m} \{f(\mathbf{b}) + g(\mathbf{a}+\mathbf{x}) \neq g(\mathbf{b}+\mathbf{a}+\mathbf{x})\} \leq \delta,$$

and

$$\Pr_{\mathbf{x} \in Z_2^m} \{g(\mathbf{a}+\mathbf{b}+\mathbf{x}) \neq f(\mathbf{a}+\mathbf{b}) + g(\mathbf{x})\} \leq \delta.$$

Hence,

$$\Pr_{\mathbf{x} \in Z_2^m} \{f(\mathbf{a}) + f(\mathbf{b}) + g(\mathbf{x}) = f(\mathbf{a}+\mathbf{b}) + g(\mathbf{x})\} \geq 1 - 3\delta > 0$$

where the last inequality is due to the fact that $\delta < 1/3$. Since the Boolean value of the predicate in the above inequality is independent from $\mathbf{x}$, its probability is either 0 or 1. The above inequality implies that this probability is 1 so that, for any $\mathbf{a}, \mathbf{b} \in Z_2^m$, $f(\mathbf{a}) + f(\mathbf{b}) = f(\mathbf{a}+\mathbf{b})$, that is, f is a linear function.

QED

The previous lemma allows us to obtain a simple probabilistic test to check whether a given function is (almost) linear whose failure probability can be made arbitrarily close to 1. Indeed, let us consider Program 7.1. In this (probabilistic) algorithm and in the following ones, we use the keyword **oracle** to denote information to which the program has random access. Moreover, the specified output corresponds to what we desire to obtain with probability at least 1/2. Finally, we assume that random choices are independently done with respect to the uniform distribution.

Theorem 7.2 ▶ *Given a function $g: Z_2^m \mapsto Z_2$ and a rational value $\delta < 1/3$, Program 7.1, with input δ and oracle the table of values of g, behaves as follows:*

1. *If g is a linear function, then it returns* YES *with probability 1.*

2. *If g is not δ-close to a linear function, then it returns* NO *with probability at least 1/2.*

PROOF

Clearly, if g is a linear function then the algorithm returns YES for any sequence of k random choices of $\mathbf{x}, \mathbf{y} \in Z_2^m$. Otherwise, if g is not δ-close to a linear function, then from Lemma 7.1 it follows that the algorithm returns YES with probability at most

$$\left(1 - \frac{\delta}{2}\right)^k = \left(1 - \frac{\delta}{2}\right)^{\lceil 2/\delta \rceil} \leq \frac{1}{2}.$$

QED The theorem is, hence, proved.

Program 7.2: Self-correcting Function

input $\mathbf{x} \in Z_2^m$;
oracle Function g, which is δ-close to linear function f;
output $f(\mathbf{x})$;
begin
   Randomly pick $\mathbf{y} \in Z_2^m$;
   return $g(\mathbf{x}+\mathbf{y}) - g(\mathbf{y})$
end.

The self-correcting functions

Another simple but important property of linear functions is that they can be easily corrected (recall that, according to the output of Program 7.1, we are not querying a linear function but a function g which is δ-close to a linear function f). The following result formally states this property, since it shows that g can be *self-corrected* in the sense that the wrong values of g (with respect to f) can be (probabilistically) corrected by using other values of g.

◀ **Theorem 7.3**

Given a function $g : Z_2^m \mapsto Z_2$, which is δ-close to a linear function f for some rational value $\delta < 1/3$, and given $\mathbf{x} \in Z_2^m$, Program 7.2, with input $\mathbf{x}$ and oracle the table of values of g, returns the value $f(\mathbf{x})$ with probability at least $1 - 2\delta$.

PROOF

It is sufficient to show that, for any $\mathbf{x} \in Z_2^m$,

$$\Pr_{\mathbf{y} \in Z_2^m} \{f(\mathbf{x}) = g(\mathbf{x}+\mathbf{y}) - g(\mathbf{y})\} \geq 1 - 2\delta.$$

To this aim, observe that both $\mathbf{y}$ and $\mathbf{x}+\mathbf{y}$ are uniformly distributed in Z_2^m (even though they are not independent). Hence, from the fact that g is δ-close to f, it follows that

$$\Pr_{\mathbf{y} \in Z_2^m} \{g(\mathbf{y}) \neq f(\mathbf{y})\} \leq \delta$$

and

$$\Pr_{\mathbf{y} \in Z_2^m} \{g(\mathbf{x}+\mathbf{y}) \neq f(\mathbf{x}+\mathbf{y})\} \leq \delta.$$

Since f is linear, then $f(\mathbf{x}) = f(\mathbf{x}+\mathbf{y}) - f(\mathbf{y})$, and the theorem is thus proved.

QED

Notice that the bound on the error probability stated by the above theorem, that is, 2δ, is larger than the bound which can be obtained by simply evaluating function g on input $\mathbf{x}$: indeed, from the fact that g is δ-close to f, it

Chapter 7

THE PCP THEOREM

follows that the error probability, in this case, is bounded by δ. However, Program 7.2 introduces some additional randomization due to the choice of vector **y**: this additional randomization will turn out to be useful in order to make accesses to the values of f independent from each other.

7.1.2 Arithmetization

Let φ be a Boolean formula in conjunctive normal form with exactly three literals per clause and let us associate with φ a polynomial p_φ over Z_2 inductively defined as follows:

1. The polynomial associated with a variable u is $p_u = 1 - x_u$, while the polynomial associated with the negation of a variable u is $p_{\bar{u}} = x_u$.

2. The polynomial associated with a clause $C = l_1 \vee l_2 \vee l_3$ is $p_C = p_{l_1} p_{l_2} p_{l_3}$.

3. The polynomial associated with a formula $\varphi = C_1 \wedge \ldots \wedge C_m$ is $p_\varphi = \sum_{i=1}^{m} p_{C_i}$.

The reason we associate the multiplication and the addition with the disjunction and the conjunction, respectively, is that, in this way, we obtain a polynomial of total degree 3: as we will see in Theorem 7.5, this is important for limiting the number of queries made by the verifier.

It is easy to see that, for any satisfying truth assignment $a_1, \ldots, a_n$ to the n variables of φ, $p_\varphi(a_1, \ldots, a_n) = 0$ where, for the sake of simplicity, we are identifying the Boolean value TRUE with 1 and the Boolean value FALSE with 0. However, p_φ does not satisfy the property that if φ is not satisfied by $a_1, \ldots, a_n$, then $p_\varphi(a_1, \ldots, a_n) = 1$. Indeed, the value of p_φ will depend on the parity of the number of clauses that are not satisfied.

Example 7.1 ▶ Let $\varphi = C_1 \wedge C_2 \wedge C_3$ where $C_1 = u_1 \vee \bar{u_3} \vee \bar{u_4}$, $C_2 = \bar{u_1} \vee u_2 \vee \bar{u_4}$, and $C_3 = u_2 \vee u_3 \vee \bar{u_4}$. The polynomial $p_\varphi(x_1, x_2, x_3, x_4)$ associated with this formula is (recall that, for any $x \in Z_2$, $x + x = 0$)

$$x_4 + x_1 x_4 + x_2 x_4 + x_1 x_2 x_4 + x_1 x_3 x_4 + x_2 x_3 x_4.$$

The truth assignment which assigns the value FALSE to all variables but x_4 does not satisfy C_3: indeed, the value of $p_\varphi(0,0,0,1)$ is 1. However, the truth assignment which assigns the value TRUE to x_1 and x_4, and the value FALSE to the other two variables satisfies neither the second nor the third clause and the value of $p_\varphi(1,0,0,1)$ is 0.

r	$p_\varphi^{\mathbf{r}}$	(1,0,0,1)
000	0	0
001	$x_4 + x_2x_4 + x_3x_4 + x_2x_3x_4$	1
010	$x_1x_4 + x_1x_2x_4$	1
011	$x_4 + x_1x_4 + x_2x_4 + x_3x_4 + x_1x_2x_4 + x_2x_3x_4$	0
100	$x_3x_4 + x_1x_3x_4$	0
101	$x_4 + x_2x_4 + x_1x_3x_4 + x_2x_3x_4$	1
110	$x_1x_4 + x_3x_4 + x_1x_2x_4 + x_1x_3x_4$	1
111	$x_4 + x_1x_4 + x_2x_4 + x_1x_2x_4 + x_1x_3x_4 + x_2x_3x_4$	0

Table 7.1 The sequence of polynomials associated with a Boolean formula

To deal with this problem, we will now introduce a sequence of polynomials such that if $a_1,\ldots,a_n$ does not satisfy φ, then the value of half of them on input $a_1,\ldots,a_n$ is 1 (hence, randomly choosing one polynomial in the sequence yields a polynomial evaluating to 1, with probability 1/2). The definition of this sequence is based on the following lemma, whose simple proof is left to the reader (see Exercise 7.2).

Let $\mathbf{v} \in Z_2^n$ be such that $\mathbf{v} \neq \mathbf{0}$. Then ◀ Lemma 7.4

$$\Pr_{\mathbf{r} \in Z_2^n} \left\{ \sum_{i=1}^{n} r_i v_i = 1 \right\} = \frac{1}{2}.$$

The sequence of polynomials associated with the Boolean formula $\varphi = C_1 \wedge \ldots \wedge C_m$ is defined as follows: for any $\mathbf{r} \in Z_2^m$, $p_\varphi^{\mathbf{r}} = \sum_{i=1}^{m} r_i p_{C_i}$. Clearly, if the truth assignment $a_1,\ldots,a_n$ satisfies φ, then $p_\varphi^{\mathbf{r}}(a_1,\ldots,a_n) = 0$ for any $\mathbf{r}$. Otherwise, the sequence

$$(p_{C_1}(a_1,\ldots,a_n),\ldots,p_{C_m}(a_1,\ldots,a_n))$$

is not identically zero and, from Lemma 7.4, it follows that

$$\Pr_{\mathbf{r} \in Z_2^m} \left\{ p_\varphi^{\mathbf{r}}(a_1,\ldots,a_n) = 1 \right\} = \frac{1}{2}.$$

Let us consider again the formula of Example 7.1, that is, $\varphi = C_1 \wedge C_2 \wedge C_3$ where ◀ Example 7.2
$C_1 = u_1 \vee \overline{u_3} \vee \overline{u_4}$, $C_2 = \overline{u_1} \vee u_2 \vee \overline{u_4}$, and $C_3 = u_2 \vee u_3 \vee \overline{u_4}$. The sequence of polynomials associated with this formula is shown in Table 7.1. The last column of the table shows the value of these polynomials corresponding to the truth assignment which assigns the value TRUE to x_1 and x_4, and the value FALSE to the other two variables. This assignment does not satisfy φ and, indeed, the value of half of the polynomials in the sequence is 1.

Chapter 7
THE PCP THEOREM

According to the above arithmetization, we can now check that a truth assignment $a_1, \ldots, a_n$ satisfies φ by randomly choosing $\mathbf{r} \in Z_2^m$ and by checking that the corresponding polynomial $p_\varphi^{\mathbf{r}}$ evaluates to 0 on input $a_1, \ldots, a_n$. Apparently, this reduction does not simplify the problem, since we still have to know the value of each variable. The following result, however, states that the evaluation of *any* polynomial of total degree 3 with input $a_1, \ldots, a_n$ can be performed by evaluating three suitable linear functions, whose definition depends only on $a_1, \ldots, a_n$, on three input sequences, which depend only on the polynomial.

Theorem 7.5 ▶ Let $\mathbf{a} = (a_1, \ldots, a_n) \in Z_2^n$. There exist three linear functions $A_{\mathbf{a}} : Z_2^n \mapsto Z_2$, $B_{\mathbf{a}} : Z_2^{n^2} \mapsto Z_2$, and $C_{\mathbf{a}} : Z_2^{n^3} \mapsto Z_2$ such that, for any 3-degree polynomial p over Z_2 with n variables, the following equality holds:

$$p(a_1, \ldots, a_n) = \alpha_p + A_{\mathbf{a}}(\mathbf{q}_{p,1}) + B_{\mathbf{a}}(\mathbf{q}_{p,2}) + C_{\mathbf{a}}(\mathbf{q}_{p,3})$$

where $\alpha_p \in Z_2$, $\mathbf{q}_{p,1} \in Z_2^n$, $\mathbf{q}_{p,2} \in Z_2^{n^2}$, and $\mathbf{q}_{p,3} \in Z_2^{n^3}$ depend only on p and can be computed in polynomial time with respect to n.

PROOF Let p be a 3-degree polynomial over Z_2 with n variables $x_1, \ldots, x_n$: without loss of generality, we may assume that the polynomial is in normal form like those shown in Table 7.1. Hence, p is the sum of a constant term α_p with $\alpha_p \in Z_2$, a set of monomials, a set of binomials, and a set of trinomials. Let $I_{p,1}$ be the set of variable indices that appear in a monomial, $I_{p,2}$ the set of pairs of indices in a binomial, and $I_{p,3}$ the set of triples of indices in a trinomial. Then, p can be written as

$$p(x_1, \ldots, x_n) = \alpha_p + \sum_{i \in I_{p,1}} x_i + \sum_{(i,j) \in I_{p,2}} x_i x_j + \sum_{(i,j,k) \in I_{p,3}} x_i x_j x_k$$

and, clearly, the following equality holds:

$$p(a_1, \ldots, a_n) = \alpha_p + \sum_{i \in I_{p,1}} a_i + \sum_{(i,j) \in I_{p,2}} a_i a_j + \sum_{(i,j,k) \in I_{p,3}} a_i a_j a_k.$$

We then define the three linear functions (whose coefficients depend only on $\mathbf{a}$) as follows:

1. $A_{\mathbf{a}} : Z_2^n \mapsto Z_2$ is defined as

$$A_{\mathbf{a}}(x_1, \ldots, x_n) = \sum_{i=1}^{n} a_i x_i.$$

2. $B_\mathbf{a}: Z_2^{n^2} \mapsto Z_2$ is defined as

$$B_\mathbf{a}(y_1,\ldots,y_{n^2}) = \sum_{i=1}^n \sum_{j=1}^n a_i a_j y_{ij}$$

where, for the sake of simplicity, we have denoted with y_{ij} the $((i-1)n+j)$-th input variable.

3. $C_\mathbf{a}: Z_2^{n^3} \mapsto Z_2$ is defined as

$$C_\mathbf{a}(z_1,\ldots,z_{n^3}) = \sum_{i=1}^n \sum_{j=1}^n \sum_{k=1}^n a_i a_j a_k z_{ijk}$$

where, for the sake of simplicity, we have denoted with z_{ijk} the $((i-1)n^2 + (j-1)n + k)$-th input variable.

Clearly, $A_\mathbf{a}$, $B_\mathbf{a}$, and $C_\mathbf{a}$ are linear functions. Moreover, from their definitions it follows that the value of p with input $\mathbf{a}$ can be computed by accessing only three values of $A_\mathbf{a}$, $B_\mathbf{a}$, and $C_\mathbf{a}$. Indeed,

$$p(a_1,\ldots,a_n) = \alpha_p + A_\mathbf{a}(\mathbf{q}_{p,1}) + B_\mathbf{a}(\mathbf{q}_{p,2}) + C_\mathbf{a}(\mathbf{q}_{p,3})$$

where $\mathbf{q}_{p,i}$ is the sequence denoting the characteristic function of set $I_{p,i}$, for $i = 1,2,3$ (for example, $\mathbf{q}_{p,2}$ is the binary sequence of length n^2 whose $((i-1)n+j)$-th element is 1 if and only if $(i,j) \in I_{p,2}$). It is easy to verify that, given p, α_p, $\mathbf{q}_{p,1}$, $\mathbf{q}_{p,2}$, and $\mathbf{q}_{p,3}$ can be computed in time polynomial with respect to n. The theorem is hence proved. QED

◀ Example 7.3

Let us consider the following polynomial:

$$p(x_1,x_2,x_3,x_4) = x_4 + x_1 x_4 + x_2 x_4 + x_1 x_2 x_4 + x_1 x_3 x_4 + x_2 x_3 x_4.$$

In this case, we have that $\alpha_p = 0$, $I_{p,1} = \{4\}$, $I_{p,2} = \{(1,4),(2,4)\}$, and $I_{p,3} = \{(1,2,4),(1,3,4),(2,3,4)\}$. Given the sequence $\mathbf{a} = (1,0,0,1)$, we have that:

1. $\mathbf{q}_{p,1} = (0,0,0,1)$ and $A_\mathbf{a}(\mathbf{q}_{p,1}) = a_4 = 1$.
2. $\mathbf{q}_{p,2} = (0,0,0,1,0,0,0,1,0,\ldots,0)$ and $B_\mathbf{a}(\mathbf{q}_{p,2}) = a_1 a_4 + a_2 a_4 = 1$.
3. $C_\mathbf{a}(\mathbf{q}_{p,3}) = a_1 a_2 a_4 + a_1 a_3 a_4 + a_2 a_3 a_4 = 0$.

Hence,
$$p(1,0,0,1) = 0+1+1+0 = 0$$
(according to what we have previously seen in Example 7.1).

As we will see in the next section, the three functions $A_\mathbf{a}$, $B_\mathbf{a}$, and $C_\mathbf{a}$ will form the encoding of the truth assignment $\mathbf{a}$ which is assumed to be a membership proof for E3-SATISFIABILITY.

Chapter 7

THE PCP THEOREM

7.1.3 The first PCP result

Let us now turn the above ideas into a precise description of a verifier for E3-SATISFIABILITY which uses $O(n^3)$ random bits and makes a constant number of queries to the proof. The verifier interprets every proof as the concatenation of three strings A', B', and C' (see Fig. 7.2) where A' (respectively, B' and C') has length 2^n (respectively, 2^{n^2} and 2^{n^3}) and is considered as a table description of a function from Z_2^n (respectively, $Z_2^{n^2}$ and $Z_2^{n^3}$) to Z_2. Ideally, A', B', and C' correspond to the three functions $A_\mathbf{a}$, $B_\mathbf{a}$, and $C_\mathbf{a}$ defined in the previous section with respect to an assignment $\mathbf{a}$, which is supposed to satisfy the Boolean formula given in input (see Fig. 7.1).

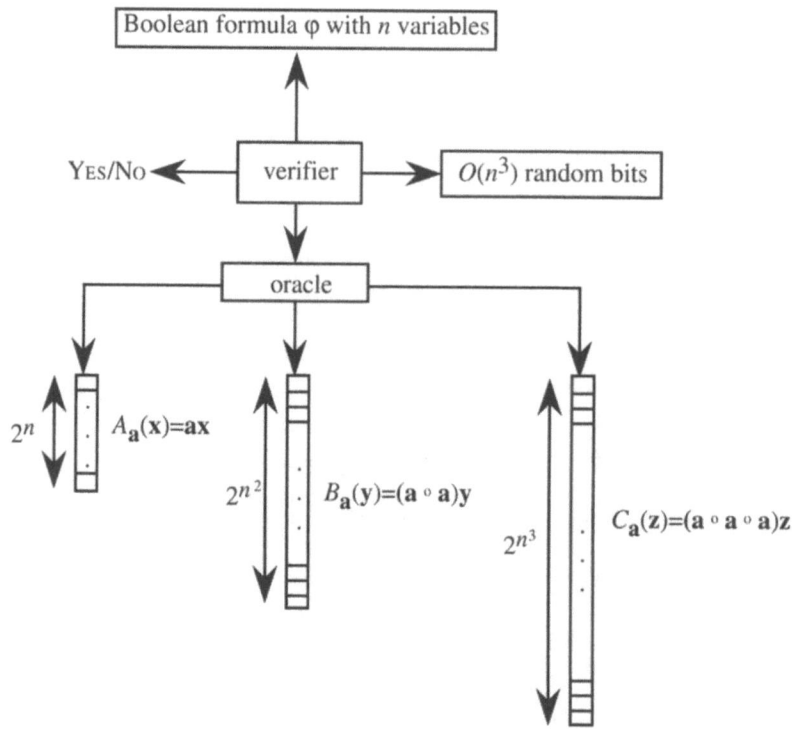

Figure 7.1
The proof of
NP $\subseteq$ PCP$[O(n^3), O(1)]$: the ideal situation

The verifier has both to check that A', B', and C' are indeed the table description of three linear functions, which are defined with respect to the same truth assignment, and to verify that this assignment is a satisfying one.

Checking linearity and consistency
From Theorem 7.2, it follows that, for any $\delta < 1/3$, if A' (respectively, B' and C') is not δ-close to a linear function, then with probability at least 1/2

Program 7.1, with input δ and oracle A' (respectively, B' and C'), returns NO.

However, even though the three applications of Program 7.1 return YES (that is, with probability at least 1/8, A', B', and C' are δ-close to three linear functions $\tilde{A}$, $\tilde{B}$, $\tilde{C}$, respectively), the verifier still has to check that the coefficients of the three functions are consistent.

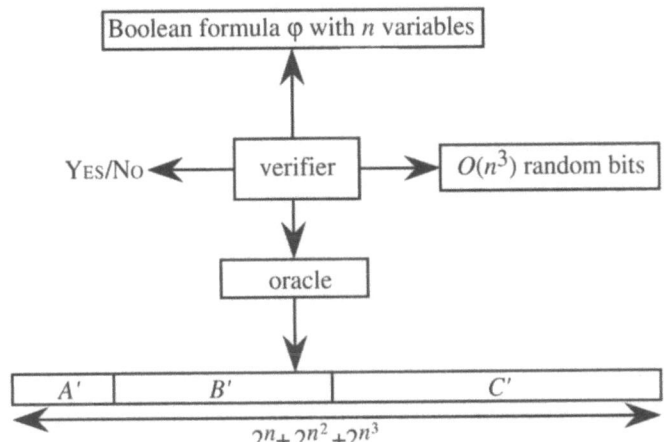

Figure 7.2
The proof of $\text{NP} \subseteq \text{PCP}[O(n^3), O(1)]$: the real situation

That is, if $\tilde{a}_1, \ldots, \tilde{a}_n$ (respectively, $\tilde{b}_1, \ldots, \tilde{b}_{n^2}$ and $\tilde{c}_1, \ldots, \tilde{c}_{n^3}$) denote the coefficients of $\tilde{A}$ (respectively, $\tilde{B}$ and $\tilde{C}$) the following equalities must be verified:

$$\tilde{b}_{(i-1)n+j} = \tilde{a}_i \tilde{a}_j,$$

and

$$\tilde{c}_{(i-1)n^2+(j-1)n+k} = \tilde{a}_i \tilde{a}_j \tilde{a}_k.$$

Observe that if, for example, we verify the first of the above two equalities by directly accessing A' and B' in correspondence of positions i, j, and $(i-1)n+j$, respectively, where i and j are randomly chosen, the three events $a'_i = \tilde{a}_i$, $a'_j = \tilde{a}_j$, and $b'_{(i-1)n+j} = \tilde{b}_{(i-1)n+j}$ are not independent from each other. For this reason, the verifier makes use of the additional randomization introduced by the self-correcting procedure previously defined.

In particular, let us consider Program 7.3 where, for any $\mathbf{x}, \mathbf{x}' \in Z_2^n$, $\mathbf{x} \circ \mathbf{x}'$ denotes the element $\mathbf{y} \in Z_2^{n^2}$ such that $y_{(i-1)n+j} = x_i x'_j$, and, for any $\mathbf{x} \in Z_2^n$ and $\mathbf{y} \in Z_2^{n^2}$, $\mathbf{x} \circ \mathbf{y}$ denotes the element $\mathbf{z} \in Z_2^{n^3}$ such that $z_{(i-1)n^2+(j-1)n+k} = x_i y_{(j-1)n+k}$. It is then possible to prove the following result (see Exercise 7.3).

◀ Lemma 7.6

Let δ < 1/24 be a constant. There exists a constant k (depending on δ) such that, if there does not exist $\tilde{\mathbf{a}} \in Z_2^n$ so that A', B', and C' are δ-close to the linear functions with coefficients specified by $\tilde{\mathbf{a}}$, $\tilde{\mathbf{a}} \circ \tilde{\mathbf{a}}$, and $\tilde{\mathbf{a}} \circ \tilde{\mathbf{a}} \circ$

Chapter 7
THE PCP THEOREM

Program 7.3: Consistency Test

oracle Functions A', B', C' δ-close to linear functions $\tilde{A}, \tilde{B}, \tilde{C}$;
output YES if $\tilde{A}, \tilde{B}, \tilde{C}$ are consistent;
begin
  Randomly pick $\mathbf{x}, \mathbf{x}' \in Z_2^n$;
  $a :=$ output of Program 7.2 with input $\mathbf{x}$ and oracle function A';
  $a' :=$ output of Program 7.2 with input $\mathbf{x}'$ and oracle function A';
  $b :=$ output of Program 7.2 with input $\mathbf{x} \circ \mathbf{x}'$ and oracle function B';
  if $aa' \neq b$ **then return** NO;
  Randomly pick $\mathbf{x} \in Z_2^n$ and $\mathbf{y} \in Z_2^{n^2}$;
  $a :=$ output of Program 7.2 with input $\mathbf{x}$ and oracle function A';
  $b :=$ output of Program 7.2 with input $\mathbf{y}$ and oracle function B';
  $c :=$ output of Program 7.2 with input $\mathbf{x} \circ \mathbf{y}$ and oracle function C';
  if $ab \neq c$ **then return** NO;
  return YES
end.

$\tilde{\mathbf{a}}$, respectively, then with probability at least $1 - \delta$ one among k calls of Programs 7.1 and 7.3 returns NO.

Checking satisfiability

It now remains to design an algorithm to verify that the coefficients $\tilde{a}_1, \ldots, \tilde{a}_n$ of the linear function δ-close to A' indeed specify a truth assignment that satisfies the original Boolean formula. This algorithm is an immediate consequence of Theorem 7.5 and is formally described in Program 7.4.

Fixed a constant $\delta < 1/24$, if A' (respectively, B' and C') is δ-close to a linear function with coefficients specified by $\tilde{\mathbf{a}}$ (respectively, $\tilde{\mathbf{a}} \circ \tilde{\mathbf{a}}$ and $\tilde{\mathbf{a}} \circ \tilde{\mathbf{a}} \circ \tilde{\mathbf{a}}$), then there exists a constant k (depending on δ) such that with probability at least $1 - \delta$ one among k calls of Program 7.4 returns NO if $\tilde{\mathbf{a}}$ is not a satisfying truth assignment (recall that if $\tilde{\mathbf{a}}$ does not satisfy φ, then, with probability $1/2$, Program 7.4 chooses a polynomial that evaluates to 1 on input $\tilde{\mathbf{a}}$.

Combining all the above results, we thus obtain that there exists a constant k such that repeating Programs 7.1, 7.3, and 7.4 k times and rejecting whenever one of these algorithms returns NO yields a verifier for E3-SATISFIABILITY which uses $O(n^3)$ random bits and makes a constant number of queries to the proof.

Indeed, each of Programs 7.1, 7.3, and 7.4 uses $O(n^3)$ random bits (note that a boolean formula with n variables has at most n^3 distinct clauses with 3 literals per clause), requires a constant number of queries, and has to be

Program 7.4: Satisfiability Test

input Boolean formula φ with n clauses;
oracle Functions A', B', C' δ-close to consistent linear functions $\tilde{A}, \tilde{B}, \tilde{C}$;
output YES if φ is satisfied by truth assignment encoded by $\tilde{A}, \tilde{B}, \tilde{C}$;
begin
    Randomly pick $\mathbf{r} \in Z_2^m$;
    Compute $\alpha_{p_\varphi^\mathbf{r}}, I_{p_\varphi^\mathbf{r},1}, I_{p_\varphi^\mathbf{r},2}$, and $I_{p_\varphi^\mathbf{r},3}$ relative to $p_\varphi^\mathbf{r}$;
    $a :=$ output of Program 7.2 with input $\mathbf{q}_{p_\varphi^\mathbf{r},1}$ and oracle function A';
    $b :=$ output of Program 7.2 with input $\mathbf{q}_{p_\varphi^\mathbf{r},2}$ and oracle function B';
    $c :=$ output of Program 7.2 with input $\mathbf{q}_{p_\varphi^\mathbf{r},3}$ and oracle function C';
    if $\alpha_{p_\varphi^\mathbf{r}} + a + b + c = 1$ **then return** NO **else return** YES
end.

Section 7.2
ALMOST TRANSPARENT SHORT PROOFS

iterated a constant number of times in order to achieve the desired error probability. In summary, we have proved the following result.

$$\text{NP} \subseteq \text{PCP}[O(n^3), O(1)].$$

◀ Theorem 7.7

7.2 Almost transparent short proofs

THE RESULT of the previous section is not only the first example of a verifier for a language in NP which uses a constant number of query bits, but it also contains the main ingredients required in the rest of the proof of the second PCP result. The major drawback of that result, however, is that the proof obtained has exponential size (see Fig. 7.2) and, therefore, requires a polynomial number of random bits (see the observation following Def. 6.9). The aim of this section is to introduce a better way of encoding a truth assignment based on polynomials whose total degree depends on the length of the assignment itself. While making use of this encoding, we will substitute the linearity test and the satisfiability test with a *low-degree test* and a *sum-check test*, respectively.

In the following, we will use the notations summarized in Table 7.2, whose last column specifies the values which will be justified during the proof of our tests (most of the algebraic results we will present, however, hold also in the general case). Moreover, we will repeatedly refer to the following definition.

For any $W \subseteq F$ and for any $w \in W$, S_w^W is the univariate polynomial of degree $|W| - 1$ defined as

◀ Definition 7.3
Selector polynomial

Chapter 7

THE PCP THEOREM

Symbol	Meaning	Value
n	number of Boolean variables	≥ 3
q	prime number	$\geq 100\lceil \log^4 n \rceil$
F	finite field $Z_q = \{0,\ldots,q-1\}$	$\lvert F \rvert = q$
H	subset $\{0,\ldots,\lvert H \rvert - 1\}$ of F	$\lvert H \rvert = \lceil \log n \rceil$
k	number of variables of a polynomial	$= \lceil \frac{\log n}{\log \log n} \rceil$
d	total degree of a polynomial	$O(k\lvert H \rvert)$
$F_{d,k}$	set of k-variate polynomials of degree d	

Table 7.2 The notations used while proving $\text{NP} \subseteq \text{PCP}[O(\log n), O(\log^4 n)]$

$$S_w^W(x) = \prod_{y \in W : y \neq w} \frac{x-y}{w-y}.$$

Clearly, $S_w^W(w) = 1$ and $S_w^W(x) = 0$ for any $x \in W$ with $x \neq w$.

7.2.1 Low-degree polynomials

The main goal of this section is to prove that polynomials over the field F whose total degree is small compared to the size of F (that is, *low-degree polynomials*) satisfy properties similar to the ones we have seen for linear functions in Sect. 7.1.1.

Definition 7.4 ▶ *A k-variate polynomial of total degree d over F is a sum of terms of the form $a x_1^{j_1} x_2^{j_2} \cdots x_k^{j_k}$ with $a \in F$ and $j_1 + \ldots + j_k \leq d$.*
k-variate polynomial

Observe that any k-variate polynomial over F (indeed, any function from F^k to F) can be represented by means of a table with q^k elements of size $\lceil \log q \rceil$.

The following result, which is a generalization of Exercise 6.21, will play a basic role in the rest of the presentation.

Lemma 7.8 ▶ *Two distinct polynomials in $F_{d,k}$ agree on at most dq^{k-1} elements of F^k.*

PROOF We will actually prove the equivalent statement that a k-variate polynomial p of total degree d over F, which is not identically zero, evaluates to 0 at no more than dq^{k-1} elements of F^k. The proof is by induction on k. The case $k=1$ is clear, since a univariate polynomial of degree d, which is not identically zero, has at most d roots.

Assume $k > 1$. Then p can be represented as

$$\sum_{i=0}^{d} x_1^i p_i(x_2, \ldots, x_k)$$

where each p_i is a polynomial in $F_{d-i,k-1}$. Since p is not identically zero, there exists p_i which is not identically zero: let m be the maximum index such that p_m is not identically zero. Let n_m be the number of elements of F^{k-1} on which p_m evaluates to zero. By the inductive hypothesis, $n_m \leq (d-m)q^{k-2}$. For any $(b_2, \ldots, b_k) \in F^{k-1}$ such that $p_m(b_2, \ldots, b_k) \neq 0$, $p(x_1, b_2, \ldots, b_k)$ is a univariate polynomial of degree m which is zero for at most m values of x_1.

Hence, the number of elements of F^k where p is zero is at most

$$\begin{aligned} qn_m + m(q^{k-1} - n_m) &= mq^{k-1} + n_m(q-m) \\ &\leq mq^{k-1} + (d-m)q^{k-2}(q-m) \\ &= dq^{k-1} - (d-m)q^{k-2}m \leq dq^{k-1} \end{aligned}$$

where the last inequality is due to the fact that $d \geq m$. The lemma is thus proved. QED

According to the values of the last column of Table 7.2, in our case we have that the fraction of elements of F^k on which two distinct polynomials in $F_{d,k}$ agree is at most

$$\frac{dq^{k-1}}{q^k} < \frac{1}{2}.$$

The above lemma thus implies that if $f : F^k \mapsto F$ is a function δ-close to a polynomial in $F_{d,k}$ with $\delta < 1/4$, then this polynomial is unique (see Fig. 7.3). In other words, polynomials are useful as encoding objects.

The way polynomials will be used as encoding objects is based on the so-called *low-degree extension* of a function which is similar to the classical Lagrange polynomial interpolation.

◀ **Theorem 7.9** *Let f be a function from H^t to $\{0,1\}$. There is a unique polynomial $p_f \in F_{t|H|,t}$ such that $p_f(\mathbf{y}) = f(\mathbf{y})$ for all $\mathbf{y} \in H^t$.*

PROOF For any $\mathbf{h} \in H^t$, we define $S_\mathbf{h}(\mathbf{x}) = \prod_{i=1}^{t} S_{h_i}^H(x_i)$. Then, $S_\mathbf{h}(\mathbf{h}) = 1$ and $S_\mathbf{h}$ evaluates to zero for all other elements of H^t. Moreover, its degree is at most $t|H|$. Finally, for any $\mathbf{x} \in F^t$, we define

$$p_f(\mathbf{x}) = \sum_{\mathbf{h} \in H^t} S_\mathbf{h}(\mathbf{x}) f(\mathbf{h}).$$

The proof of the correctness of p_f and of its uniqueness is left to the reader (see Exercise 7.4). QED

Chapter 7

THE PCP THEOREM

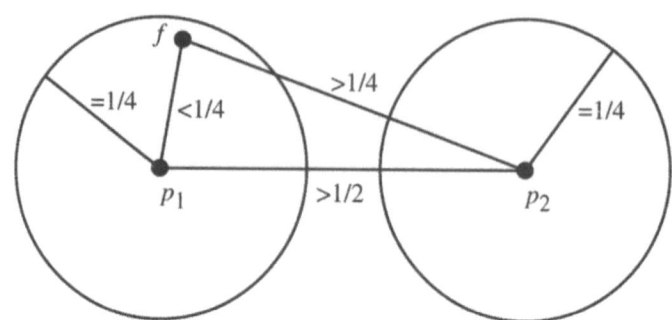

Figure 7.3
The polynomial code

Example 7.4 ▶ Let $t = 2$ and $H = \{0,1\}$. In this case, $S_0^H(x) = 1 - x$ and $S_1^H(x) = x$. Moreover, $S_{(0,0)}(x,y) = (1-x)(1-y)$, $S_{(0,1)}(x,y) = (1-x)y$, $S_{(1,0)}(x,y) = x(1-y)$, and $S_{(1,1)}(x,y) = xy$. Thus, the polynomial extension p_f of any function $f : \{0,1\}^2 \mapsto \{0,1\}$ is defined as

$$p_f(x,y) = (1-x)(1-y)f(0,0) + (1-x)yf(0,1) + x(1-y)f(1,0) + xyf(1,1).$$

To see the usefulness of the low-degree extension of a function, let us consider a truth assignment **a**, that is, a sequence of n bits (once again, we are identifying the value TRUE and FALSE with 1 and 0, respectively). According to the bounds shown in Table 7.2, we have that $|H^k| \geq n$ so that we can interpret **a** as a function from H^k to $\{0,1\}$: the membership proof will then consist of the low-degree extension of this function.

Note that we are encoding sequences of n bits by sequences of q^k words of length $\log q$: according to the bounds of Table 7.2, this increase in size is polynomially bounded with respect to n.

Example 7.5 ▶ Let $\varphi = C_1 \wedge C_2 \wedge C_3$ where $C_1 = u_1 \vee \overline{u_3} \vee \overline{u_4}$, $C_2 = \overline{u_1} \vee u_2 \vee \overline{u_4}$, and $C_3 = u_2 \vee u_3 \vee \overline{u_4}$. In this case, $n = 4$ and $k = |H| = 2$. According to the previous example, the polynomial extension of a truth assignment **a** is given by

$$p_{\mathbf{a}}(x,y) = (1-x)(1-y)a_1 + (1-x)ya_2 + x(1-y)a_3 + xya_4.$$

For example, the polynomial extension of the satisfying truth assignment (0,1,1,0) is $x + y - xy$, while the polynomial extension of the non-satisfying truth assignment (1,0,0,1) is $1 - x - y + 2xy$.

In order to verify that the membership proof is correct, we will thus need to check that it is indeed a *low-degree* polynomial encoding a *satisfying* truth assignment.

Section 7.2
ALMOST TRANSPARENT SHORT PROOFS

The low-degree test
The analogue of the linearity test presented in the proof of our first PCP result is the so-called low-degree test, that is, an algorithm that is able to check whether a given function g is indeed a polynomial of degree d. To construct such an algorithm we need to add extra information.

◀ Definition 7.5
Line

For any $\mathbf{b}, \mathbf{s} \in F^k$, *the set of q elements of F^k defined as*

$$l_{\mathbf{b},\mathbf{s}} = \{\mathbf{b} + \mathbf{s}t \mid t \in F\}$$

is said to be a line *in* F^k.

The low-degree test will require an extra table T which is supposed to describe the restrictions of g on all lines in F^k. The use of such table T is justified by the following result.

◀ Lemma 7.10

A function $g : F^k \mapsto F$ *is in* $F_{d,k}$ *if and only if, for any* $\mathbf{b}, \mathbf{s} \in F^k$, *the function*

$$g_{\mathbf{b},\mathbf{s}}(t) = g(\mathbf{b} + \mathbf{s}t)$$

belongs to $F_{d,1}$ ($g_{\mathbf{b},\mathbf{s}}$ *is also said to be the* restriction *of g to $l_{\mathbf{b},\mathbf{s}}$*).

PROOF

The necessity follows from the fact that the restriction of a multivariate polynomial of total degree d over F^k to any line in F^k is a univariate polynomial of degree d over F.

We will now prove the sufficiency by induction on k. The case $k = 1$ is trivial. Assume $k > 1$ and fix a set C of $d+1$ distinct elements of F, say $C = \{c_0, c_1, \ldots, c_d\}$. By the inductive hypothesis, for $i = 0, \ldots, d$, we have that

$$g_i(x_2, \ldots, x_k) = g(c_i, x_2, \ldots, x_k)$$

belongs to $F_{d,k-1}$: indeed, for any $\mathbf{b}, \mathbf{s} \in F^{k-1}$, $g_i(\mathbf{b} + \mathbf{s}t)$ is equal to $g(\mathbf{b}' + \mathbf{s}'t)$, where $\mathbf{b}' = (c_i, b_1, \ldots, b_{k-1})$ and $\mathbf{s}' = (0, s_1, \ldots, s_{k-1})$, and thus belongs to $F_{d,1}$.

Let us define

$$p(x_1, x_2, \ldots, x_k) = \sum_{i=0}^{d} S_{c_i}^C(x_1) g_i(x_2, \ldots, x_k).$$

We will now prove that g and p coincide on F^k and that the total degree of p is at most d. The lemma will hence follow.

In order to prove that g and p coincide on F^k, observe that, for any $\mathbf{u} \in F^{k-1}$, both $g(x, \mathbf{u})$ and $p(x, \mathbf{u})$ belong to $F_{d,1}$. In fact, $g(x, \mathbf{u})$ is equal

Chapter 7
THE PCP THEOREM

to the restriction of g to the line $l_{\mathbf{b},\mathbf{s}}$, where $\mathbf{b} = (0, u_1, \ldots, u_{k-1})$ and $\mathbf{s} = (1, 0, \ldots, 0)$ and, hence, belongs to $F_{d,1}$. On the other hand,

$$p(x_1, \mathbf{u}) = \sum_{i=0}^{d} S_{c_i}^C(x_1) g_i(\mathbf{u}),$$

which is clearly a univariate polynomial of degree d over F.

Moreover, $g(c_i, \mathbf{u}) = p(c_i, \mathbf{u})$ for any $\mathbf{u} \in F^{k-1}$ and $i = 0, 1, \ldots, d$. Hence, for any $\mathbf{u} \in F^{k-1}$, $g(x, \mathbf{u})$ and $p(x, \mathbf{u})$ coincide on F: this implies that g and p coincide on F^k.

We now prove that $p \in F_{d,k}$. Clearly, $p \in F_{2d,k}$: hence, p can be represented as

$$\sum_{\mathbf{i} \in I_{k,2d}} \alpha_{\mathbf{i}} x_1^{i_1} \cdots x_k^{i_k},$$

where $I_{k,2d}$ denotes the set of vectors of k non-negative integers whose sum is at most $2d$. Moreover, for any $\mathbf{s} \in F^k$, the restriction of p to $l_{\mathbf{0},\mathbf{s}}$, that is,

$$p_{\mathbf{0},\mathbf{s}}(t) = p(\mathbf{0} + \mathbf{s}t) = p(s_1 t, \ldots, s_k t)$$

can be represented as

$$\sum_{j=0}^{2d} t^j \sum_{\mathbf{i} \in I_{k,2d}: i_1 + \ldots + i_k = j} \alpha_{\mathbf{i}} s_1^{i_1} \cdots s_k^{i_k}.$$

Since g and p coincide and because of the hypothesis of the lemma, this restriction belongs to $F_{d,1}$, for any $\mathbf{s} \in F^k$. That is,

$$\sum_{\mathbf{i} \in I_{k,2d}: i_1 + \ldots + i_k = j} \alpha_{\mathbf{i}} s_1^{i_1} \cdots s_k^{i_k}$$

must be zero for any $j > d$ and for any $\mathbf{s} \in F^k$. This in turn implies that $\alpha_{\mathbf{i}} = 0$ for all $\mathbf{i} \in I_{k,2d}$ such that $i_1 + \ldots + i_k > d$: hence, p has no term of total degree greater than d.

QED

Example 7.6 ▶ Assume $F = \{0, 1\}$ and let us consider the bivariate function g such that $g(0,0) = 0$, $g(0,1) = 1$, $g(1,0) = 1$, and $g(1,1) = 0$. It is easy (even though tedious) to check that, for any line $l_{\mathbf{b},\mathbf{s}}$ in F^2, the restriction of g to $l_{\mathbf{b},\mathbf{s}}$ belongs to $F_{1,1}$. For example, fix $\mathbf{b} = (0,1)$ and $\mathbf{s} = (1,0)$. Then $g_{\mathbf{b},\mathbf{s}}(0) = g(0,1) = 1$ and $g_{\mathbf{b},\mathbf{s}}(1) = g(1,1) = 0$: this function coincides with the univariate linear function $1 + t$.

According to the proof of the previous lemma, let $C = \{0, 1\}$. We then have that $g_0(x_2) = g(0, x_2) = x_2$ and $g_1(x_2) = g(1, x_2) = 1 + x_2$. Hence

$$\begin{aligned} p(x_1, x_2) &= S_0^C(x_1) g_0(x_2) + S_1^C(x_1) g_1(x_2) \\ &= (1 + x_1) x_2 + x_1 (1 + x_2) = x_1 + x_2. \end{aligned}$$

That is, function g is a bivariate polynomial of degree 1.

x_1	x_2	$g(x_1,x_2)$
0	0	0
0	1	1
0	2	2
1	0	0
1	1	1
1	2	2
2	0	0
2	1	1
2	2	0

Table 7.3 An example of function $g : \{0,1,2\}^2 \mapsto \{0,1,2\}$

Section 7.2

ALMOST TRANSPARENT SHORT PROOFS

The low-degree test is based on a probabilistic version of Lemma 7.10, which states that if on most lines, most values of a function f are described by a univariate polynomial of degree d, then f is close to a k-variate polynomial of degree d. To formally state this property we need to introduce some concepts.

Let $g : F^k \mapsto F$ be a function and let $l = l_{\mathbf{b},\mathbf{s}}$ be a line. The degree d line polynomial for g on l is the univariate polynomial P_l^g of degree d which maximizes the number of elements $t \in F$ such that $P_l^g(t) = g(\mathbf{b}+\mathbf{s}t)$.

◀ Definition 7.6
Line polynomial

Note that, in the above definition, several univariate polynomials of degree d may exist that describe g equally well on l: in this case, we arbitrarily select one of them as the line polynomial.

Assume $F = \{0,1,2\}$ and let us consider the bivariate function g described by Table 7.3. Fix $\mathbf{b} = (0,1)$ and $\mathbf{s} = (1,2)$. Then $g(\mathbf{b}+\mathbf{s}0) = g(0,1) = 1$, $g(\mathbf{b}+\mathbf{s}1) = g(1,0) = 0$, and $g(\mathbf{b}+\mathbf{s}2) = g(2,2) = 0$. In this case there are two polynomials of degree 1 that describe g equally well on $l = l_{\mathbf{b},\mathbf{s}}$, that is, $p_1(t) = 1+t$ and $p_2(t) = 0$.

◀ Example 7.7

Indeed, both p_1 and p_2 agree with the restriction of g to l on two among three possible elements. We can arbitrarily set P_l^g equal to p_1.

Note also that a line has more than one parametric representation of the form given in Def. 7.5: for example, for any $\mathbf{b}, \mathbf{s} \in F^k$ and for any $c \in F - \{0\}$, the two lines $l = l_{\mathbf{b},\mathbf{s}}$ and $l = l_{\mathbf{b},c\mathbf{s}}$ coincide. Clearly, the selection of the corresponding line polynomial (among several candidates) must be done consistently.

Let $g : F^k \mapsto F$ be a function and $l = l_{\mathbf{b},\mathbf{s}}$ be a line. The success rate of g on l is defined as the fraction of elements $t \in F$ such that $P_l^g(t) = g(\mathbf{b}+\mathbf{s}t)$. The success rate of g is the average success rate of g among all lines.

◀ Definition 7.7
Success rate

Chapter 7

THE PCP THEOREM

The following theorem, whose technical and long proof is here omitted (see Bibliographical notes), is the precise statement of the probabilistic version of Lemma 7.10 previously mentioned, and is the analogue of Lemma 7.1.

Theorem 7.11 ▶ *Let δ be a constant smaller than 10^{-4}. Every function $g : F^k \mapsto F$ whose success rate is at least $1 - \delta/2$ is δ-close to a polynomial in $F_{d,k}$.*

The low-degree test (see Program 7.5) picks lines uniformly at random from all lines in F^k and checks how well the restriction of the enquired function is described by a univariate polynomial. The coefficients of this polynomial are given by a *line-table* T, which is part of the proof and has one row for each line: more formally, T is a function from F^{2k} to F^{d+1}, which, for any $\mathbf{b}$ and $\mathbf{s}$ in F^k, specifies the $d+1$ coefficients in F of a polynomial. In Program 7.5 $P_{\mathbf{b},\mathbf{s}}$ denotes this polynomial, that is, the univariate polynomial of degree d defined as $P_{\mathbf{b},\mathbf{s}}(t) = \sum_{i=1}^{d+1} T(\mathbf{b},\mathbf{s})_i t^{i-1}$ where, for $i = 1, \ldots, d+1$, $T(\mathbf{b},\mathbf{s})_i$ denotes the i-th component of the $(d+1)$-dimensional vector $T(\mathbf{b},\mathbf{s}) \in F^{d+1}$.

The low-degree test performs several repetitions of the following trial. It picks a line uniformly at random (by choosing the two parameters $\mathbf{b}, \mathbf{s} \in F^k$) and a point t uniformly at random on this line. It then reads the coefficients of the univariate polynomial provided for $l = l_{\mathbf{b},\mathbf{s}}$ in the line table T and, finally, checks whether this polynomial correctly describes the restriction of the function g to l at the chosen point t. Using Theorem 7.11 we can now prove that Program 7.5 satisfies the desired properties (the following result is the analogue of Theorem 7.2).

Theorem 7.12 ▶ *Let $\delta < 10^{-4}$ be a constant. For any function $g : F^k \mapsto F$, the following claims hold:*

1. *If g is a polynomial in $F_{d,k}$ then there exists a line-table T such that Program 7.5 with oracle g and T always accepts.*

2. *If g is not δ-close to any polynomial in $F_{d,k}$, then, for any line-table T, Program 7.5 with oracle g and T, rejects with probability at least 3/4.*

3. *Program 7.5 uses $O(k \log q)$ random bits and queries a constant number of values of g and a constant number of values of T.*

PROOF The first point is an immediate consequence of Lemma 7.10 while the third point can be easily checked by the reader. Assume now that g is not δ-close to any polynomial in $F_{d,k}$ and let p_T be the probability that the test within

Section 7.2
ALMOST TRANSPARENT SHORT PROOFS

Program 7.5: Low-Degree Test

oracle k-variate function g and line-table T;
output YES if g is δ-close to a polynomial in $F_{d,k}$;
begin
   repeat $\lceil 3/\delta \rceil$ times
   begin
      Randomly choose $\mathbf{b}, \mathbf{s} \in F^k$ and $t \in F$;
      if $P_{\mathbf{b},\mathbf{s}}(t) \neq g(\mathbf{b} + \mathbf{s}t)$ **then return** NO
   end;
   return YES
end.

the **repeat** cycle does not succeed. By definition, the success rate of g is the maximum over all line-tables T of p_T. Hence, from Theorem 7.11 we have that $p_T < 1 - \delta/2$ for every line-table. Since Program 7.5 accepts only if all tests do not succeed, after $\lceil 3/\delta \rceil$ tests, the algorithm will reject with probability at least 3/4. Thus the theorem is proved. QED

The correcting procedure

As in the case of linear functions, another important property of low-degree polynomials is that they can be easily corrected (recall that, according to the output of Program 7.5, we are not querying a low-degree polynomial but a function which is δ-close to a low-degree polynomial).

Given a function g which is δ-close to a polynomial $f \in F_{d,k}$ and an element $\mathbf{x} \in F^k$, the correcting procedure (see Program 7.6), recovers (with high probability) the correct value $f(\mathbf{x})$ (observe that the program may also return a NO answer, which implies that the verifier will reject).

The procedure uses the line-table T introduced in the description of the low-degree test, containing, for each line in F^k, a polynomial in $F_{d,1}$ that best describes f on the line. The algorithm picks a line uniformly at random from all lines in $L_\mathbf{x} = \{l_{\mathbf{x},\mathbf{s}} \mid \mathbf{s} \in F^k\}$, reads the coefficients of the corresponding polynomial in the line-table, checks whether the polynomial agrees with g on a random point on the line, and, finally, returns the values produced by this polynomial at $\mathbf{x}$.

The next theorem (which is the analogue of Theorem 7.3) shows that the probability that Program 7.6 computes a wrong value is bounded by a sufficiently small quantity.

◀ Theorem 7.13

Given a function $g : F^k \mapsto F$, which is δ-close to a polynomial $f \in F_{d,k}$ with $\delta < 10^{-4}$ and, given $\mathbf{x} \in F^k$, Program 7.6, with input $\mathbf{x}$ and oracle g and a

Chapter 7

THE PCP THEOREM

Program 7.6: Correcting Low-Degree Polynomial

input $\mathbf{x} \in F^k$;
oracle Function g which is δ-close to $f \in F_{d,k}$ and line-table T;
output $f(\mathbf{x})$;
begin
   Randomly choose in $\mathbf{s} \in F^k$;
   Randomly choose $t \in F$;
   if $P_{\mathbf{x},\mathbf{s}}(t) \neq g(\mathbf{x}+s\mathbf{t})$ **then**
      return NO
   else
      return $P_{\mathbf{x},\mathbf{s}}(0)$
end.

line-table T, returns the value $f(\mathbf{x})$ with probability at least $1-2\sqrt{\delta}-d/q$ (no matter what the line-table contains).

PROOF   For any line l in $L_{\mathbf{x}}$, recall that P_l^f denotes the univariate polynomial of degree d that best describes f on l. The only case in which Program 7.6 returns a wrong value occurs when the polynomial $P_{\mathbf{x},\mathbf{s}}$ in T corresponding to the randomly chosen line $l = l_{\mathbf{x},\mathbf{s}}$ is different from P_l^f. We now show that, in this case, $P_{\mathbf{x},\mathbf{s}}$ does not agree with g at most elements of l. Hence, Program 7.6 returns NO with high probability.

To prove that $P_{\mathbf{x},\mathbf{s}}$ does not agree with g at most elements of l, we use the fact that, for at least $(1-\sqrt{\delta})q^k$ lines in $L_{\mathbf{x}}$, P_l^f agrees with g at $(1-\sqrt{\delta})q$ elements of l (see Exercise 7.5). Let L_{good} be the set of such lines and let $l \in L_{\text{good}}$. Since two distinct polynomials in $F_{d,1}$ agree at no more than d points in F, then every polynomial in $F_{d,1}$ distinct from P_l^f agrees with g at no more than $d + (1-(1-\sqrt{\delta}))q = d+\sqrt{\delta}q$ elements of l. In particular, this is true for $P_{\mathbf{x},\mathbf{s}}$ (see Fig. 7.4, where the gray region denotes elements of l at which P_l^f and g agree, the dotted region denotes elements of l at which P_l^f, $P_{\mathbf{x},\mathbf{s}}$, and g agree, and, finally, the white region denotes elements of l at which $P_{\mathbf{x},\mathbf{s}}$ and g agree).

Hence, the probability that Program 7.6 returns a wrong value is bounded by the probability that $l_{\mathbf{x},\mathbf{s}}$ does not belong to L_{good} plus the probability that t is an element at which $P_{\mathbf{x},\mathbf{s}}$ and g agree. This probability is at most

$$1-(1-\sqrt{\delta})+\sqrt{\delta}+d/q = 2\sqrt{\delta}+d/q$$

QED   and the theorem is proved.

As in the case of linear functions, the bound on the error probability stated by the above theorem is larger than the bound immediately implied by the δ-closeness of *g* and *f*: however, Program 7.6 introduces some additional randomization that makes different accesses to the values of *f* independent from each other.

Section 7.2
ALMOST TRANSPARENT SHORT PROOFS

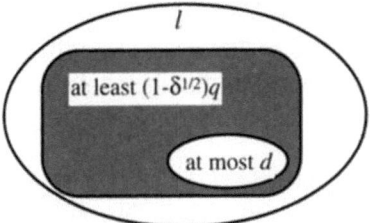

Figure 7.4
The proof of Theorem 7.13

7.2.2 Arithmetization (revisited)

In Sect. 7.1.2 we have shown how any Boolean formula φ on variables $u_1, \ldots, u_n$ can be represented by a polynomial p_φ of degree 3 such that, for any truth assignment **a** to the *n* variables of φ, **a** satisfies φ if and only if each term of p_φ evaluates to 0 with input **a**. Moreover, each term of p_φ (which corresponds to a clause of φ) can only be one of the following four polynomials (depending on the number of negated variables in the clause):

$$p_0(x,y,z) = (1-x)(1-y)(1-z) \quad \text{(no negated variables)},$$

$$p_1(x,y,z) = x(1-y)(1-z) \quad \text{(1 negated variable)},$$

$$p_2(x,y,z) = xy(1-z) \quad \text{(2 negated variables)},$$

and

$$p_3(x,y,z) = xyz \quad \text{(3 negated variables)}.$$

Observe that we may always assume that the negated variables appear at the beginning of the clause: we will then say that a clause is of type *j* if the first *j* variables are negated. We also assume that the negated (respectively, positive) variables in a clause appear in increasing order with respect to their indices (for example, the clause $u_3 \vee u_1 \vee \bar{u}_2$ is written as $\bar{u}_2 \vee u_1 \vee u_3$).

Let χ_φ^j, for $j = 0, 1, 2, 3$, be four *clause-characteristic* Boolean functions such that, for any triple of variable indices i_1, i_2, and i_3, $\chi_\varphi^j(i_1, i_2, i_3) = 1$ if and only if φ contains a clause of type *j* on the variables $u_{i_1}, u_{i_2}, u_{i_3}$.

Chapter 7
THE PCP THEOREM

We then have that a truth assignment **a** satisfies φ if and only if, for any $j = 0, 1, 2, 3$ and for any $i_1, i_2, i_3 \in \{1, \ldots, n\}$,

$$f_\varphi^j(i_1, i_2, i_3) = \chi_\varphi^j(i_1, i_2, i_3) p_j(a_{i_1}, a_{i_2}, a_{i_3}) = 0.$$

Thus, the problem of verifying that **a** is a satisfying truth assignment for φ is reduced to the task of checking that the above four functions are identically zero.

Example 7.8 ▶ Let us consider again the formula $\varphi = C_1 \wedge C_2 \wedge C_3$ where $C_1 = \overline{u_3} \vee \overline{u_4} \vee u_1$, $C_2 = \overline{u_1} \vee \overline{u_4} \vee u_2$, and $C_3 = \overline{u_4} \vee u_2 \vee u_3$. In this case we have no clauses of type 0 and 3, one clause of type 1, and two clauses of type 2. Thus the two functions to be checked are the following: $f_\varphi^1(i_1, i_2, i_3) = \chi_\varphi^1(i_1, i_2, i_3) a_{i_1}(1 - a_{i_2})(1 - a_{i_3})$, and $f_\varphi^2(i_1, i_2, i_3) = \chi_\varphi^2(i_1, i_2, i_3) a_{i_1} a_{i_2}(1 - a_{i_3})$. Let **a** = 0001. Since χ_φ^1 is equal to 1 if and only if $i_1 = 4$, $i_2 = 2$, and $i_3 = 3$ and $p_1(1, 0, 0) = 1$, we have that **a** is not a satisfying truth assignment. On the contrary, it is easy to verify that **a** = 0101 makes the above two functions identically zero.

Notice that each function χ_φ^j, for $j = 0, 1, 2, 3$ can be seen as a function from H^{3k} to $\{0, 1\}$: indeed, since $|H^k| \geq n$, each value in $\{1, \ldots, n\}$ can be encoded by means of a k-tuple of elements of H.

The zero-tester polynomials

To verify whether functions f_φ^j are identically zero, we first replace functions $\chi_\varphi^j : H^{3k} \mapsto \{0, 1\}$ and $p_j : \{0, 1\}^3 \mapsto \{0, 1\}$ with their low-degree extensions (see Theorem 7.9): let g_φ^j be the resulting polynomial. Our goal is now to verify that g_φ^j is 0 at every point in H^{3k}. The next lemma basically asserts that this latter condition can be checked by performing a so-called sum-check test (see Lemma 7.15).

Lemma 7.14 ▶ *There exists a family $\mathcal{R}$ of q^{3k} polynomials in $F_{3k|H|, 3k}$ (called zero-testers) such that, for any function $f : H^{3k} \mapsto F$ which is not identically zero, the following inequality holds:*

$$\Pr_{R \in \mathcal{R}} \left\{ \sum_{\mathbf{h} \in H^{3k}} R(\mathbf{h}) f(\mathbf{h}) = 0 \right\} \leq 3/100.$$

The family is constructible in time $q^{O(k)}$ (hence, in polynomial time with respect to n).

PROOF Consider the following $3k$-variate polynomial with coefficients in F:

$$g(t_1, \ldots, t_{3k}) = \sum_{\mathbf{h} \in H^{3k}} f(\mathbf{h}) \prod_{i=1}^{3k} t_i^{h_i}$$

where, for the sake of simplicity, we have identified each element of H with a number between 0 and $|H|-1$ to be used at the exponent.

Clearly, the degree of g is at most $3k|H|$. Moreover, g is the zero polynomial if and only if f is identically zero in H^{3k}.

From Lemma 7.8 it follows that if g is not the zero polynomial, then the fraction of all elements in F^{3k} at which g evaluates to zero is at most $3|H|k/q$. According to the values of the last column of Table 7.2, this fraction is less than $3/100$.

We will now construct, for any $\mathbf{b} \in F^{3k}$, a polynomial $R_\mathbf{b} \in F_{3k|H|,3k}$ such that

$$\sum_{\mathbf{h} \in H^{3k}} R_\mathbf{b}(\mathbf{h}) f(\mathbf{h}) = 0 \Leftrightarrow g(\mathbf{b}) = 0.$$

Let

$$I_{t_i}(x) = \sum_{h \in H} t_i^h S_h^H(x).$$

Clearly, for any $h \in H$, $I_{t_i}(h) = t_i^h$. Moreover, let s be the following $6k$-variate polynomial:

$$s(t_1, \ldots, t_{3k}, x_1, \ldots, x_{3k}) = \prod_{i=1}^{3k} I_{t_i}(x_i).$$

Then, for any $\mathbf{h} \in H^{3k}$,

$$s(t_1, \ldots, t_{3k}, \mathbf{h}) = \prod_{i=1}^{3k} I_{t_i}(h_i) = \prod_{i=1}^{3k} t_i^{h_i},$$

and

$$\sum_{\mathbf{h} \in H^{3k}} f(\mathbf{h}) s(t_1, \ldots, t_{3k}, \mathbf{h}) = \sum_{\mathbf{h} \in H^{3k}} f(\mathbf{h}) \prod_{i=1}^{3k} t_i^{h_i} = g(t_1, \ldots, t_{3k}).$$

Finally, for any $\mathbf{b} \in F^{3k}$, we define

$$R_\mathbf{b}(x_1, \ldots, x_{3k}) = s(\mathbf{b}, x_1, \ldots, x_{3k}).$$

Clearly, the degree of $R_\mathbf{b}$ is $3k|H|$. Moreover, $R_\mathbf{b}$ satisfies the desired property, that is,

$$\sum_{\mathbf{h} \in H^{3k}} R_\mathbf{b}(\mathbf{h}) f(\mathbf{h}) = 0 \Leftrightarrow g(\mathbf{b}) = 0,$$

Section 7.2

ALMOST
TRANSPARENT
SHORT PROOFS

Chapter 7
THE PCP THEOREM

since

$$\sum_{\mathbf{h} \in H^{3k}} R_{\mathbf{b}}(\mathbf{h}) f(\mathbf{h}) = \sum_{\mathbf{h} \in H^{3k}} f(\mathbf{h}) \prod_{i=1}^{3k} b_i^{h_i} = g(\mathbf{b}).$$

The family $\mathcal{R}$ of zero-tester polynomials is then defined as $\mathcal{R} = \bigcup_{\mathbf{b} \in F^{3k}} R_{\mathbf{b}}$. This family clearly satisfies the required property, that is,

$$\Pr_{R \in \mathcal{R}} \left\{ \sum_{\mathbf{h} \in H^{3k}} R(\mathbf{h}) f(\mathbf{h}) = 0 \right\} \leq 3/100.$$

QED   The lemma is thus proved.

Example 7.9 ▶ Let $|H| = 2$. In this case, we have that $I_{t_i}(x) = 1 - x + t_i x$. Hence, a polynomial $s(t_1, t_2, x_1, x_2)$ for checking the sums of functions on H^2 is

$$(1 - x_1 + t_1 x_1)(1 - x_2 + t_2 x_2) = (1 - x_1(1 - t_1))(1 - x_2(1 - t_2)).$$

Then, we have

$$\sum_{\mathbf{h} \in H^2} f(\mathbf{h}) s(t_1, t_2, \mathbf{h}) = f(0,0) + f(0,1)t_2 + f(1,0)t_1 + f(1,1)t_1 t_2.$$

Clearly, this polynomial is identically zero if and only if all its four terms are equal to zero.

As a consequence of Lemma 7.14, checking whether a function is 0 at every element of H^{3k} can be reduced to verifying that the sum over H^{3k} of the function multiplied by one of the zero-testers (chosen uniformly at random) equals to zero. The remainder of this section describes how this can be done efficiently.

The sum-check procedure

Let $f : F^{3k} \mapsto F$ be a polynomial of degree at most d and let H be any subset of F. We want to check that

$$\sum_{\mathbf{h} \in H^{3k}} f(\mathbf{h}) = 0.$$

To this aim we will need some additional information: in particular, we will need the *partial-sum polynomials* corresponding to f. For $i = 1, \ldots, 3k$, the i-th partial-sum polynomial $g_i : F^i \mapsto F$ is defined as follows:

$$g_i(x_1, \ldots, x_i) = \sum_{y_{i+1} \in H} \sum_{y_{i+2} \in H} \cdots \sum_{y_{3k} \in H} f(x_1, \ldots, x_i, y_{i+1}, y_{i+2}, \ldots, y_{3k}).$$

Clearly, for any i with $1 \leq i \leq 3k - 1$, we have that

Section 7.2
ALMOST TRANSPARENT SHORT PROOFS

Program 7.7: Sum-Check

oracle $f \in F_{d,3k}$ and table T of partial-sum polynomials;
output YES if f sums to 0 in H^{3k};
begin
  Read the coefficients of $g'_1(x)$;
  if $\sum_{x \in H} g'_1(x) \neq 0$ **then return** NO;
  Randomly choose $r_i \in F$ for $i = 1, \ldots 3k$;
  for $i := 2$ **to** $3k$ **do**
  begin
    Read the coefficients of $g'_i(r_1, \ldots, r_{i-1}, x)$;
    if $\sum_{x \in H} g'_i(x) \neq g'_{i-1}(r_{i-1})$ **then return** NO
  end;
  if $f(r_1, \ldots, r_{3k}) \neq g'_{3k}(r_{3k})$ **then**
    return NO
  else
    return YES
end.

$$g_i(x_1, \ldots, x_i) = \sum_{x \in H} g_{i+1}(x_1, \ldots, x_i, x).$$

Moreover,

$$\sum_{\mathbf{h} \in H^{3k}} f(\mathbf{h}) = \sum_{x_1 \in H} g_1(x_1).$$

This latter equality suggests that in order to perform our original sum-check (over $|H|^{3k}$ elements) we can perform a sum-check relative to g_1 (over $|H|$ elements). However, the problem still remains open of how to verify that the polynomial given in the proof is indeed equal to g_1.

The solution of this problem basically consists of pretending that the proof contains *all* the partial-sum polynomials and then sequentially checking whether these polynomials are consistent each with the next one (that is, evaluating the $(i-1)$-th partial sum polynomial at a randomly chosen element of F^{i-1} and comparing the result with the corresponding sum of $|H|$ values of the i-th partial sum polynomial). Actually, the consistency test can be modified so that the partial-sum polynomials will be seen as univariate polynomials in the last variable.

More formally, assume that the verifier has access both to the values of f and to a table T containing $\sum_{i=1}^{3k}(d+1)q^{i-1}$ rows of size $\log q$: this table is supposed to contain, for $i = 1, \ldots, 3k$ and for all $b_1, \ldots, b_{i-1} \in F$, the coefficients of the univariate polynomials $g_i(b_1, \ldots, b_{i-1}, x)$ (see Fig. 7.5). In the real situation, the proof is a string of $(q^{3k} + \sum_{i=1}^{3k}(d+1)q^{i-1})\log q$

235

Chapter 7

THE PCP THEOREM

bits, whose first $q^{3k} \log q$ bits are interpreted by the verifier as the values of f, and the remaining $\sum_{i=1}^{3k}(d+1)q^{i-1}\log q$ bits are interpreted as the $d+1$ coefficients in F of $\sum_{i=1}^{3k} q^{i-1}$ univariate polynomials $g'_i(b_1,\ldots,b_{i-1};x)$ (see Fig. 7.6).

With this notation the sum-check procedure is then described in Program 7.7. The following lemma shows that this algorithm does indeed have the desired properties.

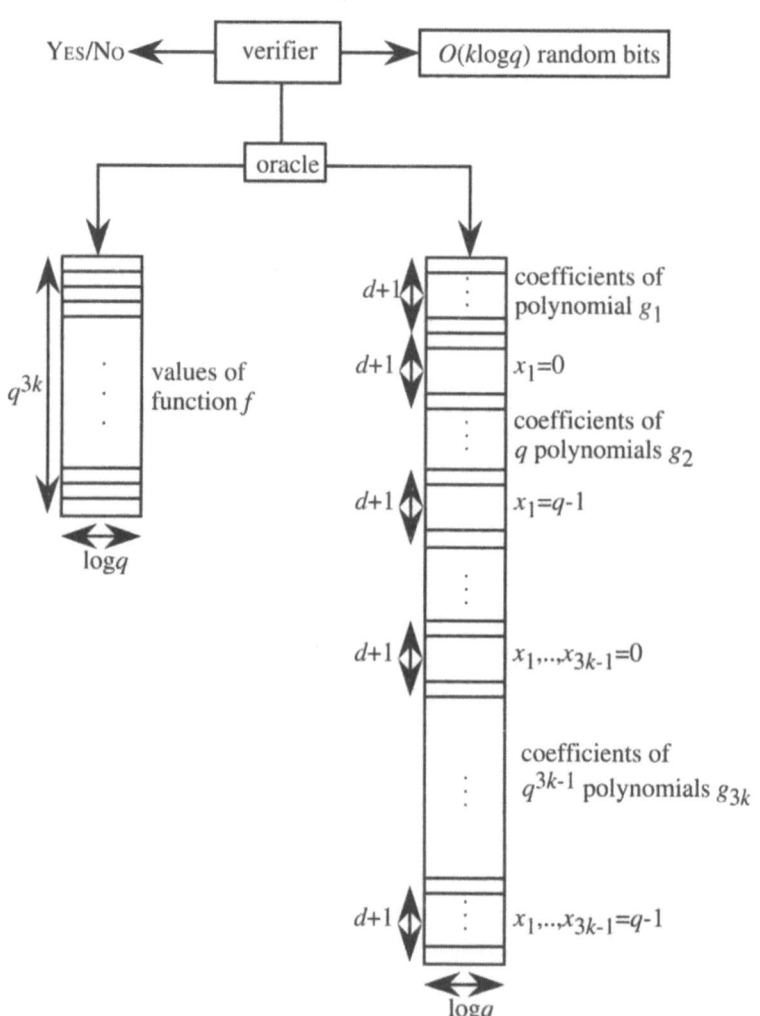

Figure 7.5
The proof for the sum-check procedure: the ideal situation

Lemma 7.15 ▶ Let $f : F^k \mapsto F$ be a polynomial in $F_{d,3k}$. The following claims hold:

1. If $\sum_{\mathbf{h}\in H^{3k}} f(\mathbf{h}) = 0$, then there exists a table T such that Program 7.7, with oracle f and T, always accepts.

2. If $\sum_{\mathbf{h} \in H^{3k}} f(\mathbf{h}) \neq 0$, then, for all tables T, Program 7.7, with oracle f and T, rejects with probability at least 3/4.

3. Program 7.7 uses $O(k \log q)$ random bits and reads only one value of f and $3k(d+1)$ rows of T.

The first point follows from the previous discussion while the third point can be easily checked by the reader. Now, suppose that

$$\sum_{\mathbf{h} \in H^{3k}} f(\mathbf{h}) \neq 0.$$

If $g_1(x)$ is identical to $g'_1(x)$ then the first test, that is, $\sum_{x \in H} g'_1(x) \neq 0$ immediately detects an error in the proof and the procedure rejects. Thus, assume that $g_1(x)$ is not identical to $g'_1(x)$: this implies that, with probability at least $1 - d/q$, $g_1(r_1) \neq g'_1(r_1)$ (see Lemma 7.8). If this latter event occurs and $g_2(r_1, x)$ is identical to $g'_2(r_1, x)$, then the second test detects an error and the procedure rejects. Otherwise, if $g_1(r_1) \neq g'_1(r_1)$ and $g_2(r_1, x)$ is not identical to $g'_2(r_1, x)$, then, with probability at least $1 - d/q$, $g_2(r_1, r_2) \neq g'_2(r_1, r_2)$.

By iterating this argument, we have that, with probability at least $(1 - d/q)^{3k-1}$, either one among the first $3k$ tests has detected an error or $g_{3k}(r_1, \ldots, r_{3k-1}, x)$ is not identical to $g'_{3k}(r_1, \ldots, r_{3k-1}, x)$. In the latter case, the final test will succeed with probability at least $1 - d/q$. In conclusion, Program 7.7 rejects with probability at least $(1 - d/q)^{3k} \geq 1 - 3kd/q \geq 3/4$ where the last inequality is due to the bounds of the last column of Table 7.2. Hence, the lemma is proved.

QED

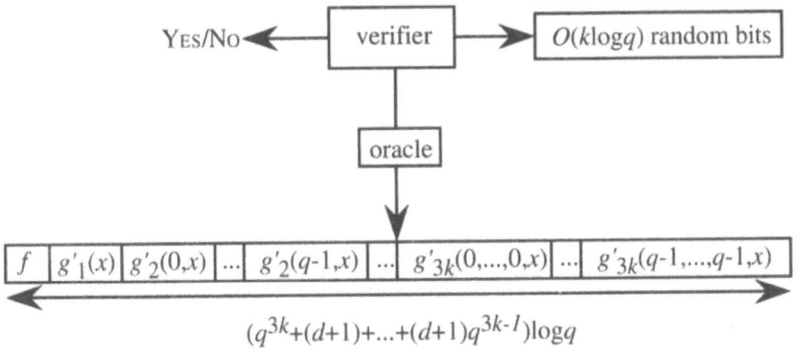

Figure 7.6
The proof for the sum-check procedure: the real situation

The combination of Lemmas 7.14 and 7.15 (which is the analogue of Theorem 7.5) yields an efficient way of checking whether the polynomial extension of the functions f_φ^j, for $j = 0, 1, 2, 3$, are identically zero on H^{3k}.

Chapter 7

THE PCP THEOREM

7.2.3 The second PCP result

The construction of a verifier which uses a logarithmic number of random bits and queries a polylogarithmic number of elements (of polylogarithmic size) of the proof can now be completed. The verifier interprets the proof as the low-degree extension $f_\mathbf{a}$ of a truth assignment plus five tables $T_\mathbf{a}$, T_0, T_1, T_2, and T_3 (see Fig. 7.7).

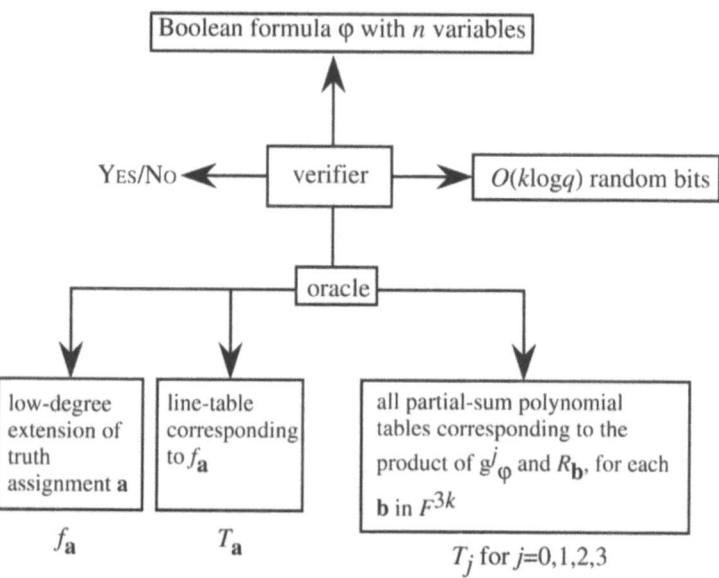

Figure 7.7
The overall verifier for proving NP $\subseteq$ PCP$[O(\log n), O(\log^4 n)]$

By using Theorem 7.12 and Program 7.5 with table $T_\mathbf{a}$ the verifier first checks that $f_\mathbf{a}$ is δ-close to a polynomial of total degree at most d. This test can be performed by querying only a constant number of values of $f_\mathbf{a}$ and a constant number of elements of $T_\mathbf{a}$, which is supposed to be the line-table corresponding to $f_\mathbf{a}$, and by using $O(k \log q)$ random bits.

Successively, the verifier tests, for $j = 0, 1, 2, 3$, whether the low-degree extension g_φ^j of function f_φ^j (with respect to the truth assignment contained in the proof) is identically zero. To this aim, the verifier randomly chooses a zero-tester polynomial R and checks that the product of R by g_φ^j sums to zero over H^{3k}. By using Program 7.7, this can be done by querying table $f_\mathbf{a}$ and T_j, which is supposed to contain all the partial-sum polynomial tables corresponding to the product of g_φ^j and R, for any zero-tester polynomial R. Indeed, since function χ_φ^j depends only on the Boolean formula φ, this function can be computed by the verifier: thus, each query to g_φ^j can be substituted by three queries to $f_\mathbf{a}$. By choosing a sufficiently small value of δ, we know that if the low-degree test accepts, then with high probability (say, 3/4) Program 7.6 returns the correct values. Since Program 7.7

detects an error with probability at least 3/4, we have that the probability that the verifier accepts a wrong proof is at most 1/2. By repeating the whole process a suitable number of times this probability can be arbitrarily reduced.

Finally, let us evaluate the number of used random bits and the number of queries to the proof: to this aim, we will refer to the values shown in the last column of Table 7.2. The low-degree test, the correcting procedure, and the sum-check procedure use $O(k\log q) = O(\log n)$ random bits. Moreover, Programs 7.5 and 7.6 query a constant number of values of $f_\mathbf{a}$ of length $O(\log q) = O(\log\log n)$ and a constant number of elements of $T_\mathbf{a}$ of length $O(k|H|\log q) = O(\log^2 n)$ while Program 7.7 (when applied to the product of g_φ^j with a zero-tester polynomial) queries a constant number of values of $f_\mathbf{a}$ of length $O(\log q) = O(\log\log n)$ and $O(k(d+1)) = O(\log^2(n))$ rows of T_j of length $O(k\log q) = O(\log^2(n))$. Thus, the overall number of queried bits is $O(\log^4(n))$.

In conclusion, we have proved the following theorem.

$\text{NP} \subseteq \text{PCP}[O(\log n), O(\log^4 n)]$. ◀ Theorem 7.16

Section 7.3
THE FINAL PROOF

7.3 The final proof

IN THIS section, we will finally prove the PCP theorem. To this aim, instead of giving other verifiers with different performances with respect to the two described in the previous two sections, we will show how these two verifiers can be combined in order to obtain the desired randomness and query complexity.

In the following we denote with $\text{poly}(n)$ (respectively, $\text{polylog}(n)$ and $\text{polyloglog}(n)$) a function of the type n^k (respectively, $\log^k n$ and $\log^k(\log n)$), for some constant k. The final proof of the PCP theorem can be summarized into the following five main steps (see Fig. 7.8):

1. **Composition lemma:** Prove that if there is an $(r_2(n), q_2(n))$-normal form verifier for E3-SATISFIABILITY (see Def. 7.9) and

   $$\text{NP} \subseteq \text{PCP}[O(r_1(n)), O(q_1(n))],$$

   then

   $$\text{NP} \subseteq \text{PCP}[O(r_1(n) + r_2(kq_1(n))), O(q_2(kq_1(n)))]$$

   where k is a constant.

Chapter 7
THE PCP THEOREM

2. Prove that

$$NP \subseteq PCP[O(\text{poly}(n)), O(1)].$$

Substep: modify the proof of the above result in order to obtain a $(\text{poly}(n), 1)$-normal form verifier for E3-SATISFIABILITY.

3. Prove that

$$NP \subseteq PCP[O(\log n), O(\text{polylog}(n))].$$

Substep: modify the proof of the above result in order to obtain a $(\log n, \text{polylog}(n))$-normal form verifier for E3-SATISFIABILITY.

4. From the first step, the third step, and the third substep it follows that NP is contained in

$$PCP[O(\log n + \log(k\text{polylog}(n))), O(\text{polylog}(k\text{polylog}(n)))].$$

Clearly,

$$O(\log n + \log(k\text{polylog}(n))) \text{ is } O(\log n)$$

and

$$O(\text{polylog}(k\text{polylog}(n))) \text{ is } O(\text{polyloglog}(n)).$$

Thus,

$$NP \subseteq PCP[O(\log n), O(\text{polyloglog}(n))].$$

5. From the first step, the second substep, and the fourth step it follows that

$$NP \subseteq PCP[O(\log n + \text{poly}(k\text{polyloglog}(n))), O(1)].$$

Since $O(\log n + \text{poly}(k\text{polyloglog}(n)))$ is $O(\log n)$, we have

$$NP \subseteq PCP[O(\log n), O(1)].$$

Section 7.3
THE FINAL PROOF

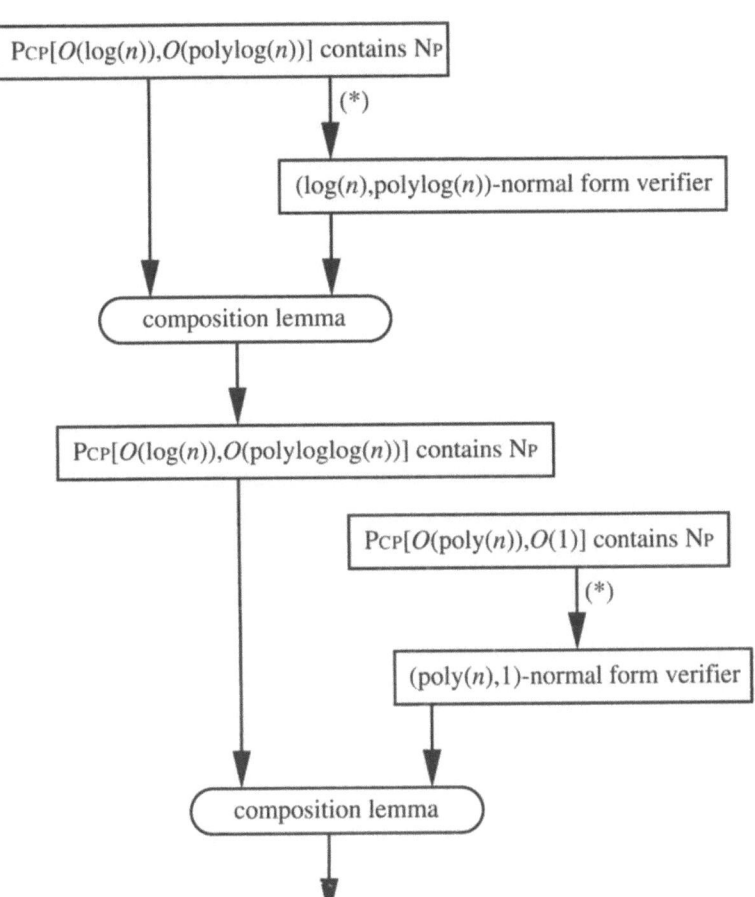

Figure 7.8
The structure of the proof of the PCP theorem (the starred arrows are not proved here)

The first step of the proof will be proved in the remainder of this section. The second and the third steps of the proof of the PCP theorem, on the other hand, have been proved in the previous two sections. However, we will not prove their corresponding substeps. We leave to the reader the task of appropriately modifying our proofs in order to obtain the results with respect to the normal-form model (see Exercises 7.8 and 7.9 and Bibliographical notes): observe that these modifications are necessary in order to apply the composition lemma.

7.3.1 Normal form verifiers

Recall that in the original definition of a probabilistically checkable proof (see Chap. 6), it is assumed that the proof is a sequence of bits to which

241

Chapter 7
THE PCP THEOREM

the verifier has random access. In this section, this will still be true but we will assume that the verifier treats this sequence of bits as words of a specified size: the verifier either reads all the bits in a word or none at all. We will also assume that the verifier is allowed to read only a constant number of such words, which implies that the number of bits it reads is proportional to the length of a word. We will call this kind of verifier *restricted*: the verifiers we have described in the previous two sections are indeed restricted (see Exercise 7.6). According to this restriction, the second parameter in the definition of PCP, that is, function q, will no longer denote the number of bits that are read but instead the length of a single word.

The last main ingredient of the proof of the PCP theorem is the so-called *composition lemma*. This lemma gives a technique to reduce the number of bits read from the proof, provided there exists a *reasonably efficient normal form verifier*. Roughly speaking, a normal form verifier is a verifier with an associated encoding scheme and it is able to check whether, given the *split encoding* of a truth assignment, this assignment satisfies a Boolean formula.

Definition 7.8 ▶ (l,c)-encoding scheme

Let l and c be two functions from $\mathbf{N}$ to $\mathbf{N}$ and let Σ be an alphabet. An (l,c)-encoding scheme is a function $E : \Sigma^* \mapsto \Sigma^*$ such that, for any $x \in \Sigma^*$ with $|x| = n$, $|E(x)| = l(n)c(n)$. For any x, $E(x)$ is said to be the codeword associated to x.

Given an (l,c)-encoding scheme E we will denote with E^{-1} a corresponding decoding scheme, that is, a function $E^{-1} : \Sigma^* \mapsto \Sigma^*$ such that, for any $y \in \Sigma^*$ with $|y| = l(n)c(n)$ for some n, $|E^{-1}(y)| = n$ and, for any $x \in \Sigma^*$ with $|x| = n$, $E^{-1}(E(x)) = x$. Observe that there may exist strings of length $l(n)c(n)$ for some n which are not codewords but are still mapped by E^{-1} to a string of length n. Note also that there exist encoding schemes which do not admit the corresponding decoding scheme: however, in the following we will consider only encoding schemes for which a decoding scheme exists.

The output of an (l,c)-encoding scheme with input a string of length n can be better visualized as a binary table with $l(n)$ rows and $c(n)$ columns: we will make use of this interpretation in the following.

An encoding scheme is said to have *minimum distance* $\delta_{\min}$ if, for any n, the distance between the encoding of any two strings of length n is at least $\delta_{\min}$ where the distance between two strings is the fraction of positions in which they differ. Note that, if E has *minimum distance* $\delta_{\min}$ and a string y has distance less than $\delta_{\min}/2$ from a codeword, then there is a unique codeword whose distance from y is less than $\delta_{\min}/2$.

We are now ready to give a formal definition of normal form verifiers.

Section 7.3

The final proof

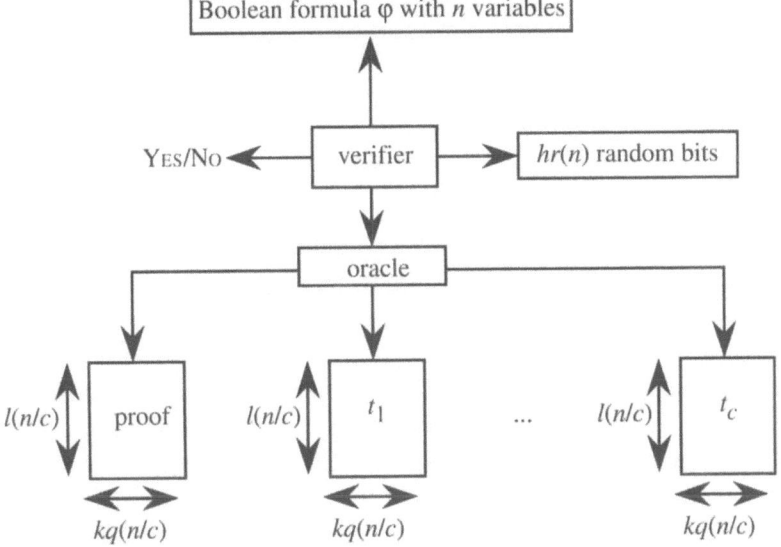

Figure 7.9
A normal-form verifier for
E3-Satisfiability

A verifier $\mathcal{A}$ (for E3-Satisfiability) is in $(r(n), q(n))$-normal form if (a) it has an associated $(l, k \cdot q)$-encoding scheme E computable in polynomial time of minimum distance $\delta_{\min}$ with $l(n) \leq 2^{hr(n)}$ for some constant h, and (b) given a Boolean formula φ in conjunctive normal form with three literals per clause containing n variables, and an integer constant c that divides n, the verifier has the following behavior (see also Fig. 7.9):

◀ Definition 7.9
$(r(n), q(n))$-normal form verifier

1. It has access to an oracle containing one proof binary table π with $l(n/c)$ rows and $kq(n/c)$ columns and c encoding tables $t_1, \ldots, t_c$ of the same size.

2. If $t_1, \ldots, t_c$ are codewords and $E^{-1}(t_1), \ldots, E^{-1}(t_c)$ is a satisfying truth-assignment for φ, then there exists π such that $\mathcal{A}$ accepts with probability 1.

3. If there exists t_i whose distance from a codeword is at least $\delta_{\min}/3$, then, for any π, $\mathcal{A}$ accepts with probability less than $1/2$.

4. If every t_i has distance from a codeword less than $\delta_{\min}/3$ but $E^{-1}(t_1), \ldots, E^{-1}(t_c)$ is not a satisfying truth assignment for φ, then, for any π, $\mathcal{A}$ accepts with probability less than $1/2$.

5. $\mathcal{A}$ makes use of exactly $hr(n)$ random bits.

243

Chapter 7
THE PCP THEOREM

6. *For any random string of length $hr(n)$, $\mathcal{A}$ reads exactly d rows of the tables in the oracle, whose indices depend only on the random string, where d is a constant.*

7. *For any d rows of the tables in the oracle which have been read by $\mathcal{A}$ with a given random string of length $hr(n)$, it is decidable in polynomial time whether $\mathcal{A}$ accepts.*

Note that, according to the previous definition, the verifier checks a *c-part split-encoded assignment*, that is, an assignment which is split into c groups of n/c variables and such that each group is then encoded by means of the encoding scheme E: on the ground of this information, the verifier decides whether the assignment satisfies the Boolean formula. If the verifier is observed to accept some proof with probability greater than 1/2, we can thus conclude that the formula is satisfiable. This restriction implies the following result whose easy proof is left to the reader (see Exercise 7.7) and is mainly based on the fact that a normal-form verifier can be used to check membership proofs for E3-SATISFIABILITY (in the standard sense) and that E3-SATISFIABILITY is NP-complete.

Theorem 7.17 ▶ *If an $(r(n), q(n))$-normal form verifier (for E3-SATISFIABILITY) exists, then $\mathrm{NP} \subseteq \mathrm{PCP}[O(r(n)), O(q(n))]$.*

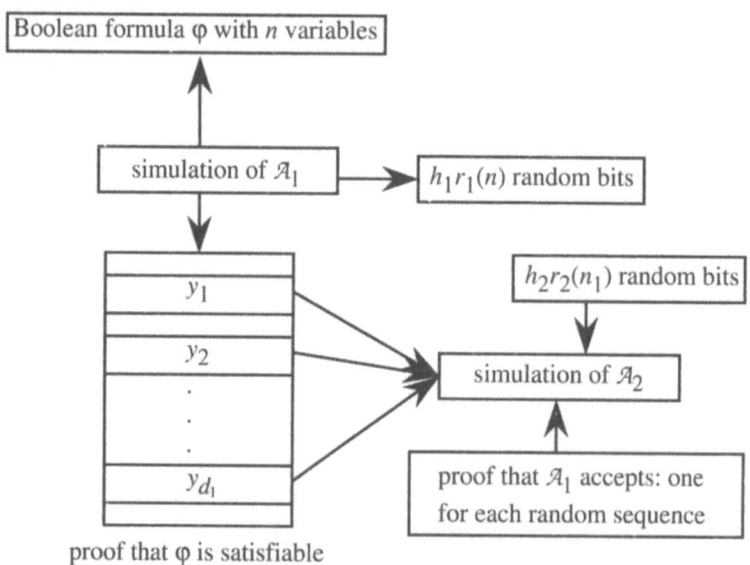

Figure 7.10
The proof of the composition lemma

7.3.2 The composition lemma

Section 7.3

THE FINAL PROOF

We are finally able to formally state and prove the composition lemma, thus concluding the proof of the PCP theorem. The basic idea behind the proof of this lemma is quite simple, and is shown in Fig. 7.10. Intuitively, the *composed* verifier first simulates verifier $\mathcal{A}_1$ and successively checks that $\mathcal{A}_1$ reading a certain set of positions of the proof π_1 accepts. This latter test could be made in polynomial time since $\mathcal{A}_1$ is a polynomial-time machine. However, one problem occurs: the composed verifier cannot read the words of proof π_1 since otherwise no reduction in the number of read bits is obtained. The solution to this problem consists of making use of verifier $\mathcal{A}_2$ which has access to one proof for each random sequence of bits received by $\mathcal{A}_1$: $\mathcal{A}_2$, however, has to be a normal form verifier, which has access to a split encoding of the proof (instead of an encoding of the entire proof).

If $\text{NP} \subseteq \text{PCP}[O(r_1(n)), O(q_1(n))]$ *and there exists an* $(r_2(n), q_2(n))$-*normal form verifier (for* E3-SATISFIABILITY*), then*

◂ **Lemma 7.18**

$$\text{NP} \subseteq \text{PCP}[O(r_1(n) + r_2(kq_1(n))), O(q_2(kq_1(n)))]$$

where k is a constant.

Let h_1, k_1 and d_1 be three constants, and let $\mathcal{A}_1$ be a verifier witnessing that E3-SATISFIABILITY belongs to $\text{PCP}[O(r_1(n)), O(q_1(n))]$. That is, for any input of length n, $\mathcal{A}_1$ uses $h_1 r_1(n)$ random bits and reads d_1 words in the proof of length $k_1 q_1(n)$.

PROOF

First, we define how to compose the proof tables. Let φ be a Boolean formula of length n and let π_1 be a proof table with $l_1(n)$ rows and $k_1 q_1(n)$ columns. Observe that, without loss of generality, we may assume $l_1(n) \leq 2^{h_1 r_1(n)}$ (see the comments following Def. 6.9). The *composed proof* table is formed by two parts: intuitively, the first part should be a proof that the second part should be a proof that φ is satisfiable (see Fig. 7.11).

Formally, for any random string r of length $h_1 r_1(n)$, let $y_1^r, \ldots, y_{d_1}^r$ be the contents of the rows of the proof table π_1 which are read by $\mathcal{A}_1$ whenever its random tape contains r. Consider the language L' consisting of the sequences $(y_1, \ldots, y_{d_1})$ such that $|y_i| = k_1 q_1(n)$, for $i = 1, \ldots, d_1$, and $\mathcal{A}_1$ accepts whenever it reads $y_1, \ldots, y_{d_1}$.

According to the definition of a verifier, $\mathcal{A}_1$'s *decision time*, that is, the running time necessary for the verifier to decide whether or not to accept (after having read the words from the proof) must be polynomial in the length of the input. It thus follows that $L' \in \text{P}$: let $\mathcal{A}_{L'}$ be a polynomial-time Turing machine that decides L', and let $\varphi_{\mathcal{A}_{L'}}$ be the corresponding

Chapter 7
THE PCP THEOREM

Boolean formula obtained by means of the proof of Cook-Levin's theorem. This formula contains $d_1 k_1 q_1(n)$ variables corresponding to the y_is and a polynomial number of additional variables z_i: moreover, $y_1 \cdots y_{d_1} \in L'$ if and only if there exists a truth assignment to the additional variables which along with the values of the ys satisfies $\varphi_{\mathcal{A}_{L'}}$.

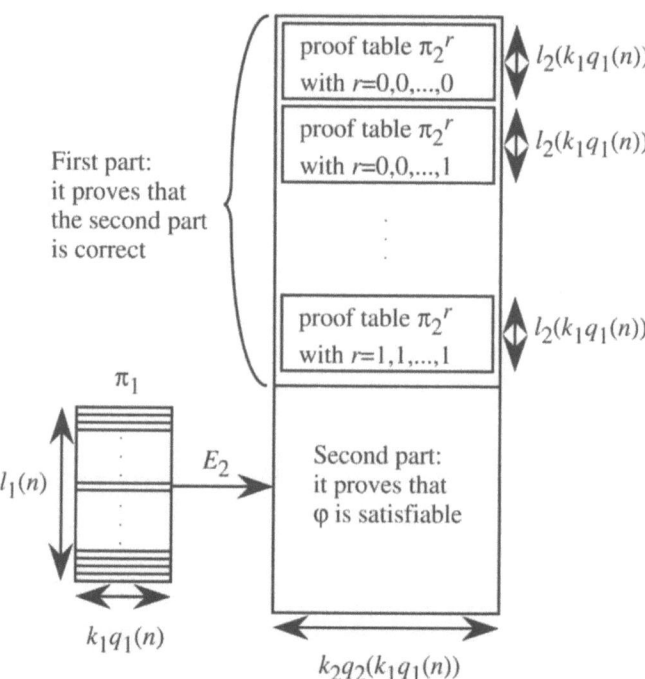

Figure 7.11
The composed proof table

Since there exists an $(r_2(n), q_2(n))$-restricted normal form verifier (for E3-SATISFIABILITY), relatively to $\varphi_{\mathcal{A}_{L'}}$ and d_1, there exist three constants h_2, k_2 and d_2, an $(l_2(n), k_2 q_2(n))$-encoding scheme of distance $\delta_{\min}$ with $l_2(n) \leq 2^{h_2 r_2(n)}$, and a verifier $\mathcal{A}_2$ which satisfies the conditions of the definition of a normal form verifier. If $y_1^r \cdots y_{d_1}^r \in L'$, then there exists a proof table π_2^r with $l_2(k_1 q_1(n))$ rows and $k_2 q_2(k_1 q_1(n))$ columns such that $\mathcal{A}_2$ with oracle $E_2(y_1^r), \ldots, E_2(y_{d_1}^r)$, and π_2^r accepts with probability 1. The first part of the composed proof table contains a table π_2^r for each random string r.

The second part of the proof table, instead, contains the encoding of each row of π_1 by making use of the encoding scheme E_2: thus, this part is a table of $l_1(n) l_2(k_1 q_1(n))$ rows and $k_2 q_2(k_1 q_1(n))$ columns.

Hence, the composed proof table contains

$$l(n) = 2^{h_1 r_1(n)} l_2(k_1 q_1(n)) + l_1(n) l_2(k_1 q_1(n))$$

rows and

$$kq(n) = k_2 q_2(k_1 q_1(n))$$

columns (see Fig. 7.11). Since

$$l_1(n) \leq 2^{h_1 r_1(n)}$$

and

$$l_2(k_1 q_1(n)) \leq 2^{h_2 r_2(k_1 q_1(n))},$$

we have that

$$l(n) \leq 2 \cdot 2^{h_1 r_1(n)} \cdot 2^{h_2 r_2(k_1 q_1(n))} \leq 2^{h(r_1(n) + r_2(k_1 q_1(n)))}$$

where h is an appropriate constant.

Finally, we have to define the composed verifier $\mathcal{A}$. First, $\mathcal{A}$ simulates $\mathcal{A}_1$ with a random string r of length $h_1 r_1(n)$. To this aim, it computes the indices of the rows of π_1 to be read and, consequently, the starting positions of their corresponding tables in the second part of the composed proof table. Successively, $\mathcal{A}$ simulates $\mathcal{A}_2$ with a random string r' of length $h_2 r_2(k_1 q_1(n))$ and with oracle the above tables and the proof table π_2^r which lies in the first part of the composed proof table. Finally, $\mathcal{A}$ accepts if and only if $\mathcal{A}_2$ accepts. Clearly, $\mathcal{A}$ makes use of *exactly* $h_1 r_1(n) + h_2 r_2(k_1 q_1(n))$ random bits.

To see the correctness of $\mathcal{A}$, assume first that φ is satisfiable. By assumption, there exists a proof table π_1 with $l_1(n)$ rows and $k_1 q_1(n)$ columns such that $\mathcal{A}_1$ with oracle π_1 accepts with probability 1. In particular, this implies that, for any random string r of length $h_1 r_1(n)$, there exists a d_1-tuple of tables $E_2(y_1), \ldots, E_2(y_{d_1})$ with $l_2(k_1 q_1(n))$ rows and $k_2 q_2(k_1 q_1(n))$ columns (where $y_1, \ldots, y_{d_1}$ denote the rows that are read by $\mathcal{A}_1$ with random string r and are contained in the proof table π_1) such that $\mathcal{A}_2$ with oracle π_2^r, $E_2(y_1), \ldots, E_2(y_{d_1})$ accepts with probability 1. We thus have that $\mathcal{A}$ with oracle the composed proof table accepts with probability 1.

On the contrary, assume that φ is not satisfiable. We will now show that the probability that, for a given proof table π with $l(n)$ rows and $kq(n)$ columns, $\mathcal{A}$ fails to reject is bounded by a suitable constant: by iterating a constant number of times the protocol of the verifier we can obtain that this probability is less than $1/2$.

Observe that the second part of π can be viewed as $l_1(n)$ tables with $l_2(k_1 q_1(n))$ rows and $k_2 q_2(k_1 q_1(n))$ columns. From the definition of a normal form verifier, we may assume that these tables are codewords (according to E_2): let us apply E_2^{-1} to them thus obtaining a table π' with $l_1(n)$

rows and $k_1q_1(n)$ columns. The probability that $\mathcal{A}_1$ with oracle π' accepts is less than 1/2 (since φ is not satisfiable). Equivalently, the probability that $\mathcal{A}_1$ with oracle π' rejects is at least 1/2. Let r be a random string of length $h_1r_1(n)$ that leads $\mathcal{A}_1$ to rejection and let $y_1^r, \ldots, y_{d_1}^r$ be the contents of the rows of π' which are read by $\mathcal{A}_1$. Then, $y_1^r \cdots y_{d_1}^r \notin L'$. Since $y_1^r, \ldots, y_{d_1}^r$ are the decoding by means of E_2^{-1} of d_1 tables $\pi_1'', \ldots, \pi_{d_1}''$, $\mathcal{A}_2$ with oracle any given table π'' with $l_2(k_1q_1(n))$ rows and $k_2q_2(k_1q_1(n))$ columns, and $\pi_1'', \ldots, \pi_{d_1}''$ rejects with probability at least 1/2. Hence, the probability that $\mathcal{A}$ rejects is at least the probability that $\mathcal{A}_1$ has rejected times the probability that $\mathcal{A}_2$ rejects: in summary, this probability is at least 1/4.

QED

7.4 Exercises

Exercise 7.1 Show that if $L_1 \in \text{PCP}[O(\log n), O(1)]$ and L_2 is Karp-reducible to L_1, then $L_2 \in \text{PCP}[O(\log n), O(1)]$.

Exercise 7.2 Prove Lemma 7.4. (Hint: use induction on n.)

Exercise 7.3 Prove Lemma 7.6. (Hint: Lemma 7.4 implies that if $\mathbf{x}, \mathbf{y} \in Z_2^n$ are distinct, then

$$\Pr_{r \in Z_2^n}\left\{\sum_{i=1}^n r_i x_i \neq \sum_{i=1}^n r_i y_i\right\} = \frac{1}{2}.$$

Use this fact to obtain a similar result on matrices in $Z_2^{n^2}$ and, then, observe that if $\tilde{\mathbf{a}}$ and $\tilde{\mathbf{b}}$ are the coefficients of the linear functions δ-close to A' and B', respectively, then $\tilde{\mathbf{a}} \circ \tilde{\mathbf{a}}$ and $\tilde{\mathbf{b}}$ can be seen as matrices.)

Exercise 7.4 Complete the proof of Theorem 7.9.

Exercise 7.5 This exercise refers to the proof of Theorem 7.13. Prove that, for at least $(1 - \sqrt{\delta})q^k$ lines l in $L_\mathbf{x}$, P_l^f correctly describes g at $(1 - \sqrt{\delta})q$ elements of l. (Hint: prove that, if S is the set of elements of F^k at which g and f disagree, then, when we pick a random line l in $L_\mathbf{x}$, the expected number of elements of l that are in S is $|S|/q^k$.)

Exercise 7.6 Check that the verifiers of Sect. 7.1 and of Sect. 7.2 are restricted according to the definition given at the beginning of Sect. 7.3.

Exercise 7.7 Prove Theorem 7.17.

Exercise 7.8 By making use of the results of Sect. 7.1, prove that there exists a $(\text{poly}(n), 1)$-normal form verifier for E3-SATISFIABILITY.

Exercise 7.9 (**) By making use of the results of Sect. 7.2, prove that a $(\log n, \text{polylog}(n))$-normal form verifier for E3-SATISFIABILITY exists.

Exercise 7.10 (**) A challenging open problem is to determine the minimum number q such that $\text{NP} = \text{PCP}[O(\log n), q]$. By Exercise 6.22 we know that at least 3 bits are needed. Prove that 9 bits are sufficient.

Exercise 7.11 (*) Recall from Def. 6.8 that the query complexity is the number of bits of the proof that the verifier accesses in a computation. In its computation on an input x and random string ρ the verifier gets a q-bit sequence from the proof and either accepts or rejects. Since the number of possible patterns of q-bit sequences is 2^q we can equivalently define the query complexity as the logarithm of the number of possible patterns. The *free bit complexity* is the logarithm of the number of *accepting* patterns. For example, consider a verifier that first asks for f bits of the proof, then decides exactly which values the remaining $q-f$ bits must have, and then just verifies that the remaining bits have these values and accepts in that case. The number of accepting patterns for this verifier is 2^f, which means that the free bit complexity is f. The free bit complexity is another PCP parameter that one can try to optimize. The class $\text{FPCP}_{c,s}[r(n), f(n)]$ is defined as $\text{PCP}_{c,s}[r(n), q(n)]$ (see Exercise 6.23) with the difference that there is no bound on the query complexity, but instead the free bit complexity is bounded by $f(n)$. Show that if $\text{NP} = \text{FPCP}_{c,s}[\log n, f]$, then the MINIMUM VERTEX COVER problem is not approximable within $1 + (c-s)/(2^f - c)$, unless $\text{P} = \text{NP}$. (Hint: use the relation between MINIMUM VERTEX COVER and MAXIMUM CLIQUE and your solution to Exercise 6.24.)

Exercise 7.12 Here are some free bit complexity results that are known:

$$\begin{aligned}\text{NP} &= \text{FPCP}_{1,0.5}[\log n, 7] = \text{FPCP}_{1,0.76}[\log n, 2] \\ &= \text{FPCP}_{0.99, 0.51}[\log n, 2] = \text{FPCP}_{0.25, 0.22125}[\log n, 0].\end{aligned}$$

Which is the best non-approximability result we can get for MINIMUM VERTEX COVER using the previous exercise and the characterizations of NP above?

7.5 Bibliographical notes

AN EXCELLENT historical account of the evolution of probabilistic proof systems, along with several open problems, is presented in [Goldreich, 1999]. We would like to start these notes quoting from this book:

Chapter 7
THE PCP THEOREM

The glory attached to the creativity involved in finding proofs, makes us forget that it is the less glorified procedure of verification which gives proofs their value.

Probabilistically checkable proof systems were first introduced as interactive proof systems between a verifier and one or more provers, who try to convince the verifier of a certain fact. Interactive proof systems first appeared in [Goldwasser, Micali, and Rackoff, 1989] (in the context of cryptography) and [Babai and Moran, 1988] (as a game theoretic extension of NP). The extension to multi-prover interactive proofs was introduced by [Ben-Or et al., 1988]: in this model the verifier interacts with two infinitely powerful provers who cannot communicate with each other during the protocol. An equivalent formulation of this latter model in terms of proof verification was suggested in [Fortnow, Rompel, and Sipser, 1988]. At the beginning, interactive protocols were not thought to be enough powerful to decide all languages in the complement class of NP. It thus came as a surprise when in [Lund et al., 1991] and in [Shamir, 1992] it was shown, by using techniques developed for program checking [Blum and Kannan, 1989], that all languages in PSPACE admit an interactive protocol. Shortly after, in [Babai, Fortnow, and Lund, 1991] it was shown that any language that can be decided by a nondeterministic exponential-time Turing machine admits a multi-prover interactive protocol while in [Babai et al., 1991] the notion of *transparent* membership proofs (i.e. proofs that can be checked in polylogarithmic time) was introduced and it was shown that NP languages admit such proofs.

The first paper showing that there is a strong connection between interactive proofs and approximation was [Feige et al., 1991]. In this paper it was first shown, by using a scaled-down version of the result of [Babai, Fortnow, and Lund, 1991], that

$$\text{NP} \subseteq \text{PCP}[O(\log n \log \log n), O(\log n \log \log n)]$$

and, subsequently, that MAXIMUM CLIQUE is not approximable unless $\text{NP} \subseteq \text{TIME}(n^{O(\log \log n)})$. This result drew the attention to the area of probabilistically checkable proofs. After a year of intense research in probabilistic proof systems the PCP theorem was shown as the result of [Arora and Safra, 1998] and [Arora et al., 1992]. Our presentation is partly inspired by [Arora and Safra, 1998], [Arora, 1994], [Hougardy, Prömel, and Steger, 1994], and [Sudan, 1995].

In the years immediately after the proof of the PCP theorem, other variants of probabilistically checkable proofs have been studied in order to obtain better inapproximability results. The PCP results and the characterizations of NP in Exercises 7.10 and 7.11 are from [Bellare, Goldreich, and Sudan, 1998], [Håstad, 1996], and [Håstad, 1997].

Several surveys of all the above developments appeared in the literature: the most recent one is [Arora, 1998]. We would like to conclude these notes quoting from this paper:

> Simplifying the proofs of the results on PCP's is an extremely important task. Current proofs fill several dozen pages each. As has happened in many other areas in the past, the process of simplifying the proofs will very likely lead to new insights.

Section 7.5

BIBLIOGRAPHICAL NOTES

Approximation Preserving Reductions

WE HAVE seen in the preceding chapters that, even though the decision problems corresponding to most NP-hard optimization problems are polynomial-time Karp-reducible to each other, the optimization problems do not share the same approximability properties. The main reason of this fact is that Karp-reductions not always preserve the measure function and, even if this happens, they rarely preserve the quality of the solutions. It is then clear that a stronger kind of reducibility has to be used that not only maps instances of a problem $\mathcal{P}_1$ to instances of a problem $\mathcal{P}_2$, but it also maps back good solutions for $\mathcal{P}_2$ to good solutions for $\mathcal{P}_1$.

In this chapter, after briefly reviewing and extending the formal framework of approximation classes introduced in Chap. 3, we provide a formal definition of approximation preserving reducibility, called AP-reducibility, that allows us to compare the approximability properties of two problems and to formally state that one optimization problem is not harder to approximate than another.

Recall that the class PTAS is contained in the class APX and that this latter class is contained in the class NPO. Moreover, we have seen that if $P \neq NP$, then these inclusions are strict, i.e., there exists an optimization problem which can be approximated within a factor r_1 but cannot be approximated within a factor $r_2 < r_1$ (e.g., MINIMUM BIN PACKING) and there exists an optimization problem that cannot be approximated within any constant factor (e.g., MINIMUM TRAVELING SALESPERSON). It is then natural to look for other optimization problems that are included in APX − PTAS or in NPO − APX. In analogy to the P versus NP question,

Chapter 8

APPROXIMATION PRESERVING REDUCTIONS

one of the primary methods for doing so makes use of the notion of approximation preserving reducibility, which orders optimization problems with respect to their difficulty of being approximated. It is, in fact, interesting to study the maximal elements in APX (respectively, NPO) with respect to this order: these problems are called APX-complete (respectively, NPO-complete) and are important for two main reasons. On the one hand, we may try to solve, for example, the APX versus PTAS question by focusing on an APX-complete problem $\mathcal{P}$, since we can either prove that APX = PTAS by showing that $\mathcal{P}$ belongs to PTAS or prove that APX $\neq$ PTAS by showing that $\mathcal{P}$ does not belong to PTAS. On the other hand, an APX-completeness result may prevent wasting time searching for a polynomial-time approximation scheme for a specific problem.

In this chapter, we use the AP-reducibility to identify the hardest problems within NPO and APX, respectively: that is, we prove several completeness results both in NPO and in APX. Since approximation preserving reducibility is an essential tool to prove non-approximability results, we also give some hints on how one should proceed in developing new reductions and, at the same time, we prove negative results for several optimization problems.

8.1 The World of NPO Problems

IN THIS preliminary section, we will review the world of NPO problems according to their approximability properties. In particular, we first generalize the definition of approximation classes given in Chap. 3 on the basis of the input-dependent approximability introduced in Chap. 4.

Definition 8.1 ▶ *Given an optimization problem $\mathcal{P}$ in NPO, an approximation algorithm $\mathcal{A}$*
$r(n)$-*approximate algorithm*   *for $\mathcal{P}$, and a function $r : \mathbf{N} \mapsto (1, \infty)$, we say that $\mathcal{A}$ is an $r(n)$-approximate algorithm for $\mathcal{P}$ if, for any instance x such that $\text{SOL}(x) \neq \emptyset$, the performance ratio of the feasible solution $\mathcal{A}(x)$ with respect to x verifies the following inequality:*

$$R(x, \mathcal{A}(x)) \leq r(|x|).$$

Observe that, in the above definition, the behavior of an approximation algorithm is not specified whenever the input instance does not admit any feasible solution.

Definition 8.2 ▶ *Given a class of functions F, F-APX is the class of all NPO problems $\mathcal{P}$*
Class F-APX   *such that, for some function $r \in F$, there exists a polynomial-time $r(n)$-approximate algorithm for $\mathcal{P}$.*

> **Problem 8.1:** Minimum $\{0,1\}$-Linear Programming
>
> INSTANCE: Matrix $A \in Z^{m \cdot n}$, vector $b \in Z^m$, vector $w \in \mathbb{N}^n$.
>
> SOLUTION: Vector $x \in \{0,1\}^n$ such that $Ax \geq b$.
>
> MEASURE: The scalar product of w and x, i.e., $\sum_{i=1}^{n} w_i x_i$.

Section 8.1

THE WORLD OF NPO PROBLEMS

In particular, APX, log-APX, poly-APX, and exp-APX will denote the classes F-APX with F equal to the set of constant functions, to the set $O(\log n)$, to the set $\bigcup_{k>0} O(n^k)$, and to the set $\bigcup_{k>0} O(2^{n^k})$, respectively. Clearly, the following inclusions hold:

$$\text{PTAS} \subseteq \text{APX} \subseteq \text{log-APX} \subseteq \text{poly-APX} \subseteq \text{exp-APX} \subseteq \text{NPO}. \quad (8.1)$$

A puzzled reader could object that there is no difference between NPO and exp-APX since the polynomial bound on the computation time of the measure function, holding for each problem in NPO, implies that such a problem is $h2^{n^k}$-approximable for some h and k.

In fact, let $\mathcal{P} = (I, \text{SOL}, m, \text{goal})$ be a problem in NPO: then, since m is computable in polynomial time, there exist h and k such that, for any $x \in I$ with $|x| = n$ and for any $y \in \text{SOL}(x)$, $m(x,y) \leq h2^{n^k}$. This implies that any feasible solution has a performance ratio bounded by $h2^{n^k}$. However, NPO problems exist for which it is hard even to decide whether any feasible solution exists (and thus to find such a feasible solution), unless $P = NP$.

Let us consider Problem 8.1. We have already seen in Example 1.10 that, given an integer matrix A and an integer vector b, deciding whether a binary vector x exists such that $Ax \geq b$ is NP-hard. This implies that, if $P \neq NP$, then MINIMUM $\{0,1\}$-LINEAR PROGRAMMING, which clearly belongs to NPO, does not belong to exp-APX.

◀ Example 8.1

The previous example shows that if $P \neq NP$, then exp-APX is strictly contained in NPO. We have also seen, in the previous chapters, examples of problems in APX that are not in PTAS and examples of problems in poly-APX that are not in APX (unless $P = NP$). In general, it is possible to show that if $P \neq NP$, then the inclusions of Eq. (8.1) are all strict (see Exercise 8.1).

The main goal of this chapter is to yield a powerful tool to prove that a problem in one of the above classes does not belong to a smaller class. In particular, paraphrasing the approach used in the theory of NP-hardness,

Chapter 8

APPROXIMATION PRESERVING REDUCTIONS

we introduce an appropriate notion of reducibility between optimization problems that allows us to formally state that an optimization problem is as hard to approximate as another optimization problem.

8.2 AP-reducibility

AS STATED in Chap. 1, a reduction from a problem $\mathcal{P}_1$ to a problem $\mathcal{P}_2$ allows us to solve $\mathcal{P}_1$ by means of an algorithm solving $\mathcal{P}_2$. In the context of approximation algorithms, the reduction should guarantee that an approximate solution of $\mathcal{P}_2$ can be used to obtain an approximate solution for $\mathcal{P}_1$. In other words, the many-to-one polynomial-time reducibility, which has been defined to study decision problems, must be modified, since we need not only a function f mapping instances of $\mathcal{P}_1$ into instances of $\mathcal{P}_2$ but also a function g mapping back solutions of $\mathcal{P}_2$ into solutions of $\mathcal{P}_1$ (see Fig. 8.1). As a matter of fact, most of the known reductions between NP-complete problems explicitly define such a correspondence between the solutions.

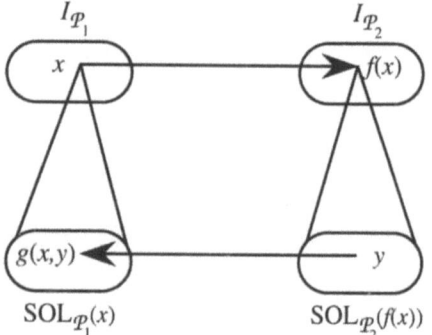

Figure 8.1 Reduction between two optimization problems

Example 8.2 ▶ Let us consider MAXIMUM CLIQUE and MINIMUM VERTEX COVER. Given a graph $G = (V,E)$, we consider the complement graph $G^c = (V,E^c)$ where E^c contains all edges that do not belong to E. For any $V' \subseteq V$, V' is a vertex cover in G^c if and only if $V - V'$ is a clique in G. Indeed, let $u,v \in V - V'$. If $(u,v) \notin E$ then $(u,v) \in E^c$ and this edge is not covered, thus contradicting the hypothesis on V'. Conversely, let $(u,v) \in E^c$. If $u,v \notin V'$ then $u,v \in V - V'$ and these two nodes are not connected, thus contradicting the hypothesis on $V - V'$. Then, for any $k \leq |V|$, G contains a clique of size k if and only if G^c contains a vertex cover of size $|V| - k$, thus proving that the corresponding two decision problems are many-to-one reducible to each other. In this case we have that $f(G) = G^c$ and, for any $V' \subseteq V$, $g(G,V') = V - V'$.

Section 8.2
AP-REDUCIBILITY

In a very broad sense, the reduction used in the above example is an approximation preserving reduction: in fact, any approximation algorithm for MINIMUM VERTEX COVER might be used to obtain an approximation algorithm for MAXIMUM CLIQUE. However, in order to preserve *guaranteed* approximation properties, the reduction from $\mathcal{P}_1$ to $\mathcal{P}_2$ has to satisfy the following basic fact: for any instance x of $\mathcal{P}_1$, if the performance ratio of the solution y of the corresponding instance $f(x)$ of $\mathcal{P}_2$ is at most r, then the performance ratio of the solution $g(x,y)$ of x is at most r', where r' may depend only on r. This is not the case for the reduction from MAXIMUM CLIQUE to MINIMUM VERTEX COVER.

◀ Example 8.3

For any n, let us consider the graph G_n containing two cliques of n nodes each and such that the i-th node of one clique is connected to all the nodes of the other clique but the i-th one. Any maximum clique of G contains n nodes. The complement graph G_n^c is formed by n disjoint pairs of connected nodes (see Fig. 8.2). The trivial solution $V' = V$ of G_n^c has performance ratio equal to 2. However, going back to G, the solution $V - V'$ is empty: clearly, this solution does not guarantee any constant performance ratio.

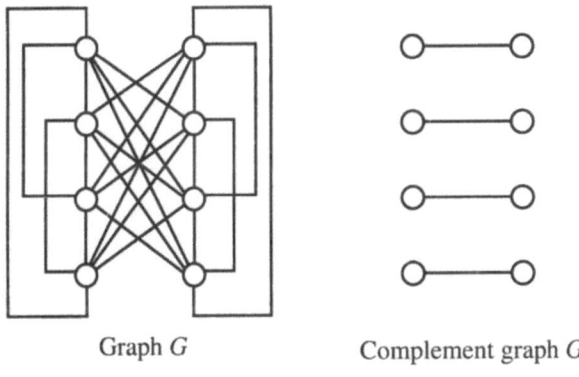

Graph G Complement graph G^c

Figure 8.2
An example of bad reduction

We are now ready to give the definition of an approximation preserving reducibility called AP-reducibility. Although different kind of reducibilities have been defined in the literature, AP reducibility is sufficiently general to encompass almost all known reducibilities while maintaining the property of establishing a linear relation between performance ratios: this is important in order to preserve membership in all approximation classes (see Bibliographical notes).

◀ Definition 8.3
AP-reducibility

Let $\mathcal{P}_1$ and $\mathcal{P}_2$ be two optimization problems in NPO. $\mathcal{P}_1$ is said to be AP-reducible *to* $\mathcal{P}_2$, in symbols $\mathcal{P}_1 \leq_{AP} \mathcal{P}_2$, if two functions f and g and a positive constant $\alpha \geq 1$ exist such that:

1. For any instance $x \in I_{\mathcal{P}_1}$ and for any rational $r > 1$, $f(x,r) \in I_{\mathcal{P}_2}$.

Chapter 8
APPROXIMATION PRESERVING REDUCTIONS

2. For any instance $x \in I_{\mathcal{P}_1}$ and for any rational $r > 1$, if $\text{SOL}_{\mathcal{P}_1}(x) \neq \emptyset$ then $\text{SOL}_{\mathcal{P}_2}(f(x,r)) \neq \emptyset$.

3. For any instance $x \in I_{\mathcal{P}_1}$, for any rational $r > 1$, and for any $y \in \text{SOL}_{\mathcal{P}_2}(f(x,r))$, $g(x,y,r) \in \text{SOL}_{\mathcal{P}_1}(x)$.

4. f and g are computable by two algorithms $\mathcal{A}_f$ and $\mathcal{A}_g$, respectively, whose running time is polynomial for any fixed rational r.

5. For any instance $x \in I_{\mathcal{P}_1}$, for any rational $r > 1$, and for any $y \in \text{SOL}_{\mathcal{P}_2}(f(x,r))$,

$$R_{\mathcal{P}_2}(f(x,r),y) \leq r \text{ implies } R_{\mathcal{P}_1}(x,g(x,y,r)) \leq 1+\alpha(r-1).$$

In the following we will refer to this condition as the AP-condition.

The triple (f,g,α) is said to be an AP-reduction from $\mathcal{P}_1$ to $\mathcal{P}_2$.

The acronym "AP" stands for *approximation preserving* and is justified by the following result.

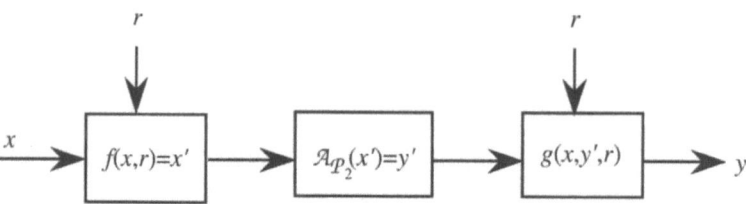

Figure 8.3
How to use an AP-reduction

Lemma 8.1 ▶ If $\mathcal{P}_1 \leq_{\text{AP}} \mathcal{P}_2$ and $\mathcal{P}_2 \in \text{APX}$ (respectively, $\mathcal{P}_2 \in \text{PTAS}$), then $\mathcal{P}_1 \in \text{APX}$ (respectively, $\mathcal{P}_1 \in \text{PTAS}$).

PROOF Let (f,g,α) be an AP-reduction from $\mathcal{P}_1$ to $\mathcal{P}_2$. If $\mathcal{P}_2 \in \text{APX}$ and $\mathcal{A}_{\mathcal{P}_2}$ is an algorithm for $\mathcal{P}_2$ with performance ratio at most r, then (see Fig. 8.3)

$$\mathcal{A}_{\mathcal{P}_1}(x) = g(x, \mathcal{A}_{\mathcal{P}_2}(f(x,r)), r)$$

is an algorithm for $\mathcal{P}_1$ with performance ratio at most $(1+\alpha(r-1))$. Analogously, if $\mathcal{P}_2 \in \text{PTAS}$ and $\mathcal{A}_{\mathcal{P}_2}$ is a polynomial-time approximation scheme for $\mathcal{P}_2$, then

$$\mathcal{A}_{\mathcal{P}_1}(x,r) = g(x, \mathcal{A}_{\mathcal{P}_2}(f(x,r'),r'),r')$$

is a polynomial-time approximation scheme for $\mathcal{P}_1$, where $r' = 1+(r-1)/\alpha$.

QED

Section 8.2
AP-REDUCIBILITY

Observe that Def. 8.3 actually introduces the notion of an approximation preserving reducibility *scheme* since, according to Fig. 8.1, for any r, two functions f_r and g_r exist that reduce $\mathcal{P}_1$ to $\mathcal{P}_2$. This extension is somewhat natural since there is no reason to ignore the quality of the solution we are looking for when reducing one optimization problem to another. However, in most (but not all) reductions that have appeared in the literature this knowledge is not required. For this reason, in the following whenever functions f and g do not use the dependency on r, we will avoid specifying this dependency. In other words, we will write $f(x)$ and $g(x,y)$ instead of $f(x,r)$ and $g(x,y,r)$, respectively.

◀ Example 8.4

The proof of Lemma 6.4 is an example of AP-reduction where f and g do not depend on the performance ratio and $\alpha = 1$.

◀ Example 8.5

Let us consider MAXIMUM CLIQUE and MAXIMUM INDEPENDENT SET and let us define the following reduction from the former problem to the latter one. Given a graph $G = (V,E)$, $f(G)$ is the complement graph G^c. Clearly, for any $V' \subseteq V$, V' is an independent set in G^c if and only if V' is a clique in G. If we set $g(G,V') = V'$, we then have that f and g satisfy the AP-condition with $\alpha = 1$. As an immediate consequence of this reduction, of Lemma 8.1, and of Theorem 6.5, we have that MAXIMUM INDEPENDENT SET does not belong to APX.

The following definition introduces a different notion of approximation preserving reducibility that is not as powerful as AP-reducibility (see Bibliographical notes) but will turn out to be very useful to prove the existence of AP-reductions. Intuitively, this notion is based on showing both a *linear* relation between the optimal measures and a *linear* relation between the absolute errors (this is why this reducibility is called L-reducibility).

◀ Definition 8.4
L-reducibility

Let $\mathcal{P}_1$ and $\mathcal{P}_2$ be two NPO problems. $\mathcal{P}_1$ is said to be L-reducible to $\mathcal{P}_2$, in symbols $\mathcal{P}_1 \leq_L \mathcal{P}_2$, if two functions f and g and two positive constants β and γ exist such that:

1. For any $x \in I_{\mathcal{P}_1}$, $f(x) \in I_{\mathcal{P}_2}$ is computable in polynomial time.

2. For any $x \in I_{\mathcal{P}_1}$, if $\text{SOL}_{\mathcal{P}_1}(x) \neq \emptyset$ then $\text{SOL}_{\mathcal{P}_2}(f(x)) \neq \emptyset$.

3. For any $x \in I_{\mathcal{P}_1}$ and for any $y \in \text{SOL}_{\mathcal{P}_2}(f(x))$, $g(x,y) \in \text{SOL}_{\mathcal{P}_1}(x)$ is computable in polynomial time.

4. For any $x \in I_{\mathcal{P}_1}$, $m^*_{\mathcal{P}_2}(f(x)) \leq \beta m^*_{\mathcal{P}_1}(x)$.

5. For any $x \in I_{\mathcal{P}_1}$ and for any $y \in \text{SOL}_{\mathcal{P}_2}(f(x))$,

$$|m^*_{\mathcal{P}_1}(x) - m_{\mathcal{P}_1}(x,g(x,y))| \leq \gamma |m^*_{\mathcal{P}_2}(f(x)) - m_{\mathcal{P}_2}(f(x),y)|.$$

The quadruple (f,g,β,γ) is said to be an L-reduction *from* $\mathcal{P}_1$ *to* $\mathcal{P}_2$.

Chapter 8
APPROXIMATION PRESERVING REDUCTIONS

Observe that both reductions considered in Examples 8.4 and 8.5 are L-reductions with $\beta = \gamma = 1$. The following lemma, whose proof is left to the reader (see Exercise 8.2), states that, whenever we restrict our attention to problems in APX, the existence of L-reductions implies the existence of AP-reductions: this lemma will be repeatedly used in the rest of the chapter.

Lemma 8.2 ▶ Let $\mathcal{P}_1$ and $\mathcal{P}_2$ be two NPO problems such that $\mathcal{P}_1 \leq_L \mathcal{P}_2$. If $\mathcal{P}_1 \in$ APX, then $\mathcal{P}_1 \leq_{AP} \mathcal{P}_2$.

Example 8.6 ▶ Let us prove the existence of an L-reduction from MAXIMUM CUT to MAXIMUM 3-SATISFIABILITY. Given a graph $G = (V, E)$, the function f is defined as $f(G) = (U_G, C_G)$ where

1. The set U_G of Boolean variables includes, for any node u, a variable x_u and, for any edge e, a variable x_e.

2. The set C_G of clauses is defined as follows. For any edge $e = (u, v)$, let

$$C_e = \{\bar{x}_u \vee x_v \vee x_e,\ x_u \vee \bar{x}_v \vee x_e,\ x_u \vee x_v \vee \bar{x}_e,\ \bar{x}_u \vee \bar{x}_v \vee \bar{x}_e,\ x_e\}.$$

Let $C_G = \bigcup_{e \in E} C_e$.

For any truth assignment τ, let us define a corresponding cut $g(G, \tau)$ as follows. First, τ is transformed into a truth assignment τ' satisfying at least four clauses of each C_e: this can be done by letting, for any node u, $\tau'(x_u)$ be equal to $\tau(x_u)$ and, for any edge $e = (u,v)$, $\tau'(x_e)$ be equal to the exclusive-or between $\tau(x_u)$ and $\tau(x_v)$. Clearly, τ' satisfies at least as many clauses as τ. Then, τ' is transformed into a cut (V_1, V_2) such that, for any $u \in V$, $u \in V_1$ if and only if x_u has been assigned the value TRUE.

The number of edges between V_1 and V_2 is exactly equal to the number of C_e-sets that are fully satisfied, that is, $m(G, (V_1, V_2)) = m((U_G, C_G), \tau') - 4|E|$. Hence, $m^*(U_G, C_G) \leq 4|E| + m^*(G)$. Actually, the equality holds since, for any cut (V_1, V_2) in G, the truth assignment giving value TRUE to any variable corresponding to a node in V_1 and to any variable corresponding to an edge going from V_1 to V_2 satisfies exactly $m(G, (V_1, V_2)) + 4|E|$ clauses.

From the proof of Theorem 2.14, it follows that $m^*(G) \geq |E|/2$. Then

$$m^*(U_G, C_G) \leq 9m^*(G),$$

and

$$\begin{aligned} m^*(G) - m(G, (V_1, V_2)) &= m^*(U_G, C_G) - m((U_G, C_G), \tau') \\ &\leq m^*(U_G, C_G) - m((U_G, C_G), \tau). \end{aligned}$$

In conclusion, the quadruple $(f, g, 9, 1)$ is an L-reduction from MAXIMUM CUT to MAXIMUM 3-SATISFIABILITY. Since MAXIMUM CUT belongs to APX (see Theorem 2.14), Lemma 8.2 implies that the above reduction is also an AP-reduction.

8.2.1 Complete problems

It is possible to see that the AP-reducibility is transitive, that is, if $\mathcal{P}_1 \leq_{AP} \mathcal{P}_2$ and $\mathcal{P}_2 \leq_{AP} \mathcal{P}_3$, then $\mathcal{P}_1 \leq_{AP} \mathcal{P}_3$ (see Exercise 8.3). It follows that this reducibility induces a partial order among problems in the same approximation class. The notion of a complete problem is formally equivalent to that of a maximum problem with respect to this order. Complete problems are important for two main reasons. On the one hand, we may try to solve, for example, the APX versus PTAS question by focusing on an APX-complete problem $\mathcal{P}$, since we can either prove that APX = PTAS by showing that $\mathcal{P}$ belongs to PTAS or prove that APX $\neq$ PTAS by showing that $\mathcal{P}$ does not belong to PTAS. On the other hand, a completeness result may prevent wasting time searching for a better approximation algorithm for a specific problem.

◀ Definition 8.5
C-hard problem

Given a class C of NPO *problems, a problem* $\mathcal{P}$ *is C-hard (with respect to the AP-reducibility) if, for any* $\mathcal{P}' \in C$, $\mathcal{P}' \leq_{AP} \mathcal{P}$. *A C-hard problem is C-complete (with respect to the AP-reducibility) if it belongs to C.*

According to Lemma 8.1 and to the above definition, we have that, for any class C of NPO problems such that $C \not\subseteq$ APX (respectively, $C \not\subseteq$ PTAS), if $\mathcal{P}$ is a C-complete problem, then $\mathcal{P} \notin$ APX (respectively, $\mathcal{P} \notin$ PTAS). In other words, C-complete problems are indeed the hardest ones in class C.

In the following two sections, we focus our attention on classes NPO and APX and show the existence of complete problems in both classes (similar results can be obtained for the other classes; see Bibliographical notes). If we show that a problem $\mathcal{P}$ is NPO-complete, then we have that $\mathcal{P} \notin$ APX, unless P = NP. Moreover, if we show that a problem $\mathcal{P}$ is APX-complete, then we have that $\mathcal{P} \notin$ PTAS, unless P = NP. Hence, the completeness results we will prove automatically yield non-approximability results.

Section 8.3
NPO-COMPLETENESS

8.3 NPO-completeness

IN ORDER to prove the first NPO-completeness result we consider Problem 8.2. This problem clearly belongs to NPO but does not belong to exp-APX (unless P = NP) since finding a feasible solution is as hard as SATISFIABILITY. This observation is consistent with the next theorem, which states that MAXIMUM WEIGHTED SATISFIABILITY is one of the hardest problems in the class of NPO maximization problems.

◀ Theorem 8.3

MAXIMUM WEIGHTED SATISFIABILITY *is complete for the class of maximization problems in* NPO.

Chapter 8

APPROXIMATION PRESERVING REDUCTIONS

> **Problem 8.2: Maximum Weighted Satisfiability**
>
> INSTANCE: Boolean formula φ with variables $x_1, \ldots, x_n$ of nonnegative weights $w_1, \ldots, w_n$.
>
> SOLUTION: Truth assignment τ to the variables that satisfies φ.
>
> MEASURE: $\max(1, \sum_{i=1}^{n} w_i \tau(x_i))$, where the Boolean values TRUE and FALSE are identified with 1 and 0, respectively.

The proof of the above theorem is left to the reader (see Exercise 8.4) and proceeds as follows: first, associate with any NPO problem a suitably defined nondeterministic Turing machine and, then, apply a variation of Cook-Levin's theorem to show that MAXIMUM WEIGHTED SATISFIABILITY is complete for the class of maximization problems in NPO.

Theorem 8.3, however, does not imply the NPO-completeness of MAXIMUM WEIGHTED SATISFIABILITY since we have restricted this class to maximization problems. However, we can consider the minimization version of MAXIMUM WEIGHTED SATISFIABILITY, called MINIMUM WEIGHTED SATISFIABILITY, and we can prove that MINIMUM WEIGHTED SATISFIABILITY is complete for the class of minimization problems in NPO. The following theorem shows that these two problems are reducible to each other via AP-reductions and, therefore, they are both NPO-complete.

Theorem 8.4 ▶ MAXIMUM WEIGHTED SATISFIABILITY *is AP-reducible to* MINIMUM WEIGHTED SATISFIABILITY, *and vice versa*.

PROOF We show that an AP-reduction exists from MAXIMUM WEIGHTED SATISFIABILITY to MINIMUM WEIGHTED SATISFIABILITY (a similar proof shows the existence of the converse reduction). First, observe that it is possible to restrict MAXIMUM WEIGHTED SATISFIABILITY to instances φ such that: (1) φ is defined over variables $v_1, \ldots, v_s$ with weights $w(v_i) = 2^{s-i}$, for $i = 1, \ldots, s$, and some other variables with weight zero and (2) each truth assignment that satisfies φ must assign the value TRUE to at least one v-variable (see the hint of Exercise 8.4).

Let x be such an instance of MAXIMUM WEIGHTED SATISFIABILITY. We first give a simple reduction with α depending on r, and then modify it to obtain an AP-reduction.

Let $f(x)$ be the formula $\varphi \wedge \alpha_1 \wedge \cdots \wedge \alpha_s$ where α_i is $z_i \equiv (\bar{v}_1 \wedge \cdots \wedge \bar{v}_{i-1} \wedge v_i)$, $z_1, \ldots, z_s$ are new variables with weights $w(z_i) = 2^i$ for $i = 1, \ldots, s$, and all other variables (even the v-variables) have zero weight. If y is a satisfying truth assignment for $f(x)$, let $g(x,y)$ be the restriction of y to the variables that occur in φ. This assignment clearly satisfies φ.

Note that exactly one among the z-variables is TRUE in any satisfying truth assignment of $f(x)$. If all z-variables were FALSE, then all v-variables would be FALSE, which is not allowed. On the other hand, it is clearly not possible that two z-variables are TRUE. Hence, for any feasible solution y of $f(x)$, we have that $m(f(x),y) = 2^i$, for some i with $1 \leq i \leq s$. Since

$$m(f(x),y) = 2^i \Leftrightarrow z_i = 1$$
$$\Leftrightarrow v_1 = v_2 = \cdots = v_{i-1} = 0, v_i = 1$$
$$\Leftrightarrow 2^{s-i} \leq m(x,g(x,y)) < 2 \cdot 2^{s-i},$$

we have that, for any feasible solution y of $f(x)$,

$$\frac{2^s}{m(f(x),y)} \leq m(x,g(x,y)) < 2\frac{2^s}{m(f(x),y)}.$$

This is in particular true for the optimal solution (observe that any satisfying truth assignment for x can be easily extended to a satisfying truth assignment for $f(x)$). Thus the performance ratio of $g(x,y)$ with respect to x is

$$R(x,g(x,y)) = \frac{m^*(x)}{m(x,g(x,y))} < \frac{2\frac{2^s}{m^*(f(x))}}{\frac{2^s}{m(f(x),y)}}$$
$$= 2\frac{m(f(x),y)}{m^*(f(x))} = 2R(f(x),y).$$

We have thus described a reduction that satisfies the AP-condition with $\alpha = (2r-1)/(r-1)$. This is, however, not sufficient since we want α to be a constant that does not depend on r. In order to achieve this result, let us extend the above construction in order to obtain

$$R(x,g(x,y)) \leq (1+2^{-k})R(f_k(x),y),$$

for every nonnegative integer k (notice that the reduction described above corresponds to $k = 0$).

We use $2^k s$ new variables named $z_{i,b_1,\ldots,b_k}$, where $i = 1,\ldots,s$ and $b_j \in \{0,1\}$, for $j = 1,\ldots,k$. Let

$$f_k(x) = \varphi \wedge \bigwedge \alpha_{i,b_1,\ldots,b_k},$$

Section 8.3

NPO-COMPLETENESS

where, for $i = 1, \ldots, s$ and for any $(b_1, \ldots, b_k) \in \{0, 1\}^k$, $\alpha_{i, b_1, \ldots, b_k}$ is the formula

$$z_{i, b_1, \ldots, b_k} \equiv (\bar{v}_1 \wedge \cdots \wedge \bar{v}_{i-1} \wedge v_i \wedge (v_{i+1} \equiv b_1) \wedge \cdots \wedge (v_{i+k} \equiv b_k)).$$

Moreover, let us define

$$w(z_{i, b_1, \ldots, b_k}) = \left\lceil \frac{K \cdot 2^s}{w(v_i) + \sum_{j=1}^{k} b_j w(v_{i+j})} \right\rceil = \left\lceil \frac{K \cdot 2^i}{1 + \sum_{j=1}^{k} b_j 2^{-j}} \right\rceil$$

(observe that by choosing K large enough, i.e., greater than 2^s, we can disregard the effect of the ceiling operation in the following computations). Finally, for any truth assignment y for $f_k(x)$, we define $g(x, y)$ as above.

As in the previous reduction exactly one of the z-variables is TRUE in any satisfying truth assignment of $f_k(x)$. If, in a solution y of $f_k(x)$, we have that $z_{i, b_1, \ldots, b_k} = 1$, then $m(f_k(x), y) = w(z_{i, b_1, \ldots, b_k})$ and we know that

$$m(x, g(x, y)) \geq w(v_i) + \sum_{j=1}^{k} b_j w(v_{i+j}) = 2^{s-i} (1 + \sum_{j=1}^{k} b_j 2^{-j})$$

and that

$$m(x, g(x, y)) \leq w(v_i) + \sum_{j=1}^{k} b_j w(v_{i+j}) + \sum_{j=k+i+1}^{s} w(v_j)$$
$$< 2^{s-i} (1 + \sum_{j=1}^{k} b_j 2^{-j})(1 + 2^{-k}).$$

Thus we get

$$\frac{K \cdot 2^s}{m(f_k(x), y)} \leq m(x, g(x, y)) < \frac{K \cdot 2^s}{m(f_k(x), y)} (1 + 2^{-k}),$$

and, therefore, $R(x, g(x, y)) < (1 + 2^{-k}) R(f_k(x), y)$.

Given any $r > 1$, if we choose k such that $2^{-k} \leq (r-1)/r$, e.g. $k_r = \lceil \log r - \log(r-1) \rceil$, then $R(f_k(x), y) \leq r$ implies $R(x, g(x, y)) < (1 + 2^{-k}) R(f_k(x), y) \leq r + r2^{-k} \leq r + r - 1 = 1 + 2(r-1)$. Hence, if we define $f(x, r) = f_{k_r}(x)$, then $(f, g, 2)$ is an AP-reduction from MAXIMUM WEIGHTED SATISFIABILITY to MINIMUM WEIGHTED SATISFIABILITY: observe how, in this case, function f uses the possibility of depending on r.

QED

Corollary 8.5 ▶ *The two problems* MAXIMUM WEIGHTED SATISFIABILITY *and* MINIMUM WEIGHTED SATISFIABILITY *are* NPO-*complete*.

8.3.1 Other NPO-complete problems

In this section, we prove the NPO-completeness of other NPO problems by using reductions from MAXIMUM WEIGHTED SATISFIABILITY and MINIMUM WEIGHTED SATISFIABILITY.

◀ Example 8.7

From Example 6.5 and from Exercise 6.9 it easily follows that MAXIMUM WEIGHTED SATISFIABILITY is AP-reducible to MAXIMUM WEIGHTED 3-SATISFIABILITY, that is, the variation in which the input formula is in conjunctive normal form and each clause contains at most three literals. Hence, MAXIMUM WEIGHTED 3-SATISFIABILITY is NPO-complete. Similarly, we can prove that MINIMUM WEIGHTED 3-SATISFIABILITY is NPO-complete.

◀ Example 8.8

In this example, we give an AP-reduction with $\alpha = 1$ from MINIMUM WEIGHTED 3-SATISFIABILITY to MINIMUM $\{0,1\}$-LINEAR PROGRAMMING.

Let φ be a Boolean formula in conjunctive normal form and denote by $u_1, \ldots, u_n$ the variables and by $c_1, \ldots, c_m$ the clauses of such a formula. First, the matrix A is defined as follows: $a_{ij} = 1$ if u_j occurs in c_i while $a_{ij} = -1$ if $\overline{u_j}$ occurs in c_i. In all other cases, $a_{ij} = 0$. Then, for any $i \leq m$, b_i is defined as 1 minus the number of negated variables in c_i. Finally, for any $i \leq m$, w_i is defined as the weight of u_i.

For example, consider the formula φ with $c_1 = (u_1 \vee \overline{u_2} \vee u_3)$, $c_2 = (u_4 \vee \overline{u_5})$, and $c_3 = (u_2 \vee u_3 \vee u_4)$ and with the weights of u_1, u_2, u_3, u_4, and u_5 defined as $2, 3, 0, 1$, and 2, respectively. In the corresponding instance of MINIMUM $\{0,1\}$-LINEAR PROGRAMMING we are asked to minimize the value

$$2x_1 + 3x_2 + x_4 + 2x_5$$

subject to the following constraints:

$$
\begin{aligned}
+x_1 - x_2 + x_3 & \geq 0 \\
+x_4 - x_5 & \geq 0 \\
+x_2 + x_3 + x_4 & \geq 1
\end{aligned}
$$

and $x_i \in \{0,1\}$.

It is easy to see that any solution of the instance of MINIMUM $\{0,1\}$-LINEAR PROGRAMMING corresponds to a truth assignment for φ having the same measure. Indeed, it is sufficient to interpret a 1 as the Boolean value TRUE and a 0 as the Boolean value FALSE.

This proves that MINIMUM $\{0,1\}$-LINEAR PROGRAMMING is NPO-complete. Similarly, we can prove that MAXIMUM $\{0,1\}$-LINEAR PROGRAMMING is NPO-complete.

8.3.2 Completeness in exp-APX

Recall that the main difference between NPO and exp-APX is that problems in the former class may have instances for which even deciding whether

Chapter 8
APPROXIMATION PRESERVING REDUCTIONS

the set of feasible solutions is empty is a difficult task. On the other hand, the existence of an exponentially approximate algorithm ensures that, for any problem in the latter class, it is possible to decide in polynomial time whether the set of feasible solutions is empty.

In order to prove completeness results in exp-APX we then slightly modify the definition of MINIMUM WEIGHTED SATISFIABILITY by artificially adding to the set of feasible solutions a trivial one, that is, the truth assignment that gives the value TRUE to any variable. Clearly, this assignment may not satisfy the Boolean formula but, in a certain sense, this is not important since this solution has the maximum possible measure.

It is then not difficult to show that this variant of MINIMUM WEIGHTED SATISFIABILITY is complete for the class of minimization problems in exp-APX (see Exercise 8.6). Moreover, starting from this problem it is possible to prove that MINIMUM TRAVELING SALESPERSON is also complete for the same class (see Exercise 8.7): this is not surprising since we have already seen in Chap. 3 that MINIMUM TRAVELING SALESPERSON is not approximable (unless P = NP).

8.4 APX-completeness

IN THIS section we see how the PCP-characterization of NP not only allows us to obtain non-approximability results (see Chap. 6) but it can also be applied in order to prove APX-completeness results. In particular, we will first show that MAXIMUM 3-SATISFIABILITY is APX-complete.

To this aim, let us consider again the proof of Theorem 6.3. That proof shows that there exist a constant ε and two functions f_s and g_s such that, for any Boolean formula φ in conjunctive normal form, $\psi = f_s(\varphi)$ is a Boolean formula in conjunctive normal form, with at most three literals per clause, that satisfies the following claim: for any truth assignment τ satisfying at least $1 - \varepsilon$ of the maximum number of satisfiable clauses of ψ, $g_s(\varphi, \tau)$ satisfies φ if and only if φ is satisfiable. We will make use of this fact in the proof of the following theorem which is the first step towards the APX-completeness of MAXIMUM 3-SATISFIABILITY.

Theorem 8.6 ▶ MAXIMUM 3-SATISFIABILITY *is complete for the class of maximization problems in* APX.

PROOF Let $\mathcal{P}$ be a maximization problem in APX and let $\mathcal{A}_\mathcal{P}$ be an $r_\mathcal{P}$-approximate algorithm for $\mathcal{P}$. Our goal is to define an AP-reduction (f, g, α) from $\mathcal{P}$ to MAXIMUM 3-SATISFIABILITY. First of all, we define

$$\alpha = 2(r_\mathcal{P} \log r_\mathcal{P} + r_\mathcal{P} - 1)\frac{1+\varepsilon}{\varepsilon},$$

Section 8.4
APX-COMPLETENESS

Program 8.1: APX-Nondeterministic

input Instance x of APX problem $\mathcal{P}$, integer $i \geq 0$;
output YES if there exists a feasible solution in the i-th interval;
begin
  guess y with $|y| \leq p(|x|)$;
  if $y \in \text{SOL}_\mathcal{P}(x) \wedge m_\mathcal{P}(x,y) \geq r_n^i t_\mathcal{P}(x) \wedge m_\mathcal{P}(x,y) \leq r_n^{i+1} t_\mathcal{P}(x)$ **then**
    return YES
  else
    return NO
end.

where ε is the constant of the observation preceding the theorem. In order to define the functions f and g, for any given r, we distinguish two cases.

1. $1 + \alpha(r-1) \geq r_\mathcal{P}$. In this case, for any x and for any y, we simply define
$$g(x,y,r) = \mathcal{A}_\mathcal{P}(x).$$
Hence, $R(x, g(x,y,r)) = R(x, \mathcal{A}_\mathcal{P}(x)) \leq r_\mathcal{P} \leq 1 + \alpha(r-1)$ and the AP-condition is satisfied (without having to define or use f at all).

2. $1 + \alpha(r-1) < r_\mathcal{P}$. In this case, let $r_n = 1 + \alpha(r-1)$. We then have that
$$r = \frac{r_n - 1}{\alpha} + 1 = \frac{\varepsilon}{2(1+\varepsilon)} \frac{r_n - 1}{r_\mathcal{P} \log r_\mathcal{P} + r_\mathcal{P} - 1} + 1 < \frac{\varepsilon}{2k(1+\varepsilon)} + 1$$
where $k = \lceil \log_{r_n} r_\mathcal{P} \rceil$ and the last inequality is due to the fact that
$$k \leq \frac{\log r_\mathcal{P}}{\log r_n} + 1 \leq \frac{r_n \log r_\mathcal{P}}{r_n - 1} + 1 < \frac{r_\mathcal{P} \log r_\mathcal{P} + r_\mathcal{P} - 1}{r_n - 1}.$$

Let $t_\mathcal{P}(x)$ denote the measure of $\mathcal{A}_\mathcal{P}(x)$. We then partition the interval $[t_\mathcal{P}(x), r_\mathcal{P} t_\mathcal{P}(x)]$ into the following k subintervals:
$$[t_\mathcal{P}(x), r_n t_\mathcal{P}(x)], [r_n t_\mathcal{P}(x), r_n^2 t_\mathcal{P}(x)], \ldots, [r_n^{k-1} t_\mathcal{P}(x), r_\mathcal{P} t_\mathcal{P}(x)].$$

Observe that $t_\mathcal{P}(x) \leq m_\mathcal{P}^*(x) \leq r_\mathcal{P} t_\mathcal{P}(x) \leq r_n^k t_\mathcal{P}(x)$, that is, the optimal measure for the input x belongs to one of the above subintervals. For any $i = 0, \ldots, k-1$, let us consider the nondeterministic Algorithm 8.1 where p denotes a polynomial bounding the length of the feasible solutions of $\mathcal{P}$.

Without loss of generality, we may assume that the solution y and its measure $m_\mathcal{P}(x,y)$ are written onto special cell tapes and are not cancelled. By applying the technique of the proof of the Cook-Levin's theorem (see also Exercise 8.4), it is then possible to derive k Boolean formulas $\varphi_0, \ldots, \varphi_{k-1}$ such that, given a truth assignment τ_i satisfying φ_i, for $i = 0, \ldots, k-1$, a feasible solution y of x can be computed in polynomial time with $m_{\mathcal{P}}(x, y) \geq r_n^j t_{\mathcal{P}}(x)$. We then define the instance ψ of MAXIMUM 3-SATISFIABILITY as follows:

$$\psi = f(x, r) = f_s(\varphi_0) \wedge \ldots \wedge f_s(\varphi_{k-1}),$$

where f_s is the function deriving from Theorem 6.3 (see the observation preceding the theorem). Without loss of generality, we may assume that each $f_s(\varphi_i)$ contains the same number m of clauses.

Let τ be any truth assignment to the variables of ψ whose performance ratio is at most r. Then

$$m^*(\psi) - m(\psi, \tau) \leq m^*(\psi) \frac{r-1}{r} \leq km \frac{r-1}{r}.$$

On the other hand, if r_i denotes the performance ratio of τ with respect to $f_s(\varphi_i)$, we have that

$$\begin{aligned} m^*(\psi) - m(\psi, \tau) &\geq m^*(f_s(\varphi_i)) - m(f_s(\varphi_i), \tau) \\ &= m^*(f_s(\varphi_i)) \frac{r_i - 1}{r_i} \geq \frac{m}{2} \frac{r_i - 1}{r_i}, \end{aligned}$$

where the last inequality is due to the fact that at least half of the clauses of a Boolean formula in conjunctive normal form are always satisfiable (see Theorem 2.19).

We then have that, for $i = 0, \ldots, k-1$,

$$\frac{m}{2} \frac{r_i - 1}{r_i} \leq km \frac{r-1}{r};$$

that is,

$$1 - 2k \frac{r-1}{r} \leq \frac{1}{r_i}.$$

By making use of the fact that $r < 1 + \varepsilon/(2k(1+\varepsilon))$ and by simple algebraic manipulations, we obtain that

$$r_i \leq 1 + \varepsilon.$$

From the observation preceding the theorem, it follows that, for $i = 0, \ldots, k-1$, $\tau_i = g_s(\varphi_i, \tau)$ satisfies φ_i if and only if φ_i is satisfiable.

Let j be the maximum i such that τ_i satisfies φ_i (observe that such a j must exist). Then

$$r_n^j t_{\mathcal{P}}(x) \leq m_{\mathcal{P}}^*(x) \leq r_n^{j+1} t_{\mathcal{P}}(x),$$

and from τ_j we can derive a solution y whose measure is at least $r_n^j t_{\mathcal{P}}(x)$, that is, y is r_n-approximate. In other words, $R(x,y) \leq r_n = 1 + \alpha(r-1)$ and the AP-condition turns out to be satisfied.

Since $\mathcal{P}$ was an arbitrary maximization problem in APX, it follows that MAXIMUM 3-SATISFIABILITY is complete for the class of maximization problems in APX.   QED

In order to prove that the MAXIMUM 3-SATISFIABILITY problem is APX-complete, it now suffices to show that any minimization problem in APX has a maximization counterpart which is at least as hard as the original problem with respect to the AP-reducibility.

For any minimization problem $\mathcal{P}$ in APX, a maximization problem $\mathcal{P}'$ in APX exists such that $\mathcal{P} \leq_{AP} \mathcal{P}'$. ◀ Theorem 8.7

Let $\mathcal{A}$ be an r-approximate algorithm for $\mathcal{P}$. For any x, let $t(x) = m_{\mathcal{P}}(x, \mathcal{A}(x))$. This means that $t(x) \leq r m_{\mathcal{P}}^*(x)$. Problem $\mathcal{P}'$ is identical to $\mathcal{P}$ apart from the measure function which is defined as follows: PROOF

$$m_{\mathcal{P}'}(x,y) = \begin{cases} (k+1)t(x) - k m_{\mathcal{P}}(x,y) & \text{if } m_{\mathcal{P}}(x,y) \leq t(x), \\ t(x) & \text{otherwise,} \end{cases}$$

where $k = \lceil r \rceil$. Clearly, $t(x) \leq m_{\mathcal{P}'}^*(x) \leq (k+1)t(x)$: this implies that $\mathcal{A}$ is a $(k+1)$-approximate algorithm for $\mathcal{P}'$, that is, $\mathcal{P}' \in$ APX. The reduction between $\mathcal{P}$ and $\mathcal{P}'$ is defined in the following way (observe that f and g do not depend on r).

1. For any instance x, $f(x) = x$.

2. For any instance x and for any solution y of $f(x)$,

$$g(x,y) = \begin{cases} y & \text{if } m_{\mathcal{P}}(x,y) \leq t(x), \\ \mathcal{A}(x) & \text{otherwise.} \end{cases}$$

3. $\alpha = k+1$.

To show that this reduction satisfies the AP-condition, let y be a solution such that $R_{\mathcal{P}'}(x,y) = m_{\mathcal{P}'}^*(x)/m_{\mathcal{P}'}(x,y) \leq r'$: we have to show that $R_{\mathcal{P}}(x, g(x,y)) \leq 1 + \alpha(r'-1)$. To this aim, let us distinguish the following two cases.

Chapter 8

APPROXIMATION
PRESERVING
REDUCTIONS

1. $m_{\mathcal{P}}(x,y) \leq t(x)$. In this case, $g(x,y)$ is equal to y. Moreover, since $m^*_{\mathcal{P}'}(x)/r' \leq m_{\mathcal{P}'}(x,y)$, $t(x) \leq m^*_{\mathcal{P}'}(x) \leq (k+1)t(x)$, $k \geq r$, and $t(x) \leq rm^*_{\mathcal{P}}(x)$, we have, by definition of α, that

$$\begin{aligned}
m_{\mathcal{P}}(x,y) &= \frac{(k+1)t(x) - m_{\mathcal{P}'}(x,y)}{k} \\
&\leq \frac{(k+1)t(x) - m^*_{\mathcal{P}'}(x)/r'}{k} \\
&= \frac{(k+1)t(x) - m^*_{\mathcal{P}'}(x)/(1+(r'-1))}{k} \\
&\leq \frac{(k+1)t(x) - (1-(r'-1))m^*_{\mathcal{P}'}(x)}{k} \\
&= \frac{(k+1)t(x) - m^*_{\mathcal{P}'}(x) + (r'-1)m^*_{\mathcal{P}'}(x)}{k} \\
&\leq m^*_{\mathcal{P}}(x) + \frac{r'-1}{k} m^*_{\mathcal{P}'}(x) \\
&\leq m^*_{\mathcal{P}}(x) + \frac{r'-1}{k}(k+1)t(x) \\
&\leq m^*_{\mathcal{P}}(x) + (r'-1)(k+1)\frac{t(x)}{r} \\
&\leq (1+\alpha(r'-1))m^*_{\mathcal{P}}(x);
\end{aligned}$$

that is, $R_{\mathcal{P}}(x, g(x,y)) = R_{\mathcal{P}}(x,y) \leq 1 + \alpha(r'-1)$.

2. $m_{\mathcal{P}}(x,y) > t(x)$. In this case, since $\alpha \geq 1$, we have that

$$R_{\mathcal{P}}(x, g(x,y)) = R_{\mathcal{P}}(x, \mathcal{A}(x)) = R_{\mathcal{P}'}(x,y) \leq r' \leq 1 + \alpha(r'-1).$$

QED

We then have that in both cases the AP-condition is satisfied so that the above reduction is an AP-reduction and $\mathcal{P} \leq_{AP} \mathcal{P}'$.

As a consequence of the previous two theorems, we have the following completeness result within APX.

Corollary 8.8 ▶ MAXIMUM 3-SATISFIABILITY *is* APX-*complete*.

8.4.1 Other APX-complete problems

STARTING FROM MAXIMUM 3-SATISFIABILITY, it is possible to prove the APX-completeness of other interesting problems. But how should we go about showing the APX-hardness of a given problem? The first step

Section 8.4
APX-COMPLETENESS

is to check whether it is already known to be APX-hard. For instance, the problem might be listed in the compendium contained at the end of this book. If we are lucky, we will answer the question as easily as that!

Using NP-hardness proofs

In case the problem is not known to be APX-hard, the first thing we can do is to check the NP-hardness proof of $\mathcal{P}$ (if any exists). It may well happen that this proof already corresponds to an approximation preserving reduction. Recall from Lemma 8.2 that an L-reduction from a problem in APX is also an AP-reduction. Thus, to show APX-completeness of a problem $\mathcal{P} \in$ APX it is enough to show that there is an L-reduction from some APX-complete problem to $\mathcal{P}$.

As an example, we prove the APX-completeness of MAXIMUM CUT by passing through two other intermediate problems, namely, the MAXIMUM 2-SATISFIABILITY problem and the MAXIMUM NOT-ALL-EQUAL 3-SATISFIABILITY problem. All the intermediate reductions are indeed the reductions used to prove the NP-hardness of these problems.

MAXIMUM 3-SATISFIABILITY $\leq_L$ MAXIMUM 2-SATISFIABILITY. ◂ Theorem 8.9

Given an instance φ of MAXIMUM 3-SATISFIABILITY with m clauses, let $a_i \lor b_i \lor c_i$ be the i-th clause, for $i = 1, \ldots, m$, where each a_i, b_i, and c_i represents either a variable or its negation (without loss of generality, we may assume that each clause contains exactly three literals, see Exercise 8.8). To this clause, we associate the following ten new clauses with at most two literals per clause: PROOF

$$a_i, \ b_i, \ c_i, \ d_i, \ \overline{a_i} \lor \overline{b_i}, \ \overline{a_i} \lor \overline{c_i}, \ \overline{b_i} \lor \overline{c_i}, \ a_i \lor \overline{d_i}, \ b_i \lor \overline{d_i}, \ c_i \lor \overline{d_i}$$

where d_i is a new variable. Let $f(\varphi) = \varphi'$ be the resulting instance of MAXIMUM 2-SATISFIABILITY.

First, note that any truth assignment to the variables of the new instance satisfies at most seven clauses in each group of ten. Moreover, for any truth assignment satisfying the i-th clause, a truth setting for d_i exists causing precisely seven of the above ten clauses to be satisfied: indeed, if only one (respectively, all) among a_i, b_i, and c_i is (respectively, are) TRUE, then we set d_i to FALSE (respectively, TRUE), otherwise d_i may be either TRUE or FALSE. Finally, for any truth assignment for which a_i, b_i, and c_i are FALSE, no truth setting of d_i can cause more than six of the ten clauses to be satisfied, and six clauses can be satisfied by setting d_i to FALSE.

This implies that

Chapter 8

APPROXIMATION PRESERVING REDUCTIONS

> **Problem 8.3:** Maximum Not-All-Equal 3-Satisfiability
>
> INSTANCE: Set U of variables, collection C of disjunctive clauses of at most three literals.
>
> SOLUTION: A truth assignment for U.
>
> MEASURE: The number of *not-all-equal* satisfied clauses, i.e., the number of clauses that have at least one true literal and at least one false literal.

$$m^*(\varphi') = 6m + m^*(\varphi) \leq 12m^*(\varphi) + m^*(\varphi) = 13m^*(\varphi)$$

where the inequality is due to the fact that at least half of the clauses of a Boolean formula in conjunctive normal form are always satisfiable (see Theorem 2.19). Finally, observe that, for any truth assignment τ' to the variables of φ', its restriction $g(\varphi, \tau') = \tau$ to the variables of φ is such that $m(\varphi, \tau) \geq m(\varphi', \tau') - 6m$. This implies that

$$\begin{aligned} m^*(\varphi) - m(\varphi, \tau) &= m^*(\varphi') - 6m - m(\varphi, \tau) \\ &\leq m^*(\varphi') - 6m - m(\varphi', \tau') + 6m \\ &= m^*(\varphi') - m(\varphi', \tau'). \end{aligned}$$

It follows that $(f, g, 13, 1)$ is, indeed, an L-reduction from MAXIMUM 3-SATISFIABILITY to MAXIMUM 2-SATISFIABILITY. QED

The MAXIMUM NOT-ALL-EQUAL 3-SATISFIABILITY problem is defined as Problem 8.3. It is easy to see that this problem is approximable either by AP-reducing it to MAXIMUM 2-SATISFIABILITY or by directly applying the greedy approach used for MAXIMUM SATISFIABILITY in Chap. 3 (see Exercise 8.9).

Theorem 8.10 ▶ MAXIMUM 2-SATISFIABILITY *is L-reducible to* MAXIMUM NOT-ALL-EQUAL 3-SATISFIABILITY.

PROOF Let φ be an instance of MAXIMUM 2-SATISFIABILITY, that is, a collection of m clauses over n variables with at most two literals per clause. The instance $f(\varphi) = \varphi'$ contains a new variable z which is inserted in each clause of φ. Clearly, the number of not-all-equal satisfied clauses in φ' does not depend on the value of z (since any truth assignment can be flipped without changing the number of not-all-equal satisfied clauses). We then assume that z is FALSE. For any truth assignment τ' such that $\tau'(z)$ is FALSE, we have that a clause of φ' is not-all-equal satisfied if and only if the corresponding clause in φ is satisfied by the restriction $g(\varphi,\tau') = \tau$ of τ' to the variables of φ. Then $(f,g,1,1)$ is an L-reduction from the MAXIMUM 2-SATISFIABILITY problem to the MAXIMUM NOT-ALL-EQUAL 3-SATISFIABILITY problem.

Section 8.4

APX-COMPLETENESS

QED

MAXIMUM NOT-ALL-EQUAL 3-SATISFIABILITY $\leq_L$ MAXIMUM CUT. ◀ Theorem 8.11

PROOF

Let φ be a given instance of the MAXIMUM NOT-ALL-EQUAL 3-SATISFIABILITY problem, that is, a collection of m clauses over n variables (without loss of generality, we may assume that each clause contains at least two literals). First, we construct a multigraph G, that is, a graph with parallel edges. For each variable x, G contains two nodes v_x and $v_{\bar{x}}$. For every clause of three literals a, b, and c, G contains the three *clause-edges* (v_a,v_b), (v_a,v_c), and (v_b,v_c). For every clause of two literals a and b, G contains two parallel *clause-edges* (v_a,v_b). Moreover, for every variable x, there are $2n_x$ parallel *literal-edges* $(v_x,v_{\bar{x}})$ where n_x denotes the number of occurrences of x in φ.

Let $l = \sum n_x$. Given a truth assignment to the variables of φ that not-all-equal satisfies p clauses, we define a partition of the nodes of G as follows: for any true literal a, v_a is in U_1 while $v_{\bar{a}}$ is in U_2. Clearly, for any not-all-equal satisfied clause, two edges of G belong to the cut, while edges corresponding to unsatisfied clauses do not contribute to the cut. The measure of this partition is $2l + 2p$.

Conversely, given a partition of the nodes of G whose measure is c, we may assume that nodes corresponding to complementary literals do not belong to the same set. Indeed, if v_x and $v_{\bar{x}}$ are in the same set, moving one of them to the other set does not decrease the number of edges in the cut since the possible missing clause-edges are compensated by literal-edges.

Therefore, we can define a solution for φ by setting to TRUE the literals in one set and to FALSE the other literals: this solution not-all-equal satisfies p clause, where $p = (c - 2l)/2$. Since $l \leq 3m \leq 6m^*(\varphi)$, we have that

$$m^*(G) = 2l + 2m^*(\varphi) \leq 14m^*(\varphi).$$

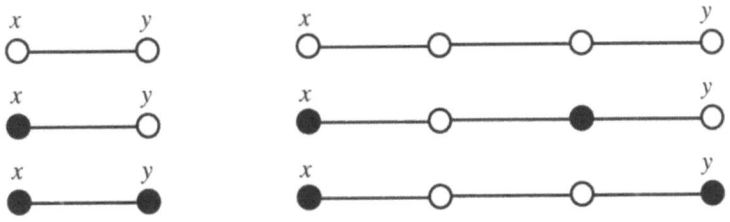

Figure 8.4
From multi-graph G to graph H

Chapter 8

APPROXIMATION
PRESERVING
REDUCTIONS

Observe that the above reduction is an L-reduction with $\beta = 14$ and $\gamma = 1$ from MAXIMUM NOT-ALL-EQUAL 3-SATISFIABILITY to MAXIMUM CUT on multi-graphs. In order to avoid multiple edges, we can simply substitute each edge of G with a 3-edge path (see the upper part of Fig. 8.4). Let $f(\varphi) = H$ be the resulting graph. Observe that any partition of the nodes of G can be extended into a partition of the nodes of H such that, for any edge of G, if the edge belongs to the cut in G then the three corresponding edges belong to the cut in H (see the middle part of Fig. 8.4) and, conversely, if the edge does not belong to the cut in G then two of the three corresponding edges belong to the cut in H (see the lower part of Fig. 8.4). From the above observation, we have that the restriction of a partition P of H to the nodes of G can then be transformed into a truth assignment $g(\varphi, P) = \tau$ whose absolute error is equal to the absolute error of P. It is then possible to prove that there exists a constant β such that that $(f, g, \beta, 1)$ is an L-reduction from MAXIMUM NOT-ALL-EQUAL 3-SATISFIABILITY to MAXIMUM CUT (see Exercise 8.10).

QED

Corollary 8.12 ▶ *The* MAXIMUM 2-SATISFIABILITY, MAXIMUM NOT-ALL-EQUAL 3-SATISFIABILITY, *and* MAXIMUM CUT *problems are* APX-*complete.*

PROOF

The corollary follows from the previous three theorems, from the APX-completeness of MAXIMUM 3-SATISFIABILITY, from Lemma 8.2, and from the the fact that the MAXIMUM 2-SATISFIABILITY, MAXIMUM NOT-ALL-EQUAL 3-SATISFIABILITY, and MAXIMUM CUT problems are approximable.

QED

Using expander graphs

Even if the original NP-hardness proof is not approximation preserving, it can still be useful. Indeed, in several cases this proof can be transformed into an approximation preserving reduction by means of some "expanding" modifications. Intuitively, these modifications try to restrict the solutions of the target problem to a subset of "good" solutions: whenever a solution which is not in this subset is computed, it can be transformed into a good solution in polynomial time and without any loss in the measure.

As an application of this technique we prove that MAXIMUM 3-SATISFIABILITY is L-reducible to MAXIMUM 3-SATISFIABILITY(B), that is, the restriction of MAXIMUM 3-SATISFIABILITY to instances in which each variable appears at most B times.

The standard NP-hardness proof of MAXIMUM 3-SATISFIABILITY(B) is based on the following idea (assume $B \geq 5$). If a variable y occurs h times with $h \geq B$, then create h new variables y_i, substitute the i-th occurrence of y with y_i, and add the following $h-1$ constraints: $y_1 \equiv y_2, y_2 \equiv y_3, \ldots,$

$y_{h-1} \equiv y_h$ (observe that each constraint $y_i \equiv y_j$ may be expressed by the two clauses $y_i \vee \overline{y_j}$ and $\overline{y_i} \vee y_j$). Graphically, the new clauses can be represented as a simple path of h nodes as shown in Fig. 8.5 in the case of $h = 5$. In the figure, each edge (y_i, y_j) corresponds to the two clauses $y_i \vee \overline{y_j}$ and $\overline{y_i} \vee y_j$.

Section 8.4

APX-COMPLETENESS

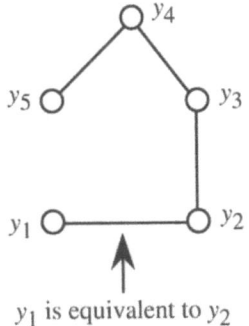

Figure 8.5
The equivalence path

This reduction is not approximation preserving since deleting just one edge of the graph may allow one to assign inconsistent truth-values to the h copies of y and to satisfy an arbitrarily large number of the clauses corresponding to the original ones (for example, consider the case in which the original formula contains $h/2$ copies of the clause containing only y and $h/2$ copies of the clause containing only $\bar{y}$). The problem is that a simple path admits "almost balanced" partitions of its nodes whose corresponding number of cut-edges is small. The solution to this problem consists of "enforcing" the bonds that link the copies of the same variable. A possible such enforcement could consist of substituting the path of Fig. 8.5 with a complete graph (see Fig. 8.6).

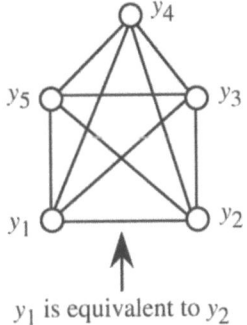

Figure 8.6
The equivalence complete graph

Observe that this new graph corresponds to the following set of equivalence constraints:
$$\{y_i \equiv y_j : 1 \leq i < j \leq h\}.$$

275

Chapter 8
APPROXIMATION PRESERVING REDUCTIONS

In this case, any proper subset T of the nodes of the graph yields a number of cut-edges equal to $|T|(h-|T|)$ which is always greater than or equal to $|T|$. If a truth assignment does not assign the same value to all copies of y, then these copies are divided into two sets according to the value they have received. Let T be the smaller of these two sets and consider the corresponding set of vertices of the complete graph. At least $|T|$ edges leave this set and each of them corresponds to an unsatisfied equivalence constraint (that is, an unsatisfied clause). Hence, by changing the truth-values of these $|T|$ copies, we gain at least $|T|$ clauses "belonging" to the equivalence complete graph. On the other hand, at most $|T|$ of the clauses corresponding to the original ones can now be unsatisfied. This implies that we do not lose anything if we impose that any truth assignment assigns the same value to all the copies of one variable.

This approach, however, has two disadvantages. First, the number of occurrences of a variable is no more bounded by a constant smaller than h (actually, we have $2h-1$ occurrences of each copy). Second, the measure of a solution for the new instance is no more bounded by a linear function of the measure of the corresponding solution for the original instance. Indeed, if a truth assignment satisfies k clauses in the original formula, then the corresponding truth assignment for the new formula satisfies $k + \sum_{i=1}^{n}(h_i - 1)h_i$ clauses where n denotes the number of variables and h_i denotes the number of occurrences of the i-th variable. This implies that this reduction would not be an L-reduction.

Actually, what we need is a graph which is an expanded version of the path of Fig. 8.5 (or, equivalently, a sparsified version of the complete graph of Fig. 8.6) with maximum degree bounded by a constant (this property is satisfied by the path) and such that any "almost balanced" partition of its nodes yields a large number of cut-edges (this property is satisfied by the complete graph). Such graphs exist and are well known: they are the *expander graphs*. The following definition is a special case of more general definitions that appeared in the literature yet sufficient for our purposes.

Definition 8.6 ▶ *Expander graph*
A d-regular graph $G = (V, E)$ is an expander if, for any $S \subset V$, the number of edges between S and $V - S$ is at least $\min(|S|, |V - S|)$.

It is well known (see Bibliographical notes) that a polynomial-time algorithm $\mathcal{A}_E$ and a fixed integer N exist such that, for any $n > N$, $\mathcal{A}_E(n)$ produces a 14-regular expander E_n with n nodes.

Theorem 8.13 ▶ MAXIMUM 3-SATISFIABILITY $\leq_L$ MAXIMUM 3-SATISFIABILITY(29).

PROOF Let φ be an instance of MAXIMUM 3-SATISFIABILITY with m clauses.

For each variable y, let h_y be the number of occurrences of y (without loss of generality, we may assume that $h_y \geq N$).

Let E_{h_y} be the 14-regular expander constructed by the above algorithm $\mathcal{A}_E$. For each node u of E_{h_y}, we create one new variable y_u and, for each edge (u,v) of E_{h_y}, we create two new clauses $y_u \vee \overline{y_v}$ and $\overline{y_u} \vee y_v$. Finally, we substitute the i-th occurrence of y in the original clauses with y_u where u is the i-th node of E_{h_y}. This concludes the definition of function f. Observe that each variable occurs at most 29 times (28 times in the clauses corresponding to edges of the expander and one time in the original clause): that is, $\varphi' = f(\varphi)$ is indeed an instance of MAXIMUM 3-SATISFIABILITY(29).

From the above discussion, it follows that we can now restrict ourselves to truth assignments for the variables of φ' that assign the same value to all copies of the same variable. Given such an assignment τ', we define a truth assignment $\tau = g(\varphi, \tau')$ for the variables of φ by assigning to y the value of its copies.

We then have that

$$m^*(\varphi') = 14 \sum_y h_y + m^*(\varphi) \leq 85 m^*(\varphi),$$

where the last inequality is due to the fact that each clause contains at most three literals and to the fact that $m^*(\varphi) \geq m/2$ (see Theorem 2.19).

Moreover, we have that, for any truth assignment τ' to the variables of φ', the corresponding truth assignment τ to the variables of φ satisfies the following equality:

$$m^*(\varphi) - m(\varphi, \tau) = m^*(\varphi') - m(\varphi', \tau').$$

That is, the quadruple $(f, g, 85, 1)$ is an L-reduction from MAXIMUM 3-SATISFIABILITY to MAXIMUM 3-SATISFIABILITY(29). QED

How few occurrences of each variable can we have in the MAXIMUM 3-SATISFIABILITY problem if it should still be hard to approximate? This is an interesting question, because MAXIMUM 3-SATISFIABILITY is a very useful problem to reduce from when showing APX-hardness results, and the fewer occurrences of each variable, the easier will hopefully the reduction be to construct.

If every variable occurs in at most two clauses it is easy to find, in polynomial time, the maximum number of clauses that can be satisfiable at the same time (see Exercise 8.11). The smallest number of occurrences that we can hope for is thus 3. Reducing MAXIMUM 3-SATISFIABILITY(29) to MAXIMUM 3-SATISFIABILITY(5) is an easier task: indeed, in this case it is sufficient to use graphs similar to that of Fig. 8.5 (see Exercise 8.12). We

Chapter 8

APPROXIMATION PRESERVING REDUCTIONS

will now construct an L-reduction from MAXIMUM 3-SATISFIABILITY(5) to MAXIMUM 3-SATISFIABILITY(3), thereby showing that there is a sharp boundary in approximability between 2 and 3 occurrences of variables.

Theorem 8.14 ▶ MAXIMUM 3-SATISFIABILITY(5) $\leq_L$ MAXIMUM 3-SATISFIABILITY(3).

PROOF Let φ be an instance of MAXIMUM 3-SATISFIABILITY(5) with n variables and m clauses. Let the function f construct a MAXIMUM 3-SATISFIABILITY(3) instance φ' with $15n$ variables and $20n + m$ clauses in the following way.

For each variable y, construct 15 variables $y_1, y_2, y_3, y_4, y_5, v_1, v_2, v_3, v_4, v_5, w_1, w_2, w_3, w_4, w_5$ and construct 20 2-SAT clauses over the 15 variables as the gadget in Fig. 8.7 shows, i.e., for each edge in the gadget from a vertex s to a vertex t, we construct the clause $\bar{s} \vee t$. This clause can thus be seen as an implication $s \Rightarrow t$, which is TRUE unless s is TRUE and t is FALSE. Finally, we substitute the i-th occurrence of y in the original clauses with y_i. In the constructed formula every variable is included in at most 3 clauses.

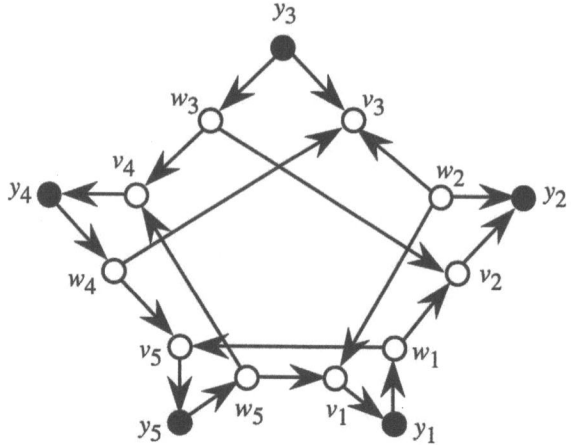

Figure 8.7
Gadget used for variable y in MAXIMUM 3-SATISFIABILITY reduction

From a solution for φ where k clauses are satisfied we get a solution to φ' by setting all variables in each gadget to the same value as the corresponding variable in φ. This will satisfy all clauses in all gadgets and k of the other clauses. Also observe that $2n \leq 3m$, since every variable occurs at least twice (otherwise we know directly which value it should have) and since the total number of occurrences of variables in φ is at most $3m$. Now we know that

$$m^*(\varphi') \leq 20n + m^*(\varphi) \leq 30m + m^*(\varphi) \leq 61m^*(\varphi).$$

Before we show how the function g constructs a solution for φ from a solution for φ' we look at some properties of the clauses created from one single gadget. If the five variables y_1, y_2, y_3, y_4, y_5 all have the same value we can satisfy all 20 clauses by assigning that value to every variable in the gadget.

If only four of the y_i variables have the same value, we will show that at most 19 clauses can be satisfied. By symmetry, we can suppose that y_1, y_2, y_3, y_4 have the same value and that y_5 has the opposite value. First assume that y_5 is TRUE. Since y_1 is FALSE there must be one clause on the path y_5, w_5, v_1, y_1 that is not satisfied (no matter which values w_5 and v_1 are assigned). If y_5 is FALSE, then y_4 is TRUE, and there must be one clause on the path y_4, w_4, v_5, y_5 that is not satisfied.

If only three of the y_i variables have the same value, we can at most satisfy 18 of the 20 clauses. We will consider two cases, namely, when the y_i variables having the same values are on one side of the gadget (for example y_1, y_2, y_3) and when they are on different sides of the gadget (for example y_1, y_3, y_4). In the second case we can use the same argument as above for each of the variables with the opposite value (for example, if y_1 is TRUE and y_2 is FALSE, then one of the clauses in the path from y_1 to y_2 is not satisfied). In the first case, we assume without loss of generality that y_4 and y_5 have the value opposite to the other y_i variables. Then, in the circuit $y_4, w_4, v_3, y_3, w_3, v_4, y_4$ there must be one clause that is not satisfied. Also, in the circuit $y_5, w_5, v_1, y_1, w_1, v_5, y_5$ there must be one clause that is not satisfied. The two circuits are disjoint, so there must be at least two clauses that are not satisfied.

Now we are ready to define the function g that, given any truth assignment τ' to the variables of φ', constructs a truth assignment τ to the variables of φ. This function first modifies τ' so that, for each gadget, all y_i variables have the same value, namely the value of the majority of the y_i variables in that gadget. This will not make fewer clauses satisfied, since we will loose at most one original clause but gain at least one gadget clause for each variable we change.

Defining the truth assignment τ to the variables of φ by setting the value of y as the value of the corresponding y_i variables yields a solution satisfying the inequality $m^*(\varphi) - m(\varphi, \tau) \leq m^*(\varphi') - m(\varphi', \tau')$. That is, $(f, g, 61, 1)$ is an L-reduction from MAXIMUM 3-SATISFIABILITY(5) to MAXIMUM 3-SATISFIABILITY(3).

QED

In conclusion, we have the following result.

MAXIMUM 3-SATISFIABILITY(B) is APX-complete for any $B \geq 3$.

◀ Corollary 8.15

Section 8.4

APX-COMPLETENESS

Chapter 8
APPROXIMATION PRESERVING REDUCTIONS

Observe that since the above constructions only add clauses with two literals they work also for MAXIMUM 2-SATISFIABILITY. Thus MAXIMUM 2-SATISFIABILITY(3) is APX-complete. This is therefore the simplest variation of the maximum satisfiability problem that is still APX-complete, and is therefore extremely useful to reduce from when showing APX-hardness results.

Using randomization

Another technique that can be useful to prove APX-hardness results is based on the following rule of thumb: whenever you do not know what to do, flip a coin. Before giving an example of this technique, we recall that randomization has already been used for developing approximation algorithms (see Chap. 5).

We will describe the technique by referring to MAXIMUM SATISFIABILITY and MAXIMUM 3-SATISFIABILITY. In particular, we will show how a 2-approximation algorithm for MAXIMUM 3-SATISFIABILITY can be used to obtain a 2.33-approximation algorithm for MAXIMUM SATISFIABILITY. This may seem somehow useless since a 2-approximation algorithm for MAXIMUM SATISFIABILITY is known (see Chap. 3). However, we have chosen this example because it is simple enough to be described in one page yet also sufficient to explain the approach: we refer to the bibliographical notes for other applications of the technique.

Recall that the standard reduction from MAXIMUM SATISFIABILITY to MAXIMUM 3-SATISFIABILITY which is based on local replacement (see Example 6.5) fails to be approximation preserving due to the possible presence of large clauses in the original instance. Large clauses, however, are easy to satisfy using a simple randomized algorithm (see Sect. 5.2). We then *combine* (in a probabilistic sense) a solution based on easy reductions and one given by the randomized algorithm, and this mixed solution will be good for any combination of small and large clauses.

Lemma 8.16 ▶ *If* MAXIMUM 3-SATISFIABILITY *admits a 2-approximation algorithm, then* MAXIMUM SATISFIABILITY *admits a 2.33-approximation algorithm.*

PROOF Let φ be an instance of MAXIMUM SATISFIABILITY with m clauses, m_3 be the number of clauses of φ with 3 or less literals, and φ_3 be the instance φ restricted to these m_3 clauses. Define $m_l = m - m_3$ and $\varphi_l = \varphi - \varphi_3$. Let τ_3 be a 2-approximate solution for φ_3 (which, by hypothesis, is computable in polynomial time), and a be the number of clauses satisfied by τ_3. It is immediate to verify that

$$m^*(\varphi) \leq 2a + m_l \ . \tag{8.2}$$

In fact, no assignment can satisfy more than $2a$ clauses in φ_3 (otherwise τ_3 would not be 2-approximate) and more then m_l clauses in φ_l. We now define a random assignment τ_R over the variables of φ according to the following distribution that depends on a parameter p with $1/2 < p < 1$: if a variable x occurs in φ_3, then $\Pr\{\tau_R(x) = \tau_3(x)\} = p$, otherwise, that is, if a variable x occurs in φ but not in φ_3, $\Pr\{\tau_R(x) = \text{TRUE}\} = \frac{1}{2}$.

Let us now estimate the average number of clauses of φ that are satisfied by τ_R. The above two conditions imply that any literal is TRUE with probability at least $(1-p)$. Hence, the probability that a clause in φ_l is not satisfied is at most p^4. On the other hand, if a clause is satisfied by τ_3, then the probability that it is still satisfied by τ_R is at least p. We can thus infer the following lower bound on the average number of clauses of φ that are satisfied by τ_R:

$$E[m(\varphi, \tau_R)] \geq pa + (1-p^4)m_l = \frac{p}{2}\left(2a + \frac{2}{p}(1-p^4)m_l\right).$$

If we choose $p = 0.86$, we have that $\frac{2}{p}(1-p^4) > 1$. Hence,

$$E[m(\varphi, \tau_R)] > \frac{p}{2}(2a + m_l) \geq \frac{p}{2}m^*(\varphi) = 0.43 m^*(\varphi),$$

where Eq. (8.2) has been applied. Using the method of conditional probabilities (see Sect. 5.4), we can find in linear time an assignment τ such that $m(\varphi, \tau) \geq E[m(\varphi, \tau_R)]$. That is, the performance ratio of τ is at most $\frac{1}{0.43} = 2.33$.   QED

The previous lemma can be generalized to obtain the following result (see Exercise 8.13).

MAXIMUM SATISFIABILITY $\leq_{AP}$ MAXIMUM 3-SATISFIABILITY.   ◀ Theorem 8.17

The above result is already known since we have previously shown that MAXIMUM 3-SATISFIABILITY is APX-complete. However, it is worth pointing out that the proof given in this paragraph does not make use of the PCP theorem.

Section 8.5

EXERCISES

8.5 Exercises

Exercise 8.1 Prove that if $P \neq NP$, then the inclusions of Eq. (8.1) are all strict. (Hint: for each pair of consecutive classes, define an artificial problem in the larger class that is not in the smaller one.)

Chapter 8

APPROXIMATION PRESERVING REDUCTIONS

Exercise 8.2 (*) Prove Lemma 8.2. (Hint: the case in which $\mathcal{P}_1$ is a minimization problem is almost trivial. The other case can be dealt with by making use of the approximation algorithm $\mathcal{A}$ for $\mathcal{P}_1$: in particular, define a new function g' which chooses the best between g and $\mathcal{A}$.)

Exercise 8.3 Show that AP-reducibility is transitive, that is, if $\mathcal{P}_1 \leq_{AP} \mathcal{P}_2$ and $\mathcal{P}_2 \leq_{AP} \mathcal{P}_3$ then $\mathcal{P}_1 \leq_{AP} \mathcal{P}_3$.

Exercise 8.4 Prove Theorem 8.3. (Hint: with any NP maximization problem $\mathcal{P}$ associate a nondeterministic Turing machine which guesses a solution y, writes y in the tape cells following the input, and writes the measure $m_\mathcal{P}(x,y)$ in the next tape cells. Successively, apply a variation of Cook-Levin's theorem to this computation model and assign a non-zero weight only to variables corresponding to the tape cells on which the machine prints $m_\mathcal{P}(x,y)$.)

Exercise 8.5 (*) Define an approximation preserving reducibility which preserves membership in FPTAS. Show a PTAS-complete problem with respect to this reducibility.

Exercise 8.6 Prove that the variation of MINIMUM WEIGHTED SATISFIABILITY in which a trivial solution is artificially added to the definition of the problem is complete for the class of minimization problems in exp-APX. (Hint: similar to Exercise 8.4.)

Exercise 8.7 (**) Prove that MINIMUM TRAVELING SALESPERSON is exp-APX-complete. (Hint: look for a proof of the NP-completeness of the corresponding decision problem.)

Exercise 8.8 In the proof of Theorem 8.9 we suppose that without loss of generality we can consider only clauses of size exactly three. Show that this is true by proving that the reduction of Example 6.5 is, indeed, an L-reduction from MAXIMUM 3-SATISFIABILITY with clauses of *at most* three literals to MAXIMUM 3-SATISFIABILITY with clauses of *exactly* three literals.

Exercise 8.9 Give an approximation algorithm for MAXIMUM NOT-ALL-EQUAL 3-SATISFIABILITY based on the greedy technique.

Exercise 8.10 Fill in the details of the proof of Theorem 8.11.

Exercise 8.11 Prove that MAXIMUM 3-SATISFIABILITY(2) is in PO.

Exercise 8.12 Prove that the MAXIMUM 3-SATISFIABILITY(29) problem is AP-reducible to the MAXIMUM 3-SATISFIABILITY(5) problem.

Exercise 8.13 (*))Prove Theorem 8.17.

Exercise 8.14 Choose any APX-complete problem that appears in the compendium at the end of the book and prove its APX-completeness.

Exercise 8.15 A problem P is in the class MAXSNP$_0$ if it is defined in terms of the expression

$$\max_S |\{(x_1, \ldots, x_k) : \varphi(G_1, \ldots, G_m, S, x_1, \ldots, x_k)\}|$$

where φ is a quantifier-free first-order expression involving the variables x_i, the structures G (which are the input), and S (which is a solution). Prove that MAXIMUM 3-SATISFIABILITY belongs to MAXSNP$_0$. (Hint: distinguish the clauses according to the number of negated variables they contain.)

Exercise 8.16 The class MAXSNP contains all the optimization problems which L-reduce to a problem in MAXSNP$_0$. Show that MINIMUM VERTEX COVER belongs to MAXSNP.

Exercise 8.17 Prove that all problems in the class MAXSNP$_0$ are approximable. (Hint: use the greedy technique.)

Exercise 8.18 (*) Without using the PCP theorem, show that any problem in MAXSNP$_0$ is L-reducible to MAXIMUM 3-SATISFIABILITY. (Hint: use the fact that, according to the definition of MAXSNP$_0$, φ can be represented as a constant-size Boolean circuit.)

Exercise 8.19 Consider the class of problems which AP-reduce to a problem in MAXSNP$_0$. Which is this class?

Exercise 8.20 (*) For any polynomial p, let MINIMUM VERTEX COVER^p be the variation of MINIMUM VERTEX COVER in which each node has associated a weight bounded by $p(n)$ where n denotes the number of nodes. Prove that MINIMUM VERTEX COVER$^p \leq_{AP}$ MINIMUM VERTEX COVER.

8.6 Bibliographical notes

COMPARING THE complexity of different combinatorial optimization problems has been an extremely active research area during the last twenty years. Actually, the need for an appropriate approximation preserving reducibility was already expressed in [Johnson, 1974a] where the author says

Chapter 8

APPROXIMATION PRESERVING REDUCTIONS

What is it that makes algorithms for different problems behave in the same way? Is there some stronger kind of reducibility than the simple polynomial reducibility that will explain these results, or are they due to some structural similarity between the problems as we define them?

One of the first answer to the above questions was given in [Ausiello, D'Atri, and Protasi, 1980] where the notion of structure preserving reduction is introduced. Successively, consistent with the general scheme shown in Fig. 8.1, several approximation preserving reducibilities have been defined in the literature which differ the way they preserve the performance ratios and/or the additional information used by functions f and g. Figure 8.8 shows a taxonomy of the main reducibilities, which appeared in [Crescenzi, 1997]: as the reader can see, AP-reducibility and L-reducibility are included in the taxonomy.

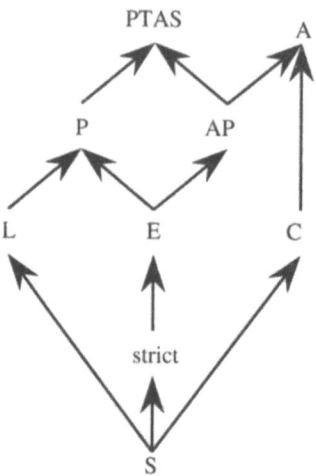

Figure 8.8
The taxonomy of approximation preserving reducibilities

These reducibilities have been introduced in the following papers: $\leq_{\text{strict}}$, $\leq_A$, and $\leq_P$ in [Orponen and Mannila, 1987], $\leq_C$ in [Simon, 1989], $\leq_L$ in [Papadimitriou and Yannakakis, 1991], $\leq_S$ in [Crescenzi, Fiorini, and Silvestri, 1991], $\leq_E$ in [Khanna, Motwani, Sudan, and Vazirani, 1999], $\leq_{\text{PTAS}}$ in [Crescenzi and Trevisan, 1994], and $\leq_{\text{AP}}$ in [Crescenzi, Kann, Silvestri, and Trevisan, 1999]. However, it is now almost wide accepted that the reducibility to be used is indeed AP-reducibility (see [Arora, 1998] where it is also suggested to refer to $\leq_{\text{PTAS}}$).

By means of these reducibilities, several notions of completeness in approximation classes have been introduced following two different approaches.

Section 8.6

BIBLIOGRAPHICAL NOTES

On one hand, attention has been focused on computationally defined classes of problems, such as NPO and APX: along this line of research, however, almost all completeness results have been shown either for artificial optimization problems or for problems for which lower bounds on the quality of the approximation were easily obtainable (see [Orponen and Mannila, 1987, Crescenzi and Panconesi, 1991]).

On the other hand, researchers focused on the logical definability of optimization problems and introduced several syntactically defined classes for which natural completeness results were obtained (see [Papadimitriou and Yannakakis, 1991, Kolaitis and Thakur, 1994, Panconesi and Ranjan, 1993] and Exercise 8.15): unfortunately, the approximability properties of the problems in these latter classes were not related to standard complexity-theoretic conjectures.

A first step toward reconciling of these two approaches consisted of proving lower bounds (modulo P $\neq$ NP or some other likely condition) on the approximability of complete problems for syntactically defined classes by means of the PCP theorem. More recently, the final step was performed in [Khanna, Motwani, Sudan, and Vazirani, 1999] where the closure of syntactically defined classes with respect to an approximation preserving reducibility was proved equal to the more familiar computationally defined classes. As a consequence of this result, for instance, any MAXSNP-completeness result appeared in the literature can be interpreted as an APX-completeness result.

The proof of the NPO-completeness of the MAXIMUM WEIGHTED SATISFIABILITY problem was given in [Ausiello, D'Atri, and Protasi, 1981, Orponen and Mannila, 1987] (even though in respect to a different kind of reducibility) while the first (non-natural) APX-complete problem was proved in [Crescenzi and Panconesi, 1991]. However, the first natural APX-completeness result is due to [Khanna, Motwani, Sudan, and Vazirani, 1999] and to [Crescenzi and Trevisan, 1994]. In the former reference, completeness results for other classes are also shown. The completeness of MINIMUM TRAVELING SALESPERSON for the class of minimization problem in exp-APX has been proved in [Orponen and Mannila, 1987] by adapting the NP-hardness proof included in [Papadimitriou, 1977] and by using the $\leq_P$-reducibility.

The short guide to approximation preserving reductions presented in Sect. 8.4.1 should be integrated with Chap. 3 of [Garey and Johnson, 1979] where general approaches, such as restriction, local replacement, and component design, for obtaining NP-completeness results are described and illustrated with a wide variety of NP-complete problems (the local replacement technique has been deeply investigated in [Trevisan, Sorkin, Sudan, and Williamson, 1996] to "automatically" obtain improved non-approximability result). Also useful to this purpose is [Arora and Lund, 1996]: this chapter also contains an interesting classification of optimization problems with respect to their hardness of approximation.

Exercise 8.20 is taken from [Crescenzi, Silvestri, and Trevisan, 1996] where, by showing approximation preserving reductions, several results are established on the difficulty of approximating weighted problems with respect to their corresponding unweighted versions.

Finally, observe that the structure of approximation classes is richer than it appears in this chapter. For example, APX-intermediate problems, that is, problems which are neither APX-complete nor in PTAS (unless P = NP) have been shown in [Crescenzi and Panconesi, 1991]. Actually, in [Crescenzi, Kann, Silvestri, and Trevisan, 1999] it has been proved that MINIMUM BIN PACKING is one such problem.

Probabilistic analysis of approximation algorithms

IN PREVIOUS chapters the performance of an approximate algorithm has been analyzed using a worst case point of view. In this chapter this condition is relaxed and the probabilistic analysis of an algorithm is performed by considering either the *average case* or the behavior in *almost all* cases.

There are two motivations for considering a weaker model. First, the empirical verification of approximate algorithms suggests that the worst case error is seldom met in practice: in most cases, the value of the solution found is much closer to the value of the optimal solution than what is suggested by the worst case analysis. Therefore we might be interested in knowing what is the average or the most likely performance of an approximate algorithm.

Moreover, we have seen that for many problems approximate algorithms with good approximation properties in the worst case cannot exist unless P = NP. For instance, in the case of MAXIMUM CLIQUE, it is known that finding an approximate solution with a worst case ratio better than $n^{1-\varepsilon}$, for any $\varepsilon > 0$, is computationally equivalent to finding the optimum. This motivates the interest to look for an algorithm that can find a good approximate solution for almost all graphs.

In this chapter, we first present some ingredients for performing the probabilistic analysis of an algorithm and then analyze several simple algorithms, showing that the bounds obtained using this approach are, in many cases, much better than the worst case bounds. To this aim, we will need to introduce a probability distribution over the set of problem instances. To yield useful results from a practical point of view, this distribution should

Chapter 9

PROBABILISTIC ANALYSIS OF APPROXIMATION ALGORITHMS

be as close as possible to the distribution of problem instances that occur in practical applications. However, we seldom know what is a realistic probability distribution and, for this reason, uniform distributions of the input instances are often considered.

9.1 Introduction

THE INPUT distribution is characterized by one or more parameters that define the size of the input. For example, the parameter n denotes the number of cities in the random instance of MINIMUM TRAVELING SALESPERSON and the number of items in the case of MINIMUM BIN PACKING; in the case of graph problems we will consider a probability distribution characterized by two parameters, which are associated to the number of nodes and edges of the graph.

Assume that the input size is characterized by a single parameter n and, for each value of n, let $D(n)$ be a probability distribution of the instances of size n; we analyze the behavior of an approximation algorithm with respect to the input distribution.

Given a problem and a probability distribution $D(n)$ defined over all problem instances, the measures of the optimum and of the solution found by an approximation algorithm are random variables depending on the input parameter n. The ideal goal of the probabilistic analysis is to characterize the above random variables for each value of the parameter n, however, this is possible in very few cases. For this reason the probabilistic analysis focuses its attention on the limit behavior of the measures of the optimum and of the approximate solutions.

There are several convergence criteria for random variables: the following definition gives those criteria that are useful in this chapter.

Definition 9.1 ▶ Let $x_1, x_2, \ldots, x_n$ be a sequence of random variables indexed by the parameter n.
Almost everywhere convergence

1. The sequence $x_1, x_2, \ldots, x_n$ converges almost everywhere (a.e.) to a value c as n goes to infinity if, for each $\varepsilon > 0$,

$$\lim_{n \to \infty} \Pr\{|x_n - c| \geq \varepsilon\} = 0.$$

2. The sequence $x_1, x_2, \ldots, x_n$ converges in expectation to a value c as n goes to infinity if the expected value $\mathrm{E}[x_n]$ satisfies the following condition

$$\lim_{n \to \infty} |\mathrm{E}[x_n] - c| = 0.$$

Section 9.1
INTRODUCTION

If $x_1, x_2, \ldots, x_n$ converges almost everywhere then we also say that $x_1, x_2, \ldots, x_n$ converges *in probability*.

The above two convergence criteria are the most used in the analysis of approximation algorithms for optimization problems; other possible convergence criteria of a sequence of random variables have also been used in the analysis of approximation algorithms, but they are omitted because such a detailed presentation is out of the scope of this book. For our purposes it is sufficient to observe that convergence almost everywhere and convergence in expectation are incomparable. In fact, there exist sequences that converge to a value c almost everywhere and to a different value c' in expectation.

The different types of convergence defined above are asymptotic and, therefore, do not provide a precise information on the rate of convergence to the limit value (clearly, we would like the convergence to be as fast as possible). We will see however that, in some cases, it is possible to prove precise bounds on the rate of convergence of a sequence of random variables.

Finally, we observe that if we consider the limit behavior of a random variable as a function of the parameter n, we actually consider a sequence of instances of the problem. Notice that there are two possible ways to model a sequence of problem instances of growing size. In the *independent model* a random instance of size n is drawn in complete independence from instances of smaller size; in the *incremental model* a random instance of size n is obtained starting from a random instance of size $n-1$ and then drawing an extra element. Clearly, in this case, problem instances are dependent each other, even if all parameter values have been drawn independently. In the following, attention is limited to the independent model because in this case the analysis is generally easier and in most cases convergence results in the independent model imply convergence in the incremental model.

9.1.1 Goals of probabilistic analysis

GIVEN A problem $\mathcal{P}$, a probability distribution $D(n)$ and an approximation algorithm $\mathcal{A}$ there are several approaches to performing a probabilistic analysis of $\mathcal{A}$. In this chapter, in order to simplify the presentation, we abuse notation and use $m^*(n)$ and $m_\mathcal{A}(n)$ to denote the random variables representing the optimal measure of a random instance of the problem of size n and the measure of the solution found by algorithm $\mathcal{A}$ when applied to a random instance of size n.

Let f and g be functions of the parameter n. If we are able to prove

Chapter 9

PROBABILISTIC ANALYSIS OF APPROXIMATION ALGORITHMS

that the optimal measure of $\mathcal{P}$ converges almost everywhere to $f(n)$ and the measure of the solution given by algorithm $\mathcal{A}$ converges almost everywhere to $g(n)$, then we can say that $\mathcal{A}$ finds almost everywhere a solution with performance ratio $f(n)/g(n)$ (equivalently that the ratio $m^*(n)/m_\mathcal{A}(n)$ converges to $f(n)/g(n)$). Analogously, if we can prove that the expected measure of the optimum converges to $f(n)$ and the expected measure of the solution given by algorithm $\mathcal{A}$ converges to $g(n)$, then we can say that the ratio of the expected measure of the optimum and of the expected measure of the approximate solution converges to $f(n)/g(n)$.

There are cases in which it is possible to obtain stronger bounds. For example, assume that the ratio $m^*(n)/m_\mathcal{A}(n)$ converges almost everywhere to 1; then it would be interesting to be able to estimate the rate of convergence that represent how fast a random variable converges to its limit. Another possibility is to study the random variable $m^*(n) - m_\mathcal{A}(n)$. For example, in some cases the difference between the optimal measure and the measure of the approximate solution returned by algorithm $\mathcal{A}$ converges almost everywhere. In other cases, we will obtain bounds on the expected value of such a difference.

Note that all the above results might depend on the considered probability distribution; in fact, it might happen that, for a given problem $\mathcal{P}$, different probability distributions of the input instances might have different characterizations of approximability.

It is also important to observe that a probabilistic analysis can be performed in order to analyze the running time of an optimal algorithm: namely, given a hard optimization problem and an algorithm that always finds its optimum, we are interested in analyzing the running time of this algorithm, assuming a probability distribution of the input instances. We expect to obtain an exponential running time in the worst case, but we might wonder whether the *expected* running time is polynomial. Here we will not address this topic and we refer to the bibliographic notes for references.

Before concluding this section we briefly discuss the criticisms of the probabilistic analysis of approximation algorithms. The probabilistic analysis of an approximation algorithm can be viewed as the analytical counterpart of an experimental analysis based on the behavior of the algorithm on a set of supposedly representative test problems. Therefore, it is important to consider probability distributions that correctly describe the input instances that arise in real cases. The main objection to this statement is that we seldom know which probability distribution is realistic.

Furthermore, the current state of the art allows one to analyze only simple solution methods for a limited class of probability distributions. In fact, for most problems the operations and the decisions taken by an algorithm during the execution of a step condition the random variable that describes the input; this dependency complicates the analysis of subsequent steps of the algorithm. For this reason, the probabilistic analysis of many algorithms of practical use is, in most cases, an open research problem.

9.2 Techniques for the probabilistic analysis of algorithms

9.2.1 Conditioning in the analysis of algorithms

THE MAIN difficulty in the probabilistic analysis of algorithms is due to dependencies between different execution steps of the algorithm. We show this with an example where the problem can be easily circumvented: the analysis of the expected measure of the solution found by the *Nearest Neighbor* algorithm for MINIMUM TRAVELING SALESPERSON (see Section 2.1.3).

Let us consider the following probability distribution D_n parameterized by the number of cities n: an instance of MINIMUM TRAVELING SALESPERSON with n vertices (cities) can be defined as a complete graph on n vertices with edge weights independently and uniformly distributed in $[0,1]$.

Recall that *Nearest Neighbor* starts from an arbitrary vertex (say vertex v_{i_1}) and then executes a loop $(n-1)$ times. In the first iteration the algorithm selects the vertex that follows v_{i_1} by choosing a vertex v_{i_2} such that edge (v_{i_1}, v_{i_2}) has minimum weight (length) among all edges with an endpoint in v_{i_1}; edge (v_{i_1}, v_{i_2}) is added to the path. In the k-th iteration of the loop, after a path through vertices $v_{i_1}, v_{i_2}, \ldots, v_{i_k}$ has been obtained, the algorithm chooses vertex $v_{i_{k+1}}$ following v_{i_k} in the path as the vertex such that $(v_{i_k}, v_{i_{k+1}})$ is the edge with minimum weight from v_{i_k} to a vertex that is not in the path; edge $(v_{i_k}, v_{i_{k+1}})$ is added to the tour. After $(n-1)$ iterations, a Hamiltonian path $v_{i_1}, v_{i_2}, \ldots, v_{i_n}$ is obtained, and the last step of the algorithm completes the tour by adding edge (v_{i_n}, v_{i_1}).

Let X_k be the random variable denoting the length of the edge chosen during the k-th iteration of the algorithm (i.e., X_k is the minimum weight of an edge from v_{i_k} to the $n-k$ unvisited vertices); it follows that X_k is distributed as the minimum value of $n-k$ independent random variables, all uniformly distributed in $[0,1]$. By these properties, it derives that the expected value of X_k is $1/(n-k+1)$ (see also Theorem 9.5).

Notice that this way of reasoning cannot be applied to the last step of the algorithm to obtain a bound on the expected length of (v_{i_n}, v_{i_1}) because the algorithm has already considered edges incident to vertex v_{i_1}. Since (v_{i_n}, v_{i_1}) is not the edge with minimum weight, the distribution of the

Chapter 9

PROBABILISTIC ANALYSIS OF APPROXIMATION ALGORITHMS

weight of the edge is conditioned by this fact and, therefore, the weight of edge (v_{i_n}, v_{i_1}) is not uniformly distributed in $[0,1]$ after the execution of the loop. Equivalently, we say that there is a *dependency* between the last step of the algorithm and its initial iteration.

In this case, it is not necessary to perform an exact analysis of the expected weight of edge (v_{i_n}, v_{i_1}): the trivial upper bound given by the value 1 on the weight of (v_{i_n}, v_{i_1}) is sufficient for the analysis (a more refined analysis shows that the exact value of the expected length of the last edge of the tour is slightly larger than $1/2$).

Let $\mathcal{H}_n$ be the *Harmonic sum*,

$$\mathcal{H}_n = \sum_{i=1}^{n} \frac{1}{n}.$$

The expected length $E[m_{NN}(n)]$ of the solution obtained by the Nearest Neighbor algorithm in a random instance of n cities can be bounded as follows

$$E[m_{NN}(n)] \leq \sum_{i=1}^{n-1} E[w_i] + 1 = \sum_{i=1}^{n-1} \frac{1}{n-i+1} + 1 = \mathcal{H}_n$$

Since the following bounds are known on the value $\mathcal{H}_n$

$$\ln n < \mathcal{H}_n < \ln n + 1$$

$$\lim_{n \to \infty} (\mathcal{H}_n - \ln n) = \gamma,$$

where γ is Euler's constant ($\gamma = 0.57721566...$), we have proved the following theorem.

Theorem 9.1 ▶ *If the distances between the cities are uniformly and independently distributed in $[0,1]$ then the expected measure $m_{NN}(n)$) of the solution found by the Nearest Neighbor algorithm in a random instance of the* MINIMUM TRAVELING SALESMAN *problem with n cities satisfies*

$$E[m_{NN}(n))] \leq \ln n + 1.$$

The analysis of the Nearest Neighbor algorithm for the TSP problem shows how the operations performed by an algorithm might condition the values considered in successive steps. In the above example the difficulty has been easily circumvented by a trivial upper bound. However, this is not always the case: in fact, dependencies between steps represent the main difficulty in the probabilistic analysis and makes the analysis of sophisticated algorithms a formidable task.

9.2.2 The first and the second moment methods

IN THIS section we introduce two inequalities that are often useful in obtaining upper and lower bounds on the measures of the optimum and of an approximate solution.

The first inequality is the first moment method, which is also known as *Boole's inequality* and can be stated as follows.

If $E_1, E_2, ..., E_k$ are events (not necessarily independent) then ◀ Proposition 9.2

$$\Pr\{\bigcup_{i=1}^{k} E_i\} \leq \sum_{i=1}^{k} \Pr\{E_i\}.$$

Boole's inequality is used to obtain upper bounds on the measure of an optimal solution in all those cases when it is required to find an element (the optimal solution) in a set of possible candidate solutions. Namely, assume that the S is the set of candidate solutions. Let E_i be the event that a candidate solution $s_i \in S$ is a feasible solution. In many cases, $\Pr\{E_i\}$ is the same for all elements of S, and therefore the following upper bound on the existence of a feasible solution can be easily obtained.

$$\Pr\{\text{feasible solution exists}\} = \Pr\{\bigcup_{i=1}^{|S|} E_i\}$$
$$\leq |S|\Pr\{s_i \text{ is feasible}\} \quad (9.1)$$

Note that if x is the random variable denoting the number of feasible solutions then $E[x] = |S|\Pr\{s_i \text{ is feasible}\}$.

Hence, Eq. 9.1 is equivalent to:

$$\Pr\{\text{a feasible solution exists}\} = \Pr\{x > 0\} \leq E[x].$$

The above inequality can also be obtained by observing that the expected value $E[x]$ of a non-negative integer random variable x is always greater than the probability of the variable to be non-zero. Formally,

$$E[x] = \sum_{i=1}^{\infty} i\Pr\{x = i\} \geq \Pr\{x > 0\}.$$

Despite its simplicity, this method allows in many cases to obtain tight upper bounds both on the optimal measure of a problem and on the measure of the solution found by an approximation algorithm.

Chapter 9

PROBABILISTIC ANALYSIS OF APPROXIMATION ALGORITHMS

The first moment method allows us to obtain an upper bound on the measure of the optimal solution. Notice that, even if the expected value of a non-negative random variable x is much larger than zero, this is not sufficient to claim that the random variable is non-zero with positive probability; it is also required that the variance should be small. A powerful technique for obtaining such a bound is the second moment method. Given a random variable Y and a real value $t > 0$, Chebyshev's inequality implies that

$$\Pr\{|Y - E[Y]| \geq t\} \leq \frac{E[Y^2] - E[Y]^2}{t^2};$$

where $E[Y^2]$ denotes the second moment.

By choosing $t = E[Y]$ it follows that

$$\Pr\{Y = 0\} \leq \Pr\{|Y - E[Y]| \geq E[Y]\} \leq \frac{E[Y^2]}{E[Y]^2} - 1. \quad (9.2)$$

Notice that obtaining a good estimate of the second moment $E[Y]^2$ is usually more difficult than analogous computations for the expected value $E[Y]$. In Section 9.5 we will apply the second moment method to the probabilistic analysis of MAXIMUM CLIQUE.

9.2.3 Convergence of random variables

The probabilistic analysis of algorithms exploits theorems and inequalities that allow one to obtain sharp and powerful bounds on the tails of random variables. The following properties of the sum of independently and identically distributed random variables are used in this chapter (many other inequalities have been proposed and used, we refer to the bibliographical notes for reference books).

There are many results in probability theory concerning the sum of independent identically distributed (i.i.d.) random variables.

Let $x_1, x_2, \ldots, x_n$ be independent samples of a random variable x with mean μ and variance σ, and let $S_j = \sum_{i=1}^{j} x_i$. The j-th centered partial sum is defined to be

$$\hat{S}_j = \sum_{i=1}^{j} x_i - j\mu.$$

A well-known result of probability theory states that the expected value of $\hat{S}_j$ is zero. In the probabilistic analysis of algorithms it it is often necessary to show that the probability of a large deviation from the mean is exponentially small.

The most often used bounds on the tail of a sum of random variables are due to Chernoff and Hoeffding. The following theorem, that requires a common distribution for the variables x_i, does not provide the strongest bound, but it is sufficient for our purposes.

◀ **Section 9.2**

TECHNIQUES FOR THE PROBABILISTIC ANALYSIS OF ALGORITHMS

Let x_i, $i = 1, 2, \ldots, n$ be independent random variables that are uniformly distributed in the range $[-1, 1]$. Let $S_n = \sum_{i=1}^{n} x_i$, and $\hat{S}_n = S_n - E[S_n]$. Then for $y \geq 0$

◀ Theorem 9.3

$$\Pr\{\hat{S}_n \geq y\sqrt{n}\} \leq e^{-y^2/2} \quad \text{and} \quad \Pr\{\hat{S}_n \leq -y\sqrt{n}\} \leq e^{-y^2/2}.$$

The following theorem provides bounds on the probability that any of the partial sums $\hat{S}_i$ exceeds some limit. Note that the theorem does not require a common distribution for the variables x_i.

Let x_i, $i = 1, 2, \ldots, n$ be independent random variables that assume values in the range $[-1, 1]$. Let $S_j = \sum_{i=1}^{j} x_i$, and $\hat{S}_j = S_j - E[S_j]$. Then, for $y \geq 0$,

◀ Theorem 9.4

$$\Pr\{\max_{1 \leq j \leq n} \hat{S}_j \geq y\sqrt{n}\} \leq 2e^{-y^2/8}$$

and

$$E[\max_{1 \leq j \leq n} \hat{S}_j] = \Theta(\sqrt{n}).$$

Assume that a fair coin is tossed n times and that, at each toss, 1 euro is won (lost) if the outcome is head (tail). Let us define the random variable $x_i = 1$ ($x_i = -1$) if the i-th coin toss is head (tail), $i = 1, 2, \ldots, n$. Since we toss a fair coin we have that $\Pr\{x_i = +1\} = \Pr\{x_i = -1\} = 0.5$.

◀ Example 9.1

Therefore, $S_j = \sum_{i=1}^{j} x_i$ represents the total win after the j-th toss (S_j is negative if we are losing); clearly for each j, $E[S_j] = 0$. The first inequality of Theorem 9.4 states that it is very unlikely that the value that we win or lose in a game involving n tosses is much larger than $\sqrt{n}$; in fact the probability that the absolute difference between heads and tails is greater than $y\sqrt{n}$ is exponentially decreasing in y. The second equality of the theorem shows that, as the length of the game increases, the expected maximum win or loss increases as the square root of such length.

In the following we will also use convergence of an ordered sequence of identically distributed random variables. Given a sequence of random variables independently drawn from a common distribution, we are interested in knowing the value of the smallest one and, more generally, the value of the k-th smallest one, $k = 1, 2, \ldots, n$. In the special case when the random variables are uniformly distributed in $[0, 1]$ the following theorem holds.

Let x_i, $i = 1, 2, \ldots, n$ be independent random variables that assume values

◀ Theorem 9.5

Chapter 9
PROBABILISTIC ANALYSIS OF APPROXIMATION ALGORITHMS

in the range $[0,1]$. Let v_j the j-th smallest value among the n values x_i, $i = 1, 2, \ldots, n$. We have that

$$E[v_j] = \frac{j}{n+1}$$

and v_j converges almost everywhere to $j/(n+1)$ as n goes to infinity.

9.3 Probabilistic analysis and multiprocessor scheduling

IN THIS section we consider MINIMUM MULTIPROCESSOR SCHEDULING on identical machines both in the general case when there is a variable number of machines and in the special case of two machines, that is represented by MINIMUM PARTITION. Recall that an instance of the MINIMUM MULTIPROCESSOR SCHEDULING on p parallel machines is specified by n jobs and a positive rational weight associated with each job, that represents its processing time; x_i represents the processing time of job i, $i = 1, 2 \ldots, n$, and the problem requires us to partition the jobs into p sets A_i, $i = 1, 2, \ldots, p$, in such a way that $\max_i A_i$ is minimized. The probability distribution that we consider assumes that the processing times are independently and uniformly distributed in $[0, 1]$. Several approximation algorithms for this problem have been analyzed from a probabilistic point of view. We will consider here the List Scheduling algorithm (see Sect. 2.2.1). Recall that the List Scheduling algorithm receives in input the list of the items to be partitioned (in any order) and assigns the first p items to p different sets. Remaining items are assigned using the following rule: the i-th item, for $i > p$, is assigned to the set with smallest sum breaking ties arbitrarily.

Let $m_{LS}(n)$ and $m^*(n)$ be the random variables that denote respectively the measures of the solution found by the List Scheduling algorithm and of the optimal solution for MINIMUM SCHEDULING ON IDENTICAL MACHINES on p machines with n jobs.

Theorem 9.6 ▶ *If the processing times are independently and uniformly distributed in $[0, 1]$ then $m_{LS}(n)$ satisfies the following inequality:*

$$\Pr\{m_{LS}(n)/m^*(n) \geq 1 + 4/n\} \leq e^{-n/32}.$$

PROOF Let W be the sum of the processing times of of all jobs, i.e., $W = \sum_{k=1}^{n} x_k$ and let j be the index of the last job that completes the schedule and let x_j be its length. In the proof of Theorem 2.6 we have shown the following bounds on $m^*(n)$ and $m_{LS}(n)$:

$$m^*(n) \geq \frac{1}{p}\sum_{i=1}^{n} x_i \text{ and } m_{LS}(n) \leq \frac{W}{p} + \frac{(p-1)x_j}{p}.$$

Since $x_j \leq 1$ we can bound the ratio between the solution found by the List Scheduling algorithm and the optimal solution as follows:

$$\frac{m_{LS}(n)}{m^*(n)} \leq 1 + \frac{p-1}{\sum_{i=1}^{n} x_i}.$$

By Theorem 9.3 we have that

$$\Pr\{m_{LS}(n)/m^*(n) \geq \frac{4(p-1)}{n}\} \leq \Pr\{\sum_{i=1}^{n}(x_i - 1/2) < -n/4\} \leq e^{-n/32}$$

and the theorem follows. QED

The theorem implies that if the processing times are independently and uniformly distributed in $[0,1]$ then the List Scheduling algorithm finds a solution with measure $m_{LS}(n)$ such that the ratio $m_{LS}(n)/m^*(n)$ converges to 1 almost everywhere. This result compares favorably with the worst case behavior of the List Scheduling algorithm analyzed in Sect. 2.2.1. It is also possible to obtain bounds on the expected difference $m_{LS}(n) - m^*(n)$ showing that, if the processing times are uniformly and independently distributed in $[0,1]$, then the expected difference between the measure of the solution returned by the List Scheduling algorithm and the optimal measure is $1/3$ if there are 2 or more jobs and 0.5 if there is only one job. We do not prove this result because a better bound can be proved for the expected difference of the solution found by the List Scheduling algorithm assuming that jobs are ordered according to the LPT rule (that is, in non-increasing order of their processing time).

Let $m_{LPT}(n)$ be the random variable that denotes the measure of the solution found by the List Scheduling algorithm with n jobs ordered according to the LPT rule on an instance of MINIMUM SCHEDULING ON IDENTICAL MACHINES on p machines.

◀ Theorem 9.7

If the processing times are independently and uniformly distributed in $[0,1]$ then $m_{LPT}(n) - m^*(n)$ converges to 0 almost everywhere.

PROOF

Let the average idle time $D_{LPT}(L_n)$ be the average, over all processors, of the difference between the makespan and $1/p$ times the sum of the processing times of all jobs. Namely,

$$D_{LPT}(L_n) = m_{LPT}(n) - \frac{1}{p}\sum_{i=1}^{n} x_i.$$

Since $\sum_{i=1}^{n} x_i/p$ is a lower bound on m^*, in order to prove the theorem it is sufficient to show that $\lim_{n\to\infty} D_{LPT}(L_n) = 0$ almost everywhere.

Since each job is assigned to the least loaded machine, the following bound holds on $D_{LPT}(L_n)$:

$$D_{LPT}(L_n) \leq \max_{1 \leq i \leq n} \left(x_i - \frac{1}{p} \sum_{k=1}^{i} x_i \right). \tag{9.3}$$

For any $\varepsilon > 0$ we separately bound the right hand side of the above inequality first with respect to values $i = 1, 2, \ldots, \lfloor \varepsilon n \rfloor$ and then to values $i = \lfloor \varepsilon n \rfloor + 1, \ldots, n$. Since jobs are ordered in nonincreasing order of their processing time and the processing times are bounded by 1 it follows that

$$\max_{1 \leq i \leq \lfloor \varepsilon n \rfloor} \left(x_i - \frac{1}{p} \sum_{k=1}^{i} x_k \right) \leq x_{\lfloor \varepsilon n \rfloor},$$

$$\max_{\lfloor \varepsilon n \rfloor + 1 \leq i \leq n} \left(x_i - \frac{1}{p} \sum_{k=1}^{i} x_k \right) \leq 1 - \sum_{k=1}^{\lfloor \varepsilon n \rfloor} x_k.$$

Hence

$$D_{LPT}(L_n) \leq \max \left(x_{\lfloor \varepsilon n \rfloor}, 1 - \sum_{k=1}^{\lfloor \varepsilon n \rfloor} x_k \right).$$

Since the sequence of random variables x_k, $k = \lfloor \varepsilon n/2 \rfloor \ldots, \lfloor \varepsilon n \rfloor$ is a nondecreasing sequence, it follows that

$$\begin{aligned}
D_{LPT}(L_n) &\leq \max \left(x_{\lfloor \varepsilon n \rfloor}, 1 - \sum_{k=1}^{\lfloor \varepsilon n \rfloor} x_k \right) \\
&\leq \max \left(x_{\lfloor \varepsilon n \rfloor}, 1 - \sum_{\lfloor \varepsilon n/2 \rfloor}^{\lfloor \varepsilon n \rfloor} x_k \right) \\
&\leq \max \left(x_{\lfloor \varepsilon n \rfloor}, 1 - \lfloor \varepsilon n/2 \rfloor x_{\lfloor \varepsilon n/2 \rfloor} \right).
\end{aligned}$$

Theorem 9.5 implies that $x_{\lfloor \varepsilon n \rfloor}$ converges almost everywhere to $1/\varepsilon$ and $x_{\lfloor \varepsilon n/2 \rfloor}$ converges almost everywhere to $2/\varepsilon$. The theorem thus follows.

QED

9.4 Probabilistic analysis and bin packing

IN THIS section, approximation algorithms for MINIMUM BIN PACKING are studied assuming the following probability distribution: an instance of size n is defined by n random variables x_i, $i = 1, 2, \ldots, n$, independently and uniformly distributed in $[0,1]$ (without loss of generality, we assume that the bin capacity is 1).

In the following, we identify an item with its size and $m^*(n)$ is the random variable that denotes the measure of an optimal packing of an instance of n items.

It is possible to show that $E[m^*(n)]$ satisfies the inequality $E[m^*(n)] \geq \frac{n}{2}$.

Let $W = \sum_i^n x_i$ be the random variable that denotes the total size of the items: since $\lceil W \rceil \leq m^*(n)$ and $E[\sum_{i=1}^n x_i] = n/2$ it follows that

$$E[m^*(n)] \geq E[\sum_{i=1}^n x_i] = \frac{n}{2}.$$

The same lower bound on the expected measure of $m^*(n)$ can be obtained by observing that the number of bins in any feasible solution is at least the number of items with size larger than 0.5. If the size of the items is uniformly distributed in $[0,1]$ then the expected number of items with size larger than 0.5 is $n/2$.

In the following, the expected values of the random variables denoting the number of bins used by *First Fit Decreasing* and *Best Fit Decreasing* (see Sect. 2.2.2) are evaluated.

The analysis exploits the *dominating algorithm* technique, that has been successfully applied for circumventing the dependency problem in the probabilistic analysis of algorithms.

The basic idea of the dominating algorithm technique is the following: if we are not able to analyze algorithm $\mathcal{A}$ we introduce a simpler algorithm $\mathcal{A}'$ that is more easy to analyze (that is, an algorithm either with no dependency or with tractable dependencies). If the solution returned by $\mathcal{A}$ is always better than the solution given by $\mathcal{A}'$ then probabilistic upper bounds on the measure of the solution found by $\mathcal{A}'$ hold for $\mathcal{A}$ as well.

The dominated algorithm used for the analysis of *First Fit Decreasing* and *Best Fit Decreasing* is called *PAIRMATCH*, and it is based on the following observation: if items are ordered in increasing order of their size, then the expected value of the sum of the i-item and of the $(n-i)$-item is 1. Therefore we might try to build a packing by matching together items i and $(n-i)$.

Notice that the obtained packing might not be feasible, since the sum of the i-item and of the $(n-i)$-item might be greater than 1.

Algorithm *PAIRMATCH* exploits this observation by packing at most two items per bin. At the start of iteration i the largest unpacked item is selected: let L_i be its size. Then the largest unpacked item having size s_i such that $L_i + s_i \leq 1$ is chosen. If such an item exists then the two selected

Section 9.4

PROBABILISTIC ANALYSIS AND BIN PACKING

items are assigned to a new bin, otherwise only the larger item is assigned to the new bin. It is easy to see that the running time of the algorithm is $O(n \log n)$.

Let $m_{PM}(n)$ be the random variable that denotes the number of bins used by *PAIRMATCH*: the following theorem evaluates the expected value of $m_{PM}(n)$.

Theorem 9.8 ▶ *If n items are independently and uniformly distributed in $[0,1]$, then $E[m_{PM}(n)]$ satisfies the equality*

$$E[m_{PM}(n)] = \frac{n}{2} + O(\sqrt{n}).$$

PROOF Given an instance of MINIMUM BIN PACKING an item is defined to be *small (large)* if its size is at most (at least) $1/2$. Notice that we do not consider items of size exactly $1/2$: however, the probability of having an item with size $1/2$ is 0, and, as a consequence, the expected number of bins used to pack items of size $1/2$ is negligible.

Let $u(n)$ be the random variable that denotes the number of large unmatched items in the solution found by *PAIRMATCH* (equivalently, the number of bins with only one large item). Since there could be at most one bin containing only one element of size smaller that $1/2$, it follows that

$$m_{PM}(n) = \lceil \frac{n-u(n)}{2} \rceil + u(n) = \lceil \frac{n+u(n)}{2} \rceil.$$

To find an upper bound on the expected value of $u(n)$, we represent large and small items on the segment $[0,1]$ as follows: for each small item of size x there is a symbol $\ominus$ at coordinate x and for each large item of size x there is a symbol $\oplus$ at coordinate $1-x$. Notice that a large item corresponding to a symbol $\oplus$ can be packed only with those small items that correspond to symbols $\ominus$ at its left.

It is possible to see that a large item is unmatched by the algorithm if and only if the number of $\oplus$ on its left (including the one corresponding to the item itself) is larger than the number of $\ominus$ on its left. Therefore, if $e(y)$ denotes the *excess* at coordinate y, that is the number of $\oplus$ minus the number of $\ominus$ in $[0,y]$, it follows that $u(n) = max_y(e(y))$. Since the sizes of the items are uniformly and independently distributed in $[0,1]$, it follows that all sequences of $\oplus$ and $\ominus$ are equally likely and that the maximum excess is distributed as the maximum excess of tails over heads at any point in a sequence of n tosses of a fair coin. The proof is completed by applying the bound provided by Theorem 9.4.

QED

Corollary 9.9 ▶ *If n items are independently drawn from the uniform distribution over the interval $[0,1]$ then $E[m_{PM}(n) - m^*(n)] = O(\sqrt{n})$.*

The following lemma shows that the solution found by PAIRMATCH is always worse than the solution found by *First Fit Decreasing* and *Best Fit Decreasing*.

Section 9.4
PROBABILISTIC ANALYSIS AND BIN PACKING

◀ **Lemma 9.10**

For any set of item sizes drawn from $[0,1]$, *the number of bins used by algorithm* PAIRMATCH *is at least the number of bins used by* First Fit Decreasing *or* Best Fit Decreasing.

PROOF

Given a packing algorithm $\mathcal{A}$, define a sequence of two stage algorithms for packing $\mathcal{A}_i$, $i = 1, 2, \ldots, n$ as follows. The first i items are packed in the same way as by algorithm $\mathcal{A}$; the remaining items are packed by another algorithm, denoted as *2-Optimal Completion (OC)*, defined as follows: *OC* packs the remaining items in the optimal way, with the restriction that at most 2 items are packed in a bin.

It is easy to see that the 2-optimal completion is done by the following algorithm, that considers items in order of nonincreasing size: when a new item x must be packed, *OC* packs it with the largest item that allows to obtain a feasible bin. Otherwise, that is, if x does not fit in any of the used bins with only one item, a new bin is used. Clearly the above definition implies that if $\mathcal{A}$ is *First Fit Decreasing* (or *Best Fit Decreasing*) then $\mathcal{A}_n$ coincides with *First Fit Decreasing* (or *Best Fit Decreasing*). Moreover, the solution found by PAIRMATCH is no better than the solution found by $\mathcal{A}_0$.

In the following, we consider the case that $\mathcal{A}$ is *First Fit Decreasing*, omitting the analogous proof for *Best Fit Decreasing*. Formally, we can write $m_{PM}(n) \geq m_{\mathcal{A}_0}(n)$ and $m_{\mathcal{A}_n} = m_{FFD}$. In order to complete the proof it is sufficient to show that, for $1 \leq i \leq n-1$, $m_{\mathcal{A}_i} \geq m_{\mathcal{A}_{i+1}}$.

Recall that items are ordered in nonincreasing order of their size: if algorithm $\mathcal{A}_i$ packs the $(i+1)$-th item in a bin with another item then the same assignment is performed by *First Fit Decreasing*, and therefore $m_{\mathcal{A}_i} = m_{\mathcal{A}_{i+1}}$. On the other hand, if $\mathcal{A}_i$ uses a new bin to pack the $(i+1)$-th item while *First Fit Decreasing* uses a bin containing two or more items, then that bin can be no longer used by either $\mathcal{A}_i$ or $\mathcal{A}_{i+1}$. This implies that $\mathcal{A}_i$ and $\mathcal{A}_{i+1}$, with the exception of the newly introduced bin, start with the same set of bins containing a single item. Therefore, $\mathcal{A}_{i+1}$ opens at most the number of bins used by $\mathcal{A}_i$, thus implying that $m_{\mathcal{A}_i} \geq m_{\mathcal{A}_{i+1}}$.

QED

Since $E[m^*(n)] \geq n/2$, Corollary 9.9 and Lemma 9.10 immediately imply the following theorem.

◀ **Theorem 9.11**

If n items are independently drawn from a uniform distribution over $[0,1]$ *then the expected number of bins used by* First Fit Decreasing *and* Best Fit

Chapter 9
PROBABILISTIC ANALYSIS OF APPROXIMATION ALGORITHMS

Decreasing *is given by*

$$E[m_{FFD}(n) - m^*(n)] = O(\sqrt{n})$$

$$E[m_{BFD}(n) - m^*(n)] = O(\sqrt{n}).$$

9.5 Probabilistic analysis and maximum clique

IN THIS section, an approximation algorithm for MAXIMUM CLIQUE is analyzed from a probabilistic point of view. The results of this section are twofold: a sharp characterization of the measure of the optimal solution and a simple sequential algorithm, with very poor behavior from the worst case point of view, but that achieves a solution whose measure is almost everywhere half the optimal measure.

Before proceeding, we need to define a probability distribution on the graphs that appear in the input. The simplest probability distribution is the one that is parameterized by the number of vertices n and assumes that all graphs with n vertices have the same probability to occur. However, this distribution is not completely satisfactory because it does not allow to represent sparse graphs with sufficiently large probability: if we draw a graph with n vertices from the above distribution, then, with overwhelming probability, the number of its edges is approximately $n^2/2$ edges.

Two main models of random graphs have been proposed and studied, denoted as $\mathcal{G}_{n,p}$ and $\mathcal{G}_{n,M}$. $\mathcal{G}_{n,p}$ denotes the class of random graphs on n labeled vertices where each of the $n(n-1)/2$ edges occurs independently and with probability p (and hence does not occur with probability $q = 1-p$). Clearly, if $p = 1/2$, then all graphs on n vertices are equiprobable. In the $\mathcal{G}_{n,M}$ model we assume that all graphs of n labeled vertices and M edges have the same probability to occur.

The two models are closely related: if $p = 2M/n(n-1)$, then the expected number of edges of a graph in $\mathcal{G}_{n,p}$ is M. This intuition can be formalized by showing interesting relationships between the two models: many properties of random graphs in the $\mathcal{G}_{n,p}$ model hold in the $\mathcal{G}_{n,M}$ model when $M = pn(n-1)$.

Here we focus our attention to the $\mathcal{G}_{n,p}$ model because the analysis is usually simpler. In fact, in this model, the existence of an edge does not condition the existence of any other edge. This is not true in the $\mathcal{G}_{n,M}$ model, as shown in the following example.

Example 9.2 ▶ Given a random graph of n nodes we want to estimate the probability that vertices $1,2,3$ form a clique, that is, that edges $(1,2)$, $(1,3)$ and $(2,3)$ occur.

Program 9.1: Sequential Clique

input A graph G;
output C a complete subgraph of G;
begin
    $C := \{v_1\}$;
    for $i := 2$ **to** n **do**
        if $\{v_i\} \cup C$ is a clique $C := C \cup \{v_i\}$
end.

Section 9.5
PROBABILISTIC ANALYSIS AND MAXIMUM CLIQUE

In the $\mathcal{G}_{n,p}$ model this probability is clearly p^3, since the existence of an edge does not depend on the presence of any other edge.

On the other hand, in the $\mathcal{G}_{n,M}$ model the probability of existence of edge $(1,2)$ is

$$\frac{M}{\binom{n}{2}} = \frac{2M}{n(n-1)}.$$

However, the probability of existence of edge $(1,3)$ given that $(1,2)$ is an edge of the graph is less than $2M/n(n-1)$ because the event that $(1,2)$ is an edge of the graph negatively conditions the probability of existence of edge $(1,3)$. Since there are M edges in total and we already know that $(1,2)$ is present there are $M-1$ remaining edges among $n(n-1)/2 - 1$. Therefore, the probability that $(1,3)$ is an edge conditioned by the existence of edge $(1,2)$ is $2(M-1)/(n(n-1)-2)$. By the same argument, the probability of existence of the third edge of the clique is $2(M-2)/(n(n-1)-4)$. Therefore, in the $\mathcal{G}_{n,M}$ model, the probability that vertices $1,2,3$ form a clique is

$$\frac{8M(M-1)(M-2)}{n(n-1)(n(n-1)-2)(n(n-1)-4)}.$$

Notice that if $p = 2M/n(n-1)$ this value is not too far from p^3 for large values of n and non-negligible values of p.

Let us first analyze the size of the maximum clique in a random graph drawn from $\mathcal{G}_{n,p}$ and, successively, the size of the clique obtained by the sequential algorithm. In the following we assume that p is fixed (that is, it does not depend on the number of vertices).

We now analyze the behavior of Program 9.1, a simple sequential algorithm that considers vertices in a fixed order.

Let $m_S(G_n)$ be the random variable denoting the solution found by Program 9.1 when applied to a random graph $G_n \in \mathcal{G}_{n,p}$. The following theorem shows that the above algorithm achieves a solution that, with probability 1, has performance ratio 2.

Before formally proving the theorem we heuristically evaluate the measure of the solution found by the algorithm. Assume that the algorithm has found a clique of size $j-1$: the probability that the next considered vertex v allows to obtain a clique of size j is p^{j-1}, that is, the probability that there exist $j-1$ edges between v and each vertex of the current clique.

Let δ_j, $j=1,2,\ldots$ be the number of vertices to be considered for obtaining a solution of size j starting from a solution of size $(j-1)$ ($\delta_1 = 1$ and, for $j \geq 2$, δ_j is the number of iterations of the loop necessary for adding one vertex to the solution).

Observe that δ_j, $j \geq 2$, is a random variable that is geometrically distributed with expected value

$$E[\delta_j] = (1/p)^{j-1}.$$

Therefore, it is reasonable to expect that the measure k of the solution found by the algorithm should be close to the solution of the following equation

$$\sum_{j=1}^{k} E[\delta_j] = n.$$

It is possible to verify that the solution of the above equation is given by $k \approx \log_{1/p} n$.

Let us now define two random variables, that shall be used in the proofs of the following theorems. Let $X(n,p,r)$ be the random variable that denotes the number of cliques of size r in a random graph $G_n \in \mathcal{G}_{n,p}$. Let $Y(n,p,r)$ be the random variable that denotes the number of maximal cliques of size r in a random graph $G_n \in \mathcal{G}_{n,p}$ (recall that a set of vertices I is a maximal clique if I is a clique and there does not exist a vertex $v \notin I$ such that $I \cup v$ is a clique).

Lemma 9.12 ▶ *The expected value of $X(n,p,r)$ is $\binom{n}{r} p^{\binom{r}{2}}$.*

PROOF Define S to be the set of all subsets of r vertices and let 1_S, $S \in \mathcal{S}$, be the indicator function, that is, 1 if S is a clique and 0 otherwise. Clearly, $X(n,p,r) = \sum_{S \in \mathcal{S}} 1_S$ and therefore

$$E[X(n,p,r)] = E[\sum_{S \in \mathcal{S}} 1_S] = \sum_{S \in \mathcal{S}} E[1_S].$$

Consider one specific set S of r vertices. Since each edge has probability p to occur, then the probability that such a set is a clique is $p^{\binom{r}{2}}$: since 1_S is a 0-1 function this probability is equal to $E[1_S]$. The lemma follows by observing that there are $\binom{n}{r}$ possible sets of r vertices.

QED

Analogously to Lemma 9.12, it is possible to show that the expected value of $Y(n,p,r)$ is

$$E[Y(n,p,r)] = \binom{n}{r} p^{\binom{r}{2}} (1-p^r)^{n-r} \qquad (9.4)$$

where the third term of the right hand side gives the probability that no vertex not in the clique is a neighbor of all vertices in the clique.

The sequence $m_S(G_n)/\log_{1/p} n$, where $G_n \in \mathcal{G}_{n,p}$, converges to 1 almost everywhere. ◀ **Theorem 9.13**

Section 9.5
PROBABILISTIC ANALYSIS AND MAXIMUM CLIQUE

We will show that, for each $\varepsilon > 0$, the following inequalities hold PROOF

$$\lim_{n\to\infty} \Pr\{\frac{m_S(G_n)}{\log_{1/p} n} - 1 \geq \varepsilon\} = 0 \qquad (9.5)$$

$$\lim_{n\to\infty} \Pr\{1 - \frac{m_S(G_n)}{\log_{1/p} n} \geq \varepsilon\} = 0 \qquad (9.6)$$

To prove Eq. (9.5) assume that a solution of size $\log_{1/p} n$ has already been obtained. We focus on the number of iterations necessary for adding $\varepsilon \log_{1/p} n$ more vertices to the solution. Namely, for $j = \log_{1/p} n + h$, with $h = 1, 2, \ldots, \varepsilon \log_{1/p} n$, the probability of adding a vertex in a single iteration of the loop is

$$p^{j-1} = p^{\log_{1/p} n + h} = \frac{p^h}{n}$$

It follows that, for $j = \log_{1/p} n + h$, with $h = 1, 2, \ldots, \varepsilon \log_{1/p} n$,

$$\Pr\{\delta_j \leq n\} \leq p^h$$

and, therefore, for n sufficiently large, we obtain

$$\Pr\{m_S(G_n) \geq (1+\varepsilon)\log_{1/p} n\}$$
$$\leq \Pr\{\delta_j \leq n \text{ for all } j = \log_{1/p} n + h, \ h = 1, 2, \ldots, \varepsilon \log_{1/p} n\}$$
$$\leq \prod_{h=1}^{\varepsilon \log_{1/p} n} p^h$$
$$\leq p^{(\varepsilon \log_{1/p} n)^2/2}$$
$$= n^{-(\varepsilon^2 \log_{1/p} n)/2}$$
$$\leq n^{-i}$$

for any integer i.

Chapter 9
PROBABILISTIC ANALYSIS OF APPROXIMATION ALGORITHMS

To prove that Eq. (9.6) actually holds, observe that the algorithm returns a maximal clique, that is, a complete graph that is not a subgraph of a larger clique. Therefore, it is sufficient to prove that, with probability 1, in a random graph there does not exist a maximal clique of size less than $(1-\varepsilon)\log_{1/p} n$.

The first moment method implies that, for each integer r with $1 \leq r \leq (1-\varepsilon)\log_{1/p} n$, the probability that a maximal clique of size r exists is bounded by $E[Y(n,p,r)]$.

Therefore, by Eq. (9.4), we have that

$$\Pr\{m_S(G_n) \leq (1-\varepsilon)\log_{1/p} n\}$$

$$\leq \sum_{r=1}^{(1-\varepsilon)\log_{1/p} n} \Pr\{G_n \text{ contains a maximal clique of size } r\}$$

$$\leq \sum_{r=1}^{(1-\varepsilon)\log_{1/p} n} \binom{n}{r} p^{\binom{r}{2}}(1-p^r)^{n-r}$$

$$\leq \sum_{r=1}^{(1-\varepsilon)\log_{1/p} n} \binom{n}{r}(1-p^r)^{n-r}.$$

Since $(1-x) \leq e^{-x}$ for any real value $x \geq 0$, $p^r \geq n^{-(1-\varepsilon)}$ for $1 \leq r \leq (1-\varepsilon)\log_{1/p} n$, and $\binom{n}{r} \leq \left(\frac{ne}{r}\right)^r$ we obtain, for n sufficiently large,

$$\Pr\{m_S(G_n) \leq (1-\varepsilon)\log_{1/p} n\}$$

$$\leq \sum_{r=1}^{(1-\varepsilon)\log_{1/p} n} \binom{n}{r}(1-p^r)^{n-r}$$

$$\leq \sum_{r=1}^{(1-\varepsilon)\log_{1/p} n} \binom{n}{r} e^{-p^r(n-r)}$$

$$\leq \sum_{r=1}^{(1-\varepsilon)\log_{1/p} n} \left(\frac{en}{r}\right)^r e^{-n^{-(1-\varepsilon)}(n-r)}$$

$$\leq \sum_{r=1}^{(1-\varepsilon)\log_{1/p} n} n^r e^{-n^{\varepsilon}+rn^{-(1-\varepsilon)}+r-r\ln r}$$

$$\leq (\log_{1/p} n) n^{\log_{1/p} n} e^{-n^{\varepsilon}}$$

$$\leq n^{-i}$$

QED for all integers i. This completes the proof of the theorem.

Observe that in the above proof the convergence to the limit is faster than what is strictly necessary to prove the theorem.

Section 9.5
PROBABILISTIC ANALYSIS AND MAXIMUM CLIQUE

We now analyze the size of the maximum clique in a random graph. Let $m^*(G_n)$ be the random variable that denotes the size of the maximum clique in a random graph in the $G_n \in \mathcal{G}_{n,p}$.

It turns out that the size of the largest clique is sharply concentrated around the value of r for which the expectation $E[X(n,p,r)]$ is about 1.

By applying Stirling's formula to approximate the factorial, the value of r that solves the equation $E[X(n,p,r)] = 1$ is close to the value of r that is the solution to the following equation

$$(2\pi)^{-1/2} n^{n+1/2} (n-r)^{-n+r-1/2} r^{-r-1/2} p^{r(r-1)/2} = 1.$$

Let $r(n,p)$ be the value of r that solves the above equation. It can be easily verified that

$$\begin{aligned} r(n,p) &= 2\log_{1/p} n - 2\log_{1/p} \log_{1/p} n + 2\log_{1/p}(e/2) + 1 + o(1) \\ &= 2\log_{1/p} n + O(\log\log n). \end{aligned}$$

The following theorem shows that the size of the maximum clique is sharply concentrated around $r(n,p)$: with probability 1 the clique number has one of two possible values.

◀ **Theorem 9.14** *For any edge probability p there exists a function $K: \mathcal{N} \to \mathcal{N}$ such that for every random graph $G_n \in \mathcal{G}_{n,p}$ with probability 1 either $m^*(G_n) = K(n)$ or $m^*(G_n) = K(n) + 1$.*

PROOF We will prove a weaker bound here, and refer to [Bollobas, 1985] (pp. 253–255) for a proof of the stronger bound that exploits the same approach applied here and utilizes a more refined analysis of the probabilities involved. Namely, we prove that

$$\frac{m^*(G_n)}{\log_{1/p} n}$$

converges almost everywhere to 2, that is the following inequalities hold:

$$\lim_{n\to\infty} \Pr\{\frac{m^*(G_{n,p})}{\log_{1/p} n} - 2 \geq \varepsilon\} = 0 \qquad (9.7)$$

and

$$\lim_{n\to\infty} \Pr\{2 - \frac{m^*(G_n)}{\log_{1/p} n} \geq \varepsilon\} = 0. \qquad (9.8)$$

Since $X(n,p,r)$ is a non-negative integer random variable, the first moment method implies that the probability of having a clique of size r, that is, $\Pr\{X(n,p,r) > 0\}$, is bounded above by the expected value of $X(n,p,r)$.

Therefore, if $E[X(n,p,r)]$ tends to zero it follows that it is unlikely that a random graph in G_n has a clique of size r, for n sufficiently large. Namely, the following inequality holds:

$$\Pr\{m^*(G_n) > r\} \leq E[X(n,p,r)] = \binom{n}{r} p^{\binom{r}{2}} \leq \frac{n^r}{r!} p^{r(r-1)/2}. \quad (9.9)$$

Let us define $r = 2\log_{1/p} n$; by observing that $p^{\log_{1/p} n} = n^{-1}$ it follows that

$$\Pr\{m^*(G_n) > 2\log_{1/p} n\} \leq \frac{n^{2\log_{1/p} n}}{(2\log_{1/p} n)!} p^{2\log_{1/p} n(2\log_{1/p} n - 1)/2}$$

$$\leq \frac{n}{(2\log_{1/p} n)!} = o(n^{-i}) \quad (9.10)$$

for any integer i and, therefore, Eq. (9.7) follows.

In order to prove Eq. (9.8), it is sufficient to show that

$$\Pr\{m^*(G_n) < 2(1-\varepsilon)\log_{1/p} n\} < o(n^{-3/2}). \quad (9.11)$$

Since $\Pr\{m^*(G_n) < r\} = \Pr\{X(n,p,r) = 0\}$, the second moment method implies that

$$\Pr\{m^*(G_n) < r\} = \Pr\{X(n,p,r) = 0\} \leq \frac{E[X(n,p,r)^2]}{E[X(n,p,r)]^2} - 1. \quad (9.12)$$

The computation of $E[X(n,p,r)^2]$ is the main difficulty of the proof. Denote by $\mathcal{S}$ the set of all subsets of r vertices and, again, let 1_S, $S \in \mathcal{S}$, be the indicator function that is 1 if the set of vertices S is a clique and 0 otherwise. The value of the random variable $X(n,p,r)^2$ is then given by

$$X(n,p,r)^2 = \left(\sum_{S_1 \in \mathcal{S}} 1_{S_1}\right)\left(\sum_{S_2 \in \mathcal{S}} 1_{S_2}\right)$$

Therefore, the second moment of $X(n,p,r)$ is given by

$$E[X(n,p,r)^2] = E\left[\sum_{S_1 \in \mathcal{S}} 1_{S_1} \sum_{S_2 \in \mathcal{S}} 1_{S_2}\right] = \sum_{S_1 \in \mathcal{S}} \sum_{S_2 \in \mathcal{S}} E[1_{S_1} 1_{S_2}].$$

Since 1_S is a 0-1 function it follows that

$$E[1_{S_1} 1_{S_2}] = \Pr\{\text{both } S_1 \text{ and } S_2 \text{ are clique}\}.$$

The estimation of the right hand side of the above equality depends on the size of the intersection of S_1 and S_2. Let $m = |S_1 \cap S_2|$ be the number of vertices in common between S_1 and S_2. For each value of m, $m = 0, 1, \ldots, r$, the number of different choices for S_1 and S_2 is given by

$$\binom{n}{r}\binom{r}{m}\binom{n-r}{r-m}. \tag{9.13}$$

In fact, there are $\binom{n}{r}$ ways of choosing S_1, $\binom{r}{m}$ ways of choosing the m vertices in common, and $\binom{n-r}{r-m}$ ways of choosing the remaining vertices of S_2.

We now consider two different cases depending on the value of m.

case 1: $m \leq 1$.
If m is either 0 or 1 then the two cliques are edge disjoint. Since the probability of existence of an edge is independent from the existence of any other edge, it follows that

$$\begin{aligned}\Pr\{\text{both } S_1 \text{ and } S_2 \text{ are clique}\} &= \Pr\{S_1 \text{ is a clique}\}\Pr\{S_2 \text{ is a clique}\} \\ &= p^{2\binom{r}{2}}. \end{aligned} \tag{9.14}$$

case 2: $2 \leq m \leq r$.
In this case, if $2 \leq m \leq r$ then if S_1 and S_2 are both cliques then they share $\binom{m}{2}$ edges. Therefore,

$$\Pr\{S_2 \text{ is a clique} \mid S_1 \text{ is a clique}\} = p^{\binom{r}{2}-\binom{m}{2}}. \tag{9.15}$$

It follows that

$$\begin{aligned}\Pr\{\text{both } S_1 \text{ and } S_2 \text{ are clique}\} &= \\ &= \Pr\{S_1 \text{ is a clique}\}\Pr\{S_2 \text{ is a clique} \mid S_1 \text{ is a clique}\} \\ &= p^{2\binom{r}{2}-\binom{m}{2}}.\end{aligned}$$

By Eqs. (9.13), (9.14), and (9.15), it follows that:

$$\begin{aligned}E[X(n,p,r)^2] &= \sum_{m=0}^{1}\binom{n}{r}\binom{r}{m}\binom{n-r}{r-m}p^{2\binom{r}{2}} \\ &+ \sum_{m=2}^{r}\binom{n}{r}\binom{r}{m}\binom{n-r}{r-m}p^{2\binom{r}{2}-\binom{m}{2}}.\end{aligned}$$

On the other hand, by applying Lemma 9.5, we have that

$$\begin{aligned}E[X(n,p,r)]^2 &= \left[\binom{n}{r}p^{\binom{r}{2}}\right]^2 \\ &= \sum_{m=0}^{r}\binom{n}{r}\binom{r}{m}\binom{n-r}{r-m}p^{2\binom{r}{2}}, \tag{9.16}\end{aligned}$$

Section 9.5

PROBABILISTIC ANALYSIS AND MAXIMUM CLIQUE

and, therefore,

$$E[X(n,p,r)^2] - E[X(n,p,r)]^2 =$$
$$= \sum_{m=2}^{r} \binom{n}{r}\binom{r}{m}\binom{n-r}{r-m} p^{2\binom{r}{2}}((1/p)^{\binom{m}{2}} - 1). \quad (9.17)$$

Plugging the right hand sides of Eqs. (9.16) and (9.17) into Eq. (9.12), after some algebraic manipulation we obtain

$$\Pr\{m^*(G_n) < r\} \leq \frac{E[X(n,p,r)^2]}{E[X(n,p,r)]^2} - 1$$
$$= \sum_{m=2}^{r} \frac{\binom{r}{m}\binom{n-r}{r-m}((1/p)^{\binom{m}{2}} - 1)}{\binom{n}{r}}$$

Let us denote, for $m = 2, 3, \ldots, r$,

$$F_m = \frac{\binom{r}{m}\binom{n-r}{r-m}((1/p)^{\binom{m}{2}} - 1)}{\binom{n}{r}}.$$

Notice that, since $F_r < 1/E[X(n,p,r)]$ and, for n sufficiently large and $3 \leq m \leq r-1$, $F_m < F_3 + F_{r-1}$, it follows that

$$\Pr\{m^*(G_n) < r\} \leq \sum_{m=2}^{r} F_m < F_2 + (r-3)(F_3 + F_{r-1}) + F_r$$
$$< \frac{r^4((1/p) - 1)}{2n^2}$$
$$+ (r-3)\left(\frac{r^6((1/p)^3 - 1)}{6n^3} + \frac{rnp^{r-1}}{E[X(n,p,r)]}\right)$$
$$+ \frac{1}{E[X(n,p,r)]}$$
$$< \frac{r^4((1/p) - 1)}{n^2}$$
$$+ \frac{1}{E[X(n,p,r)]}(1 + r^2np^{r-1}). \quad (9.18)$$

Now observe that, for $r = 2(1-\varepsilon)\log_{1/p} n$, $p^{r/2}$ is equal to $n^{(1-\varepsilon)}$. This implies that $E[X(n,p,r)] > n^i$ for n sufficiently large. In fact, we have that

$$E[X(n,p,r)] = \binom{n}{r} p^{\binom{r}{2}} > \left(\frac{n}{r}\right)^r p^{r^2/2}$$
$$> \left(\frac{n}{r}\right)^r n^{(1-\varepsilon)r} > \left(\frac{n^\varepsilon}{r}\right)^r = O(n^i)$$

for any integer i. It follows that, plugging $r = 2(1-\varepsilon)\log_{1/p} n$ into the right hand side of Eq. (9.18), Eq. (9.11) is verified.

As already observed, from Eq. (9.11) it follows that Eq. (9.8) is also verified.

The proof of the theorem is thus completed.

QED

Theorems 9.14 and 9.13 imply that the measure of the solution found by algorithm Sequential Clique indeed converges almost everywhere to one half of the optimal measure. This represent a significant improvement with respect to the worst case ratio of the algorithm.

9.6 Probabilistic analysis and graph coloring

THE PREVIOUS results can be used to study the approximability of MINIMUM GRAPH COLORING. First of all, observe that the probabilistic analysis performed for MAXIMUM CLIQUE can be immediately extended to MAXIMUM INDEPENDENT SET. Let $\alpha(G_n)$ be the random variable denoting the size of the maximum independent set in a random graph in the $\mathcal{G}_{n,p}$ model and let $q = 1 - p$. Observe that, for all graphs G, a set of vertices I is an independent set in G if and only if I is a complete graph in the complement graph $\overline{G}$ (that is, the graph with the same set of vertices in which an edge exists if and only if it does not exist in the original graph G).

Hence, from Theorem 9.14 we have that the size of the maximum independent set is sharply concentrated around $2\log_{1/q} n + O(\log\log n)$.

Let $\chi(G_n)$ be the random variable denoting the minimum number of colors necessary for coloring a random graph in the $\mathcal{G}_{n,p}$ model. Notice that, for any graph G, the number of colors required for coloring G is greater than the number of vertices of the graph divided by the size of the maximum independent set. It follows that, for any $\varepsilon > 0$,

$$\Pr\{\chi(G_n) \geq \frac{n}{\alpha(G_n)}\} > \Pr\{\chi(G_n) \geq \frac{n}{2\log_{1/q} n}(1-\varepsilon)\} \qquad (9.19)$$

approaches 1 as n goes to infinity.

Notice that the above discussion does not allow one to state the convergence of $\chi(G_n)$ to $n/2\log_{1/q} n$ because it does not imply that, for any $\varepsilon > 0$,

$$\Pr\{\chi(G_n) \leq \frac{n}{2\log_{1/q} n}(1+\varepsilon)\} \qquad (9.20)$$

almost everywhere. The convergence of $\chi(G_n)$ to the above value has been recently proved (see Bibliographical notes).

Chapter 9

PROBABILISTIC ANALYSIS OF APPROXIMATION ALGORITHMS

In an analogous way it is possible to extend the analysis of the sequential clique algorithm to evaluate the number of colors required by the sequential coloring algorithm considered in Sect. 2.2.3. Let $\chi_S(G_n)$ be the random variable denoting the number of colors used by the sequential algorithm when applied to a random graph drawn from the $\mathcal{G}_{n,p}$ model. We leave as an exercise the proof that

$$\chi_S(G_n) \frac{\log_{1/q} n}{n}$$

converges to 1 almost everywhere.

9.7 Probabilistic analysis and Euclidean TSP

IN THIS section we consider the approximability of the Euclidean MINIMUM TRAVELING SALESPERSON assuming that the cities correspond to points uniformly and independently drawn in the unit square. We will see that the length of the optimum tour through n cities grows as $\beta\sqrt{n}$, where β is a constant, and that a simple approximation algorithm finds a solution bounded by $2\sqrt{n} + (2+\sqrt{2})$.

The latter result is based on the STRIP algorithm, which can be described as follows. The algorithm first partitions the unit square into horizontal strips of height w (the bottom strip might have height less than w due to rounding). Then the algorithm visits all cities in the top strip from left to right; afterwards it moves down to the rightmost city in the second strip and visit all cities of the strip from the rightmost to the leftmost one. The algorithm proceeds by moving to the leftmost city of the third strip and visiting cities in the third strip from left to right; remaining strips are considered in the same way, until a Hamiltonian path is obtained. After visiting the last city of the bottom strip it returns to the first visited city.

The proof of the following lemma is left as an exercise.

Lemma 9.15 ▶ *Given an instance of Euclidean* MINIMUM TRAVELING SALESPERSON *with n cities in the unit square, the length of the tour obtained by the STRIP algorithm with strip height w, is at most*

$$nw + \frac{1}{w} + (2+\sqrt{2}).$$

The following theorem derives by observing that the bound of the above lemma is maximized for $w = 1/\sqrt{n}$.

Theorem 9.16 ▶ *Let $m_{STRIP}(n)$ be the length of the tour through n cities drawn in the unit* square returned by the STRIP algorithm with strip height $w = 1/\sqrt{n}$. Then we have that

$$m_{STRIP}(n) \leq 2\sqrt{n} + (2+\sqrt{2}).$$

This theorem shows that the length of the tour in a unit square grows as the square root of the number of cities. In the following, we will see that this growth rate holds also for the expected measure of the optimum. Such a result can be intuitively explained as follows: first observe that the length of a tour through $4n$ cities randomly drawn in a 2×2 square is about 4 times the length of a tour through n cities in an unit square. Furthermore, due to scaling, the length of a tour through $4n$ cities in a 2×2 square is twice the length of a tour through $4n$ cities in an unit square. It follows that, in the unit square, the length of a tour through $4n$ cities is about twice the length of the tour through n cities, that is, if the number of cities quadruples, then the length of the tour doubles. In a similar way it is possible to see that the length of a tour through $9n$ cities is about three times the length of a tour through n cities; the $\sqrt{n}$ growth rate follows.

Let $m^*(n)$ be the random variable denoting the length of an optimal tour when the cities are independently and uniformly distributed in the unit square. The next theorem shows that the expected measure of $m^*(n)$ grows as $\beta\sqrt{n}$, where β is a constant.

The main difficulty in the analysis is given by the fact that if we consider two disjoint subregions R_1 and R_2 of the unit square, then the distribution of cities in R_1, depends on the distribution in R_2; in fact, since the total number of cities is n, then the fact that in R_1 there are n_1 cities conditions the number of cities in R_2.

To prove the theorem it is useful to consider a different probability distribution and to assume that the cities are distributed according to a Poisson process. We first prove a convergence theorem assuming that the cities are distributed according to a Poisson process in the unit square and then extend it to the uniform distribution.

The advantage of considering a Poisson process instead of the uniform distribution is that in a Poisson process the distribution of cities in a subregion R_1 is independent of the distribution in a subregion R_2 disjoint from R_1. Notice that in a Poisson process the total number of cities is a random variable and, therefore, the convergence for a Poisson process does not directly imply the convergence for the uniform distribution. However, the distribution of n cities according to the uniform distribution and to a Poisson process with rate parameter n (that gives the expected number of cities in the region) are similar. In fact, given a Poisson process in the unit square, then the cities are uniformly distributed; furthermore, the total number of cities is concentrated around its mean (that is, the number of

Section 9.7

PROBABILISTIC ANALYSIS AND EUCLIDEAN TSP

Chapter 9

PROBABILISTIC ANALYSIS OF APPROXIMATION ALGORITHMS

cities in a Poisson process with rate parameter n is, with high probability, approximately n). In other words, the distribution of n cities according to the uniform distribution is well approximated by a Poisson process with rate parameter n.

Theorem 9.17 ▶ *There exists a constant β such that*

$$\lim_{n\to\infty} \frac{E[m^*(n)]}{\sqrt{n}} = \beta.$$

PROOF Let P be a Poisson process on $\mathbf{R}^2$ with unit intensity, and let $P([0,t]^2)$ be the set of cities that are included in the square $[0,t]^2$: notice that the expected number of cities in the square $[0,t]^2$ is t^2. Let us now compute the expected measure $E[m^*(P([0,t]^2))]$ of the length of an optimal tour through the cities in $P([0,t]^2)$, as t goes to infinity.

If we consider the square $[0,t]^2$ as decomposed into d^2 disjoint subsquares $R_1, R_2, \ldots R_{d^2}$ of side t/d, then a tour through all cities in the square $[0,t]^2$ can be obtained as follows. First, for all i, an optimal tour T_i through subsquare R_i is found and then the subtours are patched together to obtain a tour through all cities. This latter step is performed as follows: a city is chosen in each region R_i and the *STRIP* algorithm is used to find a tour J through the chosen cities. Together the tours T_i, $i = 1, 2, \ldots, d^2$, and the tour J form a closed walk through all the cities; using the triangle inequality this walk can be transformed into a tour of not greater length. The expected length of the obtained tour is bounded by

$$E[m^*(P([0,t]^2))] \leq \sum_{i=1}^{d^2} E[m^*(R_i)] + t(2d+2+\sqrt{2}),$$

where $t(2d+2+\sqrt{2})$ is the upper bound on the length of the tour J found by the *STRIP* algorithm when applied to an instance of the Euclidean TSP with d^2 cities belonging to a square of side t. Since the cities are distributed according to a Poisson process, then, by scaling, the expected length of the tour in R_i, $i = 1, 2, \ldots, d^2$, is $E[m^*(P([0,(t/d)]^2))]$ and hence

$$E[m^*(P([0,t]^2))] \leq d^2 E[m^*(P([0,(t/d)]^2))] + t(2d+2\sqrt{2}).$$

Setting $t = ds$, it follows that

$$\frac{E[m^*(P([0,t]^2))]}{t^2} \leq \frac{E[m^*(P([0,s]^2))]}{s^2} + \frac{2}{s} + \frac{\sqrt{2}}{t}, \quad d = 1, 2 \ldots$$

The above inequality, together with the facts that $E[m^*(P([0,t]^2))]$ is monotone and $E[m^*(P([0,t]^2))]/t^2$ is bounded (by Theorem 9.16), implies that

$$\lim_{t\to\infty} \frac{E[m^*(P([0,t]^2))]}{t^2} = \beta. \tag{9.21}$$

The convergence result obtained for $E[m^*(P([0,ds]^2))]$ is now used to prove the convergence of $E[m^*(n)]$. Observe that if $t = \sqrt{n}$ then the expected number of cities in $P([0,\sqrt{n}]^2)$ is n; furthermore, by Eq. (9.21) the expected length of a tour in $P([0,\sqrt{n}]^2)$ approaches βn as n goes to infinity. Therefore, by scaling the size of the square, we can expect that the length of a tour through n cities uniformly distributed in a unit square should be $P([0,\sqrt{n}]^2)/\sqrt{n} \approx \beta\sqrt{n}$. However, this is not sufficient to prove the theorem: to this aim we will show that the expected number of items in a Poisson process is close to its mean. Formally, by conditioning on the number k of cities in $P([0,t]^2)$, we can write $E[m^*(P([0,t]^2))]$ as follows

$$E[m^*(P([0,t]^2))] = \sum_{k=0}^{\infty} E[m^*(P([0,t]^2)) \mid |P([0,t]^2)| = k] \Pr\{|P([0,t]^2)| = k\} \quad (9.22)$$

If the cities are distributed according to a Poisson process with unit intensity, then the probability that there are k cities in $P([0,t]^2)$ is given by

$$\Pr\{|P([0,t]^2)| = k\} = \frac{t^{2k}}{k!}e^{-t^2}. \quad (9.23)$$

Now observe that, by scaling, the measure of $m^*(P([0,t]^2))$, conditioned by the fact that there are k cities, is equal to t times the length of the shortest tour through k cities that are uniformly and independently drawn in a unit square. Namely,

$$E[m^*(P([0,t]^2)) \text{ given } |P([0,t]^2)| = k] = tE[m^*(k)]. \quad (9.24)$$

Plugging equations (9.23) and (9.24) into equation (9.22) we obtain

$$E[m^*(P([0,t]^2))] = t \sum_{k=0}^{\infty} E[m^*(k)]e^{-t^2}\frac{t^{2k}}{k!}. \quad (9.25)$$

Since $E[m^*(P([0,t]^2))]/t^2$ converges to β as t goes to infinity we have that

$$\lim_{t \to \infty} \frac{\sum_{k=0}^{\infty} E[m^*(k)]e^{-t^2}\frac{t^{2k}}{k!}}{t} = \beta. \quad (9.26)$$

The proof is completed by observing that (see [Bingham, 1981]), for any $f(k)$ monotone increasing, if

$$\lim_{x \to \infty} \frac{\sum_{k=0}^{\infty} f(k)e^{-x}x^k/k!}{x^\alpha} = \beta,$$

then $\lim_{k \to \infty} f(k)/k^\alpha = \beta$. By applying this observation to Eq. (9.26) with $x = t^2$ and $\alpha = 1/2$, the theorem follows. QED

Section 9.7
PROBABILISTIC ANALYSIS AND EUCLIDEAN TSP

Chapter 9

PROBABILISTIC ANALYSIS OF APPROXIMATION ALGORITHMS

An interesting open question is to determine the value of β. Theorem 9.16 implies that $\beta \leq 2$. It is easy to see that $\beta \geq 1/(2\sqrt{2\pi}) \approx 0.199$. In fact we observe that the length of the tour is at least the sum of the distances from each city to its closest neighbor. By the linearity of expectation we have then that

$$E[m^*(n)] \geq \sum_{i=1}^{n} E[\text{distance between } x_i \text{ and its closest neighbor}]$$
$$= nE[\text{distance between } x_1 \text{ and its closest neighbor}].$$

A simple lower bound on the expected distance between x_1 and its closest neighbor is derived as follows: let $d(x_1, x_i)$ be the distance between x_1 and x_i. Let us consider a circle of radius r around x_1: the area of the circle is πr^2 and therefore the probability that a city x_i is at distance at most r from x_1 is at most πr^2 (it is exactly πr^2 if the circle is entirely contained in the unit square). Since there are n cities in total and, therefore, $(n-1)$ cities are potential candidates to be the nearest neighbor of x_1, then, by Boole's inequality, we obtain that

$$\Pr\{\text{there exists a city } x_i \text{ s.t. } d(x_i, y) \leq r\} \leq (n-1)\Pr\{d(x,y) \leq r\}$$
$$\leq n\pi r^2.$$

If $r = 1/\sqrt{2\pi n}$ then it follows that, with probability greater than $1/2$, the closest neighbor has a distance greater than r; therefore the expected distance between x_1 and its closest neighbor is at least $1/(2\sqrt{2\pi n})$. This implies that

$$E[m^*(n)] \geq \frac{1}{2\sqrt{2\pi}}\sqrt{n}$$

and, therefore, $\beta \geq 1/(2\sqrt{2\pi})$.

9.8 Exercises

Exercise 9.1 Show that there exists a sequence $x_1, \ldots, x_n$ of random variables that converges to c almost everywhere and to c' in expectation.

Exercise 9.2 Define a sequence of random variables $x_1, \ldots, x_n$ such that $E[x_n]$ is unbounded and x_n converges almost everywhere to 0.

Exercise 9.3 Prove Eq. (9.3).

Exercise 9.4 Consider MINIMUM MULTIPROCESSOR SCHEDULING assuming uniform machines, that is machine j has speed s_j and process job with processing time x_i in time x_i/s_j. Show that if the processing times of n jobs are uniformly distributed in $[0,1]$ then the LPT-rule (see Eq. 9.3) find a solution of measure $m_{LPT}(n)$ that converges almost everywhere to $m^*(n)$.

Exercise 9.5 Prove the bound of Theorem 9.8 assuming that the sizes of the items are independently drawn from any distribution with nonincreasing density. (Hint: first show that the theorem holds for a symmetric distribution.)

Exercise 9.6 Show that the 2-optimal completion procedure considered in Lemma 9.10 is correct.

Exercise 9.7 Assume that the sizes of the items of an instance of MINIMUM BIN PACKING are independently and uniformly distributed over $[0,1]$. Prove that
$$\frac{m_{BFD}(n)}{m^*(n)}$$
converges almost everywhere to 1.

Exercise 9.8 Let T_n be the random variable denoting the size of the largest tree induced in a random graph in the $\mathcal{G}_{n,p}$ model (i.e., the maximum cardinality of a set of vertices that are connected and do not contain a cycle). Prove that the sequence T_n converges almost everywhere to $2/\log_{1/p} n$.

Exercise 9.9 Let $\chi_S(G_{n,p})$ be the random variable denoting the number of colors used by the sequential coloring algorithm when applied to a random graph in the $\mathcal{G}_{n,p}$ model. Prove that the ratio
$$\chi_S(G_{n,p}) \frac{\log_{1/q} n}{n}$$
converges to 1 almost everywhere.

Exercise 9.10 Prove that, if n cities are independently and uniformly drawn in the unit square, then the expected length of the tour produced by the *STRIP* algorithm is less than $\beta \leq 0.93\sqrt{n}$. (Hint: compute the expected vertical distance between two successive cities in a strip).

Exercise 9.11 In the d-dimensional Euclidean MINIMUM TRAVELING SALESPERSON cities are distributed in the unit d-dimensional cube. Extend the *STRIP* algorithm to d dimensions and prove that the length of a tour through n cities is bounded by $a_d n^{d-1} d + c_d$, where a_d and c_d are constants that depend on d.

Exercise 9.12 Prove that there exist instances of the Euclidean MINIMUM TRAVELING SALESPERSON where cities are drawn in the unit square such that the length of the optimal tour is at least $\sqrt{n}$.

Exercise 9.13 Extend Theorem 9.17 to the d-dimensional case showing that if $m^*(n)$ is the random variable denoting the length of the optimal tour when cities are uniformly and independently distributed in the d-dimensional unit cube, then is $m^*(n) \approx \beta_d n^{d/d-1}$ as n goes to infinity.

Exercise 9.14 Prove Lemma 9.15.

9.9 Bibliographical notes

THE PROBABILISTIC analysis of combinatorial problems was motivated by the study of extremal properties and dates back to 1940s. More recently the probabilistic analysis of optimization algorithms was proposed by [Karp, 1976] as a tool for the study of hard optimization problems: since then there has been considerable interest and a large number of results have been derived. Moreover, books have been published that present the tools and the techniques that are useful in the probabilistic analysis and results for classes of problems. [Hofri, 1987] presents a general introduction to the probabilistic analysis of algorithms that is not limited to hard optimization problems. [Coffman and Lueker, 1991] deals with the probabilistic analysis of many packing and partitioning algorithms, including several scheduling problems. [Bollobas, 1985] concentrates on the study of random graphs: even though the attention is mainly focused on the study of graph properties in random graphs, several algorithms for NP-hard optimization problems are analyzed.

The presentation of the main convergence properties of random variables and of stochastic processes is out of the scope of this book. There are many books that present these topics in a simple way. We refer to [Feller, 1971] for a thorough presentation.

A more refined analysis of the List Scheduling algorithm with a better convergence rate is given by [Coffman and Gilbert, 1985]. The authors also consider the case when the processing times are exponentially distributed. [Boxma, 1985] studied the case of the Erlang–k distribution. The analysis of the LPT rule presented in this chapter is given by [Frenk and Rinnooy Kan, 1987], where more general probability distributions and the case of uniform machines are also considered. The analysis of the rate of convergence and the study of different probability distributions of the processing times of the jobs can be found in [Coffman and Flatto and Lueker,

1984] and [Rinnooy Kan and Frenk, 1986]. More results on scheduling and packing problems can be found in [Coffman and Lueker, 1991].

The analysis of approximation algorithms for MINIMUM BIN PACKING originated in the work of [Frederickson, 1980], where an algorithm was proposed with an expected number of bins bounded by $n/2 + O(n^{2/3})$. This result was subsequently improved in [Knödel, 1981, Lueker 1980]. The simple proof presented in the chapter is that of [Karp, 1982], (see also [Karp and Luby and Marchetti-Spaccamela, 1984]). An improved bound on the expected measure of the optimal solution was given in [Lueker 1980]. Other approximation algorithms have also been analyzed: in [Bentley et al., 1984] very strong bounds are presented on the expected difference between the optimum and the measure of the solution found by *First Fit Decreasing* and *Best Fit Decreasing* under various hypotheses. For example, it is proved that if the size of the items is uniformly distributed in $[0, a]$, then the expected difference between the optimal measure and the measure of the optimal solution found by *First Fit Decreasing* is $\Theta(1)$ if $a \leq 0.5$ and $\Theta(n^{1/3})$ if $a > 0.5$. The analysis of Best Fit is due to [Shor, 1986], where the connections between on-line Bin Packing algorithms and the right-upward matching problem are shown. [Karp and Luby and Marchetti-Spaccamela, 1984] studied two-dimensional Bin Packing algorithms from a probabilistic point of view (see also [Shor, 1986]). More results on the probabilistic analysis of scheduling and packing algorithms can be found in [Coffman and Lueker, 1991].

The study of random graphs was founded by Erdös and Rény after that Erdös showed how probabilistic methods might be applied to solve extremal problems in graph theory. In [Matula, 1970, 1972 and 1976] it was first observed that for fixed values of p the clique number is highly concentrated; these results have been strengthened in [Grimmett and Mc Diarmid, 1975] and [Bollobas and Erdös, 1976]. The book [Bollobas, 1985] gives a detailed presentation also in the case of edge probability p that depends on the number of vertices of the graph. [Bollobas, 1988] proved that the chromatic number of a graph converges almost everywhere to $n/2 \log_{1/q} n$.

The analysis of the convergence of the optimal measure of the TSP tour for points randomly distributed in the unit square, Theorem 9.17, is known as the BHH theorem from the initial of its authors [Beardwood and Halton and Hammersley, 1959]. The original proof of the theorem is long and complicated and shows that the tour converges almost everywhere to $\beta\sqrt{n}$ not only in expectation but also with probability 1; [Karp and Steele, 1977] gave the simple proof presented in Sect. 9.7. The analysis of the *STRIP* algorithm given above is not the best possible one: a more accurate estimate of the expected cost gives $\beta \leq 0.93$. This is not optimal: empirical studies show that $\beta \approx 0.765$. Other heuristics have been also analyzed for

Section 9.9

BIBLIOGRAPHICAL NOTES

Chapter 9
PROBABILISTIC ANALYSIS OF APPROXIMATION ALGORITHMS

MINIMUM TRAVELING SALESPERSON: [Karp, 1977] proposed a polynomial time algorithm such that, if n cities are uniformly and independently drawn in the unit square, then it finds a solution whose performance ratio converges to 1. Karp's algorithm is not practical; the probabilistic behavior of other heuristics is a hard task and there are not many results. In the case of local search algorithms, [Chandra, Karloff, and Tovey, 1994] considered the m-dimensional problem and proved that, if cities are independently and uniformly distributed, then 2-opt achieves a solution with a $O(1)$ worst case ratio with high probability in polynomial time. However, this interesting result is still far away from the good practical behavior of the method. More results on the probabilistic analysis of other variants of the MINIMUM TRAVELING SALESPERSON (including the asymmetric TSP and the directed TSP) are presented in [Karp and Steele, 1977].

Chapter 10

Heuristic methods

IN THE PRECEDING chapters the problem of computing good solutions for NP-hard optimization problems has always been considered in terms of deriving algorithms with a guaranteed behaviour, where such a guarantee refers both to the quality of the returned solution (in terms of either worst case or expected performance ratio) and to the running time (polynomial either in the worst or in the average case).

At the same time, it is well known that, for many problems, there exist algorithms that do not provide any guarantee about their worst case behavior in terms of both the quality of the returned solution and the efficiency of their execution, even if in many cases they return good solutions in a short time. In the following, we refer to such algorithms as *heuristic methods* (or *heuristics*).

Heuristics may behave very poorly if the worst case is considered: in fact, when applied to *bad* problem instances, they may return a solution very far from the optimum and/or may present an exponential running time. However, good heuristics may outperform the best approximation algorithms on most instances of a problem, thus providing a solution technique that is preferable in most practical cases to guaranteed approximation approaches.

In this chapter, we will first briefly review some of the main heuristic methods that have been introduced in the literature. Successively, we will focus our attention on construction heuristics and will describe, in particular, some heuristics to solve MINIMUM TRAVELING SALESPERSON. Finally, we will present local search based heuristics (such as simulated

Chapter 10

HEURISTIC METHODS

annealing, genetic algorithms, and tabu search), and we will discuss their application to MINIMUM TRAVELING SALESPERSON and to MINIMUM BALANCED CUT.

Throughout the whole chapter, we will restrict our attention to minimization problems: clearly, similar considerations apply to the maximization case. Moreover, for what concerns MINIMUM TRAVELING SALESPERSON, we will assume that the distance matrix is symmetric.

10.1 Types of heuristics

MOST HEURISTICS appearing in the literature can be intuitively classified as either *construction heuristics*, which iteratively construct a single feasible solution from scratch, or *enumeration heuristics*, which examine a set of feasible solutions and return the best one.

As we shall see, construction methods are strongly problem-dependent, and usually require polynomial time. On the other side, enumeration heuristics refer to general approaches for the generation of new solutions, and must be adapted to deal with the considered problem; moreover, they may present an exponential complexity if either the set of examined solutions is exponentially large (for example, in the case in which the whole solution space, or a significant fraction of it, is visited) or the domain of all possible solution measures has an exponential size.

Enumeration heuristics can, in turn, be distinguished according to the method applied for the generation of new solutions. In one case, new solutions to be examined are directly derived by limited modifications to the structure of already considered ones, while a different method derives new solutions by solving suitably generated partial problems of a smaller size. *Local search* heuristics behave according to the first of such methods, while the second one is applied, in particular, to all techniques based on linear programming formulations of problems, such as branch-and-bound and branch-and-cut.

However, all approaches based on linear programming, even if they can be easily adapted to return good solutions in polynomial time, actually aim at computing an optimal solution (in worst case exponential time). For such a reason we will only provide a short description of their main characteristics and refer the reader to more specific texts on this subject.

Branch-and-bound
Branch-and-bound is a divide-and-conquer method which tries to solve the problem by splitting it into smaller and smaller problems for which upper and lower bounds on their optimal measure can be efficiently computed.

It basically works by maintaining a set S of subinstances of the original instance x such that the union of the feasible solutions of such subinstances is equal to $\text{SOL}(x)$. At the beginning, S is initialized to contain only x. A decomposition operation, denoted as *branching*, is applied on problem instances to derive a set of mutually excluding subinstances.

The procedure relies on the application of backtracking for the generation of new subinstances and on maintaining the best solution generated thus far to limit the number of subinstances actually considered. For example, a lower bound $\underline{m}(x_i)$ on the optimal measure $m^*(x_i)$ is estimated for each subinstance x_i and a branching on x_i is not operated if $\underline{m}(x_i)$ is already greater than the value $m_0(x)$ of the best solution derived so far. In fact, this shows that the optimal solution is not in $\text{SOL}(x_i)$, since the value of any solution in $\text{SOL}(x_i)$ is greater than $m^*(x)$. In this case, the problem instance x_i is called *fathomed*.

The enumeration process is thus controlled by a mechanism for identifying branchings which do not lead to the generation of any solution better than the best one computed thus far: in this case, the heuristics stop generating all solutions in the current subtree and backtrack to the parent node.

Specific branch-and-bound procedures are largely characterized by the choices made with respect to:

Branching rule. Many different approaches can be applied to decide how to branch in correspondence to any problem instance, with the aim of deriving as quickly as possible problem instances which do not require further branching, such as, for example, instances whose lower bound corresponds to a feasible solution or instances with no feasible solution at all.

Search strategy. The applied search strategy determines which is the next subinstance x_i to be considered by the branch-and-bound algorithm among the ones in set S, that is, the ones which have been generated but not yet examined. All search strategies refer to a heuristic evaluation function $\mathcal{H}(x_i)$ which provides an estimate of the optimal solution $m^*(x_i)$.

Typical search strategies include:

1. Heuristic search. Extract from S the instance x_i such that $\mathcal{H}(x_i)$ is minimal in S.

2. Depth-first search. Extract from S the instance x_i such that $\mathcal{H}(x_i)$ is minimal among all instances in S with maximal depth, i.e., which have been generated by the longest sequence of branchings.

Section 10.1

TYPES OF
HEURISTICS

Chapter 10

HEURISTIC METHODS

3. Breadth-first search. Extract from S the instance x_i such that $\mathcal{H}(x_i)$ is minimal among all instances in S with minimal depth, i.e., which have been generated by the shortest sequence of branchings.

Bounding techniques. Any feasible solution of x_i provides a possible value for $\underline{m}(x_i)$. However, this can be a complex task by itself, since in some cases even the problem of deciding whether there exists any feasible solution turns out to be NP-hard.

On the other hand, a lower bound on the optimal measure $m^*(x)$ of a problem instance x is usually obtained by solving some *relaxation* x_r of such an instance, that is, by eliminating some of the constraints that appear in x. For example, as we have already seen in previous chapters, a relaxation of an integer linear program x_{ILP} consists of considering and solving the corresponding linear program x_{LP}, which is obtained by eliminating the integrality constraints on the set of variables occurring in x_{ILP}. Notice that x_{LP} can be solved efficiently and that if an optimal solution of x_{LP} happens to be integral, such a solution is also an optimal solution of x_{ILP}. A (worse) lower bound can be obtained even faster by considering the measure of any feasible solution of the dual program of x_{LP}.

A recent variant of branch-and-bound which seems to be successful in many applications is *branch-and-cut*. In such a method the efficiency is increased by using a more sophisticated estimation of the lower bounds by means of cutting plane generation.

In particular, let us reiterate that the well known approach of using cutting planes to compute an optimal solution of an integer linear program x_{ILP} by means of non-integral, unfeasible solutions is based on the following steps:

1. Let x_0 be the linear program relaxation of x_{ILP} and let $i = 0$.

2. Solve the linear program x_i. Let s_i be the derived solution.

3. If s_i is not integral, i.e., unfeasible for x_{ILP}, find a linear constraint c_i that is not satisfied by s_i, while not excluding any (integral) solution of x_{ILP}. Let x_{i+1} be the new linear program derived by adding c_i to x_i. Let $i = i+1$ and go to step 2.

4. If s_i is integral, return s_i, which is an optimal solution of x_{ILP}.

Thus, for any subinstance generated by the branching process, a lower bound $\underline{m}(x_i)$ on its optimal measure can be computed by applying a certain amount of steps of a cutting plane algorithm to the integer linear program formulation of the subinstance. This corresponds to spending some computation time for each subinstance derived by the branching process, with the hope that better lower bound estimates will increase the number of fathomed nodes in the branching tree.

Different types of cutting plane algorithms can be applied, ranging from classical Gomory cuts to more sophisticated polyhedral methods (see Bibliographical notes).

As said above, branch-and-bound (and branch-and-cut) algorithms are aimed at finding the optimal solution of a problem instance in a possibly exponential number of steps. The computation time of any such algorithm can be, anyway, reduced by limiting the number of subinstances examined. This can be done in at least three different ways:

1. A branching is applied to a subinstance if the lower bound on its optimal measure is sufficiently smaller than the measure of the best feasible solution derived thus far.

2. A subinstance already derived by branching is inserted into the current set S of subinstances only if the size of such a set is less than a predefined number, which specifies the maximum number of problem instances to be examined at a certain step of the computation.

3. The algorithm stops and returns the best feasible solution determined thus far, in the case that the number of branching operations performed becomes greater than a predefined number.

Limiting the number of subinstances considered may clearly result in the impossibility of finding an optimal solution. Moreover, even if the lower bound estimates make it possible to have a bound on the distance between the measure of the returned solution and the optimal measure, there is no guarantee, in general, that a solution with a given performance ratio can be computed in polynomial time. As said above, we refer the interested reader to more specific literature on branch-and-bound and branch-and-cut methods, and now turn our attention to the other types of heuristics previously introduced.

10.2 Construction heuristics

Construction heuristics operate by deriving a single feasible solution, which is either directly returned or, more often, used as a starting point for some local search heuristics (see Sect. 10.3).

Chapter 10
HEURISTIC METHODS

Notice that approximation algorithms are, by definition, construction heuristics, since they return feasible solutions. However, in many cases, construction heuristics do not provide any guarantee about the quality of the solution returned. This happens, for example, in the case of non-approximable problems or in the case of an approximable problem, if one prefers to use a very fast algorithm for generating a feasible solution, which may then be improved by means of other heuristics.

Most construction heuristics have a very simple structure and are usually greedy algorithms that start from (one or more) elementary *fragment* of a solution and iteratively build larger fragments (either by adding new items to a fragment or by merging different fragments), until a feasible solution is obtained. In the case of optimization problems for which a feasible solution is a set (or sequence) of suitable elements, this implies that such heuristics start from elementary sets (usually, singletons) and derive larger subsets by either adding new elements or merging previously obtained subsets, until a set corresponding to some feasible solution has been derived.

As an example, in this section we will consider different construction heuristics for MINIMUM TRAVELING SALESPERSON, where, as we have already seen, a feasible solution corresponds to a permutation of the set of cities in the problem instance.

A first possible construction heuristics for this problem is provided by the Christofides' approximation algorithm described in Sect. 3.1: in fact, even if this algorithm returns a solution with a guaranteed performance ratio only in the metric case, such a solution is clearly feasible for the general case.

However, since Christofides' algorithm has time complexity $O(n^3)$, it may be convenient to refer to simpler and faster construction heuristics. A possible approach consists of starting from an elementary path of just one city and iteratively adding new cities according to a predefined criterion: hence, such heuristics have the general structure given in Program 10.1, where $f(C, T)$ represents a function implementing the predefined criterion of choice of the next city to be included in the path.

Many different heuristics can be obtained in relation to different choices for what concerns:

1. The definition of $f(C, T)$, that is, how the next city to be inserted in the path is selected.

2. How the selected city is inserted in the path T: for example, it could be appended at the end of T, connected just before or after its nearest city in T, or inserted in such a way as to minimize the overall length of the resulting path.

Section 10.2
CONSTRUCTION HEURISTICS

Program 10.1: Extending Tour

input Set of cities $C = \{c_1, \ldots, c_n\}$, $n \times n$ matrix D of distances;
output Permutation $T = (c_{\pi_1}, \ldots, c_{\pi_n})$ of the cities;
begin
   Select any city $c \in C$;
   $T := \langle c \rangle$;
   $C := C - \{c\}$;
   while $C \neq \emptyset$ **do**
     begin
       Let $c = f(C, T) \in C$ be the city satisfying the predefined criterion;
       Insert c in T;
       $C := C - \{c\}$
     end
end.

Different choices with respect to points 1 and 2 above lead to different heuristics, such as:

Nearest Neighbor. This method has been considered in Sect. 2.1.3. At each step, the algorithm selects the city c_j which is nearest to the last city inserted in T, that is, for any $c_i \in C$, $D(j, \pi(k)) \leq D(i, \pi(k))$ where $T = (c_{\pi(1)}, \ldots, c_{\pi(k)})$. The city is inserted at the end of T.

Nearest Insertion. At each step, the algorithm selects the city c_j having minimal distance from any city already included in the path T, that is, for any $c_i \in C$ and for any $c_k \in T$, $D(j, k) \leq D(i, k)$. The city is inserted into the path at the point where its insertion induces the minimum increase on the path length.

Farthest Insertion. At each step, the algorithm selects the city c_j whose distance from the nearest city in T is maximal. That is, denoting as $d_T(c_i) = \min(D(i, k) \mid c_k \in T)$ the distance from a city $c_i \in C$ to the nearest city in T, city c_j is selected if, for any $c_i \in C$, $d_T(c_j) \geq d_T(c_i)$. The city is inserted into the path at the point where its insertion induces the minimum increase on the path length.

Cheapest Insertion. At each step, the algorithm selects the city whose insertion in T induces the minimum increase on the path length.

Random Insertion. At each step, a city in C is randomly selected and inserted where the minimum increase in the path length is obtained.

Chapter 10

HEURISTIC
METHODS

Program 10.2: Merging Tour

input Set of cities $C = \{c_1, \ldots, c_n\}$, $n \times n$ matrix D of distances;
output Permutation $T = (c_{\pi_1}, \ldots, c_{\pi_n})$ of the cities;
begin
    Let $\mathcal{T} = \{T_1, T_2, \ldots, T_m\}$ be a set of partial tours on C;
    while $|\mathcal{T}| > 1$ **do**
    begin
        Let $T_i, T_j \in \mathcal{T}$ be a pair of tours satisfying a predefined criterion;
        Merge T_i, T_j in a single tour T_k;
        $\mathcal{T} := (\mathcal{T} \cup \{T_k\}) - \{T_i, T_j\}$
    end;
    Return T, the unique tour in $\mathcal{T}$
end.

Smallest Sum Insertion. At each step, the algorithm selects the city c_j such that the sum of its distances from all cities in T is minimal. That is, denoting as $\Delta_T(c_i) = \sum_{c_k \in T} D(i,k)$ the sum of all distances from a city $c_i \in C$ to all cities in T, the city c_j is selected if, for any $c_i \in C$, $\Delta_T(c_j) \leq \Delta_T(c_i)$. The city is inserted at the end of the path T.

Many other heuristics can be defined in correspondence to different combinations of selection criteria and insertion strategies. Moreover, such heuristics can be further modified to consider only a fraction of the problem instance: in fact, a significant speed-up in the computation could be obtained by considering only a limited suitable subset of entries in D. This results in faster heuristics, since only a fraction of the possible city connections has to be considered; at the same time, the initial choice of such a fraction is critical in terms of both quality of the returned solution and of the existence of any feasible solution in the modified instance considered.

A different greedy approach is given in Program 10.2. In this case the construction of a tour starts from a set of simple partial tours that are iteratively merged in pairs until a single tour spanning all cities is obtained. At each step a pair of tours is selected according to the criterion of choice and such tours are then merged into a single one according to the merge operation.

In this case, different heuristics can be defined in correspondence to the criteria for the choice of different initial sets of tours $\mathcal{T}$, for the selection of the pairs of tours, and for the tour merge operations. A possible approach corresponds to the following case.

1. $\mathcal{T}$ is set up by choosing a city c and inserting in $\mathcal{T}$ the $n-1$ tours

(c,c') with $c \neq c'$.

2. Merging tours $(c, c_{i_1}, \ldots, c_{i_r})$ and $(c, c_{j_1}, \ldots, c_{j_s})$ results in the shortest among the following four tours (see Fig. 10.1):

$$(c, c_{i_1}, \ldots, c_{i_r}, c_{j_1}, \ldots, c_{j_s}), \quad (c, c_{i_1}, \ldots, c_{i_r}, c_{j_s}, \ldots, c_{j_1}),$$
$$(c, c_{j_1}, \ldots, c_{j_s}, c_{i_1}, \ldots, c_{i_r}), \quad (c, c_{i_r}, \ldots, c_{i_1}, c_{j_1}, \ldots, c_{j_s}).$$

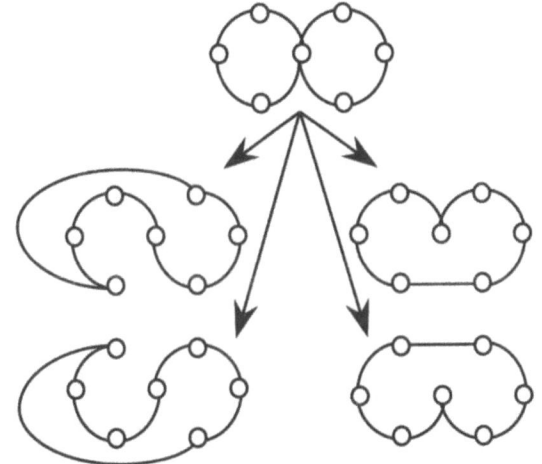

Figure 10.1
Possible tour merges

3. Let the *saving* $\sigma(T_i, T_j)$ associated to a pair of partial tours $T_i, T_j \in \mathcal{T}$ be the difference between the sum of the lengths of T_i and T_j and the length of the tour resulting from their merge: a pair with maximal saving is then selected.

Even if, in most cases, no precise estimation is known on the quality of the solutions returned by construction heuristics, bounds on the approximation ratio have been derived in some specific cases, for example, in the case of the *Nearest Neighbor* heuristics when applied to MINIMUM METRIC TRAVELING SALESPERSON, i.e., when the triangular inequality is satisfied. In this case, a logarithmic bound on the performance ratio can be proved, as shown in Sect. 2.1.3.

10.3 Local search heuristics

As we have already seen in Sect. 2.3, local search heuristics for an optimization problem $\mathcal{P}$ are algorithms that, for any instance $x \in I_{\mathcal{P}}$, visit a subset of $\text{SOL}_{\mathcal{P}}(x)$, until a solution is found which is *locally* optimal with

Chapter 10

HEURISTIC METHODS

respect to a neighborhood function $\mathcal{N}$, associating to each feasible solution $s \in \text{SOL}_\mathcal{P}(x)$ a set $\mathcal{N}(s) \subseteq \text{SOL}_\mathcal{P}(x)$ of neighbor solutions.

The topological structure induced on $\text{SOL}_\mathcal{P}(x)$ by the neighborhood function can be described, in the case of combinatorial optimization problems, also by means of a *neighborhood directed graph* $G_\mathcal{N}(x) = (V(x), A)$, where $V(x) = \text{SOL}_\mathcal{P}(x)$ and $A = \{(s_1, s_2) \mid s_2 \in \mathcal{N}(s_1)\}$.

Local search heuristics traverse the set of solutions of an instance x by visiting the associated neighborhood graph $G_\mathcal{N}(x)$ until a local optimum is found, that is, a solution $s_r \in V(x)$ with the property that $m_\mathcal{P}(x, s_r) \leq m_\mathcal{P}(x, s)$, for any s such that $(s_r, s) \in A$.

Since m may assume values from a set of exponential size, it is possible that an exponential number of steps is performed, on a worst-case instance, to reach a local optimum: for this reason, some kind of *stopping rule* should be introduced, which forces the visit to be performed only on a small subgraph of $G_\mathcal{N}(x)$.

Local search techniques are based on the generic procedure described in Program 10.3. It should be noticed that, given a problem instance x, the behavior of this algorithm depends on the following parameters:

1. *The neighborhood function $\mathcal{N}$*. The size of the neighborhood of any solution should be a compromise between the aim of obtaining a good improvement each time a new current solution s is selected and the need of limiting the cost of visiting a single neighborhood. Usually, for any solution s, the neighborhood $\mathcal{N}(s)$ is generated by applying some suitably defined local change operations on s.

2. *The starting solution s_0*. This can be derived by means of some suitable algorithm (e.g., construction heuristics), which returns a *good* feasible starting solution, or by means of some random generation procedure.

3. *The strategy of selection of new solutions*. For example, all solutions in $\mathcal{N}(s)$ could be considered by the heuristics before comparing the best one with s. This ensures that the best move from s to one of its neighbors is performed, in case s is not a local optimum. On the contrary, a move from s to a better solution $s' \in \mathcal{N}(s)$ could be performed as soon as one such neighbor is found.

10.3.1 Fixed-depth local search heuristics

Fixed-depth local search heuristics refer to neighborhoods generated by means of sequences of local change operations of bounded length. That

Section 10.3
LOCAL SEARCH HEURISTICS

Program 10.3: Local Search

input Instance x;
output Solution s;
begin
   $s :=$ initial feasible solution s_0;
   (* $\mathcal{N}$ denotes the neighborhood function *)
   repeat
      Select any $s' \in \mathcal{N}(s)$ not yet considered;
      if $m(x,s') < m(x,s)$ **then**
         $s := s'$;
   until all solutions in $\mathcal{N}(s)$ have been visited;
   return s
end.

is, given a solution $s \in \text{SOL}_\mathcal{P}(x)$ of an instance $x \in I_\mathcal{P}$ of the considered problem $\mathcal{P}$, a *k-change neighborhood* $\mathcal{N}^k(s)$ is generated, for any $k > 0$, by applying a sequence of at most k local change operations. Heuristics based on a k-change neighborhood s are also said to be *k-opt* heuristics.

Fixed-depth local search heuristics have traditionally been applied to MINIMUM TRAVELING SALESPERSON. All such heuristics refer to the same local change operation and differ from each other with respect to the maximum length of the sequence of such operations.

We now present in detail *2-opt* heuristics, which were intuitively introduced in Sect. 2.3. Recall that this method is based on the following 2-change neighborhood: for any feasible tour τ the *2-change neighborhood* $\mathcal{N}(\tau)$ is the set of all tours τ' that can be obtained from τ by deleting two edges (x,y) and (v,z) of τ and by inserting the two new edges (x,v) and (z,y) in such a way to obtain a different tour: this corresponds also to traversing some path in τ in the opposite direction (see Fig. 2.6).

The 2-opt heuristic method is described in Program 10.4 which is the instantiation of the local search approach to the 2-change neighborhood. In the algorithm, we make use of a function *Length* defined on tours, whose value is defined as $\text{Length}(\tau) = \sum_{i=1}^{n-1} D(\pi(i), \pi(i+1)) + D(\pi(n), \pi(1))$ if $\tau = \langle c_{\pi(1)}, \ldots, c_{\pi(n)} \rangle$.

The 2-opt heuristic method can be easily modified to refer to larger neighborhoods. For example, *3-opt* heuristics are based on the *3-change neighborhood* $\mathcal{N}(\tau)$ of a tour τ, which is the set of all tours which can be derived from τ by changing (at most) three edges (see Fig. 10.2). As it may be expected, 3-opt heuristics present a better behaviour than 2-opt heuristics at the expense of a higher running time.

The approach can be extended to any constant $k > 0$. Notice that the

Chapter 10

HEURISTIC METHODS

Program 10.4: 2-opt

input Set of cities $C = \{c_1, \ldots, c_n\}$, $n \times n$ matrix D of distances;
output Permutation $T = (c_{\pi_1}, \ldots, c_{\pi_n})$ of the cities;
begin
    $\tau :=$ initial tour τ_0;
    Let C be the set of pairs (p,q) of distinct integers in $\{1,\ldots,n\}$ such that $p < q$;
    $N := C$;
    repeat
        Let $\tau = (c_{i_1}, \ldots, c_{i_n})$;
        Select any pair of indices $(p,q) \in N$;
        $N := N - \{(p,q)\}$;
        $\tau' := (c_{i_1}, \ldots, c_{i_{p-1}}, c_{i_q}, c_{i_{q-1}}, \ldots, c_{i_{p+1}}, c_{i_p}, c_{i_{q+1}}, \ldots, c_{i_n})$;
        if Length$(\tau') <$ Length(τ) **then**
        begin
            $\tau := \tau'$;
            $N := C$
        end
    until $N = \emptyset$;
    return τ
end.

size of a k-change neighborhood on a set of n cities is $\binom{n}{k} = \Theta(n^k)$, which implies that $O(n^k)$ steps must be performed by the heuristic method just to check whether the current solution is a local optimum.

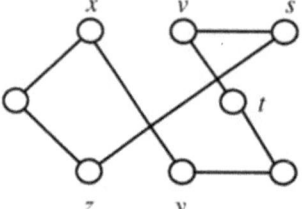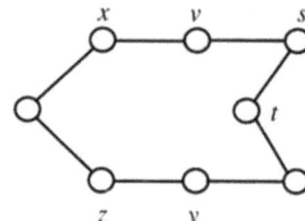

Figure 10.2
The 3-change neighborhood

It has been experimentally noticed that moving from the 2-opt to the 3-opt heuristics makes it possible to obtain better solutions at the expense of a larger processing time. The improvement on the quality of the solution provided by moving from the 3-change to the 4-change neighborhood does not sufficiently justify the additional running time.

Although quite effective in practice, fixed-depth local search heuristics for MINIMUM TRAVELING SALESPERSON may behave rather poorly in the worst case. As previously stated, the behavior of the heuristics clearly depends on the choice of the initial tour τ_0. The following theorem shows that there exist natural initial tours (such as $(c_1,\ldots,c_n)$), whose choice results in an arbitrarily poor solution.

Section 10.3
LOCAL SEARCH HEURISTICS

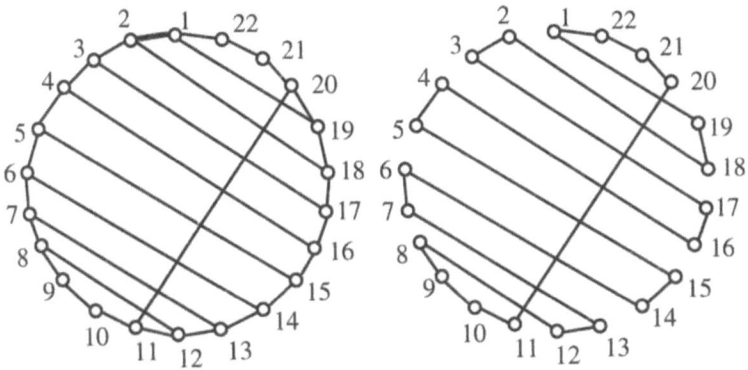

Figure 10.3
A bad instance for the 8-opt heuristics and an optimal tour

◀ Theorem 10.1

For all $k \geq 2$, for all $n \geq 2k+8$, and for all $\alpha > 1/n$, there exists an instance x of MINIMUM TRAVELING SALESPERSON with n cities $\{c_1,\ldots,c_n\}$ such that the k-opt heuristic method with input x and initial tour $\tau_0 = (c_1,\ldots,c_n)$ returns a tour whose measure $m_k(x)$ satisfies the inequality $m_k(x) > \alpha m^*(x)$.

PROOF

Assume k even and let $n \geq 2k+6$. Let us consider the following instance of MINIMUM TRAVELING SALESPERSON on a set of n cities $\{c_1,c_2,\ldots,c_n\}$ (see the left part of Fig. 10.3 where $k=8$ and the edges of weight kn are not drawn):

1. $D(1,2) = 1$;

2. $D(i,i+1) = \frac{1}{n\alpha}$, for $i = 2,\ldots,n-1$;

3. $D(n,1) = \frac{1}{n\alpha}$;

4. $D(k+3, 2k+4) = \frac{1}{n\alpha}$;

5. $D(j, 2k+4-j) = \frac{1}{n\alpha}$, for $1 \leq j \leq k$;

6. $D(i,j) = kn$, for all remaining pairs c_i, c_j.

An optimal tour can be described as follows, by using the function $\texttt{next}: \{1,\ldots,n\} \mapsto \{1,\ldots,n\}$ where $\texttt{next}(i) = j$ denotes that c_j immediately follows c_i in the tour (see the right part of Fig. 10.3):

$$\text{next}(i) = \begin{cases} 2k+4-i & \text{if } i \text{ is odd and } i < k, \\ i+1 & \text{if } i \text{ is even and } i < k, \\ i+1 & \text{if } k \leq i \leq k+2, \\ 2k+4 & \text{if } i = k+3, \\ i-1 & \text{if } i \text{ is odd and } k+3 \leq i < 2k+4, \\ 2k+4-i & \text{if } i \text{ is even and } k+3 \leq i < 2k+4, \\ i+1 & \text{if } 2k+4 \leq i < n, \\ 1 & \text{if } i = n. \end{cases}$$

As it can be immediately verified, this tour has length $m^*(x) = \frac{1}{\alpha}$, and its optimality derives from the observation that the minimum distance between two cities is $1/n\alpha$. Indeed, it is also possible to show that T^* is the unique optimal solution (see Exercise 10.4). Observe now that tour $\tau_0 = (c_1, \ldots, c_n)$ has length $1 + \frac{n-1}{n\alpha}$. By construction, T^* is the only tour of length less than $1 + \frac{n-1}{n\alpha}$, and it differs from τ_0 in $k+1$ edges. As a consequence, there is no shorter tour in the k-neighborhood of τ_0 and therefore, τ_0 might be a local optimum, which could be returned by the k-opt heuristic method. As a consequence,

$$\frac{m_k(x)}{m^*(x)} = (1 + \frac{n-1}{n\alpha})\alpha = 1 + \alpha - \frac{1}{n} > \alpha.$$

The proof in the case when k is odd is immediately derived by observing that a local optimum with respect to the $(k+1)$-change neighborhood is also a local optimum with respect to the k-change neighborhood.

QED

No upper bounds are known, in the general case, for the performance ratio of k-opt heuristics. However, the performance ratio can be upper bounded in particular cases, such as the application of 2-opt heuristics to MINIMUM EUCLIDEAN TRAVELING SALESPERSON. For any instance of this problem, a point $p_i = (u_i, v_i)$ in the plane is associated to each city c_i and, for any pair of cities c_i and c_j, $D(i,j)$ is the Euclidean distance between p_i and p_j.

Lemma 10.2 ▸ *Given an instance x of* MINIMUM EUCLIDEAN TRAVELING SALESPERSON, *let* $T = (c_{i_1}, \ldots, c_{i_n})$ *be a tour computed by Program 10.4 on input x and, for any $k \in \{1, \ldots, n\}$, let $E_k = \{(c_{i_j}, c_{i_{j+1}}) \mid D(i_j, i_{j+1}) > \frac{2m^*(x)}{\sqrt{k}}\}$. Then, $|E_k| < k$.*

PROOF

The proof is by contradiction: assume that, for some k with $1 \leq k \leq n$, $|E_k| = r \geq k$ and, for any pair $(c_{i_j}, c_{i_{j+1}}) \in E_k$, denote p_j as the *head* and p_{j+1} as the *tail* of such pair.

Let us first prove that not too many tails can be close to each other. Consider a circle $S(x,y)$ of radius $\frac{m^*(x)}{\sqrt{k}}$ centered in (x,y) and assume that, for some $p \geq \sqrt{k}$, a set $t_{s_1},\ldots,t_{s_p}$ of tails lies in $S(x,y)$: let $h_{s_1},\ldots,h_{s_p}$ be the set of corresponding heads. By hypothesis, for any pair of tails t_{s_j} and t_{s_k} in $S(x,y)$, the distance between t_{s_j} and t_{s_k} is at most $\frac{2m^*(x)}{\sqrt{k}}$: this implies that the distance between the corresponding heads h_{s_j} and h_{s_k} is at least $\frac{2m^*(x)}{\sqrt{k}}$, since otherwise deleting the two edges (h_{s_j},t_{s_j}) and (h_{s_k},t_{s_k}), and inserting the two (h_{s_j},h_{s_k}) and (t_{s_j},t_{s_k}) in T would result in a shorter tour T' (contradicting the fact that T is a local optimum). As a consequence, there exist $p \geq \sqrt{k}$ heads $\{h_{s_1},\ldots,h_{s_p}\}$ at distance at least $\frac{2m^*(x)}{\sqrt{k}}$ from each other, which implies that an optimal tour on the set of points $\{h_{s_1},\ldots,h_{s_p}\} \subseteq P$ has length at least $2m^*(x)$. This is not possible, since, for any $P' \subseteq P$, the triangle inequality implies that $m^*(P') \leq m^*(P)$. Hence, less than $\sqrt{k}$ tails can lie in $S(x,y)$.

Let us now show that if $|E_k| \geq k$ then there exists a set $\mathcal{T}$ of at least $\sqrt{k}$ tails such that their mutual distance is at least $\frac{m^*(x)}{\sqrt{k}}$. $\mathcal{T}$ is obtained as follows: starting from the empty set, as long as $P \neq \emptyset$, we choose any tail t, we add it to $\mathcal{T}$, and we delete from P all tails lying inside a circle of radius $\frac{m^*(x)}{\sqrt{k}}$ centered in t. Since, at each step, we delete at most $\sqrt{k}$ tails from P and since there are at least k tails in P, it follows that, at the end of the procedure, $|\mathcal{T}| \geq \sqrt{k}$.

Since the distance between any pair of tails in $\mathcal{T}$ is greater than $\frac{m^*(x)}{\sqrt{k}}$, an optimal tour on $\mathcal{T}$ must have a greater length than $m^*(x)$, obtaining the desired contradiction. Hence, it must be $|E_k| < k$ and the lemma is proved. QED

The proof of the following theorem easily follows from the above lemma and is omitted (see Exercise 10.5).

◀ Theorem 10.3

For any instance x of MINIMUM EUCLIDEAN TRAVELING SALESPERSON *with n cities, let $m_{2-OPT}(x)$ be the length of the tour returned by Program 10.4 with input x. Then $m_{2-OPT}(x)$ satisfies the following inequality:*

$$m_{2-OPT}(x)/m^*(x) \leq 4\sqrt{n}.$$

Local search has been applied to many different problems, with different neighborhood definitions, depending on the problem structure. As another example, we consider its application to Problem 10.1.

In this case, the local change operation we use to generate neighborhoods of solutions is the exchange of one node $v \in V_1$ with one node $u \in V_2$. That is, the 1-change neighborhood of a solution (V_1,V_2) is the set of all solutions (V'_1,V'_2) such that, there exist $v \in V_1$ and $u \in V_2$ for

Chapter 10

HEURISTIC METHODS

> **Problem 10.1: Minimum Balanced Cut**
>
> INSTANCE: Graph $G = (V,E)$, with $|V| = 2n$, edge weight function $w : E \mapsto \mathbb{Z}^+$.
>
> SOLUTION: A partition of V into subsets V_1, V_2, with $|V_1| = |V_2| = n$.
>
> MEASURE: $\sum_{\substack{(u,v) \in E: \\ u \in V_1, v \in V_2}} w(u,v)$

which $V_1' = V_1 - \{v\} \cup \{u\}$ and $V_2' = V_2 - \{u\} \cup \{v\}$. For any $k > 1$, the k-change neighborhoods can be defined analogously as the sets of all solutions which can be obtained from the current one by exchanging at most k nodes between V_1 and V_2.

As a final example, let us consider the application of local search heuristics to MAXIMUM SATISFIABILITY. In this case, we may define a local change as the operation of *flipping* the value assigned to a variable. As a consequence, given an instance (i.e., a set X of n variables and a set C of m clauses) and given a feasible solution (i.e., a truth assignment τ on X), the neighborhood of τ is given by the set of n truth assignments $\{\tau_1, \ldots, \tau_n\}$, where, for all $1 \leq i \leq n$, $\tau_i(x_j) = \tau(x_j)$ if $i \neq j$ and $\tau_i(x_i) = \overline{\tau(x_i)}$. For any k with $1 \leq k \leq n$, the k-change neighborhood of a truth assignment τ is then the set of truth assignments τ' such that the number of variables x for which $\tau(x) \neq \tau'(x)$ is at most k. Thus, the k-opt heuristic method for MAXIMUM SATISFIABILITY could start from any truth assignment and then iteratively move from the current assignment to any assignment which differs in at most k variables and which satisfies a larger set of clauses, until an assignment is found which satisfies more clauses than any other assignment in its neighborhood.

10.3.2 Variable-depth local search heuristics

Local search heuristics with variable depth were originally introduced for MINIMUM BALANCED CUT and MINIMUM TRAVELING SALESPERSON.

While any of the fixed depth local search heuristics, such as the k-opt heuristics, generate new solutions by performing sequences of local changes of bounded length (at most k) on the current solution, the number of changes performed by variable depth heuristics to derive a new solution is not bounded *a priori*.

In variable depth heuristics the next solution s' considered after solution s is derived by first applying a sequence of t changes to s (with t dependent

Section 10.3
LOCAL SEARCH HEURISTICS

Program 10.5: Variable depth local search

input Instance x;
output Solution s;
begin
   $s := $ initial feasible solution s_0;
   repeat
      $X := \varepsilon$ (the empty sequence);
      $s' := s$;
      (* begin of next candidate solution computation *)
      while the stopping rule does not apply **do**
      begin
         Let C be the set of possible change operations on s';
         Let $x \in C$ be such that $\mu(x) = \max_{c \in C} \mu(c)$;
         Append x at the end of X;
         $C := C - \{x\}$;
      end;
      Compute the change sequence X' from X;
      Apply X' on s to obtain a new solution s';
      (* end of next candidate solution computation *)
      if $m(x, s') < m(x, s)$ **then** $s := s'$
   until $s \neq s'$;
   return s
end.

from s and from the specific heuristic method), thus obtaining a sequence of solutions $s_1, \ldots, s_t$, where $s_i \in \mathcal{N}^i(s)$, and by successively choosing the best among these solutions.

In order to do that, heuristics refer to a measure μ of the gain (or profit) associated to any sequence of changes performed on the current solution. Moreover, the length of the sequence generated is determined by evaluating a suitably defined *stopping rule* at each iteration.

The basic version of the generic variable depth heuristics is shown in Program 10.5, where the external loop implements a local search among feasible solutions, while the inner loop returns a sequence of changes X (of variable length) which is used to compute the sequence X' which is actually applied to the current solution s, in order to derive a new candidate solution s' to be compared to s. Usually, X' is either X itself or a suitable prefix of X, as we will see below in the application of the technique to MINIMUM BALANCED CUT.

Several conditions must be met in order to effectively apply this local search method:

    1. A function μ must be available to evaluate the gain obtained by ap-

plying any sequence of changes. Such a function must be additive on the sequence, may take negative values, and should be such that any improvement on its value is likely to lead towards the optimal solution.

2. The change operation should be such that any sequence of any length, applied to a feasible solution, provides a new distinct feasible solution.

3. The stopping rule has to be defined by taking into account the need to avoid both generating long fruitless sequences and early stopping of the construction of very convenient new solutions.

Indeed, heuristics can be made quite more complex than the simple scheme shown here. For example, a typical extension introduces the possibility that, if the first sequence X considered does not lead to any improvement, then a certain number of different sequences are generated by introducing some limited form of backtracking.

The minimum balanced cut problem

The first application of variable depth local search was introduced by Kernighan and Lin for MINIMUM BALANCED CUT. Program 10.6, given a partition (V_1, V_2) of V, returns a new partition (V_1', V_2') which is the next solution considered in the local search. This is done under the following assumptions:

1. A change operation $x_{(a,b)}$ is an exchange of a node $a \in V_1$ with a node $b \in V_2$.

2. The profit of a single change operation $x_{(a,b)}$ is the decrease in the cut weight induced by exchanging a and b, that is,

$$\mu(x_{(a,b)}) = \sum_{(a,v) \in E : v \in V_2 - \{b\}} w(a,v) - \sum_{(a,v) \in E : v \in V_1} w(a,v) + \sum_{(b,v) \in E : v \in V_1 - \{a\}} w(b,v) - \sum_{(b,v) \in E : v \in V_2} w(b,v).$$

The profit of a sequence of change operations is the sum of their profits.

3. The next solution is generated by finding the best one among those generated by a sequence of n exchanges. Notice that, by the algorithm structure, after n exchanges, V_1 and V_2 have been interchanged and, thus, the initial solution is derived. Since there is a limited

Program 10.6: Kernighan-Lin

input Graph $G = (V, E)$ with weighted edges and partition (V_1, V_2) of V;
output Partition (V_1', V_2') of V;
begin
  $U_1 := V_1$;
  $U_2 := V_2$;
  $X := V$;
  for $i := 1$ **to** n **do**
  begin
    Choose nodes $a_i \in U_1 \cap X, b_i \in U_2 \cap X$ with maximal $\mu(x_{(a_i, b_i)})$;
    $U_1 := U_1 - \{a_i\} \cup \{b_i\}$;
    $U_2 := U_2 - \{b_i\} \cup \{a_i\}$;
    $X := X - \{a_i, b_i\}$
  end;
  Choose k with $1 \leq k \leq n$ such that $\sum_{i=1}^{k} \mu(x_{(a_i, b_i)})$ is maximal;
  Apply sequence $x_{(a_1, b_1)} \cdots x_{(a_k, b_k)}$ to (V_1, V_2) in order to obtain (V_1', V_2');
  return (V_1', V_2')
end.

upper bound on the number of changes which can be performed, it follows that the stopping rule only requires to check whether length n has been reached.

The traveling salesperson problem

The variable depth heuristics approach has been successfully applied also to MINIMUM TRAVELING SALESPERSON: in this case, indeed, it represents the most used practical line of attack for large instances of the problem.

Given a tour τ, the heuristics try to identify two sequences of k pairs of cities $C = ((c_{p_1}, c_{q_1}), \ldots, (c_{p_k}, c_{q_k}))$ and $\overline{C} = ((c_{s_1}, c_{t_1}), \ldots, (c_{s_k}, c_{t_k})\}$, with $k > 1$, such that:

1. c_{p_i} and c_{q_i} are adjacent in τ, for $1 \leq i \leq k$;
2. c_{s_i} and c_{t_i} are not adjacent in τ, for $1 \leq i \leq k$;
3. $c_{q_i} = c_{s_i}$, for $1 \leq i \leq k$;
4. $c_{t_i} = c_{p_{i+1}}$, for $1 \leq i \leq k-1$;
5. $c_{t_k} = c_{p_1}$;
6. substituting all pairs in C with all pairs in $\overline{C}$ in τ results in a tour τ' whose length is smaller than the length of τ.

Chapter 10

HEURISTIC METHODS

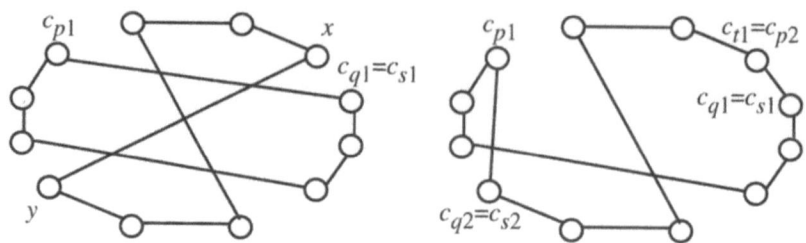

Figure 10.4
First step of LK heuristics

We present here a somewhat simplified version of the rather complicated original Lin-Kernighan heuristics, at the same time maintaining its conceptual framework. Given a tour τ, the next tour generation of the heuristic method performs the following steps:

1. Set $G = 0$ and $k = 1$. Choose any two cities c_{p_1} and c_{q_1} adjacent in τ (see the left part of Fig. 10.4). Set $C = \{c_{p_1}, c_{q_1}\}$, $c_{s_1} = c_{q_1}$, and $i = 1$.

2. Find two cities x and y such that:

    (a) x and y are adjacent in τ and do not belong to C;

    (b) substituting the $i+1$ edges $(c_{p_1}, c_{q_1}), \ldots, (c_{p_i}, c_{q_i}), (x, y)$ in τ with the $i+1$ edges $(c_{s_1}, c_{t_1}), \ldots, (c_{s_i}, x), (y, c_{p_1})$ still results in a tour;

    (c) $G_i + D(p_i, q_i) + D(x, y) - D(p_i, y) - D(s_i, x) > 0$ where $G_i = \sum_{j=1}^{i-1}[D(p_j, q_j) - D(s_j, t_j)]$;

    If x and y do not exist, then go to the next step. Otherwise, set $c_{t_i} = c_{p_{i+1}} = x$ and $c_{q_{i+1}} = c_{s_{i+1}} = y$. Set $C = C \cup \{x, y\}$. If $G_i + D(p_i, q_i) + D(x, y) - D(p_i, y) - D(s_i, x) > G$, then set $G = G_i + D(p_i, q_i) + D(x, y) - D(p_i, y) - D(s_i, x)$ and $k = i+1$. Set $i = i+1$ and repeat this step (see Figs. 10.5 and 10.6).

3. If $k > 1$, then substitute in τ the k edges $(c_{p_1}, c_{q_1}), \ldots, (c_{p_k}, c_{q_k})$ with the k edges $(c_{s_1}, c_{t_1}), \ldots, (c_{s_k}, c_{p_1})$ and return the resulting tour. Otherwise, return τ.

The whole process can be repeated, possibly starting from a different pair of adjacent cities. The procedure described here differs from the one originally introduced by Lin and Kernighan in two main points. First, the original procedure allows a same city to appear more than once in the sequence of pairs generated by the procedure, as long as no pair of cities appear in both sequences (this implies that a somewhat more sophisticated stopping rule has to be used). Second, a depth-bounded backtracking is performed with respect, in particular, to the choice of c_{t_1} and c_{t_2} (that is, the choice of x when $i = 1, 2$).

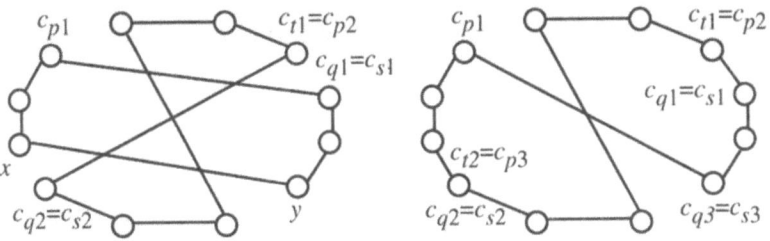

Figure 10.5
Second step of LK heuristics

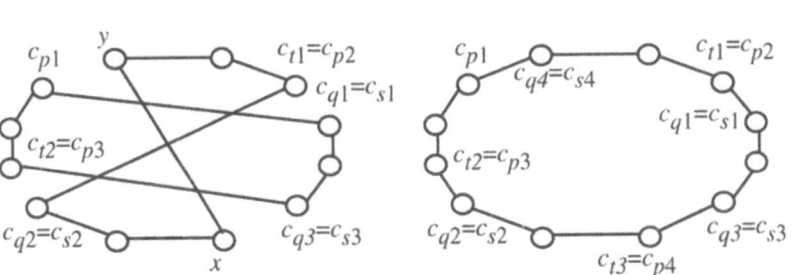

Figure 10.6
Third step of LK heuristics

10.4 Heuristics based on local search

The local search heuristic method applied to some instance x of an optimization problem moves from a current solution s to a neighbor solution s' as long as $m(x, s') < m(x, s)$. This monotone improvement of the current solution, while leading to a local optimum, does not take into account the possibility that a temporary increase of the measure of the current solution could eventually result in a better solution returned at the end of the entire search process. In this final section, we will briefly describe three approaches to overcome this limitation of local search based algorithms.

10.4.1 Simulated annealing

The simplest way of allowing a local increase in the solution measure is by defining a (possibly infinite) sequence $t_1, t_2, \ldots$, with $t_i \geq 0$ for $i \geq 1$, of *thresholds* and to allow a move from s to s' at the k-th step of the algorithm if and only if $m(x, s') - m(x, s) < t_k$. Notice that if $t_k > 0$, then s' could be selected even if it is a worse solution than s. Notice also that if $t_i = 0$ for $i \geq 1$ then the algorithm results in the above defined local search.

Chapter 10

HEURISTIC METHODS

In general, we may assume that the threshold sequence monotonically decreases to 0, that is, $t_i < t_{i-1}$, $i \geq 2$, and, for any $\varepsilon > 0$, there exists $i \geq 1$ such that $t_i \leq \varepsilon$. This implies that, while (relatively large) increases of the measure of the current solution may be accepted at the beginning of the search, only decreases or very limited increases can be performed as the algorithm proceeds. The stopping rule depends now on the number of steps performed: namely, the algorithm stops after k moves at solution s if, for all neighbor solutions s', the following inequality holds

$$m(x, s') - m(x, s) \geq t_{k+1}.$$

The method known as *Simulated annealing* corresponds to a randomized version of such an approach, where the choice of moving from s to a worse solution s' is performed with a probability inversely proportional to $m(x, s') - m(x, s)$. That is, an arbitrarily worse solution can be selected as the next current one, but this may happen only with a probability which is very small (at least after some iteration of the algorithm). In this way, simulated annealing does not deterministically preclude any search direction of the heuristics; however, the probability of selecting non-improving moves is made decreasing over time.

A particularly interesting aspect of simulated annealing is the underlying analogy between the solution of a combinatorial optimization problem and an apparently unrelated physical process such as the annealing of a solid. Annealing is the physical process where a solid which has been melt by applying a high temperature is carefully and slowly cooled with the aim of finding a low energy state for the set of particles in the solid. A typical case is the production of a crystal starting from a liquid phase.

In this framework, while in the liquid phase all particles are arranged randomly, due to the high energy of the system, the minimal energy configuration presents a regular structure in the spatial distribution of particles, which corresponds to the crystal state. Moreover, the solid may also assume any of a set of locally minimal configurations, corresponding to defective crystals. Cooling must be done slowly in order to avoid as much as possible that the solid enters one such locally minimal configuration, while leading it towards the overall minimum energy configuration.

Successful models have been introduced for the simulation of annealing which are based on probabilistic techniques. For example, the *Metropolis* algorithm simulates the cooling of the solid by generating a sequence of states as follows: given a state i with associated energy E_i, a new state j is generated by applying a small perturbation (for example, by changing the displacement of a single particle) and by computing the corresponding energy E_j. Let $\delta = E_j - E_i$: if $\delta < 0$, then the transition is accepted and

j is the next state in the sequence, otherwise, the transition from i to j is accepted with probability

$$p_T(\delta) = e^{-\frac{\delta}{k_B T}},$$

where T is the temperature of the solid and k_B is the Boltzmann constant. Without further details, it is important to observe that if the Metropolis algorithm is applied in a situation of slowly decreasing temperature, so that the solid is allowed to reach thermal equilibrium for each temperature, then the state of minimal energy of the system is approached as the temperature approaches 0.

If we consider combinatorial optimization problems, any state of the physical system corresponds to a feasible solution of the optimization problem, while the energy associated to the state corresponds to the solution measure. By the analogy between system states and feasible solutions, it is then easy to see that the Metropolis algorithm can be adapted to any optimization problem.

Unfortunately, it has been shown that, in order to eventually find the optimal solution (i.e., reaching a minimal energy state), the algorithm may require a number of transitions larger than the size of the solution space. Thus searching the optimum must be done by means of a *finite-time* implementation of the Metropolis algorithm, which results in the simulated annealing procedure used in practice.

The sequence of temperatures traversed by the algorithm, denoted as *cooling schedule*, is often generated by means of a function of the type

$$\tau := r\tau,$$

where τ is the temperature variable, initially set to value $t > 0$, and $r \in (0,1)$ is a control parameter, denoted as the cooling ratio. This results in a sequence of positive and monotonically decreasing values

$$t_1 = t, t_2 = rt, t_3 = r^2 t, \ldots$$

An integer value $l > 0$ is also defined in order to determine the number of transitions performed at each temperature. At a given temperature t_k, the move from s to s' is selected with the following probability $P(s, s', k)$:

$$P(s, s', k) = \begin{cases} 1 & \text{if } m(x, s') < m(x, s), \\ e^{-\frac{m(x,s') - m(x,s)}{t_k}} & \text{otherwise.} \end{cases}$$

A description of the generic simulated annealing heuristics is given in Program 10.7, whose behaviour is determined by the neighborhood structure $\mathcal{N}$, by the initial solution s_0, and by the values of the parameters r, t, l. The stopping rule is represented by a predicate FROZEN.

Section 10.4

HEURISTICS BASED ON LOCAL SEARCH

Chapter 10

HEURISTIC METHODS

Program 10.7: Simulated Annealing

input Instance x;
output Solution s;
begin
    $\tau := t$;
    $s :=$ initial feasible solution s_0;
    repeat
      for l **times do**
      (* local search at temperature τ *)
      begin
        Select any unvisited $s' \in \mathcal{N}(s)$;
        if $m(x,s') < m(x,s)$ **then**
          $s := s'$
        else
        begin
          $\delta := m(x,s') - m(x,s)$;
          $s := s'$ with probability $e^{-\frac{\delta}{\tau}}$
        end
      end;
    (* update of temperature *)
    $\tau := r\tau$
    until FROZEN;
    return s
end.

It is worth emphasizing that, once a neighborhood structure and an initial feasible solution have been fixed, a simulated annealing heuristic method has some additional tuning capabilities determined by the possibility of using different values for r, t and l, and of defining different stopping rule conditions.

This makes simulated annealing quite more powerful and flexible than the simple local search approach, at the same time introducing the need of reliable rules for determining a good tuning of the method.

10.4.2 Genetic algorithms

Genetic algorithms represent a type of randomized local search heuristics inspired by natural selection in biology. Roughly speaking, a genetic algorithm behaves by maintaining a *population* of solutions which evolves through a set of *generations*; at each step, the algorithm generates a new population of maximum size N (where N is a prespecified constant) from the current one by applying a certain set of (randomly applied) operators which make it possible to eliminate old solutions and generate new ones. Such a procedure presents an inherent parallelism, due to the fact that many different solutions can be tested and modified at the same time.

In particular, a new generation is usually derived through the following three phases.

Evaluation of fitness. The quality (or *fitness*) of each solution in the current population is evaluated by means of a suitably defined function. In many cases, the fitness of a solution corresponds to its measure.

Selection. Solutions in the current population are probabilistically selected to survive into the new population according to their relative fitness (better solutions have a higher probability of surviving).

Generation of new individuals. As long as the population has size less than N, new individuals are randomly created either by recombining some pair of solutions (*crossover*) or by modifying the structure of a single solution (*mutation*).

New generations are derived as long as some prespecified stopping criterion is verified, such as when there is no improvement of the best solution between two consecutive generations or after a given number of generations.

In many cases, solutions are encoded by fixed length binary strings, and both crossover and mutation are defined in terms of such strings, that is, as operators from a pair of strings into a pair of strings or from string to string, respectively. In particular, the crossover operator derives a new pair of strings σ_3 and σ_4 from a pair σ_1 and σ_2 by randomly selecting a value i, with $1 \leq i \leq |\sigma_1| = |\sigma_2|$, and by concatenating the first i bits of σ_1 and the last $|\sigma_2| - i$ bits of σ_2 to obtain σ_3 and, symmetrically, the first i bits of σ_2 and the last $|\sigma_1| - i$ bits of σ_1 to obtain σ_4.

Mutation is usually performed bit by bit, that is, each bit of a string σ is randomly flipped (with a given probability) to possibly obtain a new string σ'.

It should be clear from this general description that genetic algorithms have a certain conceptual resemblance to the evolution of a species in biology, with binary strings modeling gene sequences, crossover and mutation modeling the creation of new individuals with an original gene sequence, and selection modeling the "survival of the fittest" criterion in biology.

In order to show a specific implementation of the genetic algorithm approach, let us consider MINIMUM BALANCED CUT. In this case, a solution (i.e., a partition (V_1, V_2) of the set of nodes $\{v_1, \ldots, v_{2n}\}$) can be

Chapter 10

Heuristic Methods

encoded by a binary string σ of $2n$ bits such that the i-th bit σ_i is 0 (respectively, 1) if $v_i \in V_1$ (respectively, $v_i \in V_2$).

The fitness evaluation function can be defined as follows. Let c_{ij} be the weight associated to edge (v_i, v_j), and let us assume that $c_{ij} = 0$ if there is no edge (v_i, v_j). Then, the evaluation function ϕ is defined as

$$\phi(\sigma) = C - (\sum_{i=1}^{2n} \sum_{j=1}^{2n} c_{ij}\sigma_i(1-\sigma_j) + K\left|n - \sum_{i=1}^{2n} \sigma_i\right|).$$

Observe that $\sum_{i=1}^{2n} \sum_{j=1}^{2n} c_{ij}\sigma_i(1-\sigma_j)$ is the objective function to be minimized in MINIMUM BALANCED CUT. The expression $K\left|n - \sum_{i=1}^{2n} \sigma_i\right|$, where K is a suitably large constant, is introduced to significantly decrease the fitness of unfeasible solutions. Finally, C is another suitably large constant introduced since, by definition of genetic algorithm, the fitness has to be maximized (possible values for K and C could be $\max c_{ij} 2n$ and $\max c_{ij} n^2$, respectively).

The probability of selecting a string σ in a population G is given by

$$\phi(\sigma) / \sum_{\sigma' \in G} \phi(\sigma'),$$

that is, by the relative fitness of the string itself.

Notice that it is possible to generate unfeasible solutions by applying the crossover and mutation operators. For example, assuming $n = 4$, crossover on the third bit applied on the pair of strings 00101110 and 01100011, which represent feasible solutions, generates the pair of strings 00100011 and 01101110, which encode unfeasible solutions, since $|V_1| \neq |V_2|$. The same holds also with respect to mutations, since, due to the fact that single bits can be flipped, a string with $k \neq n$ bits equal to 1 can be generated by the mutation of a string with n such bits. Notice, however, that, even if unfeasible solutions can be generated by a genetic algorithm, the choice of a good fitness function should make such solutions very unlikely to survive in subsequent generations, due to their poor fitness.

The application of genetic algorithms to the solution of MINIMUM BALANCED CUT has been relatively simple, since a solution can be easily encoded as a binary string, as shown above. The situation can be quite different in the case of more structured problems, such as, for example, MINIMUM TRAVELING SALESPERSON. If a feasible solution for such a problem is represented by a string codifying a permutation of cities, then the definitions of mutation and crossover become more complex than in the case of MINIMUM BALANCED CUT. In fact, even if a solution of MINIMUM TRAVELING SALESPERSON could be represented as a binary string by identifying the set of edges in the tour through the set of bits equal to 1 in the string, mutation and crossover operators as simple as the ones defined above could generate a large number of unfeasible solutions.

10.4.3 Tabu search

The *tabu search* technique has the same aim as simulated annealing: to generalize simple local search in order to avoid the risks induced by the *myopic* approach of constraining heuristics to perform only moves (changes of the current solution) which lead to better solutions.

Just as simulated annealing, tabu search allows moves from a solution s to a neighbor solution s' with worse measure, hoping that this will eventually lead to a better local optimum. While simulated annealing uses randomization to introduce such possibility, tabu search follows a more complex, completely deterministic approach for the selection of the next move.

In the simplest version of the heuristic method, this is attained as follows:

1. For any solution s, let us denote as $\Delta(s)$ the set of moves which can be applied on s (to derive neighbor solutions). A function δ^* is defined which, for any subset $S \subseteq \Delta(s)$, returns a *best* move $\delta^*(S) \in \Delta(s)$ (notice that computing δ^* can even imply solving an optimization problem).

2. A list $\mathcal{T}$ of *tabu* moves (that is, moves which are excluded since they are considered to lead toward bad local optima) is maintained. This expedient introduces some kind of memory in the heuristics (for example, by forbidding moves which are the inverse of recently performed moves). $\mathcal{T}$ has a maximum size t which is an important parameter of the heuristics, roughly representing the amount of memory allowed.

At each step, the heuristic method maintains knowledge of the best solution found so far and, given a current solution s, it selects a neighbor s' as the one obtained by applying the best non-tabu move (that is, $\delta^*(\Delta(s) - \mathcal{T})$). The set of tabu moves is updated in correspondence to each change of the current solution.

The heuristics stop either when all possible moves from the current solution are tabu moves (that is, $\Delta(s) = \mathcal{T}$) or if $\bar{k}$ moves have been performed without improving the current optimum (where $\bar{k}$ is a user defined constant). Clearly, we expect that, for larger values of $\bar{k}$, better solutions are

Section 10.4

HEURISTICS BASED ON LOCAL SEARCH

returned, in general, by the algorithm, at the expense of a larger running time.

Notice that, even in this simple version, tabu search shares the high flexibility of simulated annealing and genetic algorithms, due to the possibility of tuning the heuristics by using different definitions of tabu moves and of the δ^* function and by selecting different values for $\bar{k}$ and for the maximum number t of tabu moves recorded at each time.

In many cases, δ^* and $\mathcal{T}$ have the following simple definitions:

1. For any solution s and for any subset S of $\Delta(s)$, $\delta^*(S)$ is defined as the move in S which leads to the neighbor s' such that

$$m(x,s') = \min\{m(x,y) \mid y \text{ obtained by applying a move } \delta \in S\}.$$

    Notice that this does not rule out the possibility, in the case that s is a local optimum, to move to a worse solution.

2. $\mathcal{T}$ is defined in such a way to avoid traversing solutions that have been already visited, as long as possible. Since the trivial approach of maintaining knowledge of the whole sequence of current solutions consumes too much space and time, the solution applied usually consists of storing in $\mathcal{T}$ the sequence of the last t moves performed. In this case, tabu search avoids moves which are the inverse of a move in $\mathcal{T}$. Usually, $\mathcal{T}$ is implemented as a queue of t items, so that the insertion of a new item in $\mathcal{T}$ in correspondence to a new move induces the deletion of the oldest item.

It may happen that the heuristics cannot move to a good neighbor solution because the corresponding move is the inverse of an item in $\mathcal{T}$. In such a case, it may be convenient to allow this move even if it is tabu: this is made possible in tabu search by means of *aspiration levels*. Essentially, the aspiration levels mechanism is introduced by defining a function $\mathcal{A}$ which associates a value to any pair (s,m), where s is a solution and m is a move defined on s. If s is the current solution of instance x, then, even if m is a currently tabu move, tabu search allows solution s', obtained by applying move m on s, to become the current solution if the corresponding aspiration level satisfies a given condition (for example, if $m(x,s) < A(s,m)$).

In order to better explain the above concepts, let us consider a simple application of Tabu Search to MINIMUM VERTEX COVER.

1. Given a problem instance $G = (V,E)$, for any solution (i.e., a vertex cover $U \subseteq V$), $\Delta(U)$ consists of $|V|$ moves: the first $|U|$ moves remove one vertex from U, while the remaining $|V - U|$ insert a

new vertex in the vertex cover. Notice that the $|U|$ deletion moves may lead to unfeasible solutions: however, these unfeasible solutions may be avoided by suitably modifying the definition of δ^* given above.

2. $\mathcal{T}$ is implemented as a queue of t moves, where for each move both the change operation performed (addition or deletion of a node) and the corresponding node are specified.

3. A possible aspiration level mechanism can be introduced by defining $\mathcal{A}(U,m) = |W|$, where U is the current vertex cover and W is the vertex cover obtained from U by applying move m. A tabu move m is then allowed to be applied on solution U if $\mathcal{A}(U,m) < |\overline{V}|$, where $\overline{V}$ is the smallest vertex cover found so far. That is, a tabu move is allowed if the resulting vertex cover improves the measure of the best solution derived up to the current step.

10.5 Exercises

Exercise 10.1 Implement and compare the effectiveness of Programs 10.1 and 10.2 for MINIMUM TRAVELING SALESPERSON, under the hypothesis of uniform distribution of instances.

Exercise 10.2 Define, implement, and test possible 2-opt and 3-opt heuristics for MAXIMUM k-SATISFIABILITY, under the hypothesis of uniform distribution of instances.

Exercise 10.3 Define, implement, and test a 2-opt and a variable depth heuristics for MAXIMUM CUT, under the hypothesis of uniform distribution of instances.

Exercise 10.4 Show that the optimal tour in the proof of Theorem 10.1 is unique.

Exercise 10.5 Prove Theorem 10.3.

Exercise 10.6 Implement and test Program 10.6, under the hypothesis of uniform distribution of instances.

Exercise 10.7 Implement and test the Lin Kernighan heuristics for MINIMUM TRAVELING SALESPERSON, as described in Sect. 10.3.2, under the hypothesis of uniform distribution of instances.

Chapter 10
HEURISTIC METHODS

Exercise 10.8 Define, implement and test a simulated annealing heuristics for MINIMUM TRAVELING SALESPERSON, under the hypothesis of uniform distribution of instances.

Exercise 10.9 Define, implement and test a simulated annealing heuristics for MINIMUM BALANCED CUT, under the hypothesis of uniform distribution of instances.

Exercise 10.10 Implement and test the genetic algorithm for MINIMUM BALANCED CUT, as described in Sect. 10.4.2, under the hypothesis of uniform distribution of instances.

Exercise 10.11 Define, implement and test a genetic algorithm for MAXIMUM SATISFIABILITY, under the hypothesis of uniform distribution of instances.

Exercise 10.12 Implement and test the tabu search algorithm for MINIMUM VERTEX COVER, as described in Sect. 10.4.3, under the hypothesis of uniform distribution of instances.

Exercise 10.13 Define, implement, and test a tabu search heuristics for MAXIMUM SATISFIABILITY, under the hypothesis of uniform distribution of instances.

10.6 Bibliographical notes

Enumeration methods for the solution of integer linear programs, such as branch-and-bound, represent *per se* a wide and well studied area of research. A good treatment of these topics can be found in [Garfinkel and Nemhauser, 1972]. The branch-and-cut method was introduced, in the framework of the solution of MINIMUM TRAVELING SALESPERSON, in [Padberg and Rinaldi, 1987] and further refined in [Padberg and Rinaldi, 1991].

Many approaches to the approximate solution of MINIMUM TRAVELING SALESPERSON are reviewed and discussed in [Reinelt, 1994a].

Local search heuristics are widely discussed in [Arts and Lenstra (eds.), 1997]. Their efficiency in terms of the quality of returned solutions has been studied, in general, in [Grover, 1992]. Much effort has been devoted to the characterization of the efficiency of such heuristics for MINIMUM TRAVELING SALESPERSON (see, for example, [Chandra, Karloff, and Tovey, 1994], [Plesník, 1986], and [Papadimitriou and Steiglitz, 1977]).

Section 10.6

BIBLIOGRAPHICAL NOTES

In [Johnson, Papadimitriou, Yannakakis, 1988] the class PLS is defined which includes problems whose neighborhood can be visited in polynomial time. Suitable reductions are introduced to prove completeness results in such a class. The class has been further studied and other PLS-completeness results are given in [Krentel, 1989], [Krentel, 1990], [Schäffer and Yannakakis, 1991], and [Papadimitriou, 1992].

k-opt heuristics for MINIMUM TRAVELING SALESPERSON have been introduced in [Lin, 1965] and further studied in [Chandra, Karloff, and Tovey, 1994], [Krentel, 1989], [Johnson, Papadimitriou, Yannakakis, 1988], and, from a probabilistic point of view, in [Kern, 1989].

The variable depth local search heuristics for MINIMUM BALANCED CUT was introduced in [Kernighan and Lin, 1970], and modified and made faster in [Fiduccia and Mattheyses, 1982]. The Lin-Kernighan heuristics for MINIMUM TRAVELING SALESPERSON is from [Lin and Kernighan, 1973].

Simulated annealing heuristics are widely considered in [Aarts and Korst, 1989], and in [van Laarhoven and Aarts, 1987]. The approach has been introduced in [Cerny, 1985] and [Kirkpatrick, Gelatt, Vecchi, 1983]. An experimental testing of this method for some important optimization problems, with an evaluation of its effectiveness with respect to other heuristics, is reported in [Johnson, Aragon, McGeoch, Schevon, 1991a] and [Johnson, Aragon, McGeoch, Schevon, 1991b].

Genetic algorithms are extensively treated in [Goldberg, 1989], while the tabu search heuristics is extensively described, together with many of its variants, in [Glover, 1989], [Glover, 1990], and [De Werra and Hertz, 1989]. A general description of this latter method is also given in [Glover, Taillard, Laguna and de Werra, 1993].

Appendix A

Mathematical preliminaries

In this appendix we define the basic mathematical concepts, tools, and notation that are used in the book.

A.1 Sets

ANY COLLECTION of elements is a *set* A. If a, b and c are arbitrary elements, then we represent the set A consisting of elements a, b and c as $A = \{a,b,c\}$. We also define sets according to some property and we write $\{y|\text{ property about } y\}$ to express in a compact way the fact that the set consists of all elements which satisfy the specified property; this is particularly useful to denote sets with an infinite number of elements.

The symbols $\in$ and $\notin$ denote, respectively, the fact that an element belongs or does not belong to a set. Referring to the previous example, $a \in A$ and $d \notin A$.

Examples of sets are the sets of *natural numbers* $\mathbf{N}$, of *integer numbers* Z, of *rational numbers* $\mathbf{Q}$, of *positive natural numbers* $\mathbf{N}^+$, of *positive integer numbers* Z^+, and of real points in an m-dimensional space $\mathbf{R}^m$.

Two sets A and B are *equal* (in symbols, $A = B$) if every element of A is also an element of B and vice versa. Two sets A and B are *not equal* (in symbols, $A \neq B$) when $A = B$ does not hold true. A set A is a *subset* of B (in symbols, $A \subseteq B$) if each element of A is also an element of B. A set A is a *proper subset* of B (in symbols, $A \subset B$) if $A \subseteq B$ and $A \neq B$. Two sets A and B are *disjoint* if they do not include common elements.

Appendix A

MATHEMATICAL PRELIMINARIES

A set is a *singleton* if it includes a single element (notice the difference between the singleton set $\{x\}$ and element x). A set is *empty* (in symbols, $X = \emptyset$) if it includes no element, otherwise it is *non-empty*. A set is *finite* if it includes a finite number of elements, otherwise it is *infinite*. Given a finite set X, its *cardinality* (in symbols, $|X|$) is equal to the number of its elements.

Given two sets A and B, their *union*, denoted as $A \cup B$, is the set of all elements which belong either to A or to B; their *intersection*, denoted as $A \cap B$, is the set of all elements common to A and B; finally, the *difference* set of A from B, denoted as $A - B$, is defined as the set of all elements which belong to A but not to B. In some cases, we will also consider the union of an infinite collection of sets $A_0, A_1, A_2, \ldots$ and we will denote such union as $\bigcup_{k=0}^{\infty} A_k$.

The *power set* of A (in symbols, 2^A) is a set whose elements are all possible subsets of A, including the empty set and the set A. If A is finite and consists of n elements, then the power set of A consists of 2^n elements, thus the notation 2^A.

Given a non-empty set A, a *partition* of A is a subset P of 2^A such that every element of P is non-empty, the elements of P are pairwise disjoint, and the union of the elements of P is equal to A.

A.1.1 Sequences, tuples and matrices

A *sequence* of objects is a list of these objects in some order and is designated by writing the list within parentheses. Notice that a sequence of objects is distinct from the set containing these objects because the order in which the objects appear in the sequence is significant and because the same object may appear twice in the sequence.

As with sets, sequences may be finite or infinite. Finite sequences are called *tuples*. A tuple with k elements is called *k-tuple*: a 2-tuple (respectively, 3-tuple) is also called a *pair* (respectively, *triple*).

The *Cartesian product* of k sets $A_1, \ldots, A_k$, denoted as $A_1 \times \cdots \times A_k$, is the set of all k-tuples $(a_1, \ldots, a_k)$ with $a_i \in A_i$ for $i = 1, \ldots, k$. If $A_i = A$ for $i = 1, \ldots, k$, then we denote the Cartesian product as A^k.

A (k-dimensional) *vector* v over a set A is an element of A^k (sometimes, we denote vectors with symbols in boldface, i.e., $\mathbf{v}$). The i-th component of vector v will be denoted with v_i.

A $k_1 \times k_2$ *matrix* M over a set A is an element of $(A^{k_1})^{k_2}$, that is, M is a sequence of k_2 vectors over A whose dimension is k_1 (sometimes, we denote matrices with symbols in boldface, i.e., $\mathbf{M}$). The i-th component of the j-th vector of M may be denoted with $M_{i,j}$, $M(i,j)$, or m_{ij}. Given a matrix M, M^T denotes the *transpose* of M, that is the matrix such that $M^T(i,j) = M(j,i)$.

Appendix A

FUNCTIONS AND RELATIONS

A.2 Functions and relations

GIVEN TWO sets A and B, any subset R of $A \times B$ is called a *binary relation* between A and B. The *domain* of R is the set of all x such that $(x,y) \in R$ for some y; the *codomain* or *range* of R is the set of all y such that $(x,y) \in R$ for some x. If $(x,y) \in R$, we will usually write xRy.

Given two sets A and B, a *function* f from A to B (in symbols, $f : A \mapsto B$) is a binary relation between A and B which includes, at most, one pair (a,b) for any $a \in A$. A function with domain A' and codomain B' is called a function *from A' onto B'*. When referring to functions we prefer to say that the value of f in a is b (in symbols, $f(a) = b$) instead of saying that the pair (a,b) belongs to f. Moreover, a is called the *argument* and b the *value* of f.

A function $f : A \mapsto B$ is *total* if its domain coincides with A and is *partial* if its domain is a subset of A. A function $f : A \mapsto B$ is *many-one* if the value of f may coincide on two different arguments; f is *one-one* or *injective* if, for all $a, a' \in A$ with $a \neq a'$, $f(a) \neq f(a')$, and it is *surjective* if B coincides with the codomain of f. A function is *bijective* if it is both injective and surjective; a bijective function is also called a *bijection*.

Given two functions $f, g : \mathbf{N} \mapsto \mathbf{N}$, we say that $f(n)$ is $O(g(n))$ if there exist constants c, a and n_0 such that, for all $n \geq n_0$,

$$f(n) \leq cg(n) + a.$$

We say that $f(n)$ is $\Omega(g(n))$ if there exist constants c, a and n_0 such that, for all $n \geq n_0$,

$$f(n) \geq cg(n) + a.$$

Finally, we say that $f(n)$ is $\Theta(g(n))$ if $f(n)$ is both $O(g(n))$ and $\Omega(g(n))$.

Given k sets $A_1, \ldots, A_k$, any subset R of $A_1 \times \cdots \times A_k$ is called a *k-ary relation* between the k sets $A_1, \ldots, A_k$. A *k-ary* function $f : A_1 \times A_2 \times \cdots \times A_k \mapsto B$ is a $(k+1)$-ary relation between the $k+1$ sets $A_1, \ldots, A_k, B$ which includes, at most, one $(k+1)$-tuple $(a_1, \ldots, a_k, b)$ for any $(a_1, \ldots, a_k) \in A_1 \times \cdots \times A_k$.

Given a binary relation $R \subseteq A \times B$, the *inverse relation* of R is defined as $R^{-1} = \{(y,x) : (x,y) \in R\}$. It follows from the definition that $(R^{-1})^{-1} = R$.

Similarly, given a function $f : A \mapsto B$, we say that it admits an *inverse function* $f^{-1} : B \mapsto A$ if the following holds: $f(a) = b$ if and only if $f^{-1}(b) = a$.

Appendix A
MATHEMATICAL PRELIMINARIES

Note that a binary relation R always admits a unique inverse relation R^{-1} while only injective functions admit an inverse function.

A binary relation between A and A is called a *binary relation in A*. Let R be a binary relation in A. R is called *reflexive* if $(x,x) \in R$ for all x. R is called *symmetric* if, for all pairs of elements x and y, $(x,y) \in R$ implies $(y,x) \in R$. R is called *transitive* if, for all triples of elements x, y and z, $(x,y) \in R$ and $(y,z) \in R$ imply $(x,z) \in R$.

A binary relation in A which is reflexive, symmetric and transitive is called an *equivalence* relation.

A.3 Graphs

AN UNDIRECTED *graph* G (or simply a *graph*) is a pair of finite sets (V,E) such that E is a binary symmetric relation in N. The set V is the set of *vertices* (also called *nodes*) and E is the set of *edges* (in particular, an edge (u,v) identifies both the pair (u,v) and the pair (v,u)). If $(u,v) \in E$, then u and v are said to be *adjacent* and they are the *end-points* of that edge. The number of vertices adjacent to a given vertex u is called the *degree* of u. The degree of G is the maximum over all degrees of the vertices. A graph is said to be *d-regular* if all its vertices have degree d.

A *directed graph* is a pair of finite sets (N,A) such that A is a binary (not necessarily symmetric) relation in N. Usually, elements of N are said *nodes* and elements of A are said *arcs*.

We say that $G' = (V',E')$ is the *subgraph* of $G = (V,E)$ induced by V' if $V' \subseteq V$ and $E' \subseteq \{(v_i,v_j) | v_i, v_j \in V' \land (v_i,v_j) \in E\}$.

A *weighted graph* is a graph $G = (V,E)$ together with a function $w : E \mapsto \mathbf{R}$ which associates a weight with any edge so that $w(v_i,v_j) = w(v_j,v_i)$.

A graph $G = (V,E)$ is called a *complete graph* or a *clique* if $E = V \times V$, that is, if every two vertices are adjacent. A graph $G = (V,E)$ is called a *planar* if it can be drawn on the plane without crossings between edges.

Given a graph G and two vertices v_0 and v_k, a *path from v_0 to v_k* is a sequence of distinct vertices $(v_0, \ldots, v_k)$ such that (v_i, v_{i+1}) belongs to E, $i = 0, \ldots, k-1$; the *length of the path* is the number of edges of the path or the sum of the weights of the edges if G is a weighted graph.

Given a path from v_0 to v_k, if $v_0 = v_k$, then such a path is a *cycle*. If a cycle contains all the nodes of the graph, we call it a *Hamiltonian cycle*.

A graph is *connected* if, for every two distinct vertices x and y, there is a path from x to y. A *tree* is a connected graph with no cycles; a tree is *rooted* if there is one vertex designated as the *root*.

A rooted tree can also be defined recursively as follows:

1. A single vertex is a tree. This vertex is also the *root* of the tree.

2. Suppose v is a vertex and $T_1, \ldots, T_k$ are trees with roots $v_1, \ldots, v_k$. A new tree is obtained by joining v with vertices $v_1, \ldots, v_k$. In this tree v is the root and vertices $v_1, \ldots, v_k$ are the *children* of v.

The *height of a vertex* in a tree is the length of the path from the root to the vertex itself. The *height of a tree* is the maximum height of a vertex.

In a rooted tree, a vertex with no children is called a *leaf*. A rooted tree is a *binary tree* (respectively, a *complete binary tree*) if every vertex that is not a leaf has at most (respectively, exactly) two children.

We refer to [Bollobas, 1979] for a more detailed presentation of fundamental graph concepts.

A.4 Strings and languages

AN ALPHABET is a non-empty finite set $\Sigma = \{\sigma_1, \ldots, \sigma_k\}$. A *symbol* is an element of an alphabet. A *string* is a finite tuple $x = (\sigma_{i_1}, \ldots, \sigma_{i_n})$ of symbols from Σ; the *empty string* is denoted by ε. For the sake of brevity, the string $(\sigma_{i_1}, \ldots, \sigma_{i_n})$ will be denoted by $\sigma_{i_1} \cdots \sigma_{i_n}$. The infinite set of all strings over an alphabet Σ will be denoted by Σ^*.

The *length* $|x|$ of a string $x = \sigma_{i_1} \ldots \sigma_{i_n}$ is the number n of symbols that x contains. The empty string has length 0. Clearly, the number of strings of length n over a k-symbol alphabet is equal to k^n. Given two strings x and y, the *concatenation* of x and y (in symbols, xy) is defined as the string z consisting of all symbols of x followed by all symbols of y, thus $|z| = |x| + |y|$. In particular, the concatenation of a string x with itself k times will be denoted as x^k. Given two strings x and y, x is said to be a *prefix* of y if a string z exists such that $y = xz$.

Given an alphabet Σ, a *language* over Σ is a subset of Σ^*. The *complement* of a language L, in symbols L^c, is defined as $L^c = \Sigma^* - L$.

Given an alphabet Σ, any order among the symbols of Σ induces the *lexicographic order* among the strings in Σ^* defined in the following way:

1. For any n, the strings of length n precede the strings of length $n+1$.

2. Among strings of the same length, the order is alphabetical.

A.5 Boolean logic

THE VALUES TRUE and FALSE are the *Boolean values*. These values can be manipulated with the following operations also called *connec-*

Appendix A

MATHEMATICAL PRELIMINARIES

tives: the *and* operation (in symbols, ∧) which is TRUE if and only if all its arguments are TRUE; the *or* operation (in symbols, ∨) which is TRUE if and only if at least one of its arguments is TRUE; the unary *not* operation (denoted by a bar over the argument) which is TRUE if and only if its argument is FALSE.

A *Boolean variable* is a variable which can take Boolean values; a *literal* is either a Boolean variable or its negation; given a set $V = \{v_1, \ldots, v_n\}$ of Boolean variables, $T(V) = \{v_1, \bar{v}_1, \ldots, v_n, \bar{v}_n\}$ is the corresponding set of literals; a *truth assignment* on V, is a function $f : V \mapsto \{\text{TRUE}, \text{FALSE}\}$ that satisfies, for each variable $v_i \in V$, the literal v_i if $f(v_i) = \text{TRUE}$ and the literal $\bar{v}_i$ if $f(v_i) = \text{FALSE}$.

We define a (disjunctive) *clause* as a set of literals $c \subseteq T(V)$, which we assume satisfied by a truth assignment f if and only if there exists (at least) one literal in c satisfied by f. Formally, given the literals $l_1, \ldots, l_k$, the corresponding clause is the *disjunction* $l_1 \vee \cdots \vee l_k$.

A *conjunctive normal form* (CNF) formula is defined as a collection of clauses $C = \{c_1, \ldots, c_m\}$ and is assumed to be satisfied by a truth assignment f if and only if each clause c_i, $i = 1, \ldots, m$, is satisfied by f.

A *predicate* is a function whose range is the set $\{\text{TRUE}, \text{FALSE}\}$. In the propositional calculus predicates can be expressed only by means of variables and the connectives *and*, *or*, and *not*. In the predicate calculus (or first order logic), also the logical quantifiers '∀' and '∃', called *universal* and *existential quantifiers*, can be used. '∀x' is read as 'for all x' and '∃x' as 'there exists an x such that'.

Finally we note that the logical symbols above also serve as convenient abbreviations of ordinary mathematical language and they are often used in this way.

A.6 Probability

GIVEN A set of points Ω, also called a *sample space*, an *event* is a subset of the sample space. If Ω is countable, then a *discrete probability space* is defined as follows: with each point x in the sample space there is associated a real number, called the *probability of x* and denoted as $\Pr\{x\}$. This number has to be non-negative and such that the sum of the probabilities of all points is equal to 1. The *probability of an event A* (in symbols, $\Pr\{A\}$) is the sum of the probabilities of all points in it.

For any two events A_1 and A_2,

$$\Pr\{A_1 \cup A_2\} = \Pr\{A_1\} + \Pr\{A_2\} - \Pr\{A_1 \cap A_2\}.$$

The above equality can be generalized to the case of the union of k events

Appendix A

PROBABILITY

$A_1, \ldots, A_k$:

$$\Pr\{\bigcup_{i=1}^{k} A_i\} \leq \sum_{i=1}^{k} \Pr\{A_i\}$$

where the equality holds if the events A_i are pairwise disjoint.

Two events A_1 and A_2 are *independent* if the following holds:

$$\Pr\{A_1 \cap A_2\} = \Pr\{A_1\}\Pr\{A_2\}.$$

$\Pr\{A \mid B\}$ denotes the probability of event A *conditioned by the occurrence of B* and it is equal to $\Pr\{A \mid B\} = \Pr\{A \cap B\}/\Pr\{B\}$; clearly if A and B are independent then $\Pr\{A \mid B\} = \Pr\{A\}$.

In general, a *probability space* is a triple $(\Omega, \Sigma, \Pr)$, where Ω is a set, Σ is a collection of subsets of Ω and $\Pr$ is a non-negative measure on Σ such that $\Pr\{\Omega\} = 1$. It is required that Σ must be a σ-field, that is a family of subsets that includes Ω, and that it is closed under countably many set operations. If $\Omega = (0, 1]^d$ and we want that Σ includes all one point sets then Σ contains all sets that are obtained by applying a countable number of set operations starting from subintervals of $(0, 1]$; this σ-field is also known as a *Borel* σ-field and elements of Σ are called *Borel subsets*.

A.6.1 Random variables

Given a probability space a *random variable* x is a continuous real-valued function $x: \Omega \mapsto \mathbf{R}$.[1] Given a random variable x its *distribution function* is $F(y) = \Pr\{x < y\}$, for $-\infty < y < \infty$; therefore, the probability that a random variable x assumes values in the range $[y_1, y_2]$ is given by $\Pr\{y_1 \leq x \leq y_2\} = F(y_1) - F(y_2)$.

A *probability density* on the line is a non-negative function f such that

$$\int_{-\infty}^{\infty} f(x)dx = 1.$$

To each density function f there exists a corresponding distribution function defined by

$$F(y) = \int_{-\infty}^{y} f(x)dx.$$

The *expected value* (also the *mean* or the *first moment*) of a random variable x is given by

$$E[x] = \int_{-\infty}^{\infty} y dF(y).$$

---

[1] While in the more general case it would be required that the function is measurable, for the purpose of this book we restrict to the case that it is continuous

Appendix A

MATHEMATICAL PRELIMINARIES

If Ω is a discrete sample space, then the above integrals become sums.

Moreover, the *n-th moment* $E[x^n]$ of a random variable x is given by the expected value of x^n. Clearly all the above values do not necessarily exist; however they exist for all random variables considered in the book. Finally the *variance* of x is given by $\sigma^2(x) = E[x^2] - E[x]^2$.

Note that the mean is a linear operator, that is, if $x_1, \ldots, x_n$ is a sequence of random variables then $E[\sum_{i=1}^n x_i] = \sum_{i=1}^n E[x_i]$; this equality is not in general true for other moments.

Let x_1, x_2 be two discrete random variables defined over the same probability space. We say that x_1, x_2 are independent if, for any combination of values y_1 and y_2 assumed by them

$$\Pr\{x_1 = y_1, x_2 = y_2\} = \Pr\{x_1 = y_1\}\Pr\{x_2 = y_2\}.$$

The above definition can be generalized to a sequence as follows: given a sequence $x_1, \ldots, x_n$ of random variables, then we say that $x_1, \ldots, x_n$ are *mutually independent* if for any combination of $y_1, \ldots, y_n$ assumed by them

$$\Pr\{x_1 = y_1, \ldots, x_n = y_n\} = \Pr\{x_1 = y_1\} \ldots \Pr\{x_n = y_n\}.$$

If $x_1, \ldots, x_n$ are mutually independent and each x_i has finite variance then $\sigma^2[\sum_{i=1}^n x_i] = \sum_{i=1}^n \sigma^2[x_i]$.

We now consider some examples of random variables.

If x is a random variable *uniformly distributed* in $[0, 1]$ then its density function is given by $f(x) = 1$ if $x \in [0, 1]$ and 0 otherwise. In this case, we have $E[x] = 0.5$ and $\sigma^2(x) = 1/12$.

If x is a random variable that is *Poisson distributed* with parameter n then its density function is $f(x) = e^{-n}(n^x/x!)$ if x is a non-negative integer and 0 otherwise, where, as usual, $x!$ is the *factorial* of x, that is, $x! = x(x-1)\cdots 1$. In this case we have that n is both the mean and the variance of x.

A simple random experiment is one that has two possible outcomes, called *success* and *failure*. Assume that p is the probability of success (and $1-p$ the probability of failure). We can define the random variable $x = 1$ if success, $x = 0$ if failure. In this case it is easy to see that $E[x] = p$ and $E[x^2] = p$.

Suppose now that a sequence of n failure-success experiments are performed and let x_i be the random variable that is 1 (0) if the i-th experiment has success (failure). Successive and *independent* repetitions of the same success-failure experiment are called *Bernoulli trials*. Let X be the random variable that describes the number of success. The density of X is given by the probability of having exactly k successes, $k = 1, \ldots, n$, that is

$$\binom{n}{k} p^k (1-p)^{n-k},$$

where $\binom{n}{k}$ denotes the *binomial coefficient* of n over k, that is, $\binom{n}{k} = n!/(k!(n-k)!)$. The expected value of X is np and its variance is $np(1-p)$.

We now consider a sequence of Bernoulli trials having success probability p and we are interested in the random variable Y that denotes the number of trials required to obtain the first success. Y has a *geometric distribution* with parameter p and the probability that the first success is obtained after k trials is given by

$$(1-p)^{k-1}p.$$

The expected value of Y is given by $1/p$.

A *Poisson process* on a subset A of a d-dimensional Euclidean space is a probabilistic model that allows to describe a random distribution of particles in A; in this model both the number of particles and the location of each particle are random. The distribution of particles in A is described by giving the number of particles in subsets of A: for each Borel subset B we will have a non-negative integer random variable $N(B)$ that denotes the number of particles in B. To adequately represent a distribution of particles, the values of $N(B)$ for different B must satisfy the following condition: $N(\cup_i B_i) = \sum_i B_i$ where B_i is a countable sequence of disjoint regions.

If, for each bounded region B of area $a(B)$, the distribution of particles in B is Poisson distributed with parameter $na(B)$, where n is a constant, then the distribution of particles in A constitutes a *Poisson process with parameter n*.

Interesting properties of a Poisson process are that particles are uniformly distributed in A and that the distribution of particles in two disjoint regions B and B' are independent.

There are many textbooks that present main concepts of probability; we refer to [Feller, 1971] for a thorough presentation of probability theory.

Appendix A

LINEAR PROGRAMMING

A.7 Linear programming

LINEAR PROGRAMMING provides a powerful technique to describe a wide class of optimization problems, at the same time providing general techniques for their solution.

Any linear program on n real variables $x = \{x_1, \ldots, x_n\}$ requires to minimize (or maximize) the value of an *objective function* defined as a linear combination of variables in x, subject to a set of m linear inequalities or equalities.

For example, given vectors $c \in \mathbf{R}^n$ and $b \in \mathbf{R}^m$ and matrix $A \in \mathbf{R}^{m \times n}$,

Appendix A

MATHEMATICAL PRELIMINARIES

we want to minimize $c^T x$ subject to $Ax \geq b$ and $x \geq 0$. Formally we have

$$\begin{aligned}\text{minimize} \quad & c^T x \\ \text{subject to} \quad & Ax \geq b \\ & x \geq 0.\end{aligned} \qquad (A.1)$$

Several algorithms have been proposed for solving linear programs; the most used is the the *simplex method* developed by Dantzig in the late 1940s.

The simplex method can be seen as a visit of the polyhedron specified by the set of linear inequalities $Ax \geq b$ until an optimal solution is detected. The simplex method is very effective in practice but there are instances of the problem in which the method requires exponential time.

More recently, a polynomial time algorithm based on the *ellipsoid method* used in nonlinear programming has been proposed in [Khachian, 1979]. Khachian's algorithm is asymptotically better than the simplex method, but it is slower in practice. Knowledge of the simplex method and/or of Khachian's algorithm is not a prerequisite for this book.

In this section we also overview the main results of duality theory on which some approximation algorithms, presented in Chaps. 2, 4, and 10, are based. For a more detailed treatment of these topics we refer to [Karloff, 1991] and [Schrijver, 1986].

First of all we observe that the above formulation of a linear program as a minimization problem subject to a set of linear inequalities given by $Ax \geq b$ is not the only one. If we consider minimization problems, the more general formulation is the following:

$$\begin{aligned}\text{minimize} \quad & c^T x \\ \text{subject to} \quad & Ax \geq b \\ & A'x = b' \\ & x_1 \geq 0, \ldots, x_r \geq 0.\end{aligned} \qquad (A.2)$$

Notice that variables $x_{r+1}, \ldots, x_n$ are not sign constrained. The above formulation is equivalent to (A.1). In fact:

1. Any equality constraint $a'x = b'_i$ is replaced by two inequality constraints $a'x \geq b'_i$ and $-a'x \geq -b'_i$.

2. For any unconstrained variable x_j, two variables x_j^+, x_j^- are introduced that are sign constrained ($x_j^+ \geq 0$ and $x_j^- \geq 0$); any occurrence of x_j is replaced by $x_j^+ - x_j^-$.

Appendix A

LINEAR PROGRAMMING

On the other side it is also possible to see that (A.2) is equivalent to

$$\text{minimize} \quad c^T x$$
$$\text{subject to} \quad Ax = b$$
$$x \geq 0$$

where we assume that all variables are sign-constrained.

In fact:

1. Any inequality constraint $ax \geq b_i$ can be transformed in an equality constraint by adding a new *surplus* variable s_i (also called *slack variable*). The constraint is then replaced by the pair $ax - s_i = b_i$, $s_i \geq 0$.

2. For any unconstrained variable x_j, two variables x_j^+, x_j^- are introduced that are sign constrained ($x_j^+ \geq 0$ and $x_j^- \geq 0$); any occurrence of x_j is replaced by $x_j^+ - x_j^-$.

A linear program can also be used to formulate maximization problems such as,

$$\text{maximize} \quad c^T x \quad (A.3)$$
$$\text{subject to} \quad Ax \leq b$$
$$x \geq 0.$$

A correspondence between maximization and minimization problems can be easily established. In fact:

1. Any maximization problem can be transformed into a minimization one by considering the objective function $-c^T x$.

2. Any inequality constraint $ax \leq b_i$ can be transformed into an equality constraint by adding a new slack variable s_i. The constraint is then replaced by the pair $ax + s_i = b_i$, $s_i \geq 0$.

Given a minimization linear program, called *primal*, it is possible to define a maximization linear program, called *dual*, as follows: for each constraint in the primal program, there is a variable in the dual; for each variable in the primal program there is a constraint in the dual. Namely, we have:

Appendix A

MATHEMATICAL PRELIMINARIES

	Primal		**Dual**	
constr. i	$\sum_{j=1}^{n} a_{ij}x_j = b_i$		var. i	y_i not sign-constr.
constr. i	$\sum_{j=1}^{n} a_{ij}x_j \geq b_i$		var. i	$y_i \geq 0$
var. j	x_j not sign-constr.		constr. j	$\sum_{i=1}^{m} a_{ij}y_i = c_j$
var. j	$x_j \geq 0$		constr. j	$\sum_{i=1}^{m} a_{ij}y_i \leq c_j$
	minimize $c^T x$		maximize $b^T y$.	

Analogously, the dual of a maximization linear program is a minimization linear program that is obtained using the same table above (from right to left). Clearly given linear program P, the dual of the dual of P coincides with P.

Let P and D be a primal and dual pair of linear programs. The *Duality Theorem* of linear programming states that if they are both feasible then the values of the optimal solutions coincide.

For example, program (A.1) requires to determine

$$\min\{c^T x \mid Ax \geq b, x \geq 0\}.$$

In this case the *Duality Theorem* states that

$$\min\{c^T x \mid Ax \geq b, x \geq 0\} = \max\{b^T y \mid y \geq 0, y^T A \leq c^T\}$$

provided that both sets are non empty.

Another equivalent formulation of the theorem states that

$$\max\{c^T x \mid x \geq 0, Ax = b\} = \min\{y^T \mid y^T A \geq c^T\}$$

provided that both sets are non empty.

Integer linear programming is obtained from linear programming by restricting the variables to integer values, or to subsets of the integers. An example of an integer linear program is the following:

$$\begin{aligned} \text{minimize} \quad & c^T x \\ \text{subject to} \quad & Ax = b \\ & x \in \{0, 1\}. \end{aligned}$$

In such a case, that is when variables are restricted to have values in $\{0, 1\}$, the program is also called a *0-1 linear program*. Given an integer linear program x we can associate to it a linear program x', obtained by relaxing the integrality constraints on the variables. For example, in the above case, we would obtain

$$\begin{aligned} \text{minimize} \quad & c^T x \\ \text{subject to} \quad & Ax = b \\ & 0 \leq x \leq 1. \end{aligned}$$

Clearly, $m^*(x') \leq m^*(x)$ and therefore such an approach allows us to establish a lower bound on the optimal measure of a minimization integer linear program.

Clearly, an analogous relaxation can be established in the case of maximization integer linear programs.

A.8 Two famous formulas

The *Cauchy-Schwartz inequality* states that

$$\left(\sum_{k=1}^n a_k b_k\right)^2 \leq \left(\sum_{k=1}^n a_k^2\right)\left(\sum_{k=1}^n b_k^2\right).$$

The *Stirling's formula* provides an approximation of the factorial as follows:

$$n! \approx \sqrt{2\pi n}\left(\frac{n}{e}\right)^n.$$

Appendix A

TWO FAMOUS FORMULAS

A List of NP Optimization Problems

The list contains more than 200 entries. A typical entry consists of eight parts: the first four parts are mandatory while the last four parts are optional.

1. The problem name which also specifies the goal of the problem.

2. A definition of the instances of the problem.

3. A definition of the feasible solutions of the problem.

4. A definition of the measure of a feasible solution.

5. A 'good news' part that contains the best approximation result for the problem.

6. A 'bad news' part that contains the worst negative approximation result for the problem.

7. A section of additional comments.

8. A reference to the 'closest' problem appearing in the list published in [Garey and Johnson, 1979].

Observe that the problem definitions that appear in the list follow the notation of [Garey and Johnson, 1979] while, in some cases, the definitions given in the previous chapters correspond to the most widely used in the

The compendium

A List of NP Optimization Problems

literature: in any case, the differences between the two definitions are just a matter of notations.

The list is organized according to subject matter as in [Garey and Johnson, 1979]. In particular the entries are divided into the following twelve categories:

GT Graph theory: 55 entries.

ND Network design: 65 entries.

SP Sets and partitions: 14 entries.

SR Storage and retrieval: 11 entries.

SS Sequencing and scheduling: 20 entries.

MP Mathematical programming: 18 entries.

AN Algebra and number theory: 1 entry.

GP Games and puzzles: 2 entry.

LO Logic: 14 entries.

AL Automata and language theory: 5 entries.

PO Program optimization: 2 entries.

MS Miscellaneous: 19 entries.

We have ignored problems with too obscure definitions and problems for which membership in NP was not guaranteed. Furthermore we have only included results for ordinary polynomial-time approximation algorithms, not for, e.g., bicriteria approximation algorithms and on-line algorithms.

We have certainly missed many other results. We ask everybody to help us in correcting, improving, and enlarging the problem list for future editions of the book. The easiest way to do so is via the WWW. Forms for reporting new problems, new results, and errors are available at:

> http://www.nada.kth.se/theory/compendium/

Graph Theory

Covering and Partitioning

MINIMUM VERTEX COVER ◀ GT1

INSTANCE: Graph $G = (V, E)$.

SOLUTION: A vertex cover for G, i.e., a subset $V' \subseteq V$ such that, for each edge $(u, v) \in E$, at least one of u and v belongs to V'.

MEASURE: Cardinality of the vertex cover, i.e., $|V'|$.

Good News: Approximable within $2 - \frac{\log\log|V|}{2\log|V|}$ [Monien and Speckenmeyer, 1985] and [Bar-Yehuda and Even, 1985].

Bad News: APX-complete [Papadimitriou and Yannakakis, 1991].

Comment: Trasformation from bounded MAXIMUM 3-SATISFIABILITY. Not approximable within 1.1666 [Håstad, 1997]. Admits a PTAS for planar graphs [Baker, 1994] and for unit disk graphs [Hunt III et al., 1994b]. Variation in which each vertex has a nonnegative weight and the objective is to minimize the total weight of the vertex cover is as hard to approximate as the original problem [Crescenzi, Silvestri, and Trevisan, 1996] and admits a PTAS for planar graphs [Baker, 1994]. Variation in which the degree of G is bounded by a constant B for $B \geq 3$ is APX-complete [Papadimitriou and Yannakakis, 1991] and [Alimonti and Kann, 1997] and not approximable within 1.1666 for a sufficiently large B [Clementi and Trevisan, 1996]. For $B = 3$ it is approximable within $7/6+\varepsilon$ [Berman and Fujito, 1995], for general B within $2 - \log B/B(1+\varepsilon)$ [Halldórsson, 1995a], and for $B = 5$ not within 1.0029 [Berman and Karpinski, 1998]. Approximable within $3/2$ for 6-claw-free graphs (where no independent set of size 6 exists in any neighbour set to any vertex) [Halldórsson, 1995a]. Variation in which $|E| = \Theta(|V|^2)$ is APX-complete [Clementi and Trevisan, 1996] and approximable within $2/(2 - \sqrt{1-\varepsilon})$. Approximable within $2/(1+\varepsilon)$ if every vertex has degree at least $\varepsilon|V|$ [Karpinski and Zelikovsky, 1997b] and [Karpinski, 1997].

The generalization to k-hypergraphs, for $k \geq 2$, is approximable within k [Bar-Yehuda and Even, 1981] and [Hochbaum, 1982a]. If the vertex cover is required to induce a connected graph, the problem is approximable within 2 [Arkin, Halldórsson, and Hassin, 1993]. If the graph is edge-weighted, the solution is a closed walk whose vertices form a vertex cover, and the objective is to minimize the sum of the edges in the cycle, the problem is approximable within 5.5 [Arkin, Halldórsson, and Hassin, 1993].

The constrained variation in which the input is extended with a positive integer k and a subset S of V, and the problem is to find the vertex cover of size k that contains the largest number of vertices from S, is not approximable within $|V|^\varepsilon$ for some $\varepsilon > 0$ [Zuckerman, 1993]. The maximization variation in which the input is extended just with a positive integer k, and the problem is to find k vertices that cover as many edges as possible, does not admit a PTAS [Petrank, 1994].

Garey and Johnson: GT1

GT2 ▶ MINIMUM DOMINATING SET

INSTANCE: Graph $G = (V, E)$.

SOLUTION: A dominating set for G, i.e., a subset $V' \subseteq V$ such that for all $u \in V - V'$ there is a $v \in V'$ for which $(u, v) \in E$.

MEASURE: Cardinality of the dominating set, i.e., $|V'|$.

Good News: Approximable within $1 + \log|V|$ since the problem is a special instance of MINIMUM SET COVER [Johnson, 1974a].

Bad News: Not in APX [Lund and Yannakakis, 1994] and [Bellare, Goldwasser, Lund, and Russell, 1993].

Comment: Equivalent to MINIMUM SET COVER under L-reduction [Kann, 1992b] and [Bar-Yehuda and Moran, 1984]. See MINIMUM SET COVER for more comments. Not approximable within $(1 - \varepsilon)\ln|V|$ for any $\varepsilon > 0$, unless NP $\subset$ DTIME($n^{\log\log n}$) [Feige, 1996]. If it is NP-hard to approximate within $\Omega(\log n)$, then it is complete for the class of log-approximable problems [Khanna, Motwani, Sudan, and Vazirani, 1999]. Admits a PTAS for planar graphs [Baker, 1994] and for unit disk graphs [Hunt III et al., 1994b]. Variation in which the degree of G is bounded by a constant B is APX-complete [Papadimitriou and Yannakakis, 1991] and is approximable within $\sum_{i=1}^{B+1} \frac{1}{i} - 0.433$ by reduction to MINIMUM SET COVER.

If the dominating set is restricted to be connected the problem is approximable within $\ln \Delta + 3$ where Δ is the maximum degree, and within $3\ln|V|$ for the vertex weighted version [Guha and Khuller, 1996]. The problem MAXIMUM DOMATIC PARTITION, in which the graph should be partitioned into a maximum number of dominating sets, is approximable within 4 for circular-arc graphs [Marathe, Hunt III, and Ravi, 1996].

Garey and Johnson: GT2

GT3 ▶ MINIMUM EDGE DOMINATING SET

INSTANCE: Graph $G = (V, E)$.

SOLUTION: An edge dominating set for G, i.e., a subset $E' \subseteq E$ such that for all $e_1 \in E - E'$ there is an $e_2 \in E'$ such that e_1 and e_2 are adjacent.

MEASURE: Cardinality of the edge dominating set, i.e., $|E'|$.

Good News: Approximable within 2 (any maximal matching).

Comment: Admits a PTAS for planar graphs [Baker, 1994] and for λ-precision unit disk graphs [Hunt III et al., 1994b].

Garey and Johnson: GT2

MINIMUM INDEPENDENT DOMINATING SET ◀ GT4

INSTANCE: Graph $G = (V, E)$.

SOLUTION: An independent dominating set for G, i.e., a subset $V' \subseteq V$ such that for all $u \in V - V'$ there is a $v \in V'$ for which $(u, v) \in E$, and such that no two vertices in V' are joined by an edge in E.

MEASURE: Cardinality of the independent dominating set, i.e., $|V'|$.

Bad News: NPOPB-complete [Kann, 1994b]. Not approximable within $|V|^{1-\varepsilon}$ for any $\varepsilon > 0$ [Halldórsson, 1993b].

Comment: Also called *Minimum Maximal Independence Number*. Transformation from SHORTEST PATH WITH FORBIDDEN PAIRS. Variation in which the degree of G is bounded by a constant B is APX-complete [Kann, 1992b]. Approximable within 5 for unit disk graphs [Marathe et al., 1995].

The maximization variation in which the set $\{0, 1, \ldots, k\}$ for some fixed k constrains how the independent set can dominate the remaining vertices is approximable within $O(\sqrt{|V|})$ and is not approximable within $n^{1/(k+1)-\varepsilon}$ for any $\varepsilon > 0$ unless NP = ZPP [Halldórsson, Kratochvil, and Telle, 1998].

Garey and Johnson: GT2

MINIMUM GRAPH COLORING ◀ GT5

INSTANCE: Graph $G = (V, E)$.

SOLUTION: A vertex coloring of G, i.e., a partition of V into disjoint sets $V_1, V_2, \ldots, V_k$ such that each V_i is an independent set for G.

MEASURE: Cardinality of the coloring, i.e., the number of disjoint independent sets V_i.

Good News: Approximable within $O\left(|V| \frac{(\log \log |V|)^2}{(\log |V|)^3}\right)$ [Halldórsson, 1993a].

Bad News: Not approximable within $|V|^{1/7-\varepsilon}$ for any $\varepsilon > 0$ [Bellare, Goldreich, and Sudan, 1998].

Comment: Also called *Minimum Chromatic Number*. Not approximable within $|V|^{1-\varepsilon}$ for any $\varepsilon > 0$, unless CO-RP=NP [Feige and Kilian, 1996]. If the graph is 3-colorable the problem is approximable within $O(|V|^{1/4} \log |V|)$ [Karger, Motwani, and Sudan, 1998] and [Mahajan and Ramesh, 1995], but it is not approximable within $5/3-\varepsilon$ [Khanna, Linial, and Safra, 1993]. If the graph is k-colorable the problem is approximable within

$O(|V|^{1-3/(k+1)} \log|V|)$ [Karger, Motwani, and Sudan, 1998]. Approximable with an absolute error guarantee of 1 on planar graphs, see, e.g., [Horowitz and Sahni, 1978]. Approximable within 3 but not within 1.33 for unit disk graphs [Marathe et al., 1995]. Approximable within $O(|V|/(\log^{k-1}|V|)^2)$ for k-uniform hypergraphs [Hofmeister and Lefmann, 1998]. MINIMUM COLORING WITH DEFECT d, the variation where the color classes V_i may induce graphs of degree at most d, is not approximable within $|V|^\varepsilon$ for some $\varepsilon > 0$ [Cowen, Goddard, and Jesurum, 1997].

MINIMUM FRACTIONAL CHROMATIC NUMBER, the linear programming relaxation in which the independent sets $V_1, V_2, \ldots, V_k$ do not need to be disjoint, and in the solution every independent set V_i is assigned a non-negative value λ_i such that for each vertex $v \in V$ the sum of the values assigned to the independent sets containing v is at most 1, and the measure is the sum $\sum \lambda_i$, is not approximable within $|V|^c$ for some constant c [Lund and Yannakakis, 1994].

The complementary maximization problem, where the number of "not needed colors", i.e., $|V| - k$, is to be maximized, is approximable within 4/3 [Halldórsson, 1996].

The constrained variation in which the input is extended with a positive integer k, a vertex $v_0 \in V$ and a subset S of V, and the problem is to find the k-coloring that colors the largest number of vertices from S in the same way as v_0, is not approximable within $|V|^\varepsilon$ for some $\varepsilon > 0$ [Zuckerman, 1993].

MINIMUM COLOR SUM, the variation in which the objective is to minimize $\sum_{1 \leq i \leq k} \sum_{v \in V_i} i$ is not approximable within $|V|^{1-\varepsilon}$ for any $\varepsilon > 0$ [Bar-Noy, Bellare, Halldórsson, Shachnai, and Tamir, 1998], but for bipartite graphs the problem is APX-complete and approximable within 10/9 [Bar-Noy and Kortsarz, 1997]. Generalization of MINIMUM COLOR SUM where the input is extended by positive coloring costs $k_1 \leq \ldots \leq k_{|V|}$, and i is changed to k_i in the objective function, is not approximable within $|V|^{1-\varepsilon}$ even for split, chordal, permutation and comparability graphs, but for bipartite and interval graphs the problem is approximable within $|V|^{0.5}$, but not within $|V|^{0.5-\varepsilon}$ [Jansen, 1997].

Garey and Johnson: GT4

GT6 ▶ MAXIMUM ACHROMATIC NUMBER

INSTANCE: Graph $G = (V, E)$.

SOLUTION: A complete coloring of G, i.e., a partition of V into disjoint sets $V_1, V_2, \ldots, V_k$ such that each V_i is an independent set for G and such that, for each pair of distinct sets V_i, V_j, $V_i \cup V_j$ is *not* an independent set.

MEASURE: Cardinality of the complete coloring, i.e., the number of disjoint independent sets V_i.

Good News: Approximable within factor $O(\frac{|V|}{\sqrt{\log|V|}})$ [Chaudhary and Vishwanathan, 1997].

Comment: Approximable within $O(\sqrt{|V|})$ for graphs of girth (length of shortest cycle) at least 6 [Chaudhary and Vishwanathan, 1997].

Garey and Johnson: GT5

MINIMUM EDGE COLORING ◀ GT7

INSTANCE: Graph $G = (V,E)$.

SOLUTION: An edge coloring of E, i.e., a partition of E into disjoint sets $E_1, E_2, \ldots, E_k$ such that, for $1 \leq i \leq k$, no two edges in E_i share a common endpoint in G.

MEASURE: Cardinality of the coloring, i.e., the number of disjoint sets E_i.

Good News: Approximable within 4/3, and even approximable with an absolute error guarantee of 1 [Vizing, 1964].

Bad News: Not approximable within 4/3−ε for any ε > 0 [Holyer, 1981].

Comment: Also called *Minimum Chromatic Index*. APX-intermediate unless the polynomial-time hierarchy collapses [Crescenzi, Kann, Silvestri, and Trevisan, 1999]. On multigraphs the problem is approximable within $1.1 + (0.8/m^*)$ [Nishizeki and Kashiwagi, 1990]. The maximization variation in which the input is extended with a positive integer k, and the problem is to find the maximum number of consistent vertices over all edge-colorings with k colors, is approximable within $e/(e-1)$ [Bertsimas, Teo, and Vohra, 1996], but does not admit a PTAS [Petrank, 1994].

Garey and Johnson: OPEN5

MINIMUM FEEDBACK VERTEX SET ◀ GT8

INSTANCE: Directed graph $G = (V,A)$.

SOLUTION: A feedback vertex set, i.e., a subset $V' \subseteq V$ such that V' contains at least one vertex from every directed cycle in G.

MEASURE: Cardinality of the feedback vertex set, i.e., $|V'|$.

Good News: Approximable within $O(\log|V| \log\log|V|)$ [Seymour, 1995] and [Even, Naor, Schieber, and Sudan, 1995].

Bad News: APX-hard [Kann, 1992b].

Comment: Transformations from MINIMUM VERTEX COVER and MINIMUM FEEDBACK ARC SET [Ausiello, D'Atri, and Protasi, 1980]. On undirected graphs the problem is APX-complete and approximable within 2, even if the vertices are weighted [Bafna, Berman, and Fujito, 1994]. The generalized variation in which the input is extended with a subset S of

The compendium

GRAPH THEORY

vertices and arcs, and the problem is to find a vertex set that contains at least one vertex from every directed cycle that intersects S, is approximable within $O(\min(\log |V| \log \log |V|, \log^2 |S|))$ on directed graphs [Even, Naor, Schieber, and Sudan, 1995] and within 8 on undirected graphs [Even, Naor, and Zosin, 1996]. All these problems are approximable within 9/4 for planar graphs [Goemans and Williamson, 1996]. The constrained variation in which the input is extended with a positive integer k and a subset S of V, and the problem is to find the feedback vertex set of size k that contains the largest number of vertices from S, is not approximable within $|V|^\varepsilon$ for some $\varepsilon > 0$ [Zuckerman, 1993].

Garey and Johnson: GT7

GT9 ▶ MINIMUM FEEDBACK ARC SET

INSTANCE: Directed graph $G = (V, A)$.

SOLUTION: A feedback arc set, i.e., a subset $A' \subseteq A$ such that A' contains at least one arc from every directed cycle in G.

MEASURE: Cardinality of the feedback arc set, i.e., $|A'|$.

Good News: Approximable within $O(\log |V| \log \log |V|)$ [Even, Naor, Schieber, and Sudan, 1995].

Bad News: APX-hard [Kann, 1992b].

Comment: Transformation from MINIMUM FEEDBACK VERTEX SET [Even, Naor, Schieber, and Sudan, 1995]. The generalized variation in which the input is extended with a subset S of vertices and arcs, and the problem is to find an arc set that contains at least one arc from every directed cycle that intersects S, is approximable within $O(\min(\log |V| \log \log |V|, \log^2 |S|))$ [Even, Naor, Schieber, and Sudan, 1995]. All these problems are approximable within 9/4 for planar graphs [Goemans and Williamson, 1996]. The constrained variation in which the input is extended with a positive integer k and a subset S of A, and the problem is to find the feedback arc set of size k that contains the largest number of arcs from S, is not approximable within $|E|^\varepsilon$ for some $\varepsilon > 0$ [Zuckerman, 1993]. The complementary problem of finding the maximum set of arcs A' such that $G' = (V, A')$ is acyclic is approximable within $2/(1 + \Omega(1/\sqrt{\Delta}))$ where Δ is the maximum degree [Berger and Shor, 1990] and [Hassin and Rubinstein, 1994], it is APX-complete [Papadimitriou and Yannakakis, 1991], and if $|E| = \Theta(|V|^2)$ it admits a PTAS [Arora, Frieze, and Kaplan, 1996].

Garey and Johnson: GT8

GT10 ▶ MINIMUM MAXIMAL MATCHING

INSTANCE: Graph $G = (V, E)$.

SOLUTION: A maximal matching E', i.e., a subset $E' \subseteq E$ such that no two edges in E' shares a common endpoint and every edge in $E - E'$ shares a common endpoint with some edge in E'.

MEASURE: Cardinality of the matching, i.e., $|E'|$.

Good News: Approximable within 2 (any maximal matching).

Bad News: APX-complete [Yannakakis and Gavril, 1980].

Comment: Transformation from MINIMUM VERTEX COVER.

Garey and Johnson: GT10

MAXIMUM TRIANGLE PACKING ◀ GT11

INSTANCE: Graph $G = (V, E)$.

SOLUTION: A triangle packing for G, i.e., a collection $V_1, V_2, \ldots, V_k$ of disjoint subsets of V, each containing exactly 3 vertices, such that for each $V_i = \{u_i, v_i, w_i\}$, $1 \leq i \leq k$, all three of the edges (u_i, v_i), (u_i, w_i), and (v_i, w_i) belong to E.

MEASURE: Cardinality of the triangle packing, i.e., the number of disjoint subsets V_i.

Good News: Approximable within $3/2 + \varepsilon$ for any $\varepsilon > 0$ [Hurkens and Schrijver, 1989].

Bad News: APX-complete [Kann, 1991].

Comment: Transformation from MAXIMUM 3-DIMENSIONAL MATCHING with a bounded number of occurrences. Admits a PTAS for planar graphs [Baker, 1994] and for λ-precision unit disk graphs [Hunt III et al., 1994b]. Still APX-complete when the degree of G is bounded by 4 [Kann, 1991].

Garey and Johnson: GT11

MAXIMUM H-MATCHING ◀ GT12

INSTANCE: Graph $G = (V_G, E_G)$ and a fixed graph $H = (V_H, E_H)$ with at least three vertices in some connected component.

SOLUTION: A H-matching for G, i.e., a collection $G_1, G_2, \ldots, G_k$ of disjoint subgraphs of G, each isomorphic to H.

MEASURE: Cardinality of the H-matching, i.e., the number of disjoint subgraphs G_i.

Good News: Approximable within $|V_H|/2 + \varepsilon$ for any $\varepsilon > 0$ [Hurkens and Schrijver, 1989].

Bad News: APX-hard [Kann, 1994a].

Comment: Transformation from bounded MAXIMUM 3-SATISFIABILITY. Variation in which the degree of G is bounded by a constant B is APX-complete [Kann, 1994a]. Admits a PTAS for planar graphs [Baker, 1994], but does not admit an FPTAS [Berman, Johnson, Leighton, Shor, and Snyder, 1990].

Admits a PTAS for λ-precision unit disk graphs [Hunt III et al., 1994b]. Induced MAXIMUM H-MATCHING, i.e., where the subgraphs G_i are induced subgraphs of G, has the same good and bad news as the ordinary problem, even when the degree of G is bounded.

MAXIMUM MATCHING OF CONSISTENT k-CLIQUES is the variation in which H is a k-clique and the vertices of G are partitioned into k independent sets $V_1, \ldots, V_k$, each V_i is partitioned into some sets $V_{i,j}$, and at most one vertex in each $V_{i,j}$ may be included in the matching. This problem is not in APX for $k \geq 5$ and is not approximable within 4/3 for $k \in \{2,3,4\}$. It is not in APX for any $k \geq 2$ unless NP$\subset$QP [Verbitsky, 1995].

Garey and Johnson: GT12

GT13 ▶ MINIMUM BOTTLENECK PATH MATCHING

INSTANCE: Graph $G = (V,E)$ with an even number of vertices, and a weight function $w : E \to N$.

SOLUTION: A bottleneck path matching for G, i.e., a collection $P_1, \ldots, P_{|V|/2}$ of disjoint simple paths in G with disjoint end points.

MEASURE: Weight of the heaviest path in the matching, i.e.,

$$\max_{1 \leq i \leq |V|/2} \sum_{e \in P_i} w(e).$$

Good News: Approximable within 2 [Datta and Sen, 1995].

GT14 ▶ MINIMUM CLIQUE PARTITION

INSTANCE: Graph $G = (V,E)$.

SOLUTION: A clique partition for G, i.e., a partition of V into disjoint subsets $V_1, V_2, \ldots, V_k$ such that, for $1 \leq i \leq k$, the subgraph induced by V_i is a complete graph.

MEASURE: Cardinality of the clique partition, i.e., the number of disjoint subsets V_i.

Good News: Approximable within $O\left(|V| \frac{(\log \log |V|)^2}{(\log |V|)^3}\right)$ [Halldórsson, 1993a].

Bad News: Not approximable within $|V|^{\varepsilon}$ for some $\varepsilon > 0$ [Lund and Yannakakis, 1994].

Comment: Equivalent to MINIMUM GRAPH COLORING [Paz and Moran, 1981]. The complementary maximization problem, where $|V| - k$, is to be maximized, is approximable within 4/3 [Halldórsson, 1996].

Garey and Johnson: GT15

MINIMUM k-CAPACITATED TREE PARTITION ◀ GT15

INSTANCE: Graph $G = (V, E)$, and a weight function $w : E \to N$.

SOLUTION: A k-capacitated tree partition of G, i.e., a collection of vertex disjoint subsets $E_1, \ldots, E_m$ of E such that, for each i, the subgraph induced by E_i is a tree of at least k vertices.

MEASURE: The weight of the partition, i.e., $\sum_i \sum_{e \in E_i} w(e)$.

Good News: Approximable within $2 - \frac{1}{|V|}$ [Goemans and Williamson, 1995a].

Comment: The variation in which the trees must contain exactly k vertices and the triangle inequality is satisfied is approximable within $4(1 - 1/k)(1 - 1/|V|)$. Similar results hold for the corresponding cycle and path partitioning problems with the triangle inequality [Goemans and Williamson, 1995a]. Variation in which the trees must contain exactly k vertices and the objective is to minimize $\max_i \sum_{e \in E_i} w(e)$ is not in APX, but if the triangle inequality is satisfied it is approximable within $2|V|/k - 1$ [Guttmann-Beck and Hassin, 1997]. The variation in which the number m of trees is fixed, the trees' sizes are exactly specified, and the triangle inequality is satisfied is approximable within $2p - 1$ [Guttmann-Beck and Hassin, 1998a].

MAXIMUM BALANCED CONNECTED PARTITION ◀ GT16

INSTANCE: Connected graph $G = (V, E)$, nonnegative vertex-weight function $w : V \to N$.

SOLUTION: A partition (V_1, V_2) of V into nonempty disjoint sets V_1 and V_2 such that the subgraphs of G induced by V_1 and V_2 are connected.

MEASURE: Balance of the partition, i.e., $\min\{w(V_1), w(V_2)\}$, where $w(V') = \sum_{v \in V'} w(v)$.

Good News: Approximable within 4/3 [Chlebikova, 1996].

Bad News: Not approximable with an absolute error guarantee of $|V|^{1-\varepsilon}$ for any $\varepsilon > 0$ [Chlebikova, 1996].

Comment: Variation in which the objective function is $w(V_1) \cdot w(V_2)$ is approximable within $8 - 4\sqrt{3}$ [Chlebikova, 1996].

MINIMUM CLIQUE COVER ◀ GT17

INSTANCE: Graph $G = (V, E)$.

The compendium

GRAPH THEORY

SOLUTION: A clique cover for G, i.e., a collection $V_1, V_2, \ldots, V_k$ of subsets of V, such that each V_i induces a complete subgraph of G and such that for each edge $(u, v) \in E$ there is some V_i that contains both u and v.

MEASURE: Cardinality of the clique cover, i.e., the number of subsets V_i.

Good News: Approximable within $O(f(|V|))$ if MAXIMUM CLIQUE is approximable within $f(|V|)$ [Halldórsson, 1994].

Bad News: Not approximable within $|V|^\varepsilon$ for some $\varepsilon > 0$ [Lund and Yannakakis, 1994].

Comment: Equivalent to the MINIMUM CLIQUE PARTITION problem under ratio-preserving reduction [Kou, Stockmeyer, and Wong, 1978] and [Simon, 1990]. The complementary maximization problem, where $|E| - k$ is to be maximized, is approximable within 4/3 [Halldórsson, 1996]. The constrained variation in which the input is extended with a positive integer k, a vertex $v_0 \in V$ and a subset S of V, and the problem is to find the clique cover of size k that contains the largest number of vertices from S, is not approximable within $|V|^\varepsilon$ for some $\varepsilon > 0$ [Zuckerman, 1993].

Garey and Johnson: GT17

GT18 ▶ MINIMUM COMPLETE BIPARTITE SUBGRAPH COVER

INSTANCE: Graph $G = (V, E)$.

SOLUTION: A complete bipartite subgraph cover for G, that is, a collection $V_1, V_2, \ldots, V_k$ of subsets of V, such that each V_i induces a complete bipartite subgraph of G and such that for each edge $(u, v) \in E$ there is some V_i that contains both u and v.

MEASURE: Cardinality of the complete bipartite subgraph cover, i.e., the number of subsets V_i.

Good News: Approximable within $O(f(|V|))$ if MAXIMUM CLIQUE is approximable within $f(|V|)$ [Halldórsson, 1994].

Bad News: Not approximable within $|V|^\varepsilon$ for some $\varepsilon > 0$ [Lund and Yannakakis, 1994].

Comment: Equivalent to the MINIMUM CLIQUE PARTITION problem under ratio-preserving reduction [Simon, 1990].

Garey and Johnson: GT18

GT19 ▶ MINIMUM VERTEX DISJOINT CYCLE COVER

INSTANCE: Graph $G = (V, E)$.

SOLUTION: A family F of vertex disjoint cycles covering V.

MEASURE: Number of cycles in F.

Bad News: Not in APX [Sahni and Gonzalez, 1976].

Comment: Variation in which the graph G is directed is not in APX.

MINIMUM EDGE DISJOINT CYCLE COVER ◀ **GT20**

INSTANCE: Graph $G = (V, E)$.

SOLUTION: A family F of edge disjoint cycles covering V.

MEASURE: Number of cycles in F.

Bad News: Not in APX [Sahni and Gonzalez, 1976].

Comment: Variation in which the graph G is directed is not in APX.

MINIMUM CUT COVER ◀ **GT21**

INSTANCE: Graph $G = (V, E)$.

SOLUTION: A collection $V_1, \ldots, V_m$ of cuts, i.e., a collection of subsets $V_i \subseteq V$ such that, for each edge $(u, v) \in E$, a subset V_i exists such that either $u \in V_i$ and $v \notin V_i$ or $u \notin V_i$ and $v \in V_i$.

MEASURE: Cardinality of the collection, i.e., m.

Good News: Approximable within $1 + (\log |V| - 3 \log \log |V|)/m^*(G)$ [Motwani and Naor, 1993].

Bad News: There is no polynomial-time algorithm with relative error less than 1.5. [Motwani and Naor, 1993].

Comment: The negative result is obtained by relating the problem with the coloring problem. Solvable in polynomial time for planar graphs. Observe that any graph has a cut cover of cardinality $\lceil \log |V| \rceil$.

Subgraphs and Supergraphs

MAXIMUM CLIQUE ◀ **GT22**

INSTANCE: Graph $G = (V, E)$.

SOLUTION: A clique in G, i.e. a subset $V' \subseteq V$ such that every two vertices in V' are joined by an edge in E.

MEASURE: Cardinality of the clique, i.e., $|V'|$.

Good News: $O\left(\frac{|V|}{(\log |V|)^2}\right)$-approximable [Boppana and Halldórsson, 1992].

Bad News: Not approximable within $|V|^{1/4-\varepsilon}$ for any $\varepsilon > 0$ [Bellare, Goldreich, and Sudan, 1998].

The compendium

GRAPH THEORY

Comment: Not approximable within $|V|^{1-\varepsilon}$ for any $\varepsilon > 0$, unless CO-RP=NP [Håstad, 1996]. Equivalent to MAXIMUM INDEPENDENT SET on the complementary graph. Approximable within $m^*/O(m^{3(k+1)})$ if $m^* = |V|/k+m$ [Alon and Kahale, 1994] and [Mahajan and Ramesh, 1995]. The vertex weighted version is approximable within $O(|V|(\log\log|V|/\log|V|)^2)$ [Halldórsson, 1995b].

Garey and Johnson: GT19

GT23 ▶ MAXIMUM INDEPENDENT SET

INSTANCE: Graph $G = (V, E)$.

SOLUTION: An independent set of vertices, i.e. a subset $V' \subseteq V$ such that no two vertices in V' are joined by an edge in E.

MEASURE: Cardinality of the independent set, i.e., $|V'|$.

Good News: See MAXIMUM CLIQUE.

Bad News: See MAXIMUM CLIQUE.

Comment: The same problem as MAXIMUM CLIQUE on the complementary graph. Admits a PTAS for planar graphs [Baker, 1994] and for unit disk graphs [Hunt III et al., 1994b]. Variation in which the degree of G is bounded by a constant B for $B \geq 3$ is APX-complete [Papadimitriou and Yannakakis, 1991] and [Berman and Fujito, 1995], not approximable within B^ε for some $\varepsilon > 0$ [Alon, Feige, Wigderson, and Zuckerman, 1995], and is approximable within $(B+3)/5 + \varepsilon$ for every $\varepsilon > 0$ [Berman and Fürer, 1994] and [Berman and Fujito, 1995]. Not approximable within 1.0005 for $B = 3$, 1.0018 for $B = 4$ and 1.0030 for $B = 5$ [Berman and Karpinski, 1998]. For large values of B it is approximable within $O(B/\log\log B)$ [Halldórsson and Radhakrishnan, 1994a]. Approximable within $k/2 + \varepsilon$ for $k + 1$-claw free graphs [Halldórsson, 1995a]. The vertex weighted version is approximable within 3/2 for $B = 3$ [Hochbaum, 1983], and within $(B+2)/3$ for $B > 3$ [Halldórsson, 1995b]. Approximable within $O\left(|V|/(\log^{k-1}|V|)^2\right)$ for k-uniform hypergraphs [Hofmeister and Lefmann, 1998]. MAXIMUM INDEPENDENT SET OF k-GONS, the variation in which the number of pairwise independent k-gons (cycles of size k, $k \geq 3$) is to be maximized and where two k-gons are independent if any edge connecting vertices from different k-gons must belong to at least one of these k-gons, is not approximable within 4/3 for any $k \geq 3$. It is not in APX for any $k \geq g$ unless NP$\subset$QP [Verbitsky, 1995].

Garey and Johnson: GT20

GT24 ▶ MAXIMUM INDEPENDENT SEQUENCE

INSTANCE: Graph $G = (V, E)$.

SOLUTION: An independent sequence for G, i.e., a sequence $v_1,\ldots,v_m$ of independent vertices of G such that, for all $i < m$, a vertex $\bar{v}_i \in V$ exists which is adjacent to v_{i+1} but is not adjacent to any v_j for $j \leq i$.

MEASURE: Length of the sequence, i.e., m.

Good News: Approximable within $O(|V|/\log|V|)$ [Halldórsson, 1995b].

Bad News: Not approximable within $|V|^\varepsilon$ for some $\varepsilon > 0$. [Blundo, De Santis, and Vaccaro, 1994].

Comment: Transformation from MAXIMUM CLIQUE.

MAXIMUM INDUCED SUBGRAPH WITH PROPERTY Π   ◄ GT25

INSTANCE: Graph $G = (V,E)$. The property Π must be hereditary, i.e., every subgraph of G' satisfies Π whenever G' satisfies Π, and non-trivial, i.e., it is satisfied for infinitely many graphs and false for infinitely many graphs.

SOLUTION: A subset $V' \subseteq V$ such that the subgraph induced by V' has the property Π.

MEASURE: Cardinality of the induced subgraph, i.e., $|V'|$.

Good News: Approximable within $O(|V|/\log|V|)$ if Π is checkable in time 2^{n^c} for some constant c [Halldórsson, 1995b].

Bad News: Not approximable within $|V|^\varepsilon$ for some $\varepsilon > 0$ unless P=NP, if Π is false for some clique or independent set (e.g., planar, outerplanar, and bipartite). Not approximable within $2^{\log^{0.5-\varepsilon}|V|}$ for any $\varepsilon > 0$ unless NP⊂QP, if Π is a non-trivial hereditary graph property (e.g., comparability, permutation, and perfect) [Lund and Yannakakis, 1993].

Comment: Approximable within $O(|V|(\log\log|V|/\log|V|)^2)$ if Π is false for some clique or independent set, and approximable within $(B+1)/3$ where B is the degree of the graph. The positive results above are valid even for the vertex weighted problem [Halldórsson, 1995b]. The same problem with input directed graphs is not approximable within $2^{\log^{0.5-\varepsilon}|V|}$ for any $\varepsilon > 0$ unless NP⊂QP, if Π is a non-trivial hereditary digraph property (e.g., acyclic, transitive, symmetric, antisymmetric, tournament, degree-constrained, line digraph) [Lund and Yannakakis, 1993]. Admits a PTAS for planar graphs if Π is hereditary and determined by the connected components, i.e., G' satisfies Π whenever every connected component of G' satisfies Π [Nishizeki and Chiba, 1988]. Admits a PTAS for $K_{3,3}$-free or K_5-free graphs if Π is hereditary [Chen, 1996].

Garey and Johnson: GT21

MINIMUM VERTEX-DELETION SUBGRAPH WITH PROPERTY Π   ◄ GT26

INSTANCE: Directed or undirected graph $G = (V,E)$.

The compendium

Graph Theory

SOLUTION: A subset $V' \subseteq V$ such that the subgraph induced by $V - V'$ has the property Π.

MEASURE: Cardinality of the set of deleted vertices, i.e., $|V'|$.

Good News: Approximable within some constant for any hereditary property Π with a finite number of minimal forbidden subgraphs (e.g., transitive digraph, symmetric, antisymmetric, tournament, line graph, and interval) [Lund and Yannakakis, 1993]. Approximable within some constant for any property Π that can be expressed as a universal first order sentence over subsets of edges of the graph [Kolaitis and Thakur, 1995].

Bad News: APX-hard for any non-trivial hereditary property Π [Lund and Yannakakis, 1993].

Comment: Approximable within $O(\log |V|)$ if the subgraph has to be bipartite [Garg, Vazirani, and Yannakakis, 1994]. Approximable within some constant factor for any "matroidal" property Π and under simple and natural weighting schemes [Fujito, 1996].

GT27 ▶ Minimum Edge-Deletion Subgraph with Property Π

INSTANCE: Directed or undirected graph $G = (V, E)$.

SOLUTION: A subset $E' \subseteq E$ such that the subgraph $G = (V, E - E')$ has the property Π.

MEASURE: Cardinality of the set of deleted edges, i.e., $|E'|$.

Good News: Approximable within some constant for any property Π that can be expressed as a universal first order sentence over subsets of edges of the graph [Kolaitis and Thakur, 1995].

Comment: If Π is clique, then the problem is approximable within 4, even for the edge weighted version [Hochbaum, 1998].

GT28 ▶ Maximum Induced Connected Subgraph with Property Π

INSTANCE: Graph $G = (V, E)$.

SOLUTION: A subset $V' \subseteq V$ such that the subgraph induced by V' is connected and has the property Π.

MEASURE: Cardinality of the induced connected subgraph, i.e., $|V'|$.

Bad News: Not approximable within $|V|^{1-\varepsilon}$ for any $\varepsilon > 0$ if Π is a non-trivial hereditary graph property that is satisfied by all paths and is false for some complete bipartite graph (e.g., path, tree, planar, outerplanar, bipartite, chordal, interval) [Lund and Yannakakis, 1993].

Comment: NPOPB-complete when Π is either path or chordal [Berman and Schnitger, 1992]. NPOPB-complete and not approximable within $|V|^{1-\varepsilon}$ for any $\varepsilon > 0$ when Π is simple cycle [Kann, 1995].

Garey and Johnson: GT22 and GT23

Minimum Node-Del. Connected Subgraph with Prop. Π ◀ GT29

INSTANCE: Directed or undirected graph $G = (V,E)$.

SOLUTION: A subset $V' \subseteq V$ such that the subgraph induced by $V - V'$ is connected and has the property Π.

MEASURE: Cardinality of the set of deleted vertices, i.e., $|V'|$.

Bad News: Not approximable within $|V|^{1-\varepsilon}$ for any $\varepsilon > 0$ if Π is any non-trivial hereditary property determined by the blocks (e.g., planar, outerplanar and bipartite) [Yannakakis, 1979].

Maximum Degree-Bounded Connected Subgraph ◀ GT30

INSTANCE: Graph $G = (V,E)$, a weight function $w : E \to N$, and an integer $d \geq 2$.

SOLUTION: A subset $E' \subseteq E$ such that the subgraph $G' = (V,E')$ is connected and has no vertex with degree exceeding d.

MEASURE: The total weight of the subgraph, i.e., $\sum_{e \in E'} w(E)$.

Bad News: Not in APX for any d [Kann, —].

Comment: Transformation from LONGEST PATH.

Garey and Johnson: GT26

Maximum Planar Subgraph ◀ GT31

INSTANCE: Graph $G = (V,E)$.

SOLUTION: A subset $E' \subseteq E$ such that $G' = (V,E')$ is planar.

MEASURE: Size of the subset, i.e., $|E'|$.

Good News: Approximable within factor 2.5 [Călinescu, Fernandes, Finkler, and Karloff, 1996].

Bad News: APX-complete [Călinescu, Fernandes, Finkler, and Karloff, 1996].

Comment: Transformation from MINIMUM METRIC TRAVELING SALESPERSON with distances one and two. The dual problem is APX-hard. Variation in which the subgraph should be outerplanar is APX-complete and approximable within 1.5 [Călinescu, Fernandes, Finkler, and Karloff, 1996].

Garey and Johnson: GT27

Minimum Edge Deletion k-Partition ◀ GT32

INSTANCE: Graph $G = (V,E)$ and a weight function $w : E \to N$.

The compendium

Graph Theory

SOLUTION: An k-partition, i.e., a color assignment $c: V \to [1..k]$.

MEASURE: The weight of the monochromatic edges, i.e.,
$$\sum_{(v_1,v_2) \in E : c(v_1) = c(v_2)} w(v_1, v_2).$$

Good News: Approximable within $\log|V|$ for $k = 2$ [Garg, Vazirani, and Yannakakis, 1996] and within $\varepsilon|V|^2$ for $k = 3$ and any $\varepsilon > 0$ [Kann, Khanna, Lagergren, and Panconesi, 1997].

Bad News: APX-hard [Garg, Vazirani, and Yannakakis, 1996].

Comment: Not approximable within 1.058 for $k = 2$ [Håstad, 1997]. Not approximable within $O(|E|)$ for $k \geq 3$, even for graphs with $|E| = \Omega(|V|^{2-\varepsilon})$ for any $\varepsilon > 0$ [Kann, Khanna, Lagergren, and Panconesi, 1997]. Approximable within 9/4 for planar graphs and $k = 2$ [Goemans and Williamson, 1996].

GT33 ▶ MAXIMUM k-COLORABLE SUBGRAPH

INSTANCE: Graph $G = (V, E)$.

SOLUTION: A subset $E' \subseteq E$ such that the subgraph $G' = (V, E')$ is k-colorable, i.e., there is a coloring for G' of cardinality at most k.

MEASURE: Cardinality of the subgraph, i.e., $|E'|$.

Good News: Approximable within $1/(1 - 1/k + 2\ln k/k^2)$ [Frieze and Jerrum, 1997] and [Mahajan and Ramesh, 1995].

Bad News: APX-complete for $k \geq 2$ [Papadimitriou and Yannakakis, 1991].

Comment: Equivalent to MAXIMUM k-CUT for unweighted graphs. Admits a PTAS if $|E| = \Theta(|V|^2)$ and $k = o(|V|)$ [Arora, Karger, and Karpinski, 1995].

GT34 ▶ MAXIMUM SUBFOREST

INSTANCE: Tree $G = (V, E)$ and a set of trees H.

SOLUTION: A subset $E' \subseteq E$ such that the subgraph $G' = (V, E')$ does not contain any subtree isomorphic to a tree from H.

MEASURE: Cardinality of the subgraph, i.e., $|E'|$.

Good News: Admits a PTAS [Shamir and Tsur, 1998].

GT35 ▶ MAXIMUM EDGE SUBGRAPH

INSTANCE: Graph $G = (V, E)$, a weight function $w: E \to N$, and positive integer k.

SOLUTION: A subset $V' \subseteq V$ such that $|V'| = k$.

MEASURE: Sum of the weights of the edges in the subgraph induced by V'.

Good News: Approximable within $O(|V|^{1/3-\varepsilon})$ for some $\varepsilon > 0$ [Kortsarz and Peleg, 1993] and [Feige, Kortsarz, and Peleg, 1995].

Comment: Also called *Heaviest Subgraph*. Approximable within 2 if the weights satisfy the triangle inequality [Hassin, Rubinstein, and Tamir, 1994]. The unweighted problem admits a PTAS if $|E| = \Theta(|V|^2)$ and $k = \Theta(|V|)$ [Arora, Karger, and Karpinski, 1995].

MINIMUM EDGE 2-SPANNER ◀ GT36

INSTANCE: Connected graph $G = (V, E)$.

SOLUTION: A 2-spanner of G, i.e., a spanning subgraph G' of G such that, for any pair of vertices u and v, the shortest path between u and v in G' is at most twice the shortest path between u and v in G.

MEASURE: The number of edges in G'.

Good News: Approximable within $O(\log |V|)$ [Kortsarz and Peleg, 1992].

Comment: The variation in which the goal is to minimize the maximum degree in G' is approximable within $O(\sqrt{\log |V|} \Delta^{1/4})$ where Δ is the maximum degree in G [Kortsarz and Peleg, 1994].

MAXIMUM k-COLORABLE INDUCED SUBGRAPH ◀ GT37

INSTANCE: Graph $G = (V, E)$.

SOLUTION: A subset $V' \subseteq V$ such that the induced subgraph $G' = (V', E')$ is k-colorable, i.e., there is a coloring for G' of cardinality at most k.

MEASURE: Cardinality of the vertex set of the induced subgraph, i.e., $|V'|$.

Good News: As easy to approximate as MAXIMUM INDEPENDENT SET for $k \geq 1$ (finding k independent sets) [Halldórsson, 1994].

Bad News: As hard to approximate as MAXIMUM INDEPENDENT SET for $k \geq 1$ [Panconesi and Ranjan, 1993].

Comment: Equivalent to MAXIMUM INDEPENDENT SET for $k = 1$. Admits a PTAS for 'δ-near-planar' instances for any $\delta \geq 0$ [Hunt III et al., 1994a]. Variation in which the degree of G is bounded by a constant B is approximable within $(B/k + 1)/2$ [Halldórsson, 1995a] and APX-complete.

MINIMUM EQUIVALENT DIGRAPH ◀ GT38

INSTANCE: Directed graph $G = (V, E)$.

The compendium
Graph Theory

SOLUTION: A subset $E' \subseteq E$ such that, for every ordered pair of vertices $u, v \in V$, the graph $G' = (V, E')$ contains a directed path from u to v if and only if G does.

MEASURE: Cardinality of E', i.e., $|E'|$.

Good News: Approximable within 1.645 [Khuller, Raghavachari, and Young, 1995].

Bad News: APX-complete [Khuller, Raghavachari, and Young, 1995].

Comment: APX-complete even if restricted to strongly connected graphs with no cycle longer than 17.

Garey and Johnson: GT33

GT39 ▶ Minimum Interval Graph Completion

INSTANCE: Graph $G = (V, E)$.

SOLUTION: An interval graph $G' = (V, E')$ that contains G as a subgraph, i.e., $E \subseteq E'$. An interval graph is a graph whose vertices can be mapped to distinct intervals in the real line such that two vertices in the graph have an edge between them if and only if their corresponding intervals overlap.

MEASURE: The cardinality of the interval graph, i.e., $|E'|$.

Good News: Approximable within $O(\log |T|)$ [Rao and Richa, 1998].

Garey and Johnson: GT35

GT40 ▶ Minimum Chordal Graph Completion

INSTANCE: Graph $G = (V, E)$.

SOLUTION: A chordal graph completion, i.e., a superset E' containing E such that $G' = (V, E')$ is chordal, that is, for every simple cycle of more than 3 vertices in G', there is some edge in E' that is not involved in the cycle but that joins two vertices in the cycle.

MEASURE: The size of the completion, i.e., $|E' - E|$.

Good News: Approximable within $O(|E|^{1/4} \log^{3.5} |V|)$ [Klein, Agrawal, Ravi, and Rao, 1990].

Comment: Approximable within $O(\log^4 |V|)$ for graphs with bounded degree [Klein, Agrawal, Ravi, and Rao, 1990].

Garey and Johnson: OPEN4

GT41 ▶ Maximum Constrained Hamiltonian Circuit

INSTANCE: Graph $G = (V, E)$ and subset $S \subseteq E$ of the edges.

SOLUTION: A Hamiltonian circuit C in G, i.e., a circuit that visits every vertex in V once.

MEASURE: Cardinality of the edges in S that are used in the circuit C, i.e., $|S \cap C|$.

Bad News: Not approximable within $|E|^\varepsilon$ for some $\varepsilon > 0$ [Zuckerman, 1993].

Comment: Variation in which the graph is directed has the same bad news [Zuckerman, 1993].

Garey and Johnson: Similar to GT37 and GT38

Vertex Ordering

MINIMUM BANDWIDTH ◀ GT42

INSTANCE: Graph $G = (V, E)$.

SOLUTION: A linear ordering of V, i.e., a one-to-one function

$$f : V \to \{1, 2, \ldots, |V|\}.$$

MEASURE: The bandwidth of the ordering, i.e., $\max_{(u,v) \in E} |f(u) - f(v)|$.

Good News: Approximable within $O((\log|V|)^{4.5})$ [Feige, 1998]. Approximable within $O(\sqrt{|V|/m^*(G)} \log |V|)$ [Blum, Konjevod, Ravi, and Vempala, 1998].

Bad News: Not approximable within k for any positive integer k [Unger, 1998]. Not approximable with an absolute error guarantee of $|V|^{1-\varepsilon}$ for every $\varepsilon > 0$ [Karpinski and Wirtgen, 1997].

Comment: Approximable within 3 if every vertex has degree $\Theta(|V|)$ [Karpinski, Wirtgen, and Zelikovsky, 1997]. If G is a caterpillar, then it is approximable within $O(\log|V|)$ [Haralambides, Makedon, and Monien, 1991] and is not approximable within k for any positive integer k [Unger, 1998]. Approximable within 2 if G is asteroidal triple-free [Kloks, Kratsch, and Müller, 1996]. Not approximable within $1.332 - \varepsilon$ for any $\varepsilon > 0$ if G is a tree [Blache, Karpinski, and Wirtgen, 1998]. The related problem in which the measure is the sum of the squares of the dilations is approximable within $O((\log|V|)^{3/2})$ [Blum, Konjevod, Ravi, and Vempala, 1998].

Garey and Johnson: GT40

MINIMUM DIRECTED BANDWIDTH ◀ GT43

INSTANCE: Directed acyclic graph $G = (V, E)$.

SOLUTION: A linear ordering of V, i.e. a one-to-one function

$$f : V \to \{1, 2, \ldots, |V|\}$$

such that, for all $(u, v) \in E$, $f(u) < f(v)$.

The compendium

GRAPH THEORY

MEASURE: The bandwidth of the ordering, i.e. $\max_{(u,v)\in E} f(v) - f(u)$.

Comment: Approximable within 2 if every vertex has indegree $\Theta(|V|)$ [Karpinski, Wirtgen, and Zelikovsky, 1997].

Garey and Johnson: GT41

GT44 ▶ MINIMUM LINEAR ARRANGEMENT

INSTANCE: Graph $G = (V, E)$.

SOLUTION: A one-to-one function $f : V \to [1..|V|]$.

MEASURE: The total length of the edges in the linear arrangement, that is, $\sum_{\{u,b\}\in E} |f(u) - f(v)|$.

Good News: Approximable within $O(\log |V|)$ [Rao and Richa, 1998].

Comment: Admits a PTAS if $|E| = \Theta(|V|^2)$. Variation in which the n points are on the d-dimensional grid instead of in a line and the rectilinear metric is used also admits a PTAS if $|E| = \Theta(|V|^2)$ [Arora, Frieze, and Kaplan, 1996].

Garey and Johnson: GT42

GT45 ▶ MINIMUM CUT LINEAR ARRANGEMENT

INSTANCE: Graph $G = (V, E)$.

SOLUTION: A one-to-one function $f : V \to [1..|V|]$.

MEASURE: Maximum number of cut edges in any integer point, i.e.,

$$\max_{i \in [1..|V|]} |\{(u,v) \in E : f(u) \leq i < f(v)\}|.$$

Good News: Approximable within $O(\log |V| \log \log |V|)$ [Even, Naor, Rao, and Schieber, 1995].

Comment: Admits a PTAS if $|E| = \Theta(|V|^2)$ [Arora, Frieze, and Kaplan, 1996].

Garey and Johnson: GT44

Iso- and Other Morphisms

GT46 ▶ MAXIMUM COMMON SUBGRAPH

INSTANCE: Graphs $G_1 = (V_1, E_1)$ and $G_2 = (V_2, E_2)$.

SOLUTION: A common subgraph, i.e., subsets $E_1' \subseteq E_1$ and $E_2' \subseteq E_2$ such that the two subgraphs $G_1' = (V_1, E_1')$ and $G_2' = (V_2, E_2')$ are isomorphic.

MEASURE: Cardinality of the common subgraph, i.e., $|E'|$.

Bad News: APX-hard [Halldórsson, 1994].

Comment: Transformation from MAXIMUM CUT. Not approximable with an absolute error guarantee of $\min(|V_1|,|V_2|)^\varepsilon$ for some $\varepsilon > 0$ [Verbitsky, 1994]. Transformation to MAXIMUM CLIQUE with a quadratic vertex amplification [Kann, 1992b].

Variation in which the degree of the graphs G_1 and G_2 is bounded by the constant B is not harder to approximate than the bounded degree induced common subgraph problem and is approximable within $B+1$ [Kann, 1992a].

Garey and Johnson: GT49

MAXIMUM COMMON INDUCED SUBGRAPH ◀ GT47

INSTANCE: Graphs $G_1 = (V_1, E_1)$ and $G_2 = (V_2, E_2)$.

SOLUTION: A common induced subgraph, i.e., subsets $V_1' \subseteq V_1$ and $V_2' \subseteq V_2$ such that $|V_1'| = |V_2'|$, and the subgraph of G_1 induced by V_1' and the subgraph of G_2 induced by V_2' are isomorphic.

MEASURE: Cardinality of the common induced subgraph, i.e., $|V_1'|$.

Bad News: Not approximable within $|V|^\varepsilon$ for some $\varepsilon > 0$ [Kann, 1992a].

Comment: Transformations to and from MAXIMUM CLIQUE. Variation in which the degree of the graphs G_1 and G_2 is bounded by the constant B is APX-hard and is approximable within $B+1$.

If the induced subgraph is restricted to be connected the problem is NPOPB-complete and not approximable within $(|V_1|+|V_2|)^{1-\varepsilon}$ for any $\varepsilon > 0$ [Kann, 1992a].

MAXIMUM COMMON EMBEDDED SUB-TREE ◀ GT48

INSTANCE: Trees T_1 and T_2 with labels on the nodes.

SOLUTION: A common embedded sub-tree, i.e., a labeled tree T' that can be embedded into both T_1 and T_2. An embedding from T' to T is an injective function from the nodes of T' to the nodes of T that preserves labels and ancestorship. Note that since fathership does not need to be preserved, T' does not need to be an ordinary subtree.

MEASURE: Cardinality of the common embedded sub-tree, i.e., $|T'|$.

Good News: Approximable within $\log^2 n$, where n is the number of nodes in the trees [Halldórsson and Tanaka, 1996].

Bad News: APX-hard [Zhang and Jiang, 1994].

The compendium

GRAPH THEORY

Comment: Transformation from MAXIMUM k-SET PACKING. Variation in which the problem is to minimize the edit distance between the two trees is also APX-hard.

GT49 ▶ MINIMUM GRAPH TRANSFORMATION

INSTANCE: Graphs $G_1 = (V_1, E_1)$ and $G_2 = (V_2, E_2)$.

SOLUTION: A transformation that makes G_1 isomorphic to G_2. In the transformation a set of edges $E' \subseteq E_1$ is removed from E_1 and added to E_2.

MEASURE: Cardinality of the transformation set, i.e., $|E'|$.

Bad News: APX-hard [Lin, 1994].

Comment: Transformation from MAXIMUM 3-SATISFIABILITY.

Miscellaneous

GT50 ▶ LONGEST PATH WITH FORBIDDEN PAIRS

INSTANCE: Graph $G = (V, E)$ and a collection $C = \{(a_1, b_1), \ldots, (a_m, b_m)\}$ of pairs of vertices from V.

SOLUTION: A simple path in G containing at most one vertex of each pair in C.

MEASURE: Length of the path, i.e., the number of edges in the path.

Bad News: NPOPB-complete [Berman and Schnitger, 1992].

Comment: Transformation from LONGEST COMPUTATION.

Garey and Johnson: GT54

GT51 ▶ SHORTEST PATH WITH FORBIDDEN PAIRS

INSTANCE: Graph $G = (V, E)$, a collection $C = \{(a_1, b_1), \ldots, (a_m, b_m)\}$ of pairs of vertices from V, an initial vertex $s \in V$, and a final vertex $f \in V$.

SOLUTION: A simple path from s to f in G that contains at most one vertex from each pair in C.

MEASURE: Length of the path, i.e., the number of edges in the path.

Bad News: NPOPB-complete [Kann, 1994b].

Comment: Transformation from SHORTEST COMPUTATION.

GT52 ▶ MINIMUM POINT-TO-POINT CONNECTION

INSTANCE: Graph $G = (V, E)$, a weight function $w : E \rightarrow N$, a set $S = \{s_1, \ldots, s_p\}$ of sources, and a set $D = \{d_1, \ldots, d_p\}$ of destinations.

SOLUTION: A point-to-point connection, i.e., a subset $E' \subseteq E$ such that each source-destination pair is connected in E'.

MEASURE: The weight of the connection, i.e., $\sum_{e \in E'} w(e)$.

Good News: Approximable within $2 - \frac{1}{p}$ [Goemans and Williamson, 1995a].

MINIMUM METRIC DIMENSION ◀ GT53

INSTANCE: Graph $G = (V, E)$.

SOLUTION: A metric basis for G, i.e., a subset $V' \subseteq V$ such that for each pair $u, v \in V$ there is a $w \in V'$ such that the length of the shortest path from u to w is different from the length of the shortest path from v to w.

MEASURE: The cardinality of the metric basis, i.e., $|V'|$.

Good News: Approximable within factor $O(\log |V|)$ [Khuller, Raghavachari, and Rosenfeld, 1994].

Garey and Johnson: GT61

MINIMUM TREE WIDTH ◀ GT54

INSTANCE: Graph $G = (V, E)$.

SOLUTION: A tree decomposition, i.e., a pair $(\{X_i : i \in I\}, T)$ where $T = (I, F)$ is a tree and $\{X_i\}$ is a collection of subsets of V, such that

1. $\bigcup_{i \in I} X_i = V$,
2. for any $(v, w) \in E$, there exists an $i \in I$ with $u, v \in X_i$,
3. for any $v \in V$, the set $\{i \in I : v \in X_i\}$ forms a connected subtree of T.

MEASURE: The tree width of the tree decomposition, i.e., $\max_{i \in I} |X_i| - 1$.

Good News: Approximable within $O(\log |V|)$ [Bodlaender et al., 1995].

Bad News: There is no polynomial-time algorithm with an absolute error guarantee of $|V|^{1-\varepsilon}$ for any $\varepsilon > 0$ [Bodlaender et al., 1995].

Comment: MINIMUM PATH WIDTH, the variation in which T is a path, is approximable within $O(\log^2 |V|)$ and has the same negative bound. Similar problems with the same positive and negative results are MINIMUM ELIMINATION TREE HEIGHT and MINIMUM FRONT SIZE [Bodlaender et al., 1995].

MINIMUM GRAPH INFERENCE ◀ GT55

INSTANCE: Class C of undirected edge-colored graphs, string x of colors.

SOLUTION: A graph $G \in |C|$ and a simple path in G such that the string of colors traversed in the path is equal to x.

The compendium
Graph Theory

MEASURE: Cardinality of the edge set of G.

Bad News: APX-hard and not approximable within factor $2-\varepsilon$ for any $\varepsilon > 0$ [Maruyama and Miyano, 1995].

Comment: The same negative results are valid also if all graphs in C are caterpillars.

Network Design

Spanning Trees

MINIMUM k-SPANNING TREE ◀ ND1

INSTANCE: Graph $G = (V, E)$, an integer $k \leq n$, and a weight function $w: E \to N$.

SOLUTION: A k-spanning tree, i.e., a subtree T of G of at least k nodes.

MEASURE: The weight of the tree, i.e., $\sum_{e \in T} w(e)$.

Good News: Approximable within 3 [Garg, 1996].

Bad News: APX-complete [Kann, —].

Comment: The restriction to points in the Euclidean plane admits a PTAS [Arora, 1996]. The analogous diameter and communication-cost k-spanning tree problems are not in APX [Ravi, Sundaram, Marathe, Rosenkrantz, and Ravi, 1994].

MINIMUM DEGREE SPANNING TREE ◀ ND2

INSTANCE: Graph $G = (V, E)$.

SOLUTION: A spanning tree for G.

MEASURE: The maximum degree of the spanning graph.

Good News: Approximable with an absolute error guarantee of 1 [Fürer and Raghavachari, 1994].

Bad News: Not approximable within $3/2 - \varepsilon$ for any $\varepsilon > 0$ [Garey and Johnson, 1979].

Comment: The problem is APX-intermediate unless the polynomial-time hierarchy collapses [Crescenzi, Kann, Silvestri, and Trevisan, 1999]. The generalization of the problem to Steiner trees is also approximable with an absolute error guarantee of 1 [Fürer and Raghavachari, 1994].

Garey and Johnson: ND1

MINIMUM GEOMETRIC 3-DEGREE SPANNING TREE ◀ ND3

INSTANCE: Set $P \subseteq Z \times Z$ of points in the plane.

SOLUTION: A spanning tree T for P in which no vertex has degree larger than 3.

MEASURE: The total weight of the spanning tree, i.e., $\sum_{(u,v) \in T} d(u,v)$, where $d(u,v)$ is the Euclidean distance between u and v.

Good News: Admits a PTAS [Arora, 1996].

Comment: The 4-degree spanning tree problem also admits a PTAS, but the NP-hardness of the problem is open [Arora, 1996]. The 5-degree problem is polynomial-time solvable. In d-dimensional Euclidean space for $d \geq 3$ the 3-degree spanning tree problem is approximable within 5/3. [Khuller, Raghavachari, and Young, 1996a].

The generalization to a graph with metric weight function and for each vertex $v \in V$ a degree bound $d(v)$ is called MINIMUM BOUNDED DEGREE SPANNING TREE and is approximable within $2 - \min_{v \in V}\{(d(v) - 2)/(d_T(v) - 2) : d_T(v) > 2\}$, where $d_T(v)$ is the initial degree of v [Fekete et al., 1996].

ND4 ▶ MAXIMUM LEAF SPANNING TREE

INSTANCE: Graph $G = (V, E)$.

SOLUTION: A spanning tree for G.

MEASURE: The number of leaves of the spanning tree.

Good News: Approximable within 3 [Lu and Ravi, 1992].

Bad News: APX-complete [Galbiati, Maffioli, and Morzenti, 1994].

Comment: Other problems which aim at finding spanning trees that maximize a single objective function have been considered. In particular, the problems of finding a spanning tree that has maximum diameter, or maximum height with respect to a specified root, are not in APX, the problems of finding a spanning tree that has maximum sum of the distances between all pairs of vertices, or maximum sum of the distances from a specified root, are not in PTAS [Galbiati, Maffioli, and Morzenti, 1995]. The problem of finding a spanning tree that maximizes the number of paths that connect pairs of vertices and pass through a common arc is approximable within 1.072 [Chlebikova, 1996].

Garey and Johnson: ND2

ND5 ▶ MAXIMUM MINIMUM METRIC k-SPANNING TREE

INSTANCE: Graph $G = (V, E)$, length $l(e) \in N$ for each $e \in E$ satisfying the triangle inequality.

SOLUTION: A subset $V' \subseteq V$ such that $|V'| = k$.

MEASURE: Cost of the minimum spanning tree of the subgraph induced by V'.

Good News: Approximable within factor 4 [Halldórsson, Iwano, Katoh, and Tokuyama, 1995].

Bad News: APX-complete and not approximable within $2-\varepsilon$ for any $\varepsilon > 0$ [Halldórsson, Iwano, Katoh, and Tokuyama, 1995].

Comment: Also called *Maximum Remote Minimum Spanning Tree.* Transformation from MAXIMUM INDEPENDENT SET. As hard to approximate as MAXIMUM INDEPENDENT SET for non-metric graphs. Approximable within 2.252 in the Euclidean metric, but not known to be NP-complete. MAXIMUM MINIMUM k-STEINER TREE is approximable within 3 and not approximable within $4/3-\varepsilon$, but approximable within 2.16 in the Euclidean metric. MAXIMUM MINIMUM METRIC k-TSP is approximable within 3 and not approximable within $2-\varepsilon$.

MINIMUM DIAMETER SPANNING SUBGRAPH ◀ ND6

INSTANCE: Graph $G = (V, E)$, weight $w(e) \in Z^+$ and length $l(e) \in N$ for each $e \in E$, positive integer B.

SOLUTION: A spanning subgraph $E' \subseteq E$ for G such that the sum of the weights of the edges in E' does not exceed B.

MEASURE: The diameter of the spanning subgraph.

Bad News: Not approximable within $2-\varepsilon$ for any $\varepsilon > 0$ [Plesník, 1981].

Comment: Not approximable within $3/2-\varepsilon$ for planar graphs. Not approximable within $5/4-\varepsilon$ if $l(e) = 1$ for every edge e [Plesník, 1981]. For results on minimum Steiner trees of bounded diameter, see MINIMUM STEINER TREE.

Garey and Johnson: Similar to ND4

MINIMUM COMMUNICATION COST SPANNING TREE ◀ ND7

INSTANCE: Complete graph $G = (V, E)$, weight $w(e) \in N$ for each $e \in E$, requirement $r(\{u, v\}) \in N$ for each pair of vertices from V.

SOLUTION: A spanning tree for G.

MEASURE: The weighted sum over all pairs of vertices of the cost of the path between the pair in T, i.e., $\sum_{u,v \in V} W(u,v) \cdot r(\{u,v\})$, where $W(u,v)$ denotes the sum of the weights of the edges on the path joining u and v in T.

Good News: Approximable within $O(\log^2 |V|)$ if the weight function satisfies the triangle inequality [Peleg and Reshef, 1998].

Comment: MINIMUM ROUTING COST SPANNING TREE, the variation in which $r(\{u,v\}) = 1$ for all u, v, admits a PTAS [Wu et al., 1998].

Garey and Johnson: ND7

MINIMUM STEINER TREE ◀ ND8

INSTANCE: Complete graph $G = (V, E)$, a metric given by edge weights $s : E \to N$ and a subset $S \subseteq V$ of required vertices.

The compendium
NETWORK DESIGN

The compendium

NETWORK DESIGN

SOLUTION: A Steiner tree, i.e., a subtree of G that includes all the vertices in S.

MEASURE: The sum of the weights of the edges in the subtree.

Good News: Approximable within 1.644 [Karpinski and Zelikovsky, 1997a].

Bad News: APX-complete [Bern and Plassmann, 1989].

Comment: Variation in which the weights are only 1 or 2 is still APX-complete, but approximable within 4/3 [Bern and Plassmann, 1989]. When all weights lie in an interval $[\alpha, \alpha(1+1/k)]$ the problem is approximable within $1 + 1/ek + O(1/k^2)$ [Halldórsson, Ueno, Nakao, and Kajitani, 1992]. On directed graphs the problem is approximable within $O(|S|^\varepsilon)$ for any $\varepsilon > 0$ [Charikar et al., 1998]. The variation in which the diameter of the tree is bounded by a constant d the problem is not approximable within $(1-\varepsilon)\ln|V|$ for any $\varepsilon > 0$, unless NP$\subset$ DTIME($n^{\log\log n}$), but approximable within $\log|V|$ for $d \in \{4,5\}$ [Bar-Ilan, Kortsarz, and Peleg, 1996] and within $d\log|V|$ for any constant $d \geq 6$ [Kortsarz and Peleg, 1997]. Admits a PTAS if every element of S is adjacent to $\Theta(|V-S|)$ elements from $V-S$ [Karpinski and Zelikovsky, 1997b] and [Karpinski, 1997]. A prize-collecting variation in which a penalty is associated with each vertex and the goal is to minimize the cost of the tree and the vertices in S not in the tree is approximable within $2 - 1/(|V|-1)$ [Goemans and Williamson, 1995a]. The variation, called MINIMUM k-STEINER TREE, in which an integer $k \leq |S|$ is given in input and at least k vertices of S must be included in the subtree is approximable within 17 [Blum, Ravi, and Vempala, 1996]. Variation in which there are groups of required vertices and each group must be touched by the Steiner tree is approximable within $O(\log^3 |V| \log g)$, where g is the number of groups [Garg, Konjevod, and Ravi, 1998]; when the number of groups is unbounded the problem is harder than MINIMUM DOMINATING SET to approximate, even if all edge weights are 1 [Ihler, 1992]. The constrained variation in which the input is extended with a positive integer k and a subset T of E, and the problem is to find the Steiner tree of weight at most k that contains the largest number of edges from T, is not approximable within $|E|^\varepsilon$ for some $\varepsilon > 0$ [Zuckerman, 1993]. If the solution is allowed to be a forest with at most q trees, for a given constant q, the problem is approximable within $2(1 - 1/(|S|-q+1))$ [Ravi, 1994a]. If the topology of the Steiner tree is given as input, the problem admits a PTAS [Jiang, Lawler, and Wang, 1994]. Finally, if before computing the Steiner tree the graph can be modified by reducing the weight of the edges within a given budget the problem of looking for the best reduction strategy is APX-hard [Drangmeister, Krumke, Marathe, Noltemeier, and Ravi, 1998].

Garey and Johnson: ND12

ND9 ▶ MINIMUM GEOMETRIC STEINER TREE

INSTANCE: Set $P \subseteq Z \times Z$ of points in the plane.

SOLUTION: A finite set of Steiner points, i.e., $Q \subseteq Z \times Z$.

MEASURE: The total weight of the minimum spanning tree for the vertex set $P \cup Q$, where the weight of an edge $((x_1,y_1),(x_2,y_2))$ is the discretized Euclidean length
$$\left\lceil \sqrt{(x_1-x_2)^2 + (y_1-y_2)^2} \right\rceil.$$

Good News: Admits a PTAS [Arora, 1996].

Comment: Admits a PTAS for any geometric space of constant dimension d, for example in the rectilinear metric [Arora, 1997]. MINIMUM STEINER TREES WITH OBSTACLES, the problem where a polygonally bounded region R is given in the input, and the Steiner tree has to lie inside R, admits a FPTAS under some restrictions of the input [Provan, 1988].

Garey and Johnson: ND13

MINIMUM GENERALIZED STEINER NETWORK ◀ ND10

INSTANCE: Graph $G = (V,E)$, a weight function $w : E \to N$, a capacity function $c : E \to N$, and a requirement function $r : V \times V \to N$.

SOLUTION: A Steiner network over G that satisfies all the requirements and obeys all the capacities, i.e., a function $f : E \to N$ such that, for each edge e, $f(e) \leq c(e)$ and, for any pair of nodes i and j, the number of edge disjoint paths between i and j is at least $r(i,j)$ where, for each edge e, $f(e)$ copies of e are available.

MEASURE: The cost of the network, i.e., $\sum_{e \in E} w(e) f(e)$.

Good News: Approximable within $2 \cdot \mathcal{H}(R)$ where R is the maximum requirement and, for any n, $\mathcal{H}(n) = \sum_{i=1}^{n} \frac{1}{i}$ [Goemans, Goldberg, Plotkin, Shmoys, Tardos, and Williamson, 1994].

Comment: Also called *Minimum Survivable Network*. Approximable within 2 when all the requirements are equal [Khuller and Vishkin, 1994]. If the requirements are equal and the graph is directed the problem is approximable within $O(|V|^{2/3} \log^{1/3} |V|)$ [Charikar et al., 1998]. The variation in which there are no capacity constraints on the edges is approximable within $2 \cdot \mathcal{H}(|V|)$ [Aggarwal and Garg, 1994]. This problem belongs to the class of problems of finding a minimum-cost subgraph such that the number of edges crossing each cut is at least a specified requirement, which is a function f of the cut. If f is symmetric and maximal then any problem in this class is approximable within $2R$, where R is the maximum value of f over all cuts [Williamson, Goemans, Mihail, and Vazirani, 1995].

MINIMUM ROUTING TREE CONGESTION ◀ ND11

INSTANCE: Graph $G = (V,E)$ and a weight function $w : E \to N$.

The compendium

NETWORK DESIGN

SOLUTION: A routing tree T for G, i.e., a tree T in which each internal vertex has degree 3 and the leaves correspond to vertices of G.

MEASURE: The congestion of the routing tree, i.e., the maximum, for any edge e, of

$$\sum_{(u,v)\in E, u\in S, v\notin S} w(u,v)$$

where S is one of the two connected components obtained by deleting e from T.

Good News: Approximable within factor $\log|V|$ [Khuller, Raghavachari, and Young, 1993].

Bad News: Not in APX [Seymour and Thomas, 1994].

Comment: The algorithm extends to the case when the routing tree is allowed to have vertices of higher degree. If G is planar [Seymour and Thomas, 1994], or if T is required to be a spanning tree and G is complete [Khuller, Raghavachari, and Young, 1993], the problem is solvable in polynomial time.

ND12 ▶ MAXIMUM MINIMUM SPANNING TREE DELETING k EDGES

INSTANCE: Graph $G = (V,E)$, a weight function $w : E \to N$.

SOLUTION: Subset $E' \subseteq E$ of k edges and a minimum spanning tree T in the graph $(V, E - E')$.

MEASURE: Weight of T.

Good News: Approximable within $\log k$ [Frederickson and Solis-Oba, 1996].

Comment: The good news is valid even if a minimum λ-edge connected subgraph of G is sought, for every $\lambda \geq 1$ [Frederickson and Solis-Oba, 1996].

ND13 ▶ MINIMUM UPGRADING SPANNING TREE

INSTANCE: Graph $G = (V,E)$, three edge weight functions $d_2(e) \leq d_1(e) \leq d_0(e)$ (for each $e \in E$), where $d_i(e)$ denotes the weight of edge e if i of its endpoints are "upgraded", vertex upgrade costs $c(v)$ (for each $v \in V$), a threshold value D for the weight of a minimum spanning tree.

SOLUTION: An upgrading set $W \subseteq V$ of vertices such that the weight of a minimum spanning tree in G with respect to edge weights given by d_W is bounded by D. Here, d_W denotes the edge weight function resulting from the upgrade of the vertices in W, i.e., $d_W(u,v) = d_i(u,v)$ where $|W \cap \{u,v\}| = i$.

MEASURE: The cost of the upgrading set, i.e., $c(W) = \sum_{v \in W} c(v)$.

Good News: Approximable within $O(\log|V|)$ if the difference of the largest edge weight $D_0 = \max\{d_0(e) : e \in E\}$ and the smallest edge weight $D_2 = \min\{d_2(e) : e \in E\}$ is bounded by a polynomial in $|V|$ [Krumke et al., 1997].

Bad News: At least as hard to approximate as MINIMUM DOMINATING SET [Krumke et al., 1997].

Comment: Variation in which the upgrading set must be chosen such that the upgraded graph contains a spanning tree in which no edge has weight greater than D is approximable within $2\ln|V|$. In this case no additional assumptions about the edge weights are necessary. This variation of the problem is also at least as hard to approximate as MINIMUM DOMINATING SET [Krumke et al., 1997].

Cuts and Connectivity

MAXIMUM CUT ◄ ND14

INSTANCE: Graph $G = (V,E)$.

SOLUTION: A partition of V into disjoint sets V_1 and V_2.

MEASURE: The cardinality of the cut, i.e., the number of edges with one end point in V_1 and one endpoint in V_2.

Good News: Approximable within factor 1.1383 [Goemans and Williamson, 1995b].

Bad News: APX-complete [Papadimitriou and Yannakakis, 1991].

Comment: Not approximable within 1.0624 [Håstad, 1997]. Admits a PTAS if $|E| = \Theta(|V|^2)$ [Arora, Karger, and Karpinski, 1995]. Variation in which the degree of G is bounded by a constant B for $B \geq 3$ is still APX-complete [Papadimitriou and Yannakakis, 1991] and [Alimonti and Kann, 1997]. Variation in which each edge has a nonnegative weight and the objective is to maximize the total weight of the cut is as hard to approximate as the original problem [Crescenzi, Silvestri, and Trevisan, 1996] and admits a PTAS for dense instances [Fernandez de la Vega and Karpinski, 1998] and for metric weights [Fernandez de la Vega and Kenyon, 1998]. Variation in which some pairs of vertices are restricted to be on the same side (or on different sides) is still approximable within 1.138 [Goemans and Williamson, 1995b].

MAXIMUM BISECTION, the weighted problem with the additional constraint that the partition must cut the graph into halves of the same size, is approximable within 1.54 [Frieze and Jerrum, 1997] and [Mahajan and Ramesh, 1995].

MINIMUM BISECTION, the problem where the partition must cut the graph into halves of the same size and the number of edges between them is to be minimized, admits a PTAS if the degree of each vertex is $\Theta(|V|)$ [Arora, Karger, and Karpinski, 1995].

The compendium
NETWORK DESIGN

Garey and Johnson: ND16

ND15 ▶ MINIMUM CROSSING NUMBER

INSTANCE: Directed graph $G = (V,A)$.

SOLUTION: An embedding of G in the plane.

MEASURE: The number of pairs of edges crossing one another.

Comment: Variation in which $G = (V_1, V_2, E)$ is a bipartite graph and the objective is to minimize the bipartite crossing number, that is the number of edge crossings in the embedding where the vertices in V_1 and V_2 must lie on two parallel lines and each edge corresponds to a straight line segment, is approximable within 3 [Eades and Wormald, 1994].

Garey and Johnson: OPEN3

ND16 ▶ MAXIMUM DIRECTED CUT

INSTANCE: Directed graph $G = (V,A)$.

SOLUTION: A partition of V into disjoint sets V_1 and V_2.

MEASURE: The cardinality of the cut, i.e., the number of arcs with one end point in V_1 and one endpoint in V_2.

Good News: Approximable within 1.165 [Feige and Goemans, 1995].

Bad News: APX-complete [Papadimitriou and Yannakakis, 1991].

Comment: Not approximable within 1.083 [Håstad, 1997]. Admits a PTAS if $|E| = \Theta(|V|^2)$ [Arora, Karger, and Karpinski, 1995]. Variation in which each edge has a nonnegative weight and the objective is to maximize the total weight of the cut is as hard to approximate as the original problem [Crescenzi, Silvestri, and Trevisan, 1996].

ND17 ▶ MAXIMUM k-CUT

INSTANCE: Graph $G = (V,E)$, a weight function $w : E \to N$, and an integer $k \in [2..|V|]$.

SOLUTION: A partition of V into k disjoint sets $F = \{C_1, C_2, \ldots, C_k\}$.

MEASURE: The sum of the weight of the edges between the disjoint sets, i.e.,

$$\sum_{i=1}^{k-1} \sum_{j=i+1}^{k} \sum_{\substack{v_1 \in C_i \\ v_2 \in C_j}} w(\{v_1, v_2\}).$$

Good News: Approximable within $1/(1 - 1/k + 2\ln k/k^2)$ [Frieze and Jerrum, 1997] and [Mahajan and Ramesh, 1995].

Bad News: APX-complete.

Comment: The unweighted version of the problem is equivalent to MAXIMUM k-COLORABLE SUBGRAPH. Approximable within 1.21 for $k = 3$, 1.18 for $k = 4$, and 1.15 for $k = 5$ [Frieze and Jerrum, 1997]. Not approximable within $1 + 1/(34k)$ [Kann, Khanna, Lagergren, and Panconesi, 1997]. MAXIMUM k-SECTION, the problem with the additional constraint that the partition must cut the graph into sets of equal size, is approximable within $1/(1 - 1/k + ck^3)$ for some constant $c > 0$ [Andersson, 1999].

The constrained variation in which the input is extended with a positive integer W, a vertex $v_0 \in V$ and a subset S of V, and the problem is to find the 2-cut of weight at least W with the largest number of vertices from S on the same side as v_0, is not approximable within $|V|^\varepsilon$ for some $\varepsilon > 0$ [Zuckerman, 1993].

MINIMUM NETWORK INHIBITION ON PLANAR GRAPHS ◀ ND18

INSTANCE: Planar graph $G = (V,E)$, capacity function $c : E \to N$, destruction cost function $d : E \to N$, and budget B.

SOLUTION: An attack strategy to the network, i.e., a function $\alpha : E \to [0,1]$ such that $\sum_{e \in E} \alpha(e) d(e) \leq B$.

MEASURE: The capability left in the damaged network, i.e., the minimum cut in G with capacity c' defined as $c'(e) = \alpha(e) c(e)$.

Good News: Admits an FPTAS [Phillips, 1993].

MINIMUM k-CUT ◀ ND19

INSTANCE: Graph $G = (V,E)$, a weight function $w : E \to N$, and an integer $k \in [2..|V|]$.

SOLUTION: A partition of V into k disjoint sets $F = \{C_1, C_2, \ldots, C_k\}$.

MEASURE: The sum of the weight of the edges between the disjoint sets, i.e.,

$$\sum_{i=1}^{k-1} \sum_{j=i+1}^{k} \sum_{\substack{v_1 \in C_i \\ v_2 \in C_j}} w(\{v_1, v_2\}).$$

Good News: Approximable within $2 - \frac{2}{k}$ [Saran and Vazirani, 1991].

Comment: Solvable in polynomial time $O(|V|^{k^2})$ for fixed k [Goldschmidt and Hochbaum, 1988].

If the sets in the partition are restricted to be of equal size, the problem is approximable within $|V| \cdot (k-1)/k$ [Saran and Vazirani, 1991]. If the sets in the partition are restricted to be of specified sizes and the weight function

The compendium

NETWORK DESIGN

satisfies the triangle inequality, the problem is approximable within 3 for any fixed k [Guttmann-Beck and Hassin, 1999].

The unweighted problem admits a PTAS if every vertex has degree $\Theta(|V|)$ [Arora, Karger, and Karpinski, 1995].

ND20 ▶ MINIMUM VERTEX k-CUT

INSTANCE: Graph $G = (V,E)$, a set $S = \{s_1, t_1, \ldots, s_k, t_k\}$ of special vertices, and a weight function $w : V - S \to N$, and an integer k.

SOLUTION: A vertex k-cut, i.e., a subset $C \subseteq V - S$ of vertices such that their deletion from G disconnects each s_i from t_i for $1 \leq i \leq k$.

MEASURE: The sum of the weight of the vertices in the cut, i.e., $\sum_{v \in C} w(v)$.

Good News: Approximable within factor $O(\log |V|)$ [Garg, Vazirani, and Yannakakis, 1994].

ND21 ▶ MINIMUM MULTIWAY CUT

INSTANCE: A graph $G = (V,E)$, a set $S \subseteq V$ of terminals, and a weight function $w : E \to N$.

SOLUTION: A multiway cut, i.e., a set $E' \subseteq E$ such that the removal of E' from E disconnects each terminal from all the others.

MEASURE: The weight of the cut, i.e., $\sum_{e \in E'} w(e)$.

Good News: Approximable within $3/2 - 1/|S|$ [Călinescu, Karloff, and Rabani, 1998].

Bad News: APX-complete [Dahlhaus, Johnson, Papadimitriou, Seymour, and Yannakakis, 1994].

Comment: Admits a PTAS if every vertex has degree $\Theta(|V|)$ [Arora, Karger, and Karpinski, 1995]. It remains APX-complete even if $|S| \geq 3$ is fixed. For $|S| = 4$ and $|S| = 8$ it is approximable within $4/3$ and $12/7$, respectively. In the case of directed graphs the problem is approximable within 2 [Naor and Zosin, 1997] and is APX-hard [Garg, Vazirani, and Yannakakis, 1994]. The vertex deletion variation is approximable within $2 - 2/|S|$ and is APX-complete [Garg, Vazirani, and Yannakakis, 1994].

ND22 ▶ MINIMUM MULTI-CUT

INSTANCE: A graph $G = (V,E)$, a set $S \subseteq V \times V$ of source-terminal pairs, and a weight function $w : E \to N$.

SOLUTION: A multi-cut, i.e., a set $E' \subseteq E$ such that the removal of E' from E disconnects s_i from t_i for every pair $(s_i, t_i) \in S$.

MEASURE: The weight of the cut, i.e., $\sum_{e \in E'} w(e)$.

Good News: Approximable within $O(\log |S|)$ [Garg, Vazirani, and Yannakakis, 1996]

Bad News: APX-hard.

Comment: Generalization of MINIMUM MULTIWAY CUT. When the graph is a tree the problem is approximable within 2 and is APX-complete even for trees of height one and unit edge weight [Garg, Vazirani, and Yannakakis, 1997]. The variation in which vertices instead of edges should be deleted is also approximable within $O(\log |S|)$ [Garg, Vazirani, and Yannakakis, 1994]. The variation in which the input specifies a set of demand edges and a given node v^* and the solution must be a feasible v^*-cut, i.e., the demand edges must either be in the cut or have both endpoints on the opposite part of v^* is approximable within 2.8 [Yu and Cheriyan, 1995].

MINIMUM RATIO-CUT ◀ ND23

INSTANCE: Graph $G = (V, E)$, a capacity function $c : E \to N$, and k commodities, i.e., k pairs $(s_i, t_i) \in V^2$ and a demand d_i for each pair.

SOLUTION: A cut, i.e., a partition of V into two disjoint sets V_1 and V_2.

MEASURE: The capacity of the cut divided by the demand across the cut, i.e.,

$$\sum_{\substack{v_1 \in V_1, v_2 \in V_2 \\ (v_1, v_2) \in E}} c(v_1, v_2) \Big/ \sum_{i : |\{s_i, t_i\} \cap V_1| = 1} d_i.$$

Good News: Approximable within $O(\log k)$ [Aumann and Rabaini, 1998].

Comment: Also called *Sparsest Cut*. In the uniform-demand case the problem is in APX [Klein, Plotkin, and Rao, 1993].

MINIMUM b-BALANCED CUT ◀ ND24

INSTANCE: Graph $G = (V, E)$, a vertex-weight function $w : V \to N$, an edge-cost function $c : E \to N$, and a rational b, $0 < b \leq 1/2$.

SOLUTION: A cut C, i.e., a subset $C \subseteq V$, such that

$$\min\{w(C), w(V - C)\} \geq b \cdot w(V),$$

where $w(V')$ denotes the sum of the weights of the vertices in V'.

MEASURE: The weight of the cut, i.e.,

$$\sum_{e \in \delta(C)} c(e)$$

where $\delta(C) = \{e = \{v_1, v_2\} : e \in E, v_1 \in C, v_2 \in V - C\}$.

The compendium

NETWORK DESIGN

Bad News: Not approximable within $1+|V|^{2-\varepsilon}/m^*$ for any $\varepsilon > 0$ [Bui and Jones, 1992].

Comment: Also called *Minimum b-Edge Separator*. There is a polynomial algorithm that finds a b-balanced cut within an $O(\log|V|)$ factor of the optimal $b+\varepsilon$-balanced cut for $b \leq 1/3$ [Shmoys, 1997]. Approximable within 2 for planar graphs for $b \leq 1/3$ if the vertex weights are polynomially bounded [Garg, Saran, and Vazirani, 1994]. Not approximable within $1+|V|^{1/2-\varepsilon}/m^*$ for graphs of maximum degree 3 [Bui and Jones, 1992]. The unweighted problem admits a PTAS if every vertex has degree $\Theta(|V|)$ for $b \leq 1/3$ [Arora, Karger, and Karpinski, 1995]. MINIMUM k-MULTIWAY SEPARATOR, the variation in which the removal of the cut edges partitions the graph into at most k parts, where the sum of the vertex weights in each part is at most $2\sum_{v \in V} w(v)/k$, is approximable within $(2+o(1))\ln|V|$ [Even, Naor, Rao, and Schieber, 1997].

ND25 ▶ MINIMUM b-VERTEX SEPARATOR

INSTANCE: Graph $G = (V,E)$ and a rational b, $0 < b \leq 1/2$.

SOLUTION: A partition of V into three disjoint sets A, B, and C, such that $\max\{|A|,|B|\} \leq b \cdot |V|$ and no edge has one endpoint in A and one in B.

MEASURE: The size of the separator, i.e., $|C|$.

Bad News: Not approximable within $1+|V|^{1/2-\varepsilon}/m^*$ for any $\varepsilon > 0$, even for graphs of maximum degree 3 [Bui and Jones, 1992].

Comment: For planar graphs a separator of size $O(\sqrt{|V|})$ can be found in polynomial time [Lipton and Tarjan, 1979]. An $f(|V|)$-approximation algorithm is also a $f(16|V|^5)$ algorithm for MINIMUM b-BALANCED CUT [Bui and Jones, 1992].

ND26 ▶ MINIMUM QUOTIENT CUT

INSTANCE: Graph $G = (V,E)$, a vertex-weight function $w : V \to N$, and an edge-cost function $c : E \to N$.

SOLUTION: A cut C, i.e., a subset $C \subseteq V$.

MEASURE: The quotient of the cut, i.e.,

$$\frac{c(C)}{\min\{w(C), w(V-C)\}}$$

where $c(C)$ denotes the sum of the costs of the edges (u,v) such that either $u \in C$ and $v \notin C$ or $u \notin C$ and $v \in C$ and, for any subset $V' \subseteq V$, $w(V')$ denotes the sum of the weights of the vertices in V'.

Good News: Approximable within $O(\log|V|)$ [Leighton and Rao, 1988].

Comment: Also called *Minimum Flux Cut*. Admits a PTAS for planar graphs [Park and Phillips, 1993]. The generalization to hypergraphs, also called *Minimum Net Expansion*, is approximable within $O(\log |E|)$ in the uniform vertex-weight case [Makedon and Tragoudas, 1990]. Other approximation algorithms for hypergraph partitioning problems are contained in [Makedon and Tragoudas, 1990].

Minimum k-Vertex Connected Subgraph ◀ ND27

INSTANCE: Graph $G = (V, E)$. k is a constant, $k \geq 2$.

SOLUTION: A k-vertex connected spanning subgraph $G' = (V, E')$ for G, i.e. a spanning subgraph that cannot be disconnected by removing fewer than k vertices.

MEASURE: The cardinality of the spanning subgraph, i.e., $|E'|$.

Good News: Approximable within $1 + 1/k$ [Garg, Santosh, and Singla, 1993] and [Cheriyan and Thurimella, 1996].

Comment: On directed graphs the problem is approximable within 1.61 for $k = 1$ [Khuller, Raghavachari, and Young, 1996b], and within $1 + 1/k$ for $k \geq 2$ [Cheriyan and Thurimella, 1996]. Variation in which each edge has a nonnegative weight and the objective is to minimize the total weight of the spanning subgraph is approximable within $2 + 1/|V|$ for $k = 2$ [Khuller and Raghavachari, 1996] and within $2\sum_{i=1}^{k} \frac{1}{i}$ for $k > 2$ [Ravi and Williamson, 1997]. If the weight function satisfies the triangle inequality, the problem is approximable within $3/2$ for $k = 2$ [Frederickson and Jájá, 1982] and within $2 + 2(k-1)/|V|$ for $k > 2$ [Khuller and Raghavachari, 1996].

Garey and Johnson: GT31

Minimum k-Edge Connected Subgraph ◀ ND28

INSTANCE: Graph $G = (V, E)$. k is a constant, $k \geq 2$.

SOLUTION: A k-edge connected spanning subgraph $G' = (V, E')$ for G, i.e. a spanning subgraph that cannot be disconnected by removing fewer than k edges.

MEASURE: The cardinality of the spanning subgraph, i.e., $|E'|$.

Good News: Approximable within 1.704 [Khuller and Raghavachari, 1996], [Cheriyan and Thurimella, 1996] and [Fernandes, 1997].

Bad News: APX-complete even for $k = 2$ [Fernandes, 1997].

Comment: Approximable within $\min\{1 + 5/k, 7/4 - (3k^2 - 6k + 8)/(2k^2(3k-2))\}$ for even k and within $\min\{1 + 7/k, 7/4 - (2k-3)/(4k^2)\}$ for odd k [Cheriyan and Thurimella, 1996] and [Fernandes, 1997]. On directed graphs the problem is approximable within 1.61 for $k = 1$ [Khuller, Raghavachari, and Young, 1996b], within 2 for $2 \leq k \leq 16$, and within

$1 + 4/\sqrt{k}$ for $k \geq 17$ [Cheriyan and Thurimella, 1996]. Variation in which each edge has a nonnegative weight and the objective is to minimize the total weight of the spanning subgraph is approximable within 2 for every k [Khuller and Vishkin, 1994].

ND29 ▶ MINIMUM BICONNECTIVITY AUGMENTATION

INSTANCE: Graph $G = (V, E)$ and a symmetric weight function $w : V \times V \to N$.

SOLUTION: A connectivity augmenting set E' for G, i.e., a set E' of unordered pairs of vertices from V such that $G' = (V, E \cup E')$ is biconnected.

MEASURE: The weight of the augmenting set, i.e., $\sum_{(u,v) \in E'} w(u,v)$.

Good News: Approximable within factor 2 [Frederickson and Jájá, 1981] and [Khuller and Thurimella, 1993].

Comment: The same bound is valid also when G' must be bridge connected (edge connected) [Khuller and Thurimella, 1993]. Minimum k-Connectivity Augmentation, the problem in which G' has to be k-connected (vertex or edge connected), is also approximable within 2 [Khuller, 1997]. If the weight function satisfies the triangle inequality, the problem is approximable within 3/2 [Frederickson and Jájá, 1982]. Variation in which G is planar and the augmentation must be planar is approximable within 5/3 in the unweighted case (where $w(u,v) = 1$) [Fialko and Mutzel, 1998]. If $E = \emptyset$ the problem is the same as weighted MINIMUM k-VERTEX CONNECTED SUBGRAPH.

Garey and Johnson: ND18

ND30 ▶ MINIMUM STRONG CONNECTIVITY AUGMENTATION

INSTANCE: Directed graph $G = (V, A)$ and a weight function $w : V \times V \to N$.

SOLUTION: A connectivity augmenting set A' for G, i.e., a set A' of ordered pairs of vertices from V such that $G' = (V, A \cup A')$ is strongly connected.

MEASURE: The weight of the augmenting set, i.e., $\sum_{(u,v) \in E'} w(u,v)$.

Good News: Approximable within 2 [Frederickson and Jájá, 1981].

Comment: The unweighted problem (i.e., where all weights are 1) is approximable within 1.61 [Khuller, Raghavachari, and Young, 1996b].

Garey and Johnson: ND19

ND31 ▶ MINIMUM BOUNDED DIAMETER AUGMENTATION

INSTANCE: Graph $G = (V, E)$, positive integer $D < |V|$.

SOLUTION: An augmenting set E' for G, i.e., a set E' of unordered pairs of vertices from V, such that $G' = (V, E \cup E')$ has diameter D, i.e., the maximum distance of any pair of vertices is at most D.

MEASURE: Cardinality of the augmenting set, i.e., $|E'|$.

Bad News: As hard to approximate as MINIMUM SET COVER [Li, McCormick, and Simchi-Levi, 1992].

Comment: Variation in which the size of the augmenting set is bounded by D and the problem is to minimize the diameter is approximable within $4 + 2/m^*$ [Li, McCormick, and Simchi-Levi, 1992].

Routing Problems

MINIMUM TRAVELING SALESPERSON ◀ ND32

INSTANCE: Set C of m cities, distances $d(c_i, c_j) \in N$ for each pair of cities $c_i, c_j \in C$.

SOLUTION: A tour of C, i.e., a permutation $\pi : [1..m] \to [1..m]$.

MEASURE: The length of the tour, i.e.,

$$d(\{c_{\pi(m)}, c_{\pi(1)}\}) + \sum_{i=1}^{m-1} d(\{c_{\pi(i)}, c_{\pi(i+1)}\}).$$

Bad News: NPO-complete [Orponen and Mannila, 1987].

Comment: The corresponding maximization problem (finding the tour of maximum length) is approximable within 7/5 if the distance function is symmetric and 63/38 if it is asymmetric [Kosaraju, Park, and Stein, 1994].

Garey and Johnson: ND22

MINIMUM METRIC TRAVELING SALESPERSON ◀ ND33

INSTANCE: Set C of m cities, distances $d(c_i, c_j) \in N$ satisfying the triangle inequality.

SOLUTION: A tour of C, i.e., a permutation $\pi : [1..m] \to [1..m]$.

MEASURE: The length of the tour.

Good News: Approximable within 3/2 [Christofides, 1976].

Bad News: APX-complete [Papadimitriou and Yannakakis, 1993].

The compendium

NETWORK DESIGN

Comment: Transformation from MAXIMUM 3-SATISFIABILITY. Variation in which the distances are only 1 and 2 is approximable within 7/6 [Papadimitriou and Yannakakis, 1993] but not within $5381/5380-\varepsilon$ for any $\varepsilon > 0$ [Engebretsen, 1999]. If the distance function is asymmetric the problem is approximable within $\log m$ [Frieze, Galbiati, and Maffioli, 1982]. The asymmetric problem with distances 1 and 2 is approximable within 17/12 [Vishwanathan, 1992] but not within $2805/2804-\varepsilon$ for any $\varepsilon > 0$ [Engebretsen, 1999]. The special case in which the distances are the shortest path lengths in a given weighted planar graph admits a PTAS [Arora, Grigni, Karger, Klein, and Woloszyn, 1998]. Variation in which the triangle inequality is relaxed to $d(c_i,c_j) \leq \tau(d(c_i,c_k)+d(c_k,c_j))$ for $\tau > 1$ is approximable within $(3\tau+1)\tau/2$ [Andreae and Bandelt, 1995].

The generalization in which, for each city, a neighborhood is specified in which the salesperson can meet the client, is also approximable for a variety of neighborhood types such as unit segments, unit circles, and unit rectangles [Arkin and Hassin, 1994]. Another generalization in which the salesperson has to rearrange some objects while following the route is approximable within 2.5 [Anily and Hassin, 1992]. A prize-collecting variation in which a penalty is associated with each vertex and the goal is to minimize the cost of the tour and the vertices not in the tour is approximable within $2-1/(|V|-1)$ [Goemans and Williamson, 1995a]. A clustered generalization in which vertices are partitioned into clusters that must be traversed consecutively is approximable within 2.75 in the general case, within 2.643 if the starting vertex in each cluster is given, within 1.9091 if the starting and ending vertices of each cluster are specified, within 1.8 if in each cluster two vertices are given and we are free to choose any one as starting vertex and the other as ending vertex [Guttmann-Beck, Hassin, Khuller, and Raghavachari, 1999]. A variation in which vertices can be revisited and the goal is to minimize the sum of the latencies of all vertices, where the latency of a vertex c is the length of the tour from the starting point to c, is approximable within 29 and is APX-complete [Blum, Chalasani, Coppersmith, Pulleyblank, Raghavan, and Sudan, 1994]. A combination of this problem and the matching problem, also called *Printed Circuit Board Assembly*, is approximable within 2.5 [Michel, Schroeter, and Srivastav, 1995]. Finally, the variation in which a Hamiltonian path is looked for rather than a tour is also approximable within 1.5 in the general case [Hoogeveen, 1978] while if both end vertices are specified it is approximable within 5/3 [Guttmann-Beck, Hassin, Khuller, and Raghavachari, 1999].

ND34 ▶ MINIMUM GEOMETRIC TRAVELING SALESPERSON

INSTANCE: Set $C \subseteq Z \times Z$ of m points in the plane.

SOLUTION: A tour of C, i.e., a permutation $\pi : [1..m] \to [1..m]$.

MEASURE: The length of the tour, where the distance between (x_1,y_1) and (x_2,y_2) is the discretized Euclidean length

$$\left\lceil \sqrt{(x_1-x_2)^2+(y_1-y_2)^2}\,\right\rceil.$$

Good News: Admits a PTAS [Arora, 1996].

Comment: Generalization to the d-dimensional case for d constant also admits a PTAS [Arora, 1997]. In R^m the problem is APX-complete for any l_p metric [Trevisan, 1997]. The variation in which an integer $k \leq m$ is given in the input and only at least k of the cities must be included in the tour also admits a PTAS [Arora, 1996]. MINIMUM GEOMETRIC ANGULAR TSP, the variation in which the sum of the direction changes in the tour is minimized, is approximable within $O(\log m)$. The same bound is valid also when there may be several tours covering all the cities [Aggarwal, Coppersmith, Khanna, Motwani, and Schieber, 1997]. The MAXIMUM GEOMETRIC TRAVELING SALESPERSON problem (finding the tour of maximum length) admits a PTAS [Barvinok, 1996].

Variation in which the input is given as k possibly overlapping simple polygons in the plane, and where the objective is to find the shortest tour that visits (intersects) each polygon, is approximable within $O(\log k)$ [Mata and Mitchell, 1995].

Garey and Johnson: ND23

MINIMUM METRIC TRAVELING k-SALESPERSON ◀ ND35

INSTANCE: Set C of m cities, an initial city $s \in C$, distances $d(c_i,c_j) \in N$ satisfying the triangle inequality.

SOLUTION: A collection of k subtours, each containing the initial city s, such that each city is in at least one subtour.

MEASURE: The maximum length of the k subtours.

Good News: Approximable within $1 - 1/k$ plus the performance ratio of the MINIMUM METRIC TRAVELING SALESPERSON, i.e., within $\frac{5}{2} - \frac{1}{k}$ [Frederickson, Hecht, and Kim, 1978].

Comment: The non-metric variation in which the graph is a tree and k possibly different initial cities are given in the input is approximable within $2 - 2/(p+1)$ [Averbakh and Berman, 1997].

MINIMUM METRIC BOTTLENECK WANDERING SALESPERSON ◀ ND36

INSTANCE: Set C of m cities, an initial city $s \in C$, a final city $f \in C$, distances $d(c_i,c_j) \in N$ satisfying the triangle inequality.

SOLUTION: A simple path from the initial city s to the final city f passing through all cities in C, i.e., a permutation $\pi : [1..m] \to [1..m]$ such that $v_{\pi(1)} = s$ and $v_{\pi(m)} = f$.

The compendium
NETWORK DESIGN

MEASURE: The length of the largest distance in the path, i.e.,

$$\max_{i \in [1..m-1]} d\left(\{c_{\pi(i)}, c_{\pi(i+1)}\}\right).$$

Good News: Approximable within 2 [Hochbaum and Shmoys, 1986].

Bad News: Not approximable within $2-\varepsilon$ for any $\varepsilon > 0$ [Hochbaum and Shmoys, 1986].

Comment: The same positive and negative results hold even if X is a set of point in d-dimensional space with the L_1 or L_∞ metric. If the L_2 metric is used then the upper bound is 1.969 [Feder and Greene, 1988]. The corresponding maximization problem called MAXIMUM SCATTER TSP, where the length of the shortest distance in the path is maximized, is approximable within 2, but not approximable within $2 - \varepsilon$ for any $\varepsilon > 0$ [Arkin, Chiang, Mitchell, Skiena, and Yang, 1997].

Garey and Johnson: ND24

ND37 ▶ MINIMUM CHINESE POSTMAN FOR MIXED GRAPHS

INSTANCE: Mixed graph $G = (V, A, E)$, length $l(e) \in N$ for each $e \in A \cup E$.

SOLUTION: A cycle in G (possibly containing repeated vertices) that includes each directed and undirected edge at least once, traversing directed edges only in the specified direction.

MEASURE: The total length of the cycle.

Good News: Approximable within 5/3 [Frederickson, 1979].

Comment: Approximable within 3/2 for planar graphs [Frederickson, 1979].

Garey and Johnson: ND25

ND38 ▶ MINIMUM k-CHINESE POSTMAN PROBLEM

INSTANCE: Multigraph $G = (V, E)$, initial vertex $s \in V$, length $l(e) \in N$ for each $e \in E$.

SOLUTION: A collection of k cycles, each containing the initial vertex s, that collectively traverse every edge in the graph at least once.

MEASURE: The maximum length of the k cycles.

Good News: Approximable within $2 - 1/k$ [Frederickson, Hecht, and Kim, 1978].

ND39 ▶ MINIMUM STACKER CRANE PROBLEM

INSTANCE: Mixed graph $G = (V, A, E)$, length $l(e) \in N$ for each $e \in A \cup E$ such that for every arc there is a parallel edge of no greater length.

SOLUTION: A cycle in G (possibly containing repeated vertices) that includes each directed edge in A at least once, traversing such edges only in the specified direction.

MEASURE: The total length of the cycle.

Good News: Approximable within 9/5 [Frederickson, Hecht, and Kim, 1978].

Garey and Johnson: ND26

MINIMUM k-STACKER CRANE PROBLEM ◀ ND40

INSTANCE: Mixed graph $G = (V, A, E)$, initial vertex $s \in V$, length $l(e) \in N$ for each $e \in A \cup E$,

SOLUTION: A collection of k cycles, each containing the initial vertex s, that collectively traverse each directed edge in A at least once.

MEASURE: The maximum length of the k cycles.

Good News: Approximable within $14/5 - 1/k$ [Frederickson, Hecht, and Kim, 1978].

MINIMUM GENERAL ROUTING ◀ ND41

INSTANCE: Graph $G = (V, E)$, length $l(e) \in N$ for each $e \in E$, subset $E' \subseteq E$, subset $V' \subseteq V$.

SOLUTION: A cycle in G that visits each vertex in V' exactly once and traverses each edge in E'.

MEASURE: The total length of the cycle.

Good News: Approximable within 3/2 [Jansen, 1992].

Comment: The special case where $V' = V$ is called the RURAL POSTMAN problem.

Garey and Johnson: Generalization of ND27

LONGEST PATH ◀ ND42

INSTANCE: Graph $G = (V, E)$.

SOLUTION: Simple path in G, i.e., a sequence of distinct vertices $v_1, v_2, \ldots, v_m$ such that, for any $1 \leq i \leq m - 1$, $(v_i, v_{i+1}) \in E$.

MEASURE: Length of the path, i.e., the number of edges in the path.

Good News: Approximable within $O(|V|/\log|V|)$ [Alon, Yuster, and Zwick, 1994].

Bad News: Not in APX [Karger, Motwani, and Ramkumar, 1997].

The compendium

NETWORK DESIGN

Comment: Transformation from MINIMUM METRIC TRAVELING SALESPERSON with distances one and two: APX-hard and self-improvable. Not approximable within $2^{\log^{1-\varepsilon}|V|}$ for any $\varepsilon > 0$ unless NP$\subset$QP [Karger, Motwani, and Ramkumar, 1997]. Similar results hold for a chromatic version of the problem [Bellare, 1993]. Variation in which the path must be induced subgraph of G, LONGEST INDUCED PATH, is NPOPB-complete and not approximable within $O(|V|^{1-\varepsilon})$ for any $\varepsilon > 0$, see MAXIMUM INDUCED CONNECTED SUBGRAPH WITH PROPERTY Π.

Garey and Johnson: ND29

ND43 ▶ SHORTEST WEIGHT-CONSTRAINED PATH

INSTANCE: Graph $G = (V,E)$, length function $l : E \to N$, weight function $w : E \to N$, specified vertices $s,t \in V$, and integer W.

SOLUTION: A simple path in G with total weight at most W, i.e., a sequence of distinct vertices $s = v_1, v_2, \ldots, v_m = t$ such that, for any $1 \leq i \leq m-1$, $(v_i, v_{i+1}) \in E$ and $\sum_{i=1}^{m-1} w(v_i, v_{i+1}) \leq W$.

MEASURE: The length of the path, i.e., $\sum_{i=1}^{m_1} l(v_i, v_{i+1})$.

Good News: Admits an FPTAS [Hassin, 1992] and [Phillips, 1993].

Garey and Johnson: ND30

ND44 ▶ MINIMUM RECTILINEAR GLOBAL ROUTING

INSTANCE: $m \times n$-array of gates, collection C of nets, i.e., 3-sets of gates.

SOLUTION: Wires following rectilinear paths connecting the gates in each net.

MEASURE: The largest number of wires in the same channel between two gates in the array.

Good News: Admits a PTAS$^\infty$ if $m^* \in \omega(\ln(mn))$ [Raghavan and Thompson, 1991].

Comment: Approximable within $1 + (e-1)\sqrt{2\ln(mn)/m^*}$ if $m^* > 2\ln(mn)$. In APX if $m^* \in \Omega(\ln(mn))$. The approximation algorithm will work also for nets with more than three gates, but the running time is exponential in the number of terminals.

ND45 ▶ MINIMUM TRAVELING REPAIRMAN

INSTANCE: Graph $G = (V,E)$, start vertex $r \in V$, length $l(e) \in N$ for each $e \in E$ satisfying the triangle inequality.

SOLUTION: A walk starting in r visiting all vertices in G, i.e., a permutation $\pi : [1..|V|] \to V$ such that $\pi(1) = r$, where π describes in which order the vertices are visited for the first time.

MEASURE: $\sum_{v \in V} d_\pi(r,v)$, where $d_\pi(r,v)$ is the total distance traversed in the walk from r until we first visit v.

The compendium

NETWORK DESIGN

Good News: Approximable within 1.662 [Koutsoupias, Papadimitriou, and Yannakakis, 1996].

Comment: Also called the *Minimum Latency Problem*. Variation in which the objective is the maximum search ratio, i.e., $\max_{v \in V} d_\pi(r,v)/d(r,v)$ is approximable within 6 and is APX-complete [Koutsoupias, Papadimitriou, and Yannakakis, 1996].

Flow Problems

MAXIMUM PRIORITY FLOW ◀ ND46

INSTANCE: Directed graph $G = (V,E)$, sources $s_1, \ldots, s_k \in V$, sinks $t_1, \ldots, t_k \in V$, a capacity function $c : E \to R$, a bound function $b : V \to R$, and, for any vertex v, a partial order on the set of edges leaving v.

SOLUTION: A priority flow f, i.e., a function $f : E \to R$ such that (a) for any edge e, $f(e) \leq c(e)$, (b) for any vertex $v \in V - \{s_1, \ldots, s_k, t_1, \ldots, t_k\}$, the flow is conserved at v, (c) for any vertex v, the flow leaving v is at most $b(v)$, and (d) for any vertex v and for any pair of edges e_1, e_2 leaving v, if $f(e_1) < c(e_1)$ and e_1 is less than e_2, then $f(e_2) = 0$.

MEASURE: The amount of flow entering sink t_1, i.e.,

$$\sum_{(x,t_1) \in E} f(x,t_1).$$

Bad News: Not approximable within $2^{\theta(\log^p n)}$ for some $p > 0$ unless NP $\subseteq$ DTIME$(n^{d \log^{1/\varepsilon} n})$ [Bellare, 1993].

Comment: Does not admit a PTAS.

MAXIMUM k-MULTICOMMODITY FLOW ON TREES ◀ ND47

INSTANCE: A tree $T = (V,E)$, a capacity function $c : E \to N$ and k pairs of vertices (s_i, t_i).

SOLUTION: A flow f_i for each pair (s_i, t_i) with $f_i \in N$ such that, for each $e \in E$,

$$\sum_{i=1}^{k} f_i q_i(e) \leq c(e)$$

where $q_i(e) = 1$ if e belongs to the unique path from s_i and t_i, 0 otherwise.

MEASURE: The sum of the flows, i.e., $\sum_{i=1}^{k} f_i$.

The compendium

NETWORK DESIGN

Good News: Approximable within 2 [Garg, Vazirani, and Yannakakis, 1997].

Bad News: APX-complete [Garg, Vazirani, and Yannakakis, 1997].

Comment: Transformation from MAXIMUM 3-DIMENSIONAL MATCHING. It remains APX-complete even if the edge capacities are 1 and 2.

ND48 ▶ MAXIMUM DISJOINT CONNECTING PATHS

INSTANCE: Multigraph $G = (V,E)$, set of pairs $T = \{(s_1,t_1),\ldots,(s_k,t_k)\}$ with $s_i, t_i \in V$.

SOLUTION: A collection of edge disjoint paths in G connecting some of the pairs (s_i,t_i), i.e. sequences of vertices $u_1, u_2, \ldots, u_m$ such that, for some i, $u_1 = s_i$, $u_m = t_i$, and for any j, $(u_j, u_{j+1}) \in E$.

MEASURE: The number of vertex pairs (s_i, t_i) that are connected by the paths.

Good News: Approximable within $\min\{\sqrt{|E|}, |E|/m^*\}$ [Srinivasan, 1997].

Comment: Similar to MINIMUM UNSPLITTABLE FLOW. Approximable within $O(\text{polylog}|V|)$ for constant degree graphs and butterfly graphs [Srinivasan, 1997]. Approximable within 2 if G is a tree with parallel edges. Approximable within $O(\log|V|)$ if G is a two-dimensional mesh [Aumann and Rabani, 1995], or more generally, if G is a nearly-Eulerian uniformly high-diameter planar graph [Kleinberg and Tardos, 1995].

Variation in which the objective is to minimize the number of rounds that together connect all pairs, is approximable within $O(\sqrt{|E|}\log|V|)$ on general multigraphs and within $O(\text{polylog}|V|)$ on butterfly graphs [Srinivasan, 1997].

MINIMUM PATH COLORING is the variation in which each path is assigned a color, where only paths with the same color need to be edge disjoint, and where the objective is to minimize the number of colors that are needed to connect all vertex pairs in the input. This problem is approximable within 3/2 on trees, within 2 on cycles [Raghavan and Upfal, 1994], within $O(\text{polyloglog}|V|)$ on two-dimensional meshes [Rabani, 1996], and within $O(\log|V|)$ on nearly-Eulerian uniformly high-diameter planar graphs [Kleinberg and Tardos, 1995].

Garey and Johnson: Similar to ND40

ND49 ▶ MINIMUM MAXIMUM DISJOINT CONNECTING PATHS

INSTANCE: Graph $G = (V,E)$, length function $l : E \to N$, and a pair of vertices s,t in V.

SOLUTION: Two vertex disjoint paths in G connecting vertices s and t, that is, two sequences of vertices $u_1, u_2, \ldots, u_m$ and $v_1, v_2, \ldots, v_n$ such that $|\{u_1, u_2, \ldots, u_m, v_1, v_2, \ldots, v_n\}| = m+n$, (s,u_1), (s,v_1), (u_m,t), and (v_n,t) are included in E, and for any i, (u_i, u_{i+1}) and (v_i, v_{i+1}) are included in E.

MEASURE: The longest of the two paths, i.e., the maximum between

$$l(s,u_1) + \sum_{i=1}^{m-1} l(u_i, u_{i+1}) + l(u_m, t)$$

and

$$l(s,v_1) + \sum_{i=1}^{n-1} l(v_i, v_{i+1}) + l(v_n, t).$$

Good News: Approximable within factor 2 [Li, McCormick, and Simchi-Levi, 1990].

Comment: Approximable within 2 also if the paths should be *edge disjoint* instead of vertex disjoint. Variation in which the graph is directed and we look for two vertex (edge) disjoint directed paths is also approximable within 2, and is not approximable within $2-\varepsilon$ for any $\varepsilon > 0$ [Li, McCormick, and Simchi-Levi, 1990].

Garey and Johnson: Similar to ND41

MINIMUM SINGLE-SINK EDGE INSTALLATION ◀ ND50

INSTANCE: Graph $G = (V,E)$, length function $l : E \to N$, set of sources $S \subseteq V$, sink $t \in V$, demand function $d : S \to Z^+$, finite set of cable types where each cable type is specified by its capacity and its cost per unit length.

SOLUTION: A network of cables in the graph, consisting of an integral number of each cable type for each edge in G, that routes all the demands at the sources to the sink. The demand of each source must follow a single path from source to sink.

MEASURE: The total cost of building the network of cables.

Good News: Approximable within $O(\log^2 |V|)$ [Awerbuch and Azar, 1997].

Comment: Variations in which there are multiple sinks (a generalization of MINIMUM GENERALIZED STEINER NETWORK) or just a number k of the sources need to be connected to the sink (a generalization of MINIMUM k-STEINER TREE) is also approximable within $O(\log^2 |V|)$ [Awerbuch and Azar, 1997]. Approximable within $O(\log D)$, where $D = \sum_{s \in S} d(s)$, for points in the Euclidean plane. Restricted version where the network to be designed must be a two-level tree (so that every path from a source to the sink consists of at most two edges) is approximable within $O(\log |V|)$ [Salman, Cheriyan, Ravi, and Subramanian, 1997].

MINIMUM UNSPLITTABLE FLOW ◀ ND51

INSTANCE: Graph $G = (V,E)$, with edge capacities $c : E \to Z^+$, source vertex s, collection of sinks $t_1, \ldots, t_k$ with associated non-negative integer demands $\rho_1, \ldots, \rho_k$ such that, $\rho_i \leq c(e)$ for all $e \in E$ and $i \in [1..k]$.

SOLUTION: A single s–t_i flow path for each commodity i so that the demands are satisfied and the total flow routed across any edge e is bounded by $c(e)$.

MEASURE: $\max_{e \in E} f(e)/c(e)$, where $f(e)$ is the total flow routed across e.

Good News: Approximable within 3.23 [Kolliopoulos and Stein, 1997].

Bad News: Not approximable within $3/2 - \varepsilon$ for any $\varepsilon > 0$ [Kleinberg, 1996].

Comment: Approximable within 3.23 also on directed graphs. Variation in which the objective is to maximize the routable demand is approximable within 14. If the objective is to minimize the number of rounds in which all demands are routed, is approximable within 13. The generalized problem in which there are two source vertices is not approximable within $2 - \varepsilon$ [Kolliopoulos and Stein, 1997].

Miscellaneous

ND52 ▶ MINIMUM BROADCAST TIME

INSTANCE: Graph $G = (V, E)$ and a source node $v_0 \in V$.

SOLUTION: A broadcasting scheme. At time 0 only v_0 contains the message that is to be broadcast to every vertex. At each time step any vertex that has received the message is allowed to communicate the message to at most one of its neighbors.

MEASURE: The broadcast time, i.e., the time when all vertices have received the message.

Good News: Approximable within $O(\log^2 |V|/\log\log|V|)$ [Ravi, 1994b].

Comment: Approximable within $2B$ if the degree of G is bounded by a constant B [Ravi, 1994b]. Approximable within $O(\log|V|)$ if G is chordal, k-outerplanar [Kortsarz and Peleg, 1995]. Approximable within $O(\log|V|/\log\log|V|)$ if G has bounded tree width [Marathe et al., 1995]. MINIMUM GOSSIP TIME, the extension in which there is a message on every node and the objective is to minimize the time until every node has received every message, is approximable within twice the performance ratio of MINIMUM BROADCAST TIME [Ravi, 1994b].

Garey and Johnson: ND49

ND53 ▶ MINIMUM k-CENTER

INSTANCE: Complete graph $G = (V, E)$ and distances $d(v_i, v_j) \in N$ satisfying the triangle inequality.

SOLUTION: A k-center set, i.e., a subset $C \subseteq V$ with $|C| = k$.

MEASURE: The maximum distance from a vertex to its nearest center, i.e.,
$\max_{v \in V} \min_{c \in C} d(v,c)$.

Good News: Approximable within 2 [Hochbaum and Shmoys, 1986].

Bad News: Not approximable within $2-\varepsilon$ for any $\varepsilon > 0$ [Hsu and Nemhauser, 1979] and [Plesník, 1980].

Comment: Not in APX if the distances do not satisfy the triangle inequality [Hochbaum, 1997]. MINIMUM CAPACITATED k-CENTER, the variation in which the number of vertices each center can serve is bounded by a constant L, is approximable within 5 [Khuller and Sussmann, 1996]. The converse problem, where the maximum distance from each vertex to its center is given and the number of centers is to be minimized, is approximable within $\log L + 1$ [Bar-Ilan and Peleg, 1991]. The geometric k-center problem, where the vertices lie in the plane and the geometric metric is used, is approximable within 2, but is not approximable within 1.822 [Feder and Greene, 1988]. Variants of the geometric k-center problem are also studied in the paper. The rectilinear k-center problem, where the vertices lie in the plane and the L_∞ metric is used, is approximable within 2, but is not approximable within $2-\varepsilon$ for any $\varepsilon > 0$ [Ko, Lee, and Chang, 1990]. If the vertices lie in R^d and the Euclidean or L_∞ metric is used the problem admits a PTAS whose time is exponential in k [Agarwal and Procopiuc, 1998]. Approximable within 2 in any metric space [Gonzalez, 1985].

The vertex weighted version, where the objective is to minimize the maximum weighted distance $d(v,c)w(v)$, is approximable within 2 [Plesník, 1987]. It is not approximable within $2-\varepsilon$ even if the distances are induced by a planar graph of maximum degree 3 with edge lengths 1 and vertex weights 1 [Plesník, 1980]. MINIMUM ABSOLUTE k-CENTER, the variation in which we allow the points in C to lie in edges (considered as curves) is also approximable within 2 and is not approximable within $2-\varepsilon$ [Plesník, 1988]. MINIMUM α-ALL-NEIGHBOR k-CENTER, the variation in which we want to minimize the distance from each vertex to $\alpha \geq 2$ of the k centers, is approximable within 2 when $\alpha \in \{2,3\}$ and within 3 otherwise (even for the vertex weighted version) [Khuller, Pless, and Sussmann, 1997]; it is not approximable within $2-\varepsilon$ for any $\varepsilon > 0$ [Krumke, 1995]. The asymmetric k-center problem, where $d(v_i,v_j)$ might be different from $d(v_j,v_i)$, is approximable within $O(\log^* |V|)$ [Vishwanathan, 1996].

Garey and Johnson: Similar to ND50

MINIMUM k-CLUSTERING ◀ ND54

INSTANCE: Finite set X, a distance $d(x,y) \in N$ for each pair $x,y \in X$. The distances must satisfy the triangle inequality.

SOLUTION: A partition of X into disjoint subsets $C_1, C_2, \ldots, C_k$.

MEASURE: The largest distance between two elements in the same subset, i.e.,
$$\max_{\substack{i \in [1..k] \\ x,y \in C_i}} d(x,y).$$

Good News: Approximable within 2 [Gonzalez, 1985] and [Hochbaum and Shmoys, 1986].

Bad News: Not approximable within $2-\varepsilon$ for any $\varepsilon > 0$ [Gonzalez, 1985] and [Hochbaum and Shmoys, 1986].

Comment: The same positive and negative results are valid for the geometric (Euclidean) k-clustering problem in 3 dimensions [Gonzalez, 1985] and for rectilinear k-clustering problem in 2 dimensions. The geometric k-clustering problem in 2 dimensions is not approximable within 1.969 [Feder and Greene, 1988]. Other variants are also studied in this paper.

Garey and Johnson: MS9

ND55 ▶ MINIMUM k-CLUSTERING SUM

INSTANCE: Finite set X, a distance $d(x,y) \in N$ for each pair $x,y \in X$.

SOLUTION: A partition of X into disjoint subsets $C_1, C_2, \ldots, C_k$.

MEASURE: The sum of all distances between elements in the same subset, i.e.,

$$\sum_{i=1}^{k} \sum_{v_1, v_2 \in C_i} d(v_1, v_2).$$

Bad News: Not in APX [Sahni and Gonzalez, 1976].

Comment: Approximable within 1.7 for $k = 2$ and within 2 for $k \geq 3$ if d satisfies the triangle inequality [Guttmann-Beck and Hassin, 1998b]. The maximization variation in which the subsets $C_1, \ldots, C_k$ must be of equal size is approximable within $c - 1$ or c where c is the cluster's size and is even or odd, respectively [Feo and Khellaf, 1990]: the case $c = 3$ and $c = 4$ is approximable within 2 [Feo, Goldschmidt, and Khellaf, 1992]. Moreover, if d satisfies the triangle inequality then the bounds are $2(c-1)/c$ and $(c+1)/c$, respectively [Feo and Khellaf, 1990]. Finally, if required sizes $c_1, \ldots, c_k$ of the subsets are given and the triangle inequality is satisfied then the maximization problem is approximable within $2\sqrt{2}$ [Hassin and Rubinstein, 1998].

ND56 ▶ MINIMUM k-SUPPLIER

INSTANCE: Complete graph $G = (V, E)$, distances $d(v_i, v_j) \in N$ satisfying the triangle inequality, center construction cost $c(v) \in N$ and usage weight $w(v)$ for each $v \in V$, cost bound $L \in N$.

SOLUTION: A supplier set of legal cost, i.e., a subset $S \subseteq V$ such that $\sum_{v \in S} c(v) \leq L$.

MEASURE: The maximum weighted distance from a vertex to its nearest supplier, i.e.,
$$\max_{v \in V} \min_{s \in S} w(v) d(v,s).$$

Good News: Approximable within 3 [Wang and Cheng, 1990].

Bad News: Not approximable within $3-\varepsilon$ for any $\varepsilon > 0$ [Hochbaum and Shmoys, 1986].

MINIMUM k-MEDIAN ◂ ND57

INSTANCE: Complete graph $G = (V, E)$ and distances $d(e) \in N$.

SOLUTION: A k-median set, i.e., a subset $V' \subseteq V$ with $|V'| = k$.

MEASURE: The sum of the distances from each vertex to its nearest median, i.e.,
$$\sum_{v \in V} \min_{w \in V'} d(v,w).$$

Bad News: Not in APX [Lin and Vitter, 1992].

Comment: Transformation from MINIMUM SET COVER. The problem is in APX if a small violation of the cardinality of the median set is allowed.

Garey and Johnson: ND51

MINIMUM DIAMETERS DECOMPOSITION ◂ ND58

INSTANCE: Graph $G = (V, E)$.

SOLUTION: A decomposition of the graph into two factors, F_1 and F_2, with equal diameters.

MEASURE: The diameter of F_1.

Bad News: Not approximable within $3/2 - \varepsilon$ for any $\varepsilon > 0$ [Plesník, 1982].

Comment: Variation in which the diameters of F_1 and F_2 do not need to be equal, and where the objective is to minimize the sum of the diameters, is not approximable within $5/4 - \varepsilon$ [Plesník, 1982].

MAXIMUM k-FACILITY DISPERSION ◂ ND59

INSTANCE: Complete graph $G = (V, E)$ and distances $d(v_i, v_j) \in N$ satisfying the triangle inequality.

SOLUTION: A set of k facilities, i.e., a subset $F \subseteq V$ with $|F| = k$.

The compendium

NETWORK DESIGN

MEASURE: The minimum distance between two facilities, i.e.,

$$\min_{f_1, f_2 \in F} d(f_1, f_2).$$

Good News: Approximable within 2 [Ravi, Rosenkrantz, and Tayi, 1991].

Bad News: Not approximable within $2 - \varepsilon$ for any $\varepsilon > 0$ [Ravi, Rosenkrantz, and Tayi, 1991].

Comment: Not in APX if the distances do not satisfy the triangle inequality. MAXIMUM EDGE SUBGRAPH is the variation where the measure is the average distance between any pair of facilities [Ravi, Rosenkrantz, and Tayi, 1991]. Variation in which we allow the points in F to lie in edges (considered as curves) is also approximable within 2 and is not approximable within $3/2 - \varepsilon$ [Tamir, 1991].

ND60 ▶ MINIMUM FACILITY LOCATION

INSTANCE: Complete graph $G = (V, E)$, costs $c(v_i, v_j) \in N$ that are symmetric and satisfy the triangle inequality, $F \subseteq V$ set of locations where a facility may be built, for each $v \in F$ a nonnegative cost $f(v)$ of building a facility at v, for each location $v \in V$ a nonnegative demand $d(v)$.

SOLUTION: Locations for the facilities to be built, i.e., $F' \subseteq F$.

MEASURE:

$$\sum_{v \in F'} f(v) + \sum_{u \in F'} \sum_{v \in V} d(v) \cdot c(u, v).$$

Good News: Approximable within 2.408 [Guha and Khuller, 1998].

Bad News: APX-complete [Guha and Khuller, 1998].

Comment: Transformation from the MINIMUM VERTEX COVER problem in bounded degree graphs. Not approximable within factor 1.463 unless NP $\subset$ DTIME$(n^{\log \log n})$. If $f(v) = 1$ for every $v \in F$ the same hardness results are valid, but the problem is approximable within 2.225. If $F = V$ and $f(v) = 1$ then the problem is approximable within 2.104 but not within 1.278 unless NP $\subset$ DTIME$(n^{\log \log n})$ [Guha and Khuller, 1998]. The generalization to unrestricted cost functions c is approximable within $O(\log |V|)$ [Hochbaum, 1982b].

ND61 ▶ MAXIMUM k-FACILITY LOCATION

INSTANCE: Complete graph $G = (V, E)$ and profits $p(v_i, v_j) \in N$.

SOLUTION: A set of k facilities, i.e., a subset $F \subseteq V$ with $|F| = k$.

MEASURE: The maximum total profit, i.e.,

$$\sum_{v \in V} \max_{f \in F} p(v, f).$$

Good News: Approximable within factor $e/(e-1)$ [Cornuejols, Fisher, and Nemhauser, 1977].

MINIMUM k-SWITCHING NETWORK ◀ ND62

INSTANCE: Complete graph $G = (V, E)$ and distances $d(v_i, v_j) \in N$ satisfying the triangle inequality.

SOLUTION: A partition $F = \{A_1, A_2, \ldots, A_k, B_1, B_2, \ldots, B_k\}$ of V.

MEASURE: Maximum distance between vertices in different sets with the same index, i.e.,

$$\max_{i \in [1..k]} \max_{\substack{v_1 \in A_i \\ v_2 \in B_i}} d(v_1, v_2).$$

Good News: Approximable within 3 [Hochbaum and Shmoys, 1986].

Bad News: Not approximable within $2 - \varepsilon$ for any $\varepsilon > 0$ [Hochbaum and Shmoys, 1986].

MINIMUM BEND NUMBER ◀ ND63

INSTANCE: Directed planar graph $G = (V, E)$.

SOLUTION: A planar orthogonal drawing of G, i.e., a drawing mapping vertices of G into points in the plane and edges of G into chains of horizontal and vertical segments such that no two edges cross.

MEASURE: Number of bends in the drawing.

Bad News: Not approximable within $1 + \frac{|V|^{1-\varepsilon}}{m^*(G)}$ for any $\varepsilon > 0$ [Garg and Tamassia, 1994].

MINIMUM LENGTH TRIANGULATION ◀ ND64

INSTANCE: Collection $C = \{(a_i, b_i) : 1 \leq i \leq n\}$ of pairs of integers giving the coordinates of n points in the plane.

SOLUTION: A triangulation of the set of points represented by C, i.e., a collection E of non-intersecting line segments each joining two points in C that divides the interior of the convex hull into triangular regions.

The compendium

NETWORK DESIGN

MEASURE: The discrete-Euclidean length of the triangulation, i.e.,

$$\left\lceil \sum_{((a_i,b_i),(a_j,b_j))\in E} \sqrt{(a_i-a_j)^2+(b_i-b_j)^2} \right\rceil$$

Good News: Approximable within $24\log n$ [Plaisted and Hong, 1987].

Comment: Note that the problem is not known to be NP-complete. The Steiner variation in which the point set of E must be a superset of C is approximable within 316 [Eppstein, 1992].

Triangulating planar graphs while minimizing the maximum degree is approximable so that the maximum degree is bounded by $3\Delta/2 + 11$ for general graphs and $\Delta + 8$ for triconnected graphs, where Δ is the degree of the graph [Kant and Bodlaender, 1997].

Garey and Johnson: OPEN12

ND65 ▶ MINIMUM SEPARATING SUBDIVISION

INSTANCE: A family of disjoint polygons $P_1,\ldots,P_k$.

SOLUTION: A separating subdivision, i.e., a family of k polygons $R_1,\ldots,R_k$ with pairwise disjoint boundaries such that, for each i, $P_i \subseteq R_i$.

MEASURE: The size of the subdivision, i.e., the total number of edges of the polygons $R_1,\ldots,R_k$.

Good News: Approximable within 7 [Mitchell and Suri, 1992].

Comment: The problem of separating a family of three-dimensional convex polyhedra is approximable within $O(\log n)$ while the problem of separating two d-dimensional convex polyhedra is approximable within $O(d\log n)$ where n denotes the number of facets in the input family.

Sets and Partitions

Covering, Hitting, and Splitting

MAXIMUM 3-DIMENSIONAL MATCHING ◀ SP1

INSTANCE: Set $T \subseteq X \times Y \times Z$, where X, Y, and Z are disjoint.

SOLUTION: A matching for T, i.e., a subset $M \subseteq T$ such that no elements in M agree in any coordinate.

MEASURE: Cardinality of the matching, i.e., $|M|$.

Good News: Approximable within $3/2+\varepsilon$ for any $\varepsilon > 0$ [Hurkens and Schrijver, 1989].

Bad News: APX-complete [Kann, 1991].

Comment: Transformation from MAXIMUM 3-SATISFIABILITY. Admits a PTAS for 'planar' instances [Nishizeki and Chiba, 1988]. Variation in which the number of occurrences of any element in X, Y or Z is bounded by a constant B is APX-complete for $B \geq 3$ [Kann, 1991]. The generalized MAXIMUM k-DIMENSIONAL MATCHING problem is approximable within $k/2 + \varepsilon$ for any $\varepsilon > 0$ [Halldórsson, 1994]. The constrained variation in which the input is extended with a subset S of T, and the problem is to find the 3-dimensional matching that contains the largest number of elements from S, is not approximable within $|T|^\varepsilon$ for some $\varepsilon > 0$ [Zuckerman, 1993].

Garey and Johnson: SP1

MAXIMUM SET PACKING ◀ SP2

INSTANCE: Collection C of finite sets.

SOLUTION: A set packing, i.e., a collection of disjoint sets $C' \subseteq C$.

MEASURE: Cardinality of the set packing, i.e., $|C'|$.

Good News: See MAXIMUM CLIQUE.

Bad News: See MAXIMUM CLIQUE.

Comment: Also called *Maximum Hypergraph Matching*. Equivalent to MAXIMUM CLIQUE under PTAS-reduction, where $|C| = |V|$ [Ausiello, D'Atri, and Protasi, 1980]. Therefore approximation algorithms and nonapproximability results for MAXIMUM CLIQUE will carry over to MAXIMUM SET PACKING. The problem MAXIMUM k-SET PACKING, the variation in which the cardinality of all sets in C are bounded from above by a constant $k \geq 3$, is APX-complete [Kann, 1991], and is approximable within $k/2 + \varepsilon$ for any $\varepsilon > 0$ [Hurkens and Schrijver, 1989]. Variation of MAXIMUM k-SET PACKING in which the sets have positive weights and the objective is

The compendium

SETS AND
PARTITIONS

to maximize the total weight of the sets in the set packing is approximable within $k - 1 + \varepsilon$ for any $\varepsilon > 0$ [Arkin and Hassin, 1998].

If the number of occurrences in C of any element is bounded by a constant B for $B \geq 2$ the problem is APX-complete [Berman and Fujito, 1995] and approximable within B, even for the weighted variation of the problem [Hochbaum, 1983].

Garey and Johnson: SP3

SP3 ▶ MAXIMUM SET SPLITTING

INSTANCE: Collection C of subsets of a finite set S.

SOLUTION: A partition of S into two disjoint subsets S_1 and S_2.

MEASURE: Cardinality of the subsets in C that are not entirely contained in either S_1 or S_2.

Good News: Approximable within 1.380 [Andersson and Engebretsen, 1998a].

Bad News: APX-complete [Petrank, 1994].

Comment: Also called *Maximum Hypergraph Cut*. Transformation from MAXIMUM NOT-ALL-EQUAL 3-SATISFIABILITY. Variation in which all subsets contain the same number of elements, k, is approximable within 1.138 for $k \leq 3$ and $1/(1 - 2^{1-k})$ for $k \geq 4$, and not approximable within 1.013 for $k \leq 3$ and $1 + O(k^{-3}2^{-k})$ for any $\varepsilon > 0$ for $k \geq 4$ [Kann, Lagergren, and Panconesi, 1996]. It admits a PTAS if $|C| = \Theta(|S|^k)$ [Arora, Karger, and Karpinski, 1995].

Garey and Johnson: SP4

SP4 ▶ MINIMUM SET COVER

INSTANCE: Collection C of subsets of a finite set S.

SOLUTION: A set cover for S, i.e., a subset $C' \subseteq C$ such that every element in S belongs to at least one member of C'.

MEASURE: Cardinality of the set cover, i.e., $|C'|$.

Good News: Approximable within $1 + \ln|S|$ [Johnson, 1974a].

Bad News: Not in APX [Lund and Yannakakis, 1994] and [Bellare, Goldwasser, Lund, and Russell, 1993].

Comment: Approximable within $\ln(|S|/m^*) + O(\ln\ln(|S|/m^*)) + O(1)$ [Srinivasan, 1995]. Not approximable within $(1 - \varepsilon)\ln|S|$ for any $\varepsilon > 0$, unless NP$\subset$ DTIME$(n^{\log\log n})$ [Feige, 1996]. Equivalent to MINIMUM DOMINATING SET under L-reduction and equivalent to MINIMUM HITTING SET [Ausiello, D'Atri, and Protasi, 1980]. If it is NP-hard to approximate within $\Omega(\log|S|)$, then it is complete for the class of log-approximable problems [Khanna, Motwani, Sudan, and Vazirani, 1999]. Approximable within $c\ln|S|$ if every element of S belongs to at least $\varepsilon|C|$ sets from C for any $c > 0$ and $\varepsilon > 0$ [Karpinski and Zelikovsky, 1997b] and [Karpinski, 1997].

Variation in which the subsets have positive weights and the objective is to minimize the sum of the weights in a set cover is also approximable within $1 + \ln|S|$ [Chvátal, 1979].

The problem MINIMUM k-SET COVER, the variation in which the cardinality of all sets in C are bounded from above by a constant k is APX-complete and is approximable, for any $\varepsilon > 0$, within $1.4+\varepsilon$ for $k = 3$ and $\sum_{i=1}^{k} \frac{1}{i} - 1/3$ for $k > 3$ [Halldórsson, 1996].

Variation in which also the number of occurrences in C of any element is bounded by a constant $B \geq 2$ is APX-complete [Papadimitriou and Yannakakis, 1991] and approximable within B even for the general weighted problem [Bar-Yehuda and Even, 1981] and [Hochbaum, 1982a]. If, moreover, each element needs to be covered by at least K sets with K constant, this variation is approximable within $B - K + 1$ [Peleg, Schechtman, and Wool, 1993].

Approximable within 2 if the set system (S,C) is tree representable [Garg, Vazirani, and Yannakakis, 1997].

The constrained variation in which the input is extended with a positive integer k and a subset T of C, and the problem is to find the set cover of size k that contains the largest number of subsets from T, is not approximable within $|C|^{\varepsilon}$ for some $\varepsilon > 0$ [Zuckerman, 1993].

Variation in which a distance matrix between pairs of elements in S is given and the measure of the solution is not the cardinality of the cover but its diameter (i.e., the maximum distance of any pair of elements in C') is not approximable within any constant in the general case. If the distance matrix satisfies the triangle inequality then this variation can be approximated within 2 and no better approximation is possible. Similar results hold if the cover can be partitioned into clusters and the goal is to minimize the maximum diameter over all clusters [Arkin and Hassin, 1992] (see also MINIMUM k-CLUSTERING).

Garey and Johnson: SP5

The compendium

SETS AND PARTITIONS

MINIMUM EXACT COVER ◀ SP5

INSTANCE: Collection C of subsets of a finite set S.

SOLUTION: A set cover for S, i.e., a subset $C' \subseteq C$ such that every element in S belongs to at least one member of C'.

MEASURE: Sum of cardinalities of the subsets in the set cover, i.e., $\sum_{c \in C'} |c|$.

Good News: Approximable within $1 + \ln|S|$ [Johnson, 1974a].

The compendium

SETS AND PARTITIONS

Bad News: As hard to approximate as MINIMUM SET COVER [Lund and Yannakakis, 1994].

Comment: Transformation from MINIMUM SET COVER. The only difference between MINIMUM SET COVER and MINIMUM EXACT COVER is the definition of the objective function.

SP6 ▶ MINIMUM TEST COLLECTION

INSTANCE: Collection C of subsets of a finite set S.

SOLUTION: A subcollection $C' \subseteq C$ such that for each pair of distinct elements $x_1, x_2 \in S$ there is some set $c \in C'$ that contains exactly one of x_1 and x_2.

MEASURE: Cardinality of the subcollection, i.e., $|C'|$.

Good News: Approximable within $1 + 2\ln|S|$.

Comment: Transformation to MINIMUM SET COVER [Kann, 1992b]. Observe that every solution has cardinality at least $\lceil \log|S| \rceil$.

Garey and Johnson: SP6

SP7 ▶ MINIMUM HITTING SET

INSTANCE: Collection C of subsets of a finite set S.

SOLUTION: A hitting set for C, i.e., a subset $S' \subseteq S$ such that S' contains at least one element from each subset in C.

MEASURE: Cardinality of the hitting set, i.e., $|S'|$.

Good News: See MINIMUM SET COVER.

Bad News: See MINIMUM SET COVER.

Comment: Equivalent to MINIMUM SET COVER [Ausiello, D'Atri, and Protasi, 1980]. Therefore approximation algorithms and nonapproximability results for MINIMUM SET COVER will carry over to MINIMUM HITTING SET. The constrained variation in which the input is extended with a subset T of S, and the problem is to find the hitting set that contains the largest number of elements from T, is not approximable within $|S|^\varepsilon$ for some $\varepsilon > 0$ [Zuckerman, 1993]. Several special cases in which, given compact subsets of R^d, the goal is to find a set of straight lines of minimum cardinality so that each of the given subsets is hit by at least one line, are approximable [Hassin and Megiddo, 1991].

Garey and Johnson: SP8

SP8 ▶ MINIMUM GEOMETRIC DISK COVER

INSTANCE: Set $P \subseteq Z \times Z$ of points in the plane, a rational number $r > 0$.

SOLUTION: A set of centers $C \subseteq Q \times Q$ in the Euclidean plane, such that every point in P will covered by a disk with radius r and center in one of the points in C.

MEASURE: Cardinality of the disk cover, i.e., $|C|$.

Good News: Admits a PTAS [Hochbaum and Maass, 1985].

Comment: There is a PTAS also for other convex objects, e.g., squares, and in d-dimensional Euclidean space for $d \geq 3$ [Hochbaum and Maass, 1985]. Variation in which the objective is to cover a set of points on the line by rings of given inner radius r and width w, admits a PTAS with time complexity exponential in r/w, for arbitrary r and w the problem is approximable with relative error at most 1/2 [Hochbaum and Maass, 1987].

MAXIMUM GEOMETRIC SQUARE PACKING, where the objective is to place as many squares of a given size within a region in the plane, also admits a PTAS [Hochbaum and Maass, 1985].

Weighted Set Problems

MAXIMUM CONSTRAINED PARTITION ◂ SP9

INSTANCE: Finite set A and a size $s(a) \in Z^+$ for each $a \in A$, element $a_0 \in A$, and a subset $S \subseteq A$.

SOLUTION: A partition of A, i.e., a subset $A' \subseteq A$ such that $\sum_{a \in A'} s(a) = \sum_{a \in A - A'} s(a)$.

MEASURE: Number of elements from S on the same side of the partition as a_0.

Bad News: Not approximable within $|A|^\varepsilon$ for some $\varepsilon > 0$ [Zuckerman, 1993].

Garey and Johnson: Similar to SP12

MINIMUM 3-DIMENSIONAL ASSIGNMENT ◂ SP10

INSTANCE: Three sets X, Y, and W and a cost function $c : X \times Y \times W \to N$.

SOLUTION: An assignment Λ, i.e., a subset $A \subseteq X \times Y \times W$ such that every element of $X \cup Y \cup W$ belongs to exactly one triple in A.

MEASURE: The cost of the assignment, i.e., $\sum_{(x,y,w) \in A} c(x,y,w)$.

Bad News: Not in APX [Crama and Spieksma, 1992].

Comment: The negative result holds even if $c(x,y,w)$ is either defined as

$$d(x,y) + d(x,w) + d(y,w)$$

or as

$$\min\{d(x,y) + d(x,w), d(x,y) + d(y,w), d(x,w) + d(y,w)\}$$

where d is any distance function. In these cases, however, the problem is approximable within 4/3 if d satisfies the triangle inequality. Similar results hold for the k-dimensional problem [Bandelt, Crama, and Spieksma, 1994].

Garey and Johnson: Weighted version of SP2

SP11 ▶ MAXIMUM CAPACITY REPRESENTATIVES

INSTANCE: Disjoint sets $S_1, \ldots, S_m$ and, for any $i \neq j$, $x \in S_i$, and $y \in S_j$, a non-negative capacity $c(x,y)$.

SOLUTION: A system of representatives T, i.e., a set T such that, for any i, $|T \cap S_i| = 1$.

MEASURE: The capacity of the system of representatives, i.e., $\sum_{x,y \in T} c(x,y)$.

Bad News: Not approximable within $2^{\theta(\log^p n)}$ for some $p > 0$ unless NP $\subseteq$ DTIME($n^{d \log^{1/\varepsilon} n}$) [Bellare, 1993].

Comment: Does not admit a PTAS.

SP12 ▶ MINIMUM SUM OF SQUARES

INSTANCE: Finite set A, size $s(a) \in Z^+$ for each $a \in A$, and an integer $K \geq 2$.

SOLUTION: A partition of A into K disjoint sets $A_1, A_2, \ldots, A_K$.

MEASURE:
$$\sum_{i=1}^{K} \left(\sum_{a \in A_i} s(a) \right)^2.$$

Good News: Approximable within 25/24 [Cody and Coffman, 1976].

Comment: There are somewhat better approximation ratios for $K \bmod 5 \neq 0$ [Leung and Wei, 1995].

Garey and Johnson: SP19

SP13 ▶ MINIMUM ARRAY PARTITION

INSTANCE: An $n \times n$ array A of non-negative integers, and a positive integer p.

SOLUTION: $p-1$ horizontal dividers $0 = h_0 < h_1 < \cdots < h_p = n$ and $p-1$ vertical dividers $0 = v_0 < v_1 < \cdots < v_p = n$ partitioning A into p^2 blocks.

MEASURE:
$$\max_{\substack{1 \leq i \leq p \\ 1 \leq j \leq p}} \sum_{\substack{v_{i-1} < x \leq v_i \\ h_{j-1} < y \leq h_j}} A[x,y].$$

Good News: In APX [Khanna, Muthukrishnan, and Skiena, 1997].

Bad News: Not approximable within 2.

Comment: The good news is valid also for generalizations to higher dimensions and other measures. The dual problem, where a limit δ is given instead of p, and where the problem is to find the minimum p such that the array can be partitioned into p^2 blocks where each block's measure is bounded by δ, is approximable within $O(\log n)$ [Khanna, Muthukrishnan, and Skiena, 1997].

MINIMUM RECTANGLE TILING ◂ SP14

INSTANCE: An $n \times n$ array A of non-negative numbers, positive integer p.

SOLUTION: A partition of A into p non-overlapping rectangular subarrays.

MEASURE: The maximum weight of any rectangle in the partition. The weight of a rectangle is the sum of its elements.

Good News: Approximable within 2.5 [Khanna, Muthukrishnan, and Paterson, 1998].

Bad News: Not approximable within 5/4 [Khanna, Muthukrishnan, and Paterson, 1998].

Comment: The dual problem, where the solution is a tiling where each rectangle has weight at most p, and the objective is to minimize the number of rectangles, is approximable within 2. MAXIMUM RECTANGLE PACKING, the variation in which l weighted axis-parallel rectangles are given in the input, and the objective is to maximize the sum of weights of p disjoint rectangles is approximable within $O(\log n)$ [Khanna, Muthukrishnan, and Paterson, 1998].

Storage and Retrieval

Data Storage

SR1 ▶ MINIMUM BIN PACKING

INSTANCE: Finite set U of items, a size $s(u) \in Z^+$ for each $u \in U$, and a positive integer bin capacity B.

SOLUTION: A partition of U into disjoint sets $U_1, U_2, \ldots, U_m$ such that the sum of the items in each U_i is B or less.

MEASURE: The number of bins used, i.e., the number of disjoint sets, m.

Good News: Approximable within factor 3/2 [Simchi-Levi, 1994] and within factor $71/60 + \frac{78}{71m^*}$ [Johnson and Garey, 1985], [Yue, 1991].

Bad News: Not approximable within $3/2 - \varepsilon$ for any $\varepsilon > 0$ [Garey and Johnson, 1979].

Comment: Admits an FPTAS$^\infty$, that is, is approximable within $1 + \varepsilon$ in time polynomial in $1/\varepsilon$, where $\varepsilon = O(\log^2(m^*)/m^*)$ [Karmarkar and Karp, 1982]. APX-intermediate unless the polynomial-time hierarchy collapses [Crescenzi, Kann, Silvestri, and Trevisan, 1999]. A survey of approximation algorithms for MINIMUM BIN PACKING is found in [Coffman, Garey, and Johnson, 1997]. If a partial order on U is defined and we require the bin packing to obey this order, then the problem is approximable within 2 [Wee and Magazine, 1982], and is not in FPTAS$^\infty$ [Queyranne, 1985]. The generalization in which the cost of a bin is a monotone and concave function of the number of items in the bin is approximable within 7/4 and is not approximable within 4/3 unless some information about the cost function is used [Anily, Bramel, and Simchi-Levi, 1994]. The generalization of this problem in which a conflict graph is given such that adjacent items are assigned to different bins is approximable within 2.7 for graphs that can be colored in polynomial time [Jansen and Öhring, 1997] and not approximable within $|U|^\varepsilon$ for a given ε in the general case [Lund and Yannakakis, 1994].

Garey and Johnson: SR1

SR2 ▶ MINIMUM HEIGHT TWO DIMENSIONAL PACKING

INSTANCE: Set of rectangles $B = \{(x_i, y_i)\}$ with positive sizes (width $x_i \leq 1$ and height y_i).

SOLUTION: A packing P of the rectangles in B into a unit-width bin with infinite height. The rectangles must be packed orthogonally and may not be rotated.

MEASURE: Height of the packing P.

Good News: Approximable within 2 [Schiermeyer, 1994].

Comment: Also called *Minimum Strip Packing*. Approximable within $1+\varepsilon + O(\max y_i/\varepsilon^2)/m^*$ in time polynomial in $1/\varepsilon$ for any $\varepsilon > 0$ [Kenyon and Rémila, 1996]. Variation in which the dimensions of the rectangles are bounded below by a constant admits a PTAS [Fernandez de la Vega and Zissimopoulos, 1991]. The three-dimensional variation is approximable within 3.25 [Li and Cheng, 1990].

MAXIMUM d-VECTOR COVERING ◀ SR3

INSTANCE: Set X of vectors in $[0,1]^d$.

SOLUTION: A partition of X into m subsets $S_1,\ldots,S_m$.

MEASURE: The number of unit covers in $\{S_1,\ldots,S_m\}$ where a set S of vectors in $[0,1]^d$ is a unit cover if, for any $i \leq d$, the sum of the i-th components of the elements in S is at least 1.

Good News: Approximable within 4/3 for $d=1$ [Assmann, Johnson, Kleitman, and Leung, 1984] and within $(2\ln d)/(1+o(1))$ for $d \geq 2$ [Alon, Csirik, Sevastianov, Vestjens, and Woeginger, 1996].

MINIMUM DYNAMIC STORAGE ALLOCATION ◀ SR4

INSTANCE: Set A of items to be stored, each $a \in A$ having a size $s(a) \in Z^+$, an arrival time $r(a) \in Z^+$, and a departure time $d(a) \in Z^+$.

SOLUTION: A feasible allocation of storage for A, i.e., a function $\sigma : A \to Z^+$ such that, for all $a, a' \in A$, if $I(a) \cap I(a')$ is nonempty then either $d(a) \leq r(a')$ or $d(a') \leq r(a)$, where $I(x) = [\sigma(x), \sigma(x) + s(x) - 1]$.

MEASURE: The size of the allocation, i.e., $\max_{a \in A}(\sigma(a) + s(a) - 1)$.

Good News: Approximable within 5 [Gergov, 1996].

Comment: Approximable within 5/2 if the sizes are powers of 2 and within 2 if only two sizes are allowed [Gergov, 1996].

Compression and Representation

SHORTEST COMMON SUPERSEQUENCE ◀ SR5

INSTANCE: Finite alphabet Σ, finite set R of strings from Σ^*.

SOLUTION: A string $w \in \Sigma^*$ such that each string $x \in R$ is a subsequence of w, i.e. one can get x by taking away letters from w.

MEASURE: Length of the supersequence, i.e., $|w|$.

Good News: Approximable within $(|R|+3)4$ [Fraser and Irving, 1995].

The compendium

STORAGE AND
RETRIEVAL

Bad News: Not in APX [Jiang and Li, 1994b].

Comment: Transformation from MINIMUM FEEDBACK VERTEX SET and self-improvability. Not approximable within $\log^\delta |R|$ for some $\delta > 0$ unless NP$\subset$ DTIME($n^{\text{polylog}\,n}$) [Jiang and Li, 1994b]. APX-complete if the size of the alphabet Σ is fixed [Jiang and Li, 1994b] and [Bonizzoni, Duella, and Mauri, 1994]. Variation in which the objective is to find the longest minimal common supersequence (a supersequence that cannot be reduced to a shorter common supersequence by removing a letter) is APX-hard even over the binary alphabet, and SHORTEST MAXIMAL COMMON NON-SUPERSEQUENCE is APX-hard even over the binary alphabet [Middendorf, 1994]. Variation in which the longest common subsequence is known is also APX-hard even over the binary alphabet [D'Anzeo, 1996].

Garey and Johnson: SR8

SR6 ▶ SHORTEST COMMON SUPERSTRING

INSTANCE: Finite alphabet Σ, finite set R of strings from Σ^*.

SOLUTION: A string $w \in \Sigma^*$ such that each string $x \in R$ is a substring of w, i.e., $w = w_0 x w_1$ where $w_0, w_1 \in \Sigma^*$.

MEASURE: Length of the superstring, i.e., $|w|$.

Good News: Approximable within 2.75 [Armen and Stein, 1994].

Bad News: APX-complete [Blum, Jiang, Li, Tromp, and Yannakakis, 1994].

Comment: Transformation from MINIMUM METRIC TRAVELING SALESPERSON with distances one and two. Variation in which there are negative strings in the input and a solution cannot contain any negative string as a substring, is approximable within $O(\log m^*)$ [Li, 1990]. If the number of negative strings is constant, or if no negative strings contain positive strings as substrings, the problem is approximable within a constant [Jiang and Li, 1994a]. The complementary MAXIMUM COMPRESSION problem, where the objective is to maximize $\sum_{x \in R} |x| - |w|$, is approximable within 2 [Tarhio and Ukkonen, 1988] and [Turner, 1989].

Garey and Johnson: SR9

SR7 ▶ LONGEST COMMON SUBSEQUENCE

INSTANCE: Finite alphabet Σ, finite set R of strings from Σ^*.

SOLUTION: A string $w \in \Sigma^*$ such that w is a subsequence of each $x \in R$, i.e. one can get w by taking away letters from each x.

MEASURE: Length of the subsequence, i.e., $|w|$.

Good News: Approximable within $O(m/\log m)$, where m is the length of the shortest string in R [Halldórsson, 1995b].

Bad News: Not approximable within $n^{1/4-\varepsilon}$ for any $\varepsilon > 0$, where n is the maximum of $|R|$ and $|\Sigma|$ [Berman and Schnitger, 1992], [Jiang and Li, 1994b] and [Bellare, Goldreich, and Sudan, 1998].

Comment: Transformation from the MAXIMUM INDEPENDENT SET problem. APX-complete if the size of the alphabet Σ is fixed [Jiang and Li, 1994b] and [Bonizzoni, Duella, and Mauri, 1994]. Variation in which the objective is to find the shortest maximal common subsequence (a subsequence that cannot be extended to a longer common subsequence) is not approximable within $|R|^{1-\varepsilon}$ for any $\varepsilon > 0$ [Fraser, Irving, and Middendorf, 1996].

Garey and Johnson: SR10

MAXIMUM COMMON SUBTREE ◀ SR8

INSTANCE: Collection $T_1, \ldots, T_n$ of trees.

SOLUTION: A tree T' isomorphic to a subtree (i.e., induced connected subgraph) of each T_i.

MEASURE: Size, i.e., number of nodes, of the common subtree T'.

Good News: Approximable within $O(\log\log n / \log^2 n)$, where n is the size of the smallest T_i [Akutsu and Halldórsson, 1997].

Bad News: Not approximable within factor $n^{1/4-\varepsilon}$ for any $\varepsilon > 0$ [Akutsu and Halldórsson, 1997].

Comment: Transformation from the MAXIMUM INDEPENDENT SET problem. $n^{1-\varepsilon}$ hardness holds for the variation where the nodes are labeled and where maximum degree is bounded by a constant $B \geq 3$. Approximable within $O(n/\log^2 n)$ when the maximum degree is constant. Variation in which the nodes are labeled is approximable within $O(\log\log n / \log^2 n)$ if the number of distinct labels is $O(log^{O(1)} n)$ and within $O(n/\log n)$ in general [Akutsu and Halldórsson, 1997]. Approximable within 2 if the solution is an isomorphic edge subset of each tree. APX-hard when the number of input trees is a constant at least 3, but polynomial-time solvable if the maximum degree is additionally bounded by constant or for only two trees [Akutsu and Halldórsson, 1994].

MAXIMUM COMMON POINT SET ◀ SR9

INSTANCE: Positive integer d, collection $S_1, \ldots, S_k$ of d-dimensional point sets.

SOLUTION: A point set S' congruent to a subset of each set S_i in the collection.

MEASURE: Size, i.e., number of elements, of the common point set S'.

Good News: Approximable within $O(n/\log n)$, where n is the size of the smallest S_i [Akutsu and Halldórsson, 1994].

Bad News: Not approximable within factor n^ε for some $\varepsilon > 0$ [Akutsu and Halldórsson, 1994].

The compendium

STORAGE AND RETRIEVAL

Comment: Transformation from MAXIMUM INDEPENDENT SET. The bad news is valid even if a small gap is allowed between corresponding points.

SR10 ▶ MINIMUM RECTANGLE COVER

INSTANCE: An arbitrary polygon P.

SOLUTION: A collection of m rectangles whose union is exactly equal to the polygon P.

MEASURE: Size, i.e. number m of elements, of the collection.

Good News: Approximable within $O(\sqrt{n}\log n)$, where n denotes the number of vertices of the polygon [Levcopoulos, 1997].

Comment: If the vertices are given as polynomially bounded integer coordinates, then the problem is $O(\log n)$-approximable [Gudmundsson and Levcopoulos, 1999]. If the polygon is hole-free then it is approximable within $O(\alpha(n))$, where $\alpha(n)$ is the extremely slowly growing inverse of Ackermanns' function [Gudmundsson and Levcopoulos, 1997]. In the case of rectilinear polygons with holes, the rectangular covering is APX-hard [Berman and DasGupta, 1997]. If the solution can contain only squares, then the problem is 14-approximable [Levcopoulos and Gudmundsson, 1997].

Garey and Johnson: SR25

Miscellaneous

SR11 ▶ MINIMUM TREE COMPACT PACKING

INSTANCE: A tree $T = (V,E)$, a node-weight function $w : V \to Q^+$ such that $\sum_{v \in V} w(v) = 1$, and a page-capacity p.

SOLUTION: A compact packing of T into pages of capacity p, i.e., a function $\tau : V \to Z^+$ such that $|\tau^{-1}(i)| = p$.

MEASURE: The number of page faults of the packing, i.e., $\sum_{v \in V} c_\tau(v) w(v)$ where

$$c_\tau(v) = \sum_{i=0}^{l(v)-1} \Delta_\tau(v_i),$$

$l(v)$ denotes the number of edges in the path from the root to v, v_i denotes the ith node in this path, and $\Delta_\tau(v)$ is equal to 0 if the parent of v is assigned the same page of u, it is equal to 1 otherwise.

Good News: Approximable with an absolute error guarantee of 1 [Gil and Itai, 1995].

Comment: If all $w(v)$ are equal then it is approximable with an absolute error guarantee of 1/2.

Sequencing and Scheduling

Sequencing on One Processor

MAXIMUM JOB SEQUENCING TO MINIMIZE TASK WEIGHT ◀ SS1

INSTANCE: Set T of tasks, for each task $t \in T$ a length $l(t) \in Z^+$, a weight $w(t) \in Z^+$, and a deadline $d(t) \in Z^+$, a subset $S \subseteq T$, and a positive integer K.

SOLUTION: A one-processor schedule σ for T such that the sum of $w(t)$, taken over all $t \in T$ for which $\sigma(t) + l(t) > d(t)$ does not exceed K.

MEASURE: Cardinality of jobs in S completed by the deadline.

Bad News: Not approximable within $|T|^\varepsilon$ for some $\varepsilon > 0$ [Zuckerman, 1993].

Garey and Johnson: Similar to SS3

MINIMUM STORAGE-TIME SEQUENCING ◀ SS2

INSTANCE: Set T of tasks, a directed acyclic graph $G = (T, E)$ defining precedence constraints for the tasks, for each task a length $l(t) \in Z^+$, and for each edge in the graph a weight $w(t_1, t_2)$ measuring the storage required to save the intermediate results generated by task t_1 until it is consumed by task t_2.

SOLUTION: A one-processor schedule for T that obeys the precedence constraints, i.e., a permutation $\pi : [1..|T|] \to [1..|T|]$ such that, for each edge $(t_i, t_j) \in E$, $\pi^{-1}(i) < \pi^{-1}(j)$.

MEASURE: The total storage-time product, i.e.,

$$\sum_{(t_{\pi(i)}, t_{\pi(j)}) \in E} w(t_{\pi(i)}, t_{\pi(j)}) \sum_{k=\min(i,j)}^{\max(i,j)} l(t_{\pi(k)}).$$

Good News: Approximable within $O(\log |T|)$ [Rao and Richa, 1998].

MINIMUM CONSTRAINED SEQUENCING WITH DELAYS ◀ SS3

INSTANCE: Set T of tasks, a directed acyclic graph $G = (T, E)$ defining precedence constraints for the tasks, a positive integer D, and for each task an integer delay $0 \leq d(t) \leq D$.

SOLUTION: A one-processor schedule for T that obeys the precedence constraints and the delays, i.e., an injective function $S : T \to Z^+$ such that, for each edge $(t_i, t_j) \in E$, $S(t_j) - S(t_i) > d(t_i)$.

MEASURE: The finish time for the schedule, i.e., $\max_{t \in T} S(t)$.

Good News: Approximable within $2 - 1/(D+1)$ [Bernstein, Rodeh, and Gertner, 1989].

Comment: Approximable within $2 - 2/(D+1)$ if $d(t) \in \{0, D\}$ for all tasks t.

SS4 ▶ Minimum Sequencing with Release Times

INSTANCE: Set T of tasks, for each task $t \in T$ a release time $r(t) \in Z^+$, a length $l(t) \in Z^+$, and a weight $w(t) \in Z^+$.

SOLUTION: A one-processor schedule for T that obeys the release times, i.e., a function $f : T \to N$ such that, for all $u \geq 0$, if $S(u)$ is the set of tasks t for which $f(t) \leq u < f(t) + l(t)$, then $|S(u)| = 1$ and for each task t, $f(t) \geq r(t)$.

MEASURE: The weighted sum of completion times, i.e. $\sum_{t \in T} w(t)(f(t) + l(t))$.

Good News: Approximable within factor 1.6853 [Goemans, Queyranne, Schulz, Skutella, and Wang, 1998].

Comment: Approximable within factor $e/(e-1) \approx 1.58$ if all weights $w(t) = 1$ [Chekuri, Motwani, Natarajan, and Stein, 1997]. Variation in which there are precedence constraints on T instead of release times is approximable within 2 [Hall, Schulz, Shmoys, and Wein, 1997]. The preemptive case is approximable within $4/3$ [Schulz and Skutella, 1996]. Variation in which there are precedence constraints and release times is approximable within e [Schulz and Skutella, 1997b]. If all weights $w(t) = 1$ and the objective is to minimize the max-stretch $\max_{t \in T} (f(t) + l(t) - r(t))/l(t)$ the nonpreemptive problem is not approximable within $O(|T|^{1-\varepsilon})$ for any $\varepsilon > 0$, but the preemptive problem admits a PTAS [Bender, Chakrabarti, and Muthukrishnan, 1998].

SS5 ▶ Minimum Time-Cost Tradeoff

INSTANCE: Set J of activities, a directed acyclic graph defining precedence constraints for the activities, a positive budget B, and for each activity $j \in J$ a non-increasing cost function c_j described as a step function with l_j steps:

$$c_j(t) = \begin{cases} \infty & \text{if } 0 \leq t < a_{j,1}, \\ c_j(a_{j,i}) & \text{if } a_{j,i} \leq t < a_{j,i+1}, 1 \leq i < l_j, \\ 0 & \text{if } a_{j,l_j} \leq t, \end{cases}$$

where $\infty > c_j(a_{j,1}) > \cdots > c_j(a_{j,l_j})$.

SOLUTION: A one-processor schedule for J that obeys the precedence constraints and activities durations $t_j > 0$ for each $j \in J$ that obeys the budget, i.e. $\sum_{j \in J} c_j(t_j) \leq B$.

MEASURE: The total duration of all activities, i.e., $\sum_{j \in J} t_j$.

Good News: Approximable within $(3/2)\log f + (7/2)$, where f is the ratio of the maximum allowed duration of any activity to the minimum allowed non-zero duration of any activity [Skutella, 1997].

Comment: Variation in which $a_{j,i} \in \{0,1,2\}$ for all i,j is approximable within $3/2$ but not approximable within $3/2 - \varepsilon$ for any $\varepsilon > 0$ [Skutella, 1997].

Multiprocessor Scheduling

MINIMUM MULTIPROCESSOR SCHEDULING ◂ SS6

INSTANCE: Set T of tasks, number m of processors, length $l(t,i) \in Z^+$ for each task $t \in T$ and processor $i \in [1..m]$.

SOLUTION: An m-processor schedule for T, i.e., a function $f : T \to [1..m]$.

MEASURE: The finish time for the schedule, i.e.,

$$\max_{i \in [1..m]} \sum_{\substack{t \in T: \\ f(t)=i}} l(t,i).$$

Good News: Approximable within 2 [Lenstra, Shmoys, and Tardos, 1990].

Bad News: Not approximable within $3/2 - \varepsilon$ for any $\varepsilon > 0$ [Lenstra, Shmoys, and Tardos, 1990].

Comment: Admits an FPTAS for the variation in which the number of processors m is constant [Horowitz and Sahni, 1976]. Admits a PTAS for the uniform variation, in which $l(t,i)$ is independent of the processor i [Hochbaum and Shmoys, 1987].

A variation in which, for each task t and processor i, a cost $c(t,i)$ is given in input and the goal is to minimize a weighted sum of the finish time and the cost is approximable within 2 [Shmoys and Tardos, 1993].

Variation in which each task has a nonnegative penalty and the objective is to minimize the finish time for accepted jobs plus the sum of the penalties of rejected jobs, is approximable within $2-1/m$ and admits an FPTAS for fixed m [Bartal et al., 1996].

Garey and Johnson: Generalization of SS8

MINIMUM PRECEDENCE CONSTRAINED SCHEDULING ◂ SS7

INSTANCE: Set T of tasks, each having length $l(t) = 1$, number m of processors, and a partial order $<$ on T.

SOLUTION: An m-processor schedule for T that obeys the precedence constraints, i.e., a function $f : T \to N$ such that, for all $u \geq 0$, $|f^{-1}(u)| \leq m$ and such that $t < t'$ implies $f(t') > f(t)$.

The compendium

SEQUENCING AND
SCHEDULING

MEASURE: The finish time for the schedule, i.e., $\max_{t \in T} f(t)$.

Good News: Approximable within $2 - 2/|T|$ [Lam and Sethi, 1977].

Bad News: Not approximable within $4/3 - \varepsilon$ for any $\varepsilon > 0$ [Lenstra and Rinnooy Kan, 1978].

Comment: A variation with an enlarged class of allowable constraints is approximable within $3 - 4/(|T|+1)$ while a variation in which the partial order $<$ is substituted with a weak partial order $\leq$ is approximable within $2 - 2/(|T|+1)$ [Berger and Cowen, 1991].

Variation in which there is a communication delay of 1, i.e., if two tasks $t < t'$ are scheduled on different machines, then $f(t') > f(t) + 1$, is approximable within 3 [Rayward-Smith, 1987], and not approximable within $5/4 - \varepsilon$ [Hoogeveen, Lenstra, and Veltman, 1995]. The same problem without limit on the number of processors, i.e., with $m = \infty$, is not approximable within $7/6 - \varepsilon$ [Hoogeveen, Lenstra, and Veltman, 1995].

The variation of this problem in which the average weighted completion time has to be minimized is approximable within 10/3 (respectively, 6.143 if the tasks have arbitrary length) [Möhring, Schäffter, and Schulz, 1996].

Generalization where the tasks have lengths $l(t)$ and the processors have speed factors $s(i)$ is approximable within $O(\log m)$ [Chudak and Shmoys, 1997].

Garey and Johnson: SS9

SS8 ▶ MINIMUM RESOURCE CONSTRAINED SCHEDULING

INSTANCE: Set T of tasks each having length $l(t)$, number m of processors, number r of resources, resource bounds b_i, $1 \leq i \leq r$, and resource requirement $r_i(t)$, $0 \leq r_i(t) \leq b_i$, for each task t and resource i.

SOLUTION: An m processor schedule for T that obeys the resource constraints, i.e., a function $f : T \to Z$ such that for all $u \leq 0$, if $S(u)$ is the set of tasks t for which $f(t) \leq u < f(t) + l(t)$, then $|S(u)| \leq m$ and for each resource i

$$\sum_{t \in S(u)} r_i(t) \leq b_i.$$

MEASURE: The makespan of the schedule, i.e., $\max_{t \in T}(f(t) + l(t))$.

Good News: Approximable within 2 [Garey and Graham, 1975].

Comment: Note that the restriction in which there is only one resource, i.e., the available processors, is identical to minimizing the makespan of the schedule of parallel tasks on m processors. In this case, minimizing the average response time, i.e., $\frac{1}{|T|}\sum_{t \in T}(f(t) + l(t))$ is approximable within 32 [Turek, Schwiegelshohn, Wolf, and Yu, 1994].

The further variation in which each task can be executed by any number of processors and the length of a task is a function of the number of processors allotted to it is also approximable [Ludwig and Tiwari, 1994].

Variation in which every task is of length 1 and has an integer ready-time, which means that it cannot be processed before its ready-time, is approximable [Srivastav and Stangier, 1994].

Garey and Johnson: SS10

◀ SS9 MINIMUM PREEMPTIVE SCHEDULING WITH SET-UP TIMES

INSTANCE: Set C of compilers, set T of tasks, number m of processors, length $l(t) \in Z^+$ and compiler $c(t) \in C$ for each $t \in T$, set-up time $s(c) \in Z^+$ for each $c \in C$.

SOLUTION: An m-processor preemptive schedule for T, i.e., for each $t \in T$, a partition of t into any number of subtasks $t_1, \ldots, t_k$ such that $\sum_{i=1}^{k} l(t_i) = l(t)$ and an assignment $\sigma = (\sigma_1, \sigma_2)$ that assigns to each subtask t_i of t a pair of nonnegative integers $(\sigma_1(t_i), \sigma_2(t_i))$ such that $1 \leq \sigma_1(t_i) \leq m$ and, for $1 \leq i < k$, $\sigma_2(t_{i+1}) \geq \sigma_2(t_i) + l(t_i)$. Such a schedule must satisfy the following additional constraint: Whenever two subtasks t_i of t and t'_j of t' with $\sigma_2(t_i) < \sigma_2(t'_j)$ are scheduled consecutively on the same machine (i.e., $\sigma_1(t_i) = \sigma_1(t'_j)$ and no other subtask t''_k has $\sigma_1(t''_k) = \sigma_1(t_i)$ and $\sigma_2(t_i) < \sigma_2(t''_k) < \sigma_2(t'_j)$), then $\sigma_2(t'_j) \geq \sigma_2(t_i) + l(t_i)$ if they have the same compiler (i.e., $c(t) = c(t')$) and $\sigma_2(t'_j) \geq \sigma_2(t_i) + l(t_i) + s(c(t'))$ if they have different compilers.

MEASURE: The overall completion time, i.e., the maximum over all subtasks of $\sigma_2(t_i) + l(t_i)$.

Good News: Approximable within

$$\max\left(\frac{3m}{2m+1}, \frac{3m-4}{2m-2}\right)$$

[Chen, 1993].

Comment: If all set-up times are equal, then the problem is approximable within $3/2 - 1/2m$ for $m \geq 2$ and admits an FPTAS for $m = 2$ [Wöginger and Yu, 1992].

Garey and Johnson: SS6 and SS12

◀ SS10 MINIMUM PROCESSOR SCHEDULING WITH SPEED FACTORS

INSTANCE: Set T of tasks, number m of processors, for each task $t \in T$ a length $l(t) \in Z^+$, and for each processor $i \in [1..m]$ a speed factor $s(i) \in Q$ such that $s(1) = 1$ and $s(i) \geq 1$ for every i.

SOLUTION: An m-processor schedule for T, i.e., a function $f : T \to [1..m]$.

The compendium

SEQUENCING AND
SCHEDULING

MEASURE: The finish time for the schedule, i.e., $\max_{i\in[1..m]} \sum_{\substack{t\in T:\\ f(t)=i}} l(t)/s(i)$.

Good News: Admits a PTAS [Hochbaum and Shmoys, 1988].

Bad News: Does not admit an FPTAS [Hochbaum and Shmoys, 1988].

Comment: Admits an FPTAS for the variation in which the number of processors m is constant [Horowitz and Sahni, 1976].

SS11 ▶ MINIMUM PARALLEL PROCESSOR TOTAL FLOW TIME

INSTANCE: Set T of tasks, number m of identical processors, for each task $t \in T$ a release time $r(t) \in Z^+$ and a length $l(t) \in Z^+$.

SOLUTION: An m-processor schedule for T that obeys the resource constraints and the release times, i.e., a function $f : T \to N \times [1..m]$ such that, for all $u \geq 0$ and for each processor i, if $S(u,i)$ is the set of tasks t for which $f(t)_1 \leq u < f(t)_1 + l(t)$ and $f(t)_2 = i$, then $|S(u,i)| = 1$ and for each task t, $f(t)_1 \geq r(t)$.

MEASURE: The total flow time for the schedule, i.e., $\sum_{t\in T} (f(t)_1 - r(t))$.

Good News: Approximable within factor $O\left(\sqrt{n/m}\log(n/m)\right)$ where $n = |T|$ [Leonardi and Raz, 1997].

Bad News: Not approximable within $O(n^{1/3-\varepsilon})$ for any $\varepsilon > 0$ [Leonardi and Raz, 1997].

Comment: In the case $m = 1$, it is approximable within $\sqrt{n}$ and it is not approximable within $O(n^{1/2-\varepsilon})$ for any $\varepsilon > 0$ [Kellerer, Tautenhahn, and Woeginger, 1996]. In the preemptive case, that is, in the case a job that is running can be preempted and continue later on any machine, the problem is approximable within $O(\log(\min\{n/m,P\}))$ and it is not approximable within $O(\log(\max\{n/m,P\}))$ where $P = \max_{s,t\in T} l(s)/l(t)$ [Leonardi and Raz, 1997]. Variation in which all speed factors are 1 and the load is measured using the L_p norm, i.e., $\max_{[1..m]}(\sum l(t)^p)^{1/p}$, admits a PTAS for any $p \geq 1$ [Alon, Azar, Woeginger, and Yadid, 1997].

SS12 ▶ MINIMUM WEIGHTED COMPLETION TIME SCHEDULING

INSTANCE: Set T of tasks, number m of identical processors, for each task $t \in T$ a release time $r(t) \in Z^+$, a length $l(t) \in Z^+$, and a weight $w(t) \in Z^+$.

SOLUTION: An m-processor schedule for T that obeys the resource constraints and the release times, i.e., a function $f : T \to N \times [1..m]$ such that, for all $u \geq 0$ and for each processor i, if $S(u,i)$ is the set of tasks t for which $f(t)_1 \leq u < f(t)_1 + l(t)$ and $f(t)_2 = i$, then $|S(u,i)| = 1$ and for each task t, $f(t)_1 \geq r(t)$.

MEASURE: The weighted sum of completion times, i.e., $\sum_{t\in T} w(t)(f(t)_1 + l(t))$.

Good News: Approximable within 2 [Schulz and Skutella, 1997a].

Comment: The preemptive case is also approximable within 2 [Phillips, Stein, and Wein, 1995]. Generalization to unrelated processors (where the length of a task also depends in the processor) is approximable within $2+\varepsilon$ for any $\varepsilon > 0$ for both nonpreemptive and preemptive scheduling [Schulz and Skutella, 1997a]. Restriction (of the nonpreemptive case) where $r(t) = 0$ for all t is approximable within 3/2 [Schulz and Skutella, 1997b]. Generalization where there are precedence constraints on T that must be obeyed in the solution is approximable within 5.33 [Chakrabati, Phillips, Schulz, Shmoys, Stein, and Wein, 1996].

Garey and Johnson: SS13

MINIMUM 3-DEDICATED PROCESSOR SCHEDULING ◀ SS13

INSTANCE: Set T of tasks, set P of 3 processors, and, for each task $t \in T$, a length $l(t) \in Z^+$ and a required subset of processors $r(t) \subseteq P$.

SOLUTION: A schedule for T, i.e., a starting time function $s: T \to Z^+$ such that, for any two tasks t_1 and t_2 with $r(t_1) \cap r(t_2) \neq \emptyset$, either $s(t_1) + l(t_1) < s(t_2)$ or $s(t_2) + l(t_2) < s(t_1)$.

MEASURE: The makespan of the schedule, i.e., $\max_{t \in T}(s(t) + r(t))$.

Good News: Admits a PTAS [Amoura, Bampis, Kenyon, and Manoussakis, 1997].

Comment: Admits a PTAS even for any fixed number of processors.

Shop Scheduling

MINIMUM OPEN-SHOP SCHEDULING ◀ SS14

INSTANCE: Number $m \in Z^+$ of processors, set J of jobs, each $j \in J$ consisting of m operations $o_{i,j}$ with $1 \leq i \leq m$ (with $o_{i,j}$ to be executed by processor i), and for each operation a length $l_{i,j} \in N$.

SOLUTION: An open-shop schedule for J, i.e., a set of one-processor schedules $f_i: J \to N$, $1 \leq i \leq m$, such that $f_i(j) > f_i(j')$ implies $f_i(j) \geq f_i(j') + l_{i,j'}$, such that for each $j \in J$ the intervals $[f_i(j), f_i(j) + l_{i,j})$ are all disjoint.

MEASURE: The completion time of the schedule, i.e.,

$$\max_{1 \leq i \leq m, j \in J} f_i(j) + l_{i,j}.$$

Good News: Approximable within 2 [Lenstra and Shmoys, 1995].

Bad News: Not approximable within $5/4 - \varepsilon$ for any $\varepsilon > 0$ [Williamson et al., 1994].

The compendium

SEQUENCING AND
SCHEDULING

Comment: Approximable within 3/2 if $m = 3$ [Chen and Strusevich, 1993a]. Variation in which $m = 2$, but the two processors are replaced by respectively m_1 and m_2 identical parallel processors, is approximable within 3/2 if $\min(m_1,m_2) = 1$, 5/3 if $m_1 = m_2 = 2$ and $2 - 2/\max(m_1+m_2)^2$ otherwise [Chen and Strusevich, 1993b].

Garey and Johnson: SS14

SS15 ▶ MINIMUM FLOW-SHOP SCHEDULING

INSTANCE: Number $m \in Z^+$ of processors, set J of jobs, each $j \in J$ consisting of m operations $o_{i,j}$ with $1 \leq i \leq m$ (with $o_{i,j}$ to be executed by processor i), and for each operation a length $l_{i,j} \in N$.

SOLUTION: An open-shop schedule for J (see the definition of MINIMUM OPEN-SHOP SCHEDULING) such that, for each $j \in J$ and $1 \leq i < m$, $f_{i+1}(j) \geq f_i(j) + l_{i,j}$.

MEASURE: The completion time of the schedule, i.e.,
$$\max_{j \in J} f_m(j) + l_{m,j}.$$

Good News: Approximable within $m/2$ if m is even and within $m/2 + 1/6$ if m is odd [Chen et al., 1996].

Bad News: Not approximable within $5/4 - \varepsilon$ for any $\varepsilon > 0$ [Williamson et al., 1994].

Comment: Approximable within 5/3 if $m = 3$ [Chen et al., 1996]. Variation in which $m = 2$, but the two processors are replaced by respectively m_1 and m_2 identical parallel processors, is approximable within $2 - 1/\max(m_1+m_2)$ [Chen, 1994].

Garey and Johnson: SS15

SS16 ▶ MINIMUM FLOW-SHOP SCHEDULING WITH SET-UP TIMES

INSTANCE: Set C of compilers, set J of jobs, for each $j \in J$, which consists of two operations $o_{i,j}$, $i = 1,2$, a compiler $k(j) \in C$, for each operation $o_{i,j}$ a length $l_{i,j} \in N$ and for each $c \in C$ a pair of set-up times $(s_1(c), s_2(c))$, where $s_i(c) \in Z^+$.

SOLUTION: A two-processor flow-shop schedule for J (see the definition of MINIMUM FLOW-SHOP SCHEDULING) such that, whenever two operations $o_{i,j}$ and $o_{i,j'}$ with $f_i(j) < f_i(j')$ are scheduled consecutively (i.e., no other operation $o_{i,j''}$ has $f_i(j) < f_i(j'') < f_i(j')$) and have different compilers (i.e., $k(j) \neq k(j')$), then $f_i(j') \geq f_i(j) + l_{i,j} + s_i(k(j'))$.

MEASURE: The completion time of the schedule, i.e.,

$$\max_{j \in J} f_2(j) + l_{2,j}.$$

Good News: Approximable within 4/3 [Chen, Potts, and Strusevich, 1998].

MINIMUM JOB SHOP SCHEDULING ◀ SS17

INSTANCE: Number $m \in Z^+$ of processors, set J of jobs, each $j \in J$ consisting of a sequence of n_j operations $o_{i,j}$ with $1 \leq i \leq n_j$, for each such operation a processor $p_{i,j} \in [1..m]$ and a length $l_{i,j} \in N$.

SOLUTION: A job shop schedule for J, i.e., a collection of one-processor schedules $f_p : \{o_{i,j} : p_{i,j} = p\} \to N$ such that $f_p(o_{i,j}) > f_p(o_{i',j'})$ implies $f_p(o_{i,j}) \geq f_p(o_{i',j'}) + l_{i',j'}$ and such that $f_p(o_{i+1,j}) \geq f_p(o_{i,j}) + l_{i,j}$.

MEASURE: The completion time of the schedule, i.e., $\max_{j \in J} f_p(o_{n_j,j}) + l_{n_j,j}$.

Good News: Approximable within

$$O\left(\lceil \log(\min(m\mu, l_{\max}))/\log\log(m\mu)\rceil \log(m\mu)/\log\log(m\mu)\right),$$

where $\mu = \max_{j \in J} n_j$ and $l_{\max} = \max l_{i,j}$ [Goldberg, Paterson, Srinivasan, and Sweedyk, 1997].

Bad News: Not approximable within $5/4 - \varepsilon$ for any $\varepsilon > 0$ [Williamson et al., 1994].

Comment: Transformation from 3-PARTITION. Approximable within $O(\log^2 m)$ if each job must be processed on each machine at most once [Shmoys, Stein, and Wein, 1994]. The variation in which the operations must be processed in an order consistent to a particular partial order, and the variation in which there are different types of machines, for each type, there are a specified number of identical processors, and each operation may be processed on any processor of the appropriate type, are approximable within $O(\log^2(m\mu))$ [Shmoys, Stein, and Wein, 1994].

Garey and Johnson: SS18

Miscellaneous

MINIMUM FILE TRANSFER SCHEDULING ◀ SS18

INSTANCE: A file transfer graph, i.e., a graph $G = (V, E)$, a port constraint function $p : V \to N$ and a file length function $L : E \to N$.

SOLUTION: A file transfer schedule, i.e., a function $s : E \to N$ such that, for each vertex v and for each $t \in N$,

$$|\{u : (u,v) \in E \land s(e) \leq t \leq s(e) + L(e)\}| \leq p(v).$$

The compendium

SEQUENCING AND
SCHEDULING

MEASURE: The makespan of the schedule, i.e., $\max_{e \in E}(s(e) + L(e))$.

Good News: Approximable within 2.5 [Coffman, Garey, Johnson, and Lapaugh, 1985].

Comment: Several special cases with better guarantees are also obtainable [Coffman, Garey, Johnson, and Lapaugh, 1985].

SS19 ▸ MINIMUM SCHEDULE LENGTH

INSTANCE: A network $N = (V, E, b, c)$ where $G = (V, E)$ is a graph, $b : V \to \mathbb{N}$ is the vertex-capacity function, and $c : E \to \mathbb{N}$ is the edge-capacity function, and a set T of tokens $t = (u, v, p)$ where $u, v \in V$ and p is either a path from u to v or the empty set.

SOLUTION: A schedule S, i.e., a sequence $f_0, \ldots, f_l$ of configuration functions $f_i : T \to V$ such that

1. for any token $t = (u, v, p)$, $f_0(t) = u$ and $f_l(t) = v$,
2. for any $0 \leq i \leq l-1$ and for any token t, if $f_i(t) = v$ and $f_{i+1}(t) = w$ then (a) $(u,v) \in E$, (b) $|\{t' : f_i(t') = w\}| < b(w)$, (c) $|\{t' : f_{i+1}(t') = w\}| \leq b(w)$, and (d) $|\{t' : f_i(t') = v \wedge f_{i+1}(t') = w\}| \leq c(v, w)$.

MEASURE: The length of the schedule, i.e., l.

Bad News: Not in APX [Clementi and Di Ianni, 1996].

Comment: Transformation from MINIMUM GRAPH COLORING. It remains non-approximable even for layered graphs. The variation in which the vertex-capacity is unbounded and the delay, i.e., the maximum number of times a token is neither in the final destination nor moved, must be minimized is also non-approximable within $|T|^{1-\varepsilon}$ for any ε [Di Ianni, 1998].

SS20 ▸ MINIMUM VEHICLE SCHEDULING ON TREE

INSTANCE: Rooted tree $T = (V, E, v_0)$, a forward travel time $f : E \to \mathbb{N}$, a backward travel time $b : E \to \mathbb{N}$, a release time $r : V \to \mathbb{N}$, and an handling time $h : V \to \mathbb{N}$.

SOLUTION: A vehicle routing schedule that starts from v_0, visits all nodes of T, returns to v_0, and, for any node v_i, starts processing v_i not before the release time $r(v_i)$, i.e., a permutation π of $1, \ldots, |V|$ and a waiting function w such that, for any i,

$$d(v_0, v_{\pi(1)}) + \sum_{j=1}^{i-1}[w(v_{\pi(j)}) + h(v_{\pi(j)}) + d(v_{\pi(j)}, v_{\pi(j+1)})] \geq r(v_{\pi(i)})$$

where $d(u, v)$ denotes the length of the unique path from u to v.

MEASURE: The total completion time, i.e.,

$$d(v_0, v_{\pi(1)}) + \sum_{j=1}^{n-1} [w(v_{\pi(j)}) + h(v_{\pi(j)}) + d(v_{\pi(j)}, v_{\pi(j+1)})] + w(v_{\pi(n)}) + h(v_{\pi(n)}) + d(v_{\pi(n)}, v_0).$$

Good News: Approximable within factor 2 [Karuno, Nagamochi, and Ibaraki, 1993].

Mathematical Programming

MP1 ▶ MINIMUM $\{0,1\}$-LINEAR PROGRAMMING

INSTANCE: Integer $m \times n$-matrix $A \in Z^{m \cdot n}$, integer m-vector $b \in Z^m$, nonnegative integer n-vector $c \in N^n$.

SOLUTION: A binary n-vector $x \in \{0,1\}^n$ such that $Ax \geq b$.

MEASURE: The scalar product of c and x, i.e., $\sum_{i=1}^{n} c_i x_i$.

Bad News: NPO-complete [Orponen and Mannila, 1987].

Comment: Transformation from MINIMUM WEIGHTED SATISFIABILITY. Variation in which $c_i = 1$ for all i is NPOPB-complete and not approximable within $n^{1-\varepsilon}$ for any $\varepsilon > 0$ [Kann, 1994b]. Variation in which there are at most k non-zero entries on each row of the matrix with k constant is approximable within k [Hall and Hochbaum, 1986] and [Hochbaum, Megiddo, Naor, and Tamir, 1993].

Garey and Johnson: MP1

MP2 ▶ MAXIMUM BOUNDED $\{0,1\}$-LINEAR PROGRAMMING

INSTANCE: Integer $m \times n$-matrix $A \in Z^{m \cdot n}$, integer m-vector $b \in Z^m$, nonnegative binary n-vector $c \in \{0,1\}^n$.

SOLUTION: A binary n-vector $x \in \{0,1\}^n$ such that $Ax \leq b$.

MEASURE: The scalar product of c and x, i.e., $\sum_{i=1}^{n} c_i x_i$.

Bad News: NPOPB-complete [Berman and Schnitger, 1992].

Comment: Transformation from LONGEST PATH WITH FORBIDDEN PAIRS. Not approximable within $n^{1-\varepsilon}$ [Jonsson, 1997] and not within $m^{1-\varepsilon}$ for any $\varepsilon > 0$ [Kann, 1995].

Garey and Johnson: MP1

MP3 ▶ MAXIMUM PACKING INTEGER PROGRAMMING

INSTANCE: Rational $m \times n$-matrix $A \in [0,1]^{m \cdot n}$, rational m-vector $b \in [1,\infty)^m$, rational n-vector $c \in [0,1]^n$.

SOLUTION: A rational n-vector $x \in [0,1]^n$ such that $Ax \leq b$.

MEASURE: The scalar product of c and x, i.e., $\sum_{i=1}^{n} c_i x_i$.

Good News: Approximable within $O(m^*(cm^{1/B}/m^*)^{B/(B-1)})$ for some $c > 0$, where $B = \min_i b_i$ [Srinivasan, 1995].

Comment: Variation in which $A \in \{0,1\}^{m \cdot n}$ and $b \in Z^+$ is approximable within $O(m^*(m^{1/(B+1)}/m^*)^{(B+1)/B})$ [Srinivasan, 1995].

MINIMUM COVERING INTEGER PROGRAMMING ◀ MP4

INSTANCE: Rational $m \times n$-matrix $A \in [0,1]^{m \cdot n}$, rational m-vector $b \in [1,\infty)^m$, rational n-vector $c \in [0,1]^n$.

SOLUTION: A rational n-vector $x \in [0,1]^n$ such that $Ax \geq b$.

MEASURE: The scalar product of c and x, i.e., $\sum_{i=1}^{n} c_i x_i$.

Good News: Approximable within

$$1 + O(\max(\log(mB/m^*), \sqrt{\log(mB/m^*)/B})),$$

where $B = \min_i b_i$ [Srinivasan, 1995].

Comment: Variation in which $A \in \{0,1\}^{m \cdot n}$ and $b \in Z^+$ is approximable within $O(m^*(m^{1/(B+1)}/m^*)^{(B+1)/B})$ [Srinivasan, 1995].

MAXIMUM QUADRATIC PROGRAMMING ◀ MP5

INSTANCE: Positive integer n, set of linear constraints, given as an $m \times n$-matrix A and an m-vector b, specifying a region $S \subseteq R^n$ by $S = \{x \in [0,1]^n : Ax \leq b\}$.

SOLUTION: A multivariate polynomial $f(x_1, \ldots, x_n)$ of total degree at most 2.

MEASURE: The maximum value of f in the region specified by the linear constants, i.e., $\max_{x \in S} f(x)$.

Bad News: Does not admit a μ-approximation for any constant $0 < \mu < 1$ [Bellare and Rogaway, 1995].

Comment: A μ-approximation algorithm finds a solution that differs from the optimal solution by at most the value $\mu(\max_{x \in S} f(x) - \min_{x \in S} f(x))$. Variation in which we look for a polynomial f of any degree does not admit a μ-approximation for $\mu = 1 - n^{-\delta}$ for some $\delta > 0$ [Bellare and Rogaway, 1995]. Note that these problems are known to be solvable in polynomial space but are not known to be in NP.

Garey and Johnson: MP2

MINIMUM GENERALIZED 0-1 ASSIGNMENT ◀ MP6

INSTANCE: Integer $m \times n$-matrix $A \in Z^{m \cdot n}$, integer m-vector $b \in Z^m$, and binary $m \times n$-matrix $C \in \{0,1\}^{m \cdot n}$.

The compendium
MATHEMATICAL PROGRAMMING

SOLUTION: A binary $m \times n$-matrix $X \in \{0,1\}^{m \cdot n}$ such that there is exactly one 1 in each column of X, and $\sum_{j=1}^{n} A_{i,j} X_{i,j} \leq b_i$ for all $i \in [1..m]$.

MEASURE: $\sum_{i=1}^{m} \sum_{j=1}^{n} C_{i,j} X_{i,j}$.

Bad News: Not in APX [Sahni and Gonzalez, 1976].

MP7 ▶ MINIMUM QUADRATIC 0-1 ASSIGNMENT

INSTANCE: Nonnegative integer $n \times n$-matrix $C \in N^{n \cdot n}$, nonnegative integer $m \times m$-matrix $D \in N^{m \cdot m}$.

SOLUTION: Binary $n \times m$-matrix $X \in \{0,1\}^{n \cdot m}$ such that there is at most one 1 in each row of X and exactly one 1 in each column of X.

MEASURE: $\sum_{\substack{i,j=1 \\ i \neq j}}^{n} \sum_{\substack{k,l=1 \\ k \neq l}}^{m} C_{i,j} D_{k,l} X_{i,k} X_{j,l}$.

Bad News: Not in APX [Sahni and Gonzalez, 1976].

Comment: Not in APX even if D satisfies the triangle inequality [Queyranne, 1986].

MP8 ▶ MINIMUM PLANAR RECORD PACKING

INSTANCE: Collection C of n records, for each record $c \in C$ a probability $p(c)$ such that $0 \leq p(c) \leq 1$.

SOLUTION: For each record $c \in C$ a placement $z(c)$ in the plane, given as integer coordinates, such that all records are placed on different points in the plane.

MEASURE: $\sum_{c_1 \in C} \sum_{c_2 \in C} p(c_1) p(c_2) d(z(c_1), z(c_2))$, where $d(z(c_1), z(c_2))$ is the discretized Euclidean distance between the points $z(c_1)$ and $z(c_2)$.

Good News: Approximable with an absolute error guarantee of $\lfloor 4\sqrt{2} + 8\sqrt{\pi} \rfloor$, that is, one can in polynomial time find a solution with objective function value at most $m^* + \lfloor 4\sqrt{2} + 8\sqrt{\pi} \rfloor$ [Karp, McKellar, and Wong, 1975].

MP9 ▶ MINIMUM RELEVANT VARIABLES IN LINEAR SYSTEM

INSTANCE: Integer $m \times n$-matrix $A \in Z^{m \cdot n}$, integer m-vector $b \in Z^m$.

SOLUTION: A rational n-vector $x \in Q^n$ such that $Ax = b$.

MEASURE: The number of non-zero elements in x.

Bad News: Not in APX [Amaldi and Kann, 1998].

Comment: Not approximable within $2^{\log^{1-\varepsilon} n}$ for any $\varepsilon > 0$ unless NP$\subseteq$QP [Amaldi and Kann, 1998]. The above nonapproximability results are still true for the variation in which the solutions are restricted by $Ax \geq b$ instead of $Ax = b$. Variation in which the solution vector is restricted to contain binary numbers is NPOPB-complete and is not approximable within $n^{0.5-\varepsilon}$ for any $\varepsilon > 0$ [Amaldi and Kann, 1998]. The complementary maximization problem, where the number of zero elements in the solution is to be maximized, and the solution vector is restricted to contain binary numbers, is NPOPB-complete and is not approximable within $(n+m)^{1/3-\varepsilon}$ for any $\varepsilon > 0$ [Kann, 1995].

Garey and Johnson: MP5

MAXIMUM SATISFYING LINEAR SUBSYSTEM ◀ MP10

INSTANCE: System $Ax = b$ of linear equations, where A is an integer $m \times n$-matrix, and b is an integer m-vector.

SOLUTION: A rational n-vector $x \in Q^n$.

MEASURE: The number of equations that are satisfied by x.

Good News: Approximable within $O(m/\log m)$ [Halldórsson, 1995b].

Bad News: Not approximable within m^ε for some $\varepsilon > 0$ [Amaldi and Kann, 1995] and [Arora, Babai, Stern, and Sweedyk, 1997].

Comment: For any prime $q \geq 2$ the problem over GF[q] is approximable within q, but is not approximable within $q - \varepsilon$ for any $\varepsilon > 0$, even if the number of variables in each equation is exactly three. When there are at most two variables in each equation the problem over GF[2] is approximable within 1.383 but is not approximable within 1.0909 [Håstad, 1997]; not approximable within 1.0030 if the number of occurrences of any variable is bounded by 3 [Berman and Karpinski, 1998]. When there are at most two variables in each equation the problem over GF[q] for any prime $q \geq 2$ is approximable within $q - \delta$ for some $\delta > 0$ (dependent on q) but not approximable within $70/69 - \varepsilon$ for any $\varepsilon > 0$ [Andersson, Engebretsen, and Håstad, 1999]. When there are exactly k variables in each equation and the number of equations is $\Theta(n^k)$ the problem over GF[q] for any prime $q \geq 2$ admits a randomized PTAS [Andersson and Engebretsen, 1998b]. The variation in which the system consists of relations ($>$ or $\geq$) is APX-complete and approximable within 2 [Amaldi and Kann, 1995]. If the variables are restricted to assume only binary values, the problem is harder to approximate than MAXIMUM INDEPENDENT SET. Approximability results for even more variants of the problem can be found in [Amaldi and Kann, 1995].

MINIMUM UNSATISFYING LINEAR SUBSYSTEM ◀ MP11

INSTANCE: System $Ax = b$ of linear equations, where A is an integer $m \times n$-matrix, and b is an integer m-vector.

The compendium

MATHEMATICAL PROGRAMMING

SOLUTION: A rational n-vector $x \in Q^n$.

MEASURE: The number of equations that are *not* satisfied by x.

Bad News: Not in APX [Arora, Babai, Stern, and Sweedyk, 1997].

Comment: Not approximable within $2^{\log^{1-\varepsilon} n}$ for any $\varepsilon > 0$ unless NP$\subset$QP [Arora, Babai, Stern, and Sweedyk, 1997]. If the system consists of relations ($>$ or $\geq$) the problem is even harder to approximate; there is a transformation from MINIMUM DOMINATING SET to this problem. If the variables are restricted to assume only binary values the problem is NPOPB-complete both for equations and relations, and is not approximable within $n^{1-\varepsilon}$ for any $\varepsilon > 0$. Approximability results for even more variants of the problem can be found in [Amaldi and Kann, 1998].

MP12 ▶ MAXIMUM HYPERPLANE CONSISTENCY

INSTANCE: Finite sets P and N of integer n-vectors. P consists of positive examples and N of negative examples.

SOLUTION: A hyperplane specified by a normal vector $w \in Q^n$ and a bias w_0.

MEASURE: The number of examples that are consistent with respect to the hyperplane, i.e., $|\{x \in P : wx > w_0\}| + |\{x \in N : wx < w_0\}|$.

Good News: Approximable within 2 [Amaldi and Kann, 1995].

Bad News: APX-complete [Amaldi and Kann, 1995].

Comment: Variation in which only one type of misclassification, either positive or negative, is allowed is not approximable within n^ε for some $\varepsilon > 0$ [Amaldi and Kann, 1995]. The complementary minimization problem, where the number of misclassifications is to be minimized, is not in APX unless P=NP, and is not approximable within $2^{\log^{1-\varepsilon} n}$ for any $\varepsilon > 0$ unless NP$\subset$QP [Arora, Babai, Stern, and Sweedyk, 1997] and [Amaldi and Kann, 1998].

Garey and Johnson: Similar to MP6

MP13 ▶ MAXIMUM KNAPSACK

INSTANCE: Finite set U, for each $u \in U$ a size $s(u) \in Z^+$ and a value $v(u) \in Z^+$, a positive integer $B \in Z^+$.

SOLUTION: A subset $U' \subseteq U$ such that $\sum_{u \in U'} s(u) \leq B$.

MEASURE: Total value of the chosen elements, i.e., $\sum_{u \in U'} v(u)$.

Good News: Admits an FPTAS [Ibarra and Kim, 1975].

Comment: The special case when $s(u) = v(u)$ for all $u \in U$ is called MAX-
IMUM SUBSET SUM. The corresponding minimization problem where
$\sum_{u \in U'} s(u) \geq B$ also admits an FPTAS, as well as several other variations
of the knapsack problem [Gens and Levner, 1979].

Garey and Johnson: MP9

MAXIMUM INTEGER m-DIMENSIONAL KNAPSACK ◀ MP14

INSTANCE: Nonnegative integer $m \times n$-matrix, $A \in N^{m \cdot n}$, nonnegative integer m-vector $b \in N^m$, nonnegative integer n-vector $c \in N^n$. m is a constant, $m \in Z^+$.

SOLUTION: Nonnegative integer n-vector $x \in N^n$ such that $Ax \leq b$.

MEASURE: The scalar product of c and x, i.e., $\sum_{i=1}^{n} c_i x_i$.

Good News: Admits a PTAS [Chandra, Hirschberg, and Wong, 1976].

Garey and Johnson: Generalization of MP10

MAXIMUM INTEGER k-CHOICE KNAPSACK ◀ MP15

INSTANCE: Nonnegative integer $n \times k$-matrices $A, C \in N^{n \cdot k}$, nonnegative integer $b \in N$.

SOLUTION: Nonnegative integer vector $x \in N^n$, function $f : [1..n] \to [1..k]$ such that $\sum_{i=1}^{n} a_{i,f(i)} x_i \leq b$.

MEASURE: $\sum_{i=1}^{n} c_{i,f(i)} x_i$.

Good News: Admits an FPTAS [Chandra, Hirschberg, and Wong, 1976].

Comment: The corresponding minimization problem, where $\sum_{i=1}^{n} a_{i,f(i)} x_i \geq b$ also
admits an FPTAS as well as the variation in which there does not need to
be a chosen element for each i [Gens and Levner, 1979].

Garey and Johnson: Similar to MP11

MAXIMUM CLASS-CONSTRAINED KNAPSACK ◀ MP16

INSTANCE: n set sizes specified by an n-vector $U \in N^n$, m knapsacks with different volumes and number of compartments specified by two m-vectors $V \in N^m, C \in N^m$, such that $\sum_i U_i = \sum_j V_j$.

The compendium
MATHEMATICAL PROGRAMMING

SOLUTION: A placement of the set elements into the knapsacks, specified by two $n \times m$-matrices: $I \in \{0,1\}^{n \times m}, Q \in N^{n \times m}$, such that: $\sum_i Q_{i,j} \leq V_j$ and $\sum_i I_{i,j} \leq C_j$ for every j, $\sum_j Q_{i,j} \leq U_i$ for every i, and $I_{i,j} = 0 \Rightarrow Q_{i,j} = 0$ for every (i,j).

MEASURE: Total number of packed elements, i.e., $\sum_i \sum_j Q_{i,j}$.

Good News: Approximable with absolute error at most $\sum_{j=1}^{N} \frac{V_j}{C_j+1}$ [Shachnai and Tamir, 1998].

MP17 ▶ NEAREST LATTICE VECTOR

INSTANCE: Lattice basis $\{b_1, \ldots, b_m\}$ where $b_i \in Z^k$, a point $b_0 \in Q^k$, and a positive integer p.

SOLUTION: A vector b in the lattice, where $b \neq b_0$.

MEASURE: The distance between b_0 and b in the ℓ_p norm.

Bad News: Not in APX [Arora, Babai, Stern, and Sweedyk, 1997].

Comment: Not approximable within $2^{\log^{1-\varepsilon} mp}$ for any $\varepsilon > 0$ unless NP$\subset$QP. The special case where b_0 is the zero vector and $p = \infty$ is not approximable within $2^{\log^{0.5-\varepsilon} mp}$ for any $\varepsilon > 0$ unless NP$\subset$QP [Arora, Babai, Stern, and Sweedyk, 1997].

MP18 ▶ MINIMUM BLOCK-ANGULAR CONVEX PROGRAMMING

INSTANCE: K disjoint convex compact sets B^k called blocks, M nonnegative continuous convex functions $f_m^k : B^k \to R$.

SOLUTION: A positive number λ such that

$$\sum_{k=1}^{K} f_m^k(x^k) \leq \lambda \text{ for } 1 \leq m \leq M, \text{ and } x^k \in B^k \text{ for } 1 \leq k \leq K.$$

MEASURE: λ.

Good News: Admits an FPTAS [Grigoriadis and Khachiyan, 1994].

Algebra and Number Theory

Solvability of Equations

MAXIMUM SATISFIABILITY OF QUADRATIC EQUATIONS ◀ **AN1**

INSTANCE: Prime number q, set $P = \{p_1(x), p_2(x), \ldots, p_m(x)\}$ of polynomials of degree at most 2 over GF[q] in n variables. The polynomials may not contain any monomial x_i^2 for any i.

SOLUTION: A subset $P' \subseteq P$ of the polynomials such that there is a root common to all polynomials in P'.

MEASURE: Cardinality of the subset, i.e., $|P'|$.

Good News: Approximable within $q^2/(q-1)$ [Håstad, Phillips, and Safra, 1993].

Bad News: Not approximable within $q - \varepsilon$ for any ε [Håstad, Phillips, and Safra, 1993].

Comment: Over the rationals or over the reals the problem is not approximable within $n^{1-\varepsilon}$ for any $\varepsilon > 0$ [Håstad, Phillips, and Safra, 1993]. For linear polynomials the problem is not approximable within q^ε for some $\varepsilon > 0$ [Amaldi and Kann, 1995].

Garey and Johnson: Similar to AN9

Games and Puzzles

GP1 ▶ MINIMUM GRAPH MOTION PLANNING

INSTANCE: Graph $G = (V, E)$, start vertex $s \in V$ where the robot is initially placed, goal vertex $t \in V$, and a subset $W \subseteq V$ of vertices where the obstacles are initially placed.

SOLUTION: A motion scheme for the robot and the obstacles. In each time step either the robot or one obstacle may be moved to a neighboring vertex that is not occupied by the robot or an obstacle.

MEASURE: The number of steps until the robot has been moved to the goal vertex t.

Good News: Approximable within $O(\sqrt{|V|})$ [Papadimitriou, Raghavan, Sudan, and Tamaki, 1994].

Bad News: APX-complete [Papadimitriou, Raghavan, Sudan, and Tamaki, 1994].

Comment: Approximable within $O(l_{\max}/l_{\min})$, where $l_{\max}$ and $l_{\min}$ are the lengths of the longest and shortest paths of degree-2 vertices in G.

GP2 ▶ MINIMUM TRAVEL ROBOT LOCALIZATION

INSTANCE: A simple polygon P and the visibility polygon V of the location of a robot inside P.

SOLUTION: A motion scheme for the robot to determine its exact location.

MEASURE: The distance the robot travels according to the motion scheme.

Good News: Approximable within $|H| - 1$ where H is the set of points in P with visibility polygon V [Dudek, Romanik, and Whitesides, 1994].

Comment: The problem SHORTEST WATCHMAN ROUTE where the polygon P have holes and the problem is to find a shortest tour inside P such that the robot can see all of P is approximable within $O(\log m)$, where m is the minimum number of edges in an optimal rectilinear tour [Mata and Mitchell, 1995].

Logic

Propositional Logic

MAXIMUM SATISFIABILITY ◄ LO1

INSTANCE: Set U of variables, collection C of disjunctive clauses of literals, where a literal is a variable or a negated variable in U.

SOLUTION: A truth assignment for U.

MEASURE: Number of clauses satisfied by the truth assignment.

Good News: Approximable within 1.2987 [Asano, Hori, Ono, and Hirata, 1997].

Bad News: APX-complete [Papadimitriou and Yannakakis, 1991].

Comment: Variation in which each clause has a nonnegative weight and the objective is to maximize the total weight of the satisfied clauses is also approximable within 1.2987 [Asano, Hori, Ono, and Hirata, 1997]. Generalization in which each clause is a disjunction of conjunctions of literals and each conjunction consists of at most k literals, where k is a positive constant, is still APX-complete [Papadimitriou and Yannakakis, 1991]. Admits a PTAS for 'planar' instances [Khanna and Motwani, 1996]. The corresponding minimization problem MINIMUM SATISFIABILITY is approximable within 2 [Bertsimas, Teo, and Vohra, 1996] and its variation in which each clause has a nonnegative weight and the objective is to minimize the total weight of the satisfied clauses is as hard to approximate as the unweighted version [Crescenzi, Silvestri, and Trevisan, 1996].

Garey and Johnson: LO1

MAXIMUM k-SATISFIABILITY ◄ LO2

INSTANCE: Set U of variables, collection C of disjunctive clauses of at most k literals, where a literal is a variable or a negated variable in U. k is a constant, $k \geq 2$.

SOLUTION: A truth assignment for U.

MEASURE: Number of clauses satisfied by the truth assignment.

Good News: Approximable within $1/(1-2^{-k})$ if every clause consists of exactly k literals [Johnson, 1974a].

Bad News: APX-complete [Papadimitriou and Yannakakis, 1991].

Comment: If $k=3$, the problem is approximable within 1.249 [Trevisan, Sorkin, Sudan, and Williamson, 1996] and is approximable within 8/7 for satisfiable instances [Karloff and Zwick, 1997]. MAXIMUM k-SATISFIABILITY is not approximable within $1/(1-2^{-k}) - \varepsilon$ for any $\varepsilon > 0$ and $k \geq 3$, even if every clause consists of exactly k literals [Håstad, 1997].

The compendium

Logic

MAXIMUM 2-SATISFIABILITY is approximable within 1.0741 [Feige and Goemans, 1995], and is not approximable within 1.0476 [Håstad, 1997]. The weighted version of this problem is as hard to approximate as the unweighted version [Crescenzi, Silvestri, and Trevisan, 1996].

If every clause consists of exactly k literals, the weighted version of the problem is as hard to approximate as the unweighted version [Crescenzi, Silvestri, and Trevisan, 1996]. Variation in which the number of occurrences of any literal is bounded by a constant B for $B \geq 3$ is still APX-complete even for $k = 2$ [Papadimitriou and Yannakakis, 1991] and [Chap.8 of this book]; for $B = 6$ it is not approximable within 1.0014 [Berman and Karpinski, 1998].

Admits a PTAS if $|C| = \Theta(|U|^k)$ [Arora, Karger, and Karpinski, 1995]. Variation in which each clause is a Horn clause, i.e., contains at most one nonnegated variable, is APX-complete, even for $k = 2$ [Kohli, Krishnamurti, and Mirchandani, 1994].

Garey and Johnson: LO2 and LO5

LO3 ▶ MINIMUM k-SATISFIABILITY

INSTANCE: Set U of variables, collection C of disjunctive clauses of at most k literals, where a literal is a variable or a negated variable in U. k is a constant, $k \geq 2$.

SOLUTION: A truth assignment for U.

MEASURE: Number of clauses satisfied by the truth assignment.

Good News: Approximable within $2(1 - 1/2^k)$ [Bertsimas, Teo, and Vohra, 1996].

Bad News: APX-complete for every $k \geq 2$ [Kohli, Krishnamurti, and Mirchandani, 1994].

Comment: Transformation from MAXIMUM 2-SATISFIABILITY. Variation in which each clause is a Horn clause, i.e., contains at most one nonnegated variable, is APX-complete, even for $k = 2$ [Kohli, Krishnamurti, and Mirchandani, 1994].

Garey and Johnson: LO2

LO4 ▶ MAXIMUM NOT-ALL-EQUAL 3-SATISFIABILITY

INSTANCE: Set U of variables, collection C of disjunctive clauses of at most 3 literals, where a literal is a variable or a negated variable in U.

SOLUTION: A truth assignment for U and a subset $C' \subseteq C$ of the clauses such that each clause in C' has at least one true literal and at least one false literal.

MEASURE: $|C'|$.

Good News: Approximable within 1.138 [Kann, Lagergren, and Panconesi, 1996].

Bad News: APX-complete [Papadimitriou and Yannakakis, 1991]. Not approximable within 1.090 [Zwick, 1998].

Comment: Transformation from MAXIMUM 2-SATISFIABILITY. Approximable within 1.096 for satisfiable instances [Zwick, 1998]. MAXIMUM NOT-ALL-EQUAL SATISFYABILITY, without restrictions on the number of literals in a clause, is approximable within 1.380 [Andersson and Engebretsen, 1998a].

Garey and Johnson: LO3

MINIMUM 3DNF SATISFIABILITY ◀ LO5

INSTANCE: Set U of variables, collection C of conjunctive clauses of at most three literals, where a literal is a variable or a negated variable in U.

SOLUTION: A truth assignment for U.

MEASURE: Number of clauses satisfied by the truth assignment.

Bad News: Not in APX [Kolaitis and Thakur, 1994].

Garey and Johnson: LO8

MAXIMUM DISTINGUISHED ONES ◀ LO6

INSTANCE: Disjoint sets X, Z of variables, collection C of disjunctive clauses of at most 3 literals, where a literal is a variable or a negated variable in $X \cup Z$.

SOLUTION: Truth assignment for X and Z that satisfies every clause in C.

MEASURE: The number of Z variables that are set to true in the assignment.

Bad News: NPOPB-complete [Kann, 1992b].

Comment: Transformation from MAXIMUM NUMBER OF SATISFIABLE FORMULAS [Panconesi and Ranjan, 1993]. Not approximable within $(|X| + |Z|)^{1-\varepsilon}$ for any $\varepsilon > 0$ [Jonsson, 1997]. MAXIMUM ONES, the variation in which all variables are distinguished, i.e., $|X| = \emptyset$, is also NPOPB-complete [Kann, 1992b], and is not approximable within $|Z|^{1-\varepsilon}$ for any $\varepsilon > 0$ [Jonsson, 1997]. MAXIMUM WEIGHTED SATISFIABILITY, the weighted version, in which every variable is assigned a nonnegative weight, is NPO-complete [Ausiello, D'Atri, and Protasi, 1981].

MINIMUM DISTINGUISHED ONES ◀ LO7

INSTANCE: Disjoint sets X, Z of variables, collection C of disjunctive clauses of at most 3 literals, where a literal is a variable or a negated variable in $X \cup Z$.

SOLUTION: Truth assignment for X and Z that satisfies every clause in C.

The compendium
Logic

MEASURE: The number of Z variables that are set to true in the assignment.

Bad News: NPOPB-complete [Kann, 1994b].

Comment: Transformation from MINIMUM INDEPENDENT DOMINATING SET. Not approximable within $(|X|+|Z|)^{1-\varepsilon}$ for any $\varepsilon > 0$ [Jonsson, 1997]. MINIMUM ONES, the variation in which all variables are distinguished, i.e., $X = \emptyset$, is also NPOPB-complete [Kann, 1994b], and is not approximable within $|Z|^{1-\varepsilon}$ for any $\varepsilon > 0$ [Jonsson, 1997]. MINIMUM ONES for clauses of 2 literals is approximable within 2 [Gusfield and Pitt, 1992]. MINIMUM WEIGHTED SATISFIABILITY, the weighted version, in which every variable is assigned a nonnegative weight, is NPO-complete [Orponen and Mannila, 1987]. Variations corresponding to three- and four-valued logics have also been considered [Errico and Rosati, 1995].

LO8 ▶ MAXIMUM WEIGHTED SATISFIABILITY WITH BOUND

INSTANCE: Set U of variables, boolean expression F over U, a nonnegative bound $B \in N$, for each variable $u \in U$ a weight $w(u) \in N$ such that $B \le \sum_{u \in U} w(u) \le 2B$.

SOLUTION: A truth assignment for U, i.e., a subset $U' \subseteq U$ such that the variables in U' are set to true and the variables in $U - U'$ are set to false.

MEASURE: $\sum_{v \in U'} w(v)$ if the truth assignment satisfies the boolean expression F and B otherwise.

Good News: Approximable within 2 [Crescenzi and Panconesi, 1991].

Bad News: APX-complete [Crescenzi and Panconesi, 1991].

Comment: Variation with $\sum_{u \in U} w(u) \le (1 + 1/(|U|-1))B$ is PTAS-complete [Crescenzi and Panconesi, 1991].

LO9 ▶ MAXIMUM NUMBER OF SATISFIABLE FORMULAS

INSTANCE: Set U of variables, collection C of 3CNF formulas.

SOLUTION: A truth assignment for U.

MEASURE: Number of formulas satisfied by the truth assignment.

Bad News: NPOPB-complete [Kann, 1992b].

Comment: Transformation from LONGEST INDUCED PATH. Not approximable within $|C|$.

LO10 ▶ MINIMUM NUMBER OF SATISFIABLE FORMULAS

INSTANCE: Set U of variables, collection C of 3CNF formulas.

SOLUTION: A truth assignment for U.

MEASURE: Number of formulas satisfied by the truth assignment.

Bad News: NPOPB-complete [Kann, 1994b].

Comment: Transformation from MINIMUM DISTINGUISHED ONES. Not approximable within $|C|^{1-\varepsilon}$ for any $\varepsilon > 0$ [Kann, 1994b].

MINIMUM EQUIVALENCE DELETION ◄ LO11

INSTANCE: Set U of variables, collection C of equivalences over U.

SOLUTION: A truth assignment for U.

MEASURE: Number of equivalences not satisfied by the truth assignment.

Good News: Approximable within $O(\log|U|)$ [Garg, Vazirani, and Yannakakis, 1996].

Bad News: APX-hard [Garg, Vazirani, and Yannakakis, 1996].

Comment: The dual problem is approximable within 1.138 [Kann, 1994b].

MAXIMUM k-CONSTRAINT SATISFACTION ◄ LO12

INSTANCE: Set U of variables, collection C of conjunctive clauses of at most k literals, where a literal is a variable or a negated variable in U, and k is a constant, $k \geq 2$.

SOLUTION: A truth assignment for U.

MEASURE: Number of clauses satisfied by the truth assignment.

Good News: Approximable within 2^{k-1} [Trevisan, 1996].

Bad News: APX-complete [Berman and Schnitger, 1992].

Comment: Transformation from MAXIMUM 2-SATISFIABILITY. Approximable within 1.165 but not within 1.111 for $k = 2$, and approximable within 2 but not within $2-\varepsilon$ for $k = 3$. There are also results of specific variations of the problem for $k \leq 3$ [Zwick, 1998]. Not approximable within $2^{0.09k}$ for large enough k [Trevisan, 1996]. Not in APX when $k = \log|C|$ [Verbitsky, 1995]. When there are exactly k variables in each clause and the number of clauses is $\Theta(|U|^k)$ the problem admits a randomized PTAS [Andersson and Engebretsen, 1998b]. A complete classification of the approximability of optimization problems derived from Boolean constraint satisfaction is contained in [Khanna, Sudan, and Williamson, 1997] and in [Khanna, Sudan, and Trevisan, 1997].

MINIMUM LENGTH EQUIVALENT FREGE PROOF ◄ LO13

INSTANCE: A Frege proof π of a tautology φ.

SOLUTION: A Frege proof π' of φ shorter than π, i.e., containing at most as many symbols as π.

MEASURE: Number of symbols in π'.

Good News: Approximable within $O(n)$.

Bad News: APX-hard [Alekhnovich, Buss, Moran, and Pitassi, 1998].

Comment: The result applies to all Frege systems, to all extended Frege systems, to resolution, to Horn clause resolution, to the sequent calculus, and to the cut-free sequent calculus. Not approximable within $2^{\log^{(1-\varepsilon)} n}$ for any $\varepsilon > 0$ unless NP $\subseteq$ QP [Alekhnovich, Buss, Moran, and Pitassi, 1998].

Miscellaneous

LO14 ▶ MAXIMUM HORN CORE

INSTANCE: Set M of truth assignments on n variables.

SOLUTION: A Horn core of M, i.e., a subset $M' \subseteq M$ such that M' is equal to the set of truth assignments satisfying a Horn boolean formula.

MEASURE: The cardinality of the core, i.e., $|M'|$.

Bad News: Not in APX [Kavvadias, Papadimitriou, and Sideri, 1993].

Automata and Language Theory

Automata Theory

MINIMUM CONSISTENT FINITE AUTOMATON ◀ AL1

INSTANCE: Two finite sets P, N of binary strings.

SOLUTION: A deterministic finite automaton $(Q, \{0,1\}, \delta, q_0, F)$ accepting all strings in P and rejecting all strings in N.

MEASURE: Number of states in the automaton.

Bad News: Not approximable within $2-\varepsilon$ for any $\varepsilon > 0$ [Simon, 1990].

Comment: Transformation from MINIMUM GRAPH COLORING. Not approximable within $(|P|+|N|)^{1/14-\varepsilon}$ for any $\varepsilon > 0$ [Pitt and Warmuth, 1993].

Garey and Johnson: AL8

LONGEST COMPUTATION ◀ AL2

INSTANCE: Nondeterministic Turing machine M, binary input string x.

SOLUTION: Nondeterministic guess string c produced by M on input x.

MEASURE: The length of the shortest of the strings c and x, i.e., $\min(|c|, |x|)$.

Bad News: NPOPB-complete [Berman and Schnitger, 1992].

Comment: Variation in which the machine is oblivious is also NPOPB-complete.

SHORTEST COMPUTATION ◀ AL3

INSTANCE: Nondeterministic Turing machine M, binary input string x.

SOLUTION: Nondeterministic guess string c produced by M on input x.

MEASURE: The length of the shortest of the strings c and x, i.e., $\min(|c|, |x|)$.

Bad News: NPOPB-complete [Kann, 1994b].

Comment: Variation in which the Turing machine is oblivious is also NPOPB-complete.

Formal Languages

MINIMUM LOCALLY TESTABLE AUTOMATON ORDER ◀ AL4

INSTANCE: A locally testable language L, i.e., a language L such that, for some positive integer j, whether or not a string x is in the language depends on (a) the prefix and suffix of x of length $j-1$, and (b) the set of substrings of x of length j.

SOLUTION: An order j, i.e., a positive integer j witnessing the local testability of L.

MEASURE: The value of the order, i.e., j.

Good News: Admits a PTAS [Kim and McNaughton, 1993].

Miscellaneous

AL5 ▶ MINIMUM PERMUTATION GROUP BASE

INSTANCE: Permutation group on n letters.

SOLUTION: A base for G, i.e., a sequence of points $b_1, \ldots, b_k$ such that the only element in G fixing all of the b_i is the identity.

MEASURE: The size of the base, i.e., k.

Good News: Approximable within $\log \log n$ [Blaha, 1992].

Program Optimization

Code Generation

MINIMUM REGISTER SUFFICIENCY ◂ PO1

INSTANCE: Directed acyclic graph $G = (V, E)$.

SOLUTION: Computation for G that uses k registers, i.e., an ordering $v_1, \ldots, v_n$ of the vertices in V and a sequence $S_0, \ldots, S_n$ of subsets of V, each satisfying $|S_i| \leq k$, such that S_0 is empty, S_n contains all vertices with in-degree 0 in G, and, for $1 \leq i \leq n$, $v_i \in S_i$, $S_i - \{v_i\} \subseteq S_{i-1}$, and S_{i-1} contains all vertices u for which $(v_i, u) \in A$.

MEASURE: Number of registers, i.e., k.

Good News: Approximable within $O(\log^2 n)$ [Klein, Agrawal, Ravi, and Rao, 1990].

Garey and Johnson: PO1

MINIMUM LOCAL REGISTER ALLOCATION ◂ PO2

INSTANCE: Sequence of instructions forming a basic block (no jumps), number N of available registers, cost S_i (for $1 \leq i \leq N$) of loading or storing register i.

SOLUTION: A register allocation for the instruction sequence.

MEASURE: The total loading and storing cost for executing the instructions using the register allocation.

Good News: Approximable within 2 [Farach and Liberatore, 1998].

The compendium
MISCELLANEOUS

Miscellaneous

MS1 ▶ MAXIMUM BETWEENNESS

INSTANCE: Finite set A, collection C of ordered triples (a,b,c) of distinct elements from A.

SOLUTION: A one-to-one function $f : A \to [1..|A|]$.

MEASURE: Number of triples where either $f(a) < f(b) < f(c)$ or $f(c) < f(b) < f(a)$.

Good News: Approximable within 2 for satisfiable instances [Chor and Sudan, 1995].

Bad News: APX-complete [Chor and Sudan, 1995].

Comment: Admits a PTAS if $|C| = \Theta(|A|^3)$ [Arora, Frieze, and Kaplan, 1996].

Garey and Johnson: MS1

MS2 ▶ MINIMUM DECISION TREE

INSTANCE: Finite set X of objects, collection $T = \{T_1, T_2, \ldots, T_m\}$ of binary tests $T_i : X \to \{0,1\}$.

SOLUTION: A decision tree t for X using the tests in T, where a decision tree is a binary tree in which each non-leaf vertex is labelled by a test from T, each leaf is labelled by an object from X, the edge from a non-leaf vertex to its left son is labelled 0 and the one to its right son is labelled 1, and, if $T_{i_1}, O_{i_1}, T_{i_2}, O_{i_2}, \ldots, T_{i_k}, O_{i_k}$ is the sequence of vertex and edge labels on the path from the root to a leaf labelled by $x \in X$, then x is the unique object for which $T_{i_j}(x) = O_{i_j}$ for all j, $1 \le j \le k$.

MEASURE: Number of leaves of t.

Bad News: APX-hard [Hancock, Jiang, Li, and Tromp, 1996].

Comment: Not approximable within $2^{\log^{(1-\varepsilon)} |X|}$ unless $\text{NP} \subseteq \text{QP}$ [Hancock, Jiang, Li, and Tromp, 1996].

Garey and Johnson: MS15

MS3 ▶ NEAREST CODEWORD

INSTANCE: Linear binary code C of length n and a string x of length n.

SOLUTION: A codeword y of C.

MEASURE: The Hamming distance between x and y, i.e., $d(x,y)$.

Bad News: Not in APX [Arora, Babai, Stern, and Sweedyk, 1997].

Comment: Not approximable within $2^{\log^{1-\varepsilon} n}$ for any $\varepsilon > 0$ unless NP$\subset$QP [Arora, Babai, Stern, and Sweedyk, 1997]. The complementary maximization problem, where the number of bits that agree between x and y is to be maximized, does not admit a PTAS [Petrank, 1992].

MINIMUM TREE ALIGNMENT ◀ MS4

INSTANCE: Set S of sequences over an alphabet Σ, a score scheme $\mu: \Sigma \cup \{\Delta\} \times \Sigma \cup \{\Delta\} \to N$ satisfying the triangle inequality, and a tree T of bounded degree whose leafs are labeled with the sequences S.

SOLUTION: A sequence labeling of the interior nodes of the tree T.

MEASURE: The sum over all edges (x,y) in the tree of the edit distance $d(x,y)$ between the labels of the endpoints of the edge. The edit distance is the minimum alignment cost $\sum_i \mu(x'[i], y'[i])$ over all possible alignments x' and y' of x and y. An alignment is obtained by inserting spaces (denoted by Δ) into the original sequence, either between two characters or in the beginning or at the end. x' and y' must have the same length.

Good News: Admits a PTAS [Jiang, Lawler, and Wang, 1994].

Comment: Variation in which T is not given as an input and the objective is to find the tree with the minimum total alignment cost is called MINIMUM EVOLUTIONARY TREE and is APX-hard [Jiang, Lawler, and Wang, 1994]. The variation of this latter problem which uses the quartet paradigm belongs to PTAS [Jiang, Kearney, and Li, 1998]. MINIMUM SEQUENCE ALIGNMENT, the variation in which the objective is to minimize the sum of edit distances of all pairs of sequences in the alignment, is approximable within $2 - 3/|S|$ [Pevzner, 1992]. The generalization of this latter problem to weighted sums is approximable within $O(\log |S|)$ [Wu et al., 1998].

MAXIMUM STRING FOLDING ◀ MS5

INSTANCE: Finite alphabet Σ, finite string s from Σ^*.

SOLUTION: A folding of s in the 3-dimensional rectangular grid Z^3, that is, a mapping of s into Z^3 such that adjacent symbols of s are adjacent in the lattice, and no site in the lattice is occupied by more than one symbol.

MEASURE: The number of pairs of equal symbols that lie at adjacent lattice sites.

Bad News: APX-hard via random reductions [Nayak, Sinclair, and Zwick, 1998].

MINIMUM ATTRACTION RADIUS FOR BINARY HOPFIELD NET ◀ MS6

INSTANCE: An n-node synchronous binary Hopfield network and a stable initial vector of states $u \in \{-1, 1\}^n$. A binary Hopfield network is a complete graph where each edge has an integer weight $w(v_i, v_j)$ and each vertex has an integer threshold value $t(v_i)$. At each time step t each

The compendium

MISCELLANEOUS

vertex v_i has a state $x(t,v_i)$. $x(0,v_i)$ is given by u and $x(t+1,v_i) = \text{sgn}\left(\sum_{j=1}^{n} w(v_i,v_j) - t_i\right)$ where sgn is the sign function. An initial vector of states is stable if $x(t,v_i)$ eventually converges for all i.

SOLUTION: An initial vector of states v that either converges to a different vector than u or is not stable.

MEASURE: The Hamming distance between u and v. If v is the vector nearest to u that does not converge to the same vector as u, then this distance is the attraction radius.

Bad News: Not approximable within $n^{1-\varepsilon}$ for any $\varepsilon > 0$ [Floréen and Orponen, 1993].

Comment: Transformation from MINIMUM INDEPENDENT DOMINATING SET.

MS7 ▶ MAXIMUM CHANNEL ASSIGNMENT

INSTANCE: Net of hexagonal cells in which n cells C_i are assigned a positive load r_i, an interference radius r, and m channels F_i.

SOLUTION: A channel assignment A, i.e., a multivalued function A assigning a set of cells to a channel such that if $C_i, C_j \in A(F_k)$ then the distance between C_i and C_j is greater than $2r$.

MEASURE: The number of satisfied request, i.e., $\sum_i \min\{r_i, |\{F_j : C_i \in F_j\}|\}$.

Good News: Approximable within $1/(1-e^{-1})$ [Simon, 1989].

Comment: Admits a PTAS if the number of channels is fixed. Similar results hold in the case in which each cell has a set of forbidden channels.

MS8 ▶ MINIMUM k-LINK PATH IN A POLYGON

INSTANCE: Polygon P with n integer-coordinates vertices and two points s and t in P.

SOLUTION: A k-link path between s and t, i.e., a sequence $p_0, \ldots, p_h$ of points inside P with $h \leq k$ such that $p_0 = s$, $p_k = t$, and, for all i with $0 \leq i < h$, the segment between p_i and p_{i+1} is inside P.

MEASURE: The Euclidean length of the path, i.e., $\sum_{i=0}^{h-1} d(p_i, p_{i+1})$ where d denotes the Euclidean distance.

Good News: Admits an FPTAS [Mitchell, Piatko, and Arkin, 1992].

MS9 ▶ MINIMUM SIZE ULTRAMETRIC TREE

INSTANCE: $n \times n$ matrix M of positive integers.

SOLUTION: An ultrametric tree, i.e., an edge-weighted tree $T(V,E)$ with n leaves such that, for any pair of leaves i and j, $d_{ij}^T \geq M[i,j]$ where d_{ij}^T denotes the sum of the weights in the path between i and j.

MEASURE: The size of the tree, i.e., $\sum_{e \in E} w(e)$ where $w(e)$ denotes the weight of edge e.

Bad News: Not approximable within n^ε for some $\varepsilon > 0$ [Farach, Kannan, and Warnow, 1993].

Comment: Transformation from graph coloring. Variation in which M is a metric is approximable within $O(\log n)$ [Wu, Chao, and Tang, 1998].

MINIMUM RECTANGLE PARTITION ON INTERIOR POINTS ◀ MS10

INSTANCE: Rectangle R and finite set P of points located inside R.

SOLUTION: A set of line segments that partition R into rectangles such that every point in P is on the boundary of some rectangle.

MEASURE: The total length of the introduced line segments.

Good News: Approximable within 1.75 [Gonzalez and Zheng, 1989].

Comment: Variation in which R is a rectilinear polygon is approximable within 4 [Gonzalez and Zheng, 1990].

MINIMUM SORTING BY REVERSALS ◀ MS11

INSTANCE: Permutation π of the numbers 1 to n.

SOLUTION: A sequence $\rho_1, \rho_2, \ldots, \rho_t$ of reversals of intervals such that $\pi \cdot \rho_1 \cdot \rho_2 \cdots \rho_t$ is the identity permutation. A reversal of an interval $[i,j]$ is the permutation $(1, 2, \ldots, i-1, j, j-1, \ldots, i+1, i, j+1, \ldots, n)$.

MEASURE: The number of reversals, i.e., t.

Good News: Approximable within 3/2 [Christie, 1998].

Bad News: Not approximable within 1.0008 [Berman and Karpinski, 1998].

Comment: MINIMUM SORTING BY TRANSPOSITIONS, the problem where transpositions are used instead of reversals, is approximable within 3/2. A transposition $\rho(i,j,k)$ where $i < j < k$ is the permutation $(1, \ldots, i-1, j, \ldots, k-1, i \ldots, j-1, k, \ldots, n)$ [Bafna and Pevzner, 1995].

MAXIMUM COMPATIBLE BINARY CONSTRAINT SATISFACTION ◀ MS12

INSTANCE: Set $\{x_1, \ldots, x_n\}$ of variables taking values from the sets $D_1, \ldots, D_n$ respectively, binary compatibility constraints $C(x_i, x_j) \subseteq D_i \times D_j$ for $1 \leq i < j \leq n$.

The compendium

MISCELLANEOUS

SOLUTION: A compatible variable assignment to a subset of the variables such that for all pairs (x_i, x_j) of variables in the subset, the corresponding values $(d_i, d_j) \in C(x_i, x_j)$.

MEASURE: The number of variables in the compatible variable assignment.

Bad News: Not approximable within $(\sum_i |D_i|)^{1/4-\varepsilon}$ for any $\varepsilon > 0$ [Jagota, 1993] and [Bellare, Goldreich, and Sudan, 1998].

Comment: Equivalent to MAXIMUM CLIQUE under PTAS-reduction [Jagota, 1993].

MS13 ▶ SHORTEST PATH MOTION IN 3 DIMENSIONS

INSTANCE: Two points s, t and collection of polyhedral obstacles in three-dimensional Euclidean space.

SOLUTION: A path from s to t that avoids the obstacles.

MEASURE: The total length of the path in Euclidean metric.

Good News: Admits an FPTAS [Papadimitriou, 1985] and [Choi, Sellen, and Yap, 1997].

MS14 ▶ MAXIMUM MAP LABELING

INSTANCE: A set of n points in the plane.

SOLUTION: A collection of n axis-aligned equal-sized squares (labels) such that (a) every point is the corner of exactly one square, and (b) all squares are pairwise disjoint.

MEASURE: The area of the squares.

Good News: Approximable within 2 [Wagner and Wolff, 1995].

Bad News: Not approximable within $2 - \varepsilon$ for any $\varepsilon > 0$ [Formann and Wagner, 1991].

Comment: The generalized problem, where equal-sized rectangles or ellipses are allowed as labels, and where the labels may have any orientation and may have any point on the boundary at the point in the plane, is in APX. The generalized problem is approximable within 14 when the labels are squares and within 15 when the labels are circles [Doddi et al., 1997].

MS15 ▶ MINIMUM PHYLOGENETIC TREE DISTANCE

INSTANCE: A 3-degree (binary) tree $T = (V, E)$ with unique labels on the leaves and with edge weights $w : E \to N$. A 3-degree tree T' with the same set of unique labels as T.

SOLUTION: A sequence of edges $e_1,\ldots,e_m$ representing nearest neighbor interchanges transforming the tree T into T' (so that the labels match). A nearest neighbor interchange of $e_i = \{u,v\}$ is a transformation where one of the two subtrees adjacent to u is exchanged with one of the two subtrees adjacent to v.

MEASURE: The cost of the sequence of interchanges, i.e., $\sum_{1 \leq i \leq m} w(e_i)$.

Good News: Approximable within $6 + 6\log n$ where n is the number of leaves [Dasgupta et al., 1997].

Comment: The variation when another type of transformation called *linear-cost subtree transfer* is used is approximable within 2 [Dasgupta et al., 1997]. MAXIMUM AGREEMENT SUBTREE for three trees of unbounded degree is not approximable within $2^{\log^\delta n}$ for any $\delta < 1$, unless NP$\subset$QP [Hein, Jiang, Wang, and Zhang, 1996].

The compendium
MISCELLANEOUS

MAXIMUM FREQUENCY ALLOCATION ◀ MS16

INSTANCE: Graph $G = (V,E)$, and, for each node u, a finite set $L(u) \subset Z^+$ of available frequencies and an integer M_u.

SOLUTION: A frequency allocation for G, i.e., for each node u, a subset $S(u)$ of $L(u)$ such that, for all $u \in V$, $1 \leq |S(u)| \leq M_u$, and, for all $(u,v) \in E$, $S(u) \cap S(v) = \emptyset$.

MEASURE: The stability number of the allocation, i.e., $\sum_{u \in V} |S(u)|$.

Good News: Approximable within 2 [Malesinska and Panconesi, 1996].

Bad News: APX-complete [Malesinska and Panconesi, 1996].

Comment: Transformation from MAXIMUM 3-SATISFIABILITY.

MINIMUM RED-BLUE SEPARATION ◀ MS17

INSTANCE: Finite sets R and B of points in the Euclidean plane.

SOLUTION: A simple polygon P that separates the red points (points in R) from the blue points (points in B).

MEASURE: Euclidean length of the perimeter of P.

Good News: Approximable within $O(\log|R|+|B|)$ [Mata and Mitchell, 1995].

Comment: Variation in which the objective function is the combinatorial size of the separating polygon, that is, the number of vertices in the polygon, is approximable within $\log(m^*)$ [Agarwal and Suri, 1994].

MINIMUM NUMERICAL TAXONOMY ◀ MS18

INSTANCE: $n \times n$ distance matrix D.

The compendium

MISCELLANEOUS

SOLUTION: Tree metric d_T, i.e., an edge weighted tree T of n nodes such that $d_T(i,j)$ is the distance between i and j in T.

MEASURE: $\max_{i<j} |d_T(i,j) - D[i,j]|$.

Good News: Approximable within 3 [Agarwala et al., 1996].

Bad News: Not approximable within $9/8-\varepsilon$ for any $\varepsilon > 0$ [Agarwala et al., 1996].

MS19 ▶ MINIMUM RING LOADING

INSTANCE: Positive integer n and a sequence $\{d_{i,j}\}_{0 \leq i < j < n}$ of non-negative integers denoting the pairwise demand between nodes i and j of a ring of size n.

SOLUTION: A map $\Phi : \{(i,j) : 1 \leq i < j \leq n\} \to \{0,1\}$.

MEASURE: $\max_{0 \leq k < n} c_k$, where

$$c_k = \sum_{0 \leq i < j < n} \{d_{i,j} : \Phi(i,j) = 1 \wedge (i \leq k < j)\}$$
$$+ \sum_{0 \leq i < j < n} \{d_{i,j} : \Phi(i,j) = 0 \wedge (k < i \vee k \geq j)\}.$$

Good News: Admits a PTAS [Khanna, 1997].

Bibliography

Aarts, E. H. L., and Korst, J. (1989), *Simulated Annealing and Boltzmann Machines*, John Wiley & Sons, New York.

Agarwal, P. K., and Procopiuc, C. M. (1998), "Exact and approximation algorithms for clustering", *Proc. 9th Annual ACM-SIAM Symposium on Discrete Algorithms*, ACM-SIAM, 658–667. *(ND53)*

Agarwal, P. K., and Suri, S. (1994), "Surface approximation and geometric partitions", *Proc. 5th Annual ACM-SIAM Symposium on Discrete Algorithms*, ACM-SIAM, 34–43. *(MS17)*

Agarwala, R., Bafna, V., Farach, M., Narayanan, B., Paterson, M., and Thorup, M. (1996), "On the approximability of numerical taxonomy", *Proc. 7th Annual ACM-SIAM Symposium on Discrete Algorithms*, ACM-SIAM, 365–372. *(MS18)*

Aggarwal, A., Coppersmith, D., Khanna, S., Motwani, R., and Schieber, B. (1997), "The angular-metric traveling salesman problem", *Proc. 8th Annual ACM-SIAM Symposium on Discrete Algorithms*, ACM-SIAM, 221–229. *(ND34)*

Aggarwal, M., and Garg, N. (1994), "A scaling technique for better network design", *Proc. 5th Annual ACM-SIAM Symposium on Discrete Algorithms*, ACM-SIAM, 233–239. *(ND10)*

Aho, A. V., Hopcroft, J. E., and Ullman, J. D. (1974), *The Design and Analysis of Computer Algorithms*, Addison-Wesley, Reading MA.

Akutsu, T., and Halldórsson, M. (1994), "On the approximation of largest common point sets and largest common subtrees", *Proc. 5th Annual Inter-*

Bibliography

national Symposium on Algorithms and Computation, Lecture Notes in Computer Science 834, Springer-Verlag, Berlin, 405–413. *(SR8, SR9)*

Akutsu, T., and Halldórsson, M. (1997), "On the approximation of largest common point sets and largest common subtrees", Unpublished manuscript. *(SR8)*

Alekhnovich, M., Buss, S., Moran, S., and Pitassi, T. (1998), "Minimum propositional proof length is NP-hard to linearly approximate", *Proc. 23rd International Symposium on Mathematical Foundations of Computer Science*, Lecture Notes in Computer Science 1450, Springer-Verlag, Berlin, 176–184. *(LO13)*

Alimonti, P., and Kann, V. (1997), "Hardness of approximating problems on cubic graphs", *Proc. 3rd Italian Conference on Algorithms and Complexity*, Lecture Notes in Computer Science 1203, Springer-Verlag, Berlin, 288–298. *(GT1, ND14)*

Alizadeh, F. (1995), "Interior point methods in semidefinite programming with applications to combinatorial optimization", *SIAM J. Optimization* **5**, 13–51.

Alon, N., Azar, Y., Woeginger, G. J., and Yadid, T. (1997), "Approximation schemes for scheduling", *Proc. 8th Annual ACM-SIAM Symposium on Discrete Algorithms*, ACM-SIAM, 493–500. *(SS11)*

Alon, N., Csirik, J., Sevastianov, S. V., Vestjens, A. P. A., and Woeginger, G. J. (1996), "On-line and off-line approximation algorithms for vector covering problems", *Proc. 4th Annual European Symposium on Algorithms*, Lecture Notes in Computer Science 1136, Springer-Verlag, Berlin, 406–418. *(SR3)*

Alon, N., Feige, U., Wigderson, A., and Zuckerman, D. (1995), "Derandomized graph products", *Computational Complexity* **5**, 60–75. *(GT23)*

Alon, N., and Kahale, N. (1994), "Approximating the independence number via the θ-function", Unpublished manuscript. *(GT22)*

Alon, N., and Spencer, J. H. (1992), *The Probabilistic Method*, John Wiley & Sons, New York.

Alon, N., Yuster, R., and Zwick, U. (1994), "Color-coding: a new method for finding simple paths, cycles and other small subgraphs within large graphs", *Proc. 26th ACM Symposium on Theory of Computing*, ACM, 326–335. *(ND42)*

Amaldi, E., and Kann, V. (1995), "The complexity and approximability of finding maximum feasible subsystems of linear relations", *Theoretical Computer Science* **147**, 181–210. *(MP10, MP12, AN1)*

Amaldi, E., and Kann, V. (1998), "On the approximability of minimizing nonzero variables or unsatisfied relations in linear systems", *Theoretical Computer Science* **209**, 237–260. *(MP9, MP11, MP12)*

Bibliography

Amoura, A. K., Bampis, E., Kenyon, C., and Manoussakis, Y. (1997), "How to schedule independent multiprocessor tasks", *Proc. 5th Annual European Symposium on Algorithms*, Lecture Notes in Computer Science 1284, Springer-Verlag, Berlin, 1–12. *(SS13)*

Andersson, G. (1999), "An approximation algorithm for MAX p-SECTION", *Proc. 16th Symposium on Theoretical Aspects of Computer Science*, Lecture Notes in Computer Science 1563, Springer-Verlag, Berlin, 237–247. *(ND17)*

Andersson, G., and Engebretsen, L. (1998a), "Better approximation algorithms for set splitting and not-all-equal sat", *Information Processing Letters* **65**, 305–311. *(SP3, LO4)*

Andersson, G., and Engebretsen, L. (1998b), "Sampling methods applied to non-boolean optimization problems", *Proc. 3rd International Symposium on Randomization and Approximation Techniques in Computer Science*, Lecture Notes in Computer Science 1518, Springer-Verlag, Berlin, 357–368. *(MP10, LO12)*

Andersson, G., Engebretsen, L., and Håstad, J. (1999), "A new way to use semidefinite programming with applications to linear equations mod p", *Proc. 10th Annual ACM-SIAM Symposium on Discrete Algorithms*, ACM-SIAM, to appear. *(MP10)*

Andreae, T., and Bandelt, H. (1995), "Performance guarantees for approximation algorithms depending on parametrized triangle inequalities", *SIAM J. Discrete Mathematics* **8**, 1–16. *(ND33)*

Anily, A., Bramel, J., and Simchi-Levi, D. (1994), "Worst-case analysis of heuristics for the bin-packing problem with general cost structures", *Operations Research* **42**, 287–298. *(SR1)*

Anily, S., and Hassin, R. (1992), "The swapping problem", *Networks* **22**, 419–433. *(ND33)*

Arkin, E. M., Chiang, Y., Mitchell, J. S. B., Skiena, S. S., and Yang, T. (1997), "On the maximum scatter TSP", *Proc. 8th Annual ACM-SIAM Symposium on Discrete Algorithms*, ACM-SIAM, 211–220. *(ND36)*

Arkin, E. M., Halldórsson, M. M., and Hassin, R. (1993), "Approximating the tree and tour covers of a graph", *Information Processing Letters* **47**, 275–282. *(GT1)*

Arkin, E. M., and Hassin, R. (1992), "Multiple-choice minimum diameter problems", Unpublished manuscript. *(SP4)*

Arkin, E. M., and Hassin, R. (1994), "Approximation algorithms for the geometric covering salesman problem", *Discrete Applied Mathematics* **55**, 197–218. *(ND33)*

Arkin, E. M., and Hassin, R. (1998), "On local search for weighted packing probems", *Mathematics of Operations Research* **23**, 640–648. *(SP2)*

Bibliography

Armen, C., and Stein, C. (1994), "A $2+\frac{3}{4}$-approximation algorithm for the shortest superstring problem", Technical Report PCS-TR94-214, Department of Computer Science, Dartmouth College, Hanover, NH. *(SR6)*

Arora, S. (1994), *Probabilistic Checking of Proofs and hardness of approximation problems*, PhD thesis, U. C. Berkeley.

Arora, S. (1996), "Polynomial time approximation scheme for Euclidean TSP and other geometric problems", *Proc. 37th Annual IEEE Symposium on Foundations of Computer Science*, IEEE Computer Society, 2–11. *(ND1, ND3, ND9, ND34)*

Arora, S. (1997), "Nearly linear time approximation schemes for Euclidean TSP and other geometric problems", *Proc. 38th Annual IEEE Symposium on Foundations of Computer Science*, IEEE Computer Society, 554–563. *(ND9, ND34)*

Arora, S. (1998), "The approximability of NP-hard problems", *Proc. 30th ACM Symposium on Theory of Computing*, ACM, 337–348.

Arora, S., Babai, L., Stern, J., and Sweedyk, Z. (1997), "The hardness of approximate optima in lattices, codes, and systems of linear equation", *J. Computer and System Sciences* **54**, 317–331. *(MP10, MP11, MP12, MP17, MS3)*

Arora, S., Frieze, A., and Kaplan, H. (1996), "A new rounding procedure for the assignment problem with applications to dense graph arrangement problems", *Proc. 37th Annual IEEE Symposium on Foundations of Computer Science*, IEEE Computer Society, 21–30. *(GT9, GT44, GT45, MS1)*

Arora, S., Grigni, M., Karger, D., Klein, P., and Woloszyn, A. (1998), "A polynomial-time approximation scheme for weighted planar graph TSP", *Proc. 9th Annual ACM-SIAM Symposium on Discrete Algorithms*, ACM-SIAM, 33–41. *(ND33)*

Arora, S., Karger, D., and Karpinski, M. (1995), "Polynomial time approximation schemes for dense instances of NP-hard problems", *Proc. 27th ACM Symposium on Theory of Computing*, ACM, 284–293. *(GT33, GT35, ND14, ND16, ND19, ND21, ND24, SP3, LO2)*

Arora, S., and Lund, C. (1996), "Hardness of approximations", in *Approximation Algorithms for NP-hard Problems*, PWS, Boston, Mass., 399–446.

Arora, S., Lund, C., Motwani, R., Sudan, M., and Szegedy, M. (1992), "Proof verification and hardness of approximation problems", *Proc. 33rd Annual IEEE Symposium on Foundations of Computer Science*, IEEE Computer Society, 14–23.

Arora, S., and Safra, S. (1998), "Probabilistic checking of proofs: A new characterization of NP", *J. ACM* **45**, 70–122.

Arts, E., and Lenstra, J. K. (eds.) (1997), *Local Search in Combinatorial Optimization*, John Wiley & Sons, New York.

Bibliography

Asano, T. (1997), "Approximation algorithms for MAX SAT: Yannakakis vs. Goemans-Williamson", *Proc. 5th Israel Symposium on Theory of Computing and Systems*, IEEE Computer Society, 24–37.

Asano, T., Hori, K., Ono, T., and Hirata, T. (1997), "A theoretical framework of hybrid approaches to MAX SAT", *Proc. 8th Annual International Symposium on Algorithms and Computation*, Lecture Notes in Computer Science 1350, Springer-Verlag, Berlin, 153–162. *(LO1)*

Assmann, S. F., Johnson, D. S., Kleitman, D. J., and Leung, J. Y. (1984), "On a dual version of the one-dimensional bin packing problem", *J. Algorithms* **5**, 502–525. *(SR3)*

Aumann, Y., and Rabani, Y. (1995), "Improved bounds for all optical routing", *Proc. 6th Annual ACM-SIAM Symposium on Discrete Algorithms*, ACM-SIAM, 567–576. *(ND48)*

Aumann, Y., and Rabani, Y. (1998), "An $o(\log k)$ approximate min-cut max-flow theorem and approximation algorithm", *SIAM J. Computing* **27**, 291–301. *(ND23)*

Ausiello, G., D'Atri, A., and Protasi, M. (1980), "Structure preserving reductions among convex optimization problems", *J. Computer and System Sciences* **21**, 136–153. *(GT8, SP2, SP4, SP7)*

Ausiello, G., D'Atri, A., and Protasi, M. (1981), "Lattice theoretic ordering properties for NP-complete optimization problems", *Annales Societatis Mathematicae Polonae* **4**, 83–94. *(LO6)*

Ausiello, G., Marchetti-Spaccamela, A., and Protasi, M. (1980), "Toward a unified approach for the classification of NP-complete optimization problems", *Theoretical Computer Science* **12**, 83–96.

Averbakh, I., and Berman, O. (1997), "$(p-1)/(p+1)$-approximate algorithms for p-traveling salesmen problems on a tree with minmax objective", *Discrete Applied Mathematics* **75**, 201–216. *(ND35)*

Awerbuch, B., and Azar, Y. (1997), "Buy-at-bulk network design", *Proc. 38th Annual IEEE Symposium on Foundations of Computer Science*, IEEE Computer Society, 542–547. *(ND50)*

Babai, L., Fortnow, L., Levin, L., and Szegedy, M. (1991), "Checking computations in polylogarithmic time", *Proc. 23rd ACM Symposium on Theory of Computing*, ACM, 21–31.

Babai, L., Fortnow, L., and Lund, C. (1991), "Non-deterministic exponential time has two-prover interactive protocol", *Computational Complexity* **1**, 3–40.

Babai, L., and Moran, S. (1988), "Arthur-Merlin games: A randomized proof system, and a hierarchy of complexity classes", *J. Computer and System Sciences* **36**, 254–276.

Bafna, V., Berman, P., and Fujito, T. (1994), "Approximating feedback vertex set for undirected graphs within ratio 2", Unpublished manuscript. *(GT8)*

Bibliography

Bafna, V., and Pevzner, P. A. (1995), "Sorting permutations by transpositions", *Proc. 6th Annual ACM-SIAM Symposium on Discrete Algorithms*, ACM-SIAM, 614–621. *(MS11)*

Baker, B. S. (1983), "A new proof for the first fit decreasing algorithm", Technical report, Bell Laboratories, Murray Hill.

Baker, B. S. (1994), "Approximation algorithms for NP-complete problems on planar graphs", *J. ACM* **41**, 153–180. *(GT1, GT2, GT3, GT11, GT12, GT23)*

Balcàzar, J. L., Diaz, J., and Gabarrò, J. (1988), *Structural Complexity, Vols. 1–2*, Springer-Verlag, Berlin.

Bandelt, H., Crama, Y., and Spieksma, F. C. R. (1994), "Approximation algorithms for multi-dimensional assignment problems with decomposable costs", *Discrete Applied Mathematics* **49**, 25–50. *(SP10)*

Bar-Ilan, J., Kortsarz, G., and Peleg, D. (1996), "Generalized submodular cover problems and applications", *Proc. 4th Israel Symposium on Theory of Computing and Systems*, IEEE Computer Society, 110–118. *(ND8)*

Bar-Ilan, J., and Peleg, D. (1991), "Approximation algorithms for selecting network centers", *Proc. 2nd Workshop on Algorithms and Data Structures*, Lecture Notes in Computer Science 519, Springer-Verlag, Berlin, 343–354. *(ND53)*

Bar-Noy, A., Bellare, M., Halldórsson, M. M., Shachnai, H., and Tamir, T. (1998), "On chromatic sums and distributed resource allocation", *Information and Computation* **140**, 183–202. *(GT5)*

Bar-Noy, A., and Kortsarz, G. (1997), "The minimum color sum of bipartite graphs", *Proc. 24th International Colloquium on Automata, Languages and Programming*, Lecture Notes in Computer Science 1256, Springer-Verlag, Berlin, 738–748. *(GT5)*

Bar-Yehuda, R., and Even, S. (1981), "A linear-time approximation algorithm for the weighted vertex cover problem", *J. Algorithms* **2**, 198–203. *(GT1, SP4)*

Bar-Yehuda, R., and Even, S. (1985), "A local-ratio theorem for approximating the weighted vertex cover problem", in *Analysis and Design of Algorithms for Combinatorial Problems*, volume 25 of *Annals of Discrete Mathematics*, Elsevier Science, Amsterdam, 27–46. *(GT1)*

Bar-Yehuda, R., and Moran, S. (1984), "On approximation problems related to the independent set and vertex cover problems", *Discrete Applied Mathematics* **9**, 1–10. *(GT2)*

Bartal, Y., Leonardi, S., Marchetti-Spaccamela, A., Sgall, J., and Stougie, L. (1996), "Multiprocessor scheduling with rejection", *Proc. 7th Annual ACM-SIAM Symposium on Discrete Algorithms*, ACM-SIAM, 95–103. *(SS6)*

Bibliography

Barvinok, A. I. (1996), "Two algorithmic results for the traveling salesman problem", *Mathematics of Operations Research* **21**, 65–84. *(ND34)*

Beardwood, J., Halton, J. H., and Hammersley, J. M. (1959), "The shortest path through many points", *Proc. Cambridge Philosophical Society* **55**, 299–327.

Bellare, M. (1993), "Interactive proofs and approximation: reductions from two provers in one round", *Proc. 2nd Israel Symposium on Theory of Computing and Systems*, IEEE Computer Society, 266–274. *(ND42, ND46, SP11)*

Bellare, M., Goldreich, O., and Sudan, M. (1998), "Free bits, PCPs and non-approximability – towards tight results", *SIAM J. Computing* **27**, 804–915. *(GT5, GT22, SR7, MS12)*

Bellare, M., Goldwasser, S., Lund, C., and Russell, A. (1993), "Efficient probabilistically checkable proofs and applications to approximation", *Proc. 25th ACM Symposium on Theory of Computing*, ACM, 294–304. *(GT2, SP4)*

Bellare, M., and Rogaway, P. (1995), "The complexity of approximating a non-linear program", *Mathematical Programming* **69**, 429–441. *(MP5)*

Ben-Or, M., Goldwasser, S., Kilian, S., and Wigderson, A. (1988), "Multi prover interactive proofs: How to remove intractability assumptions", *Proc. 20th ACM Symposium on Theory of Computing*, ACM, 113–121.

Bender, M. A., Chakrabarti, S., and Muthukrishnan, S. (1998), "Flow and stretch metrics for scheduling continuous job streams", *Proc. 9th Annual ACM-SIAM Symposium on Discrete Algorithms*, ACM-SIAM, 270–279. *(SS4)*

Bentley, J. L., Johnson, D. S., Leighton, F. T., McGeoch, C. C., and McGeoch, L. A. (1984), "Some unexpected expected behavior for bin packing", *Proc. 16th ACM Symposium on Theory of Computing*, ACM, 279–288.

Berger, B., and Cowen, L. (1991), "Complexity results and algorithms for $\{<, \leq, =\}$-constrained scheduling", *Proc. 2nd Annual ACM-SIAM Symposium on Discrete Algorithms*, ACM-SIAM, 137–147. *(SS7)*

Berger, B., and Shor, P. W. (1990), "Approximation algorithms for the maximum acyclic subgraph problem", *Proc. 1st Annual ACM-SIAM Symposium on Discrete Algorithms*, ACM-SIAM, 236–243. *(GT9)*

Berman, F., Johnson, D., Leighton, T., Shor, P. W., and Snyder, L. (1990), "Generalized planar matching", *J. Algorithms* **11**, 153–184. *(GT12)*

Berman, P., and DasGupta, B. (1997), "Complexities of efficient solutions of rectilinear polygon cover problems", *Algorithmica* **17**, 331–356. *(SR10)*

Berman, P., and Fujito, T. (1995), "Approximating independent sets in degree 3 graphs", *Proc. 4th Workshop on Algorithms and Data Structures*, Lecture Notes in Computer Science 955, Springer-Verlag, Berlin, 449–460. *(GT1, GT23, SP2)*

Bibliography

Berman, P., and Fürer, M. (1994), "Approximating maximum independent set in bounded degree graphs", *Proc. 5th Annual ACM-SIAM Symposium on Discrete Algorithms*, ACM-SIAM, 365–371. *(GT23)*

Berman, P., and Karpinski, M. (1998), "On some tighter inapproximability results", Technical Report TR98-029, The Electronic Colloquium on Computational Complexity. *(GT1, GT23, MP10, LO2, MS11)*

Berman, P., and Schnitger, G. (1992), "On the complexity of approximating the independent set problem", *Information and Computation* **96**, 77–94. *(GT28, GT50, SR7, MP2, LO12, AL2)*

Bern, M., and Plassmann, P. (1989), "The Steiner problem with edge lengths 1 and 2", *Information Processing Letters* **32**, 171–176. *(ND8)*

Bernstein, D., Rodeh, M., and Gertner, I. (1989), "Approximation algorithms for scheduling arithmetic expressions on pipelined machines", *J. Algorithms* **10**, 120–139. *(SS3)*

Bertsimas, D., Teo, C-P., and Vohra, R. (1996), "On dependent randomized rounding algorithms", *Proc. 5th International Conference on Integer Programming and Combinatorial Optimization*, Lecture Notes in Computer Science 1084, Springer-Verlag, Berlin, 330–344. *(GT7, LO1, LO3)*

Bertsimas, D., and Vohra, R. (1994), "Linear programming relaxations, approximation algorithms, and randomization : a unified view of covering problems", Technical Report OR 285-94, MIT.

Bingham, N. H. (1981), "Tauberian theorems and the central limit theorem", *The Annals of Probability* **9**, 221–231.

Blache, G., Karpinski, M., and Wirtgen, J. (1998), "On approximation intractability of the bandwidth problem", Technical Report TR98-014, The Electronic Colloquium on Computational Complexity. *(GT42)*

Blaha, K. D. (1992), "Minimum bases for permutation groups: the greedy approximation", *J. Algorithms* **13**, 297–306. *(AL5)*

Blum, A., Chalasani, P., Coppersmith, D., Pulleyblank, B., Raghavan, P., and Sudan, M. (1994), "The minimum latency problem", *Proc. 26th ACM Symposium on Theory of Computing*, ACM, 163–171. *(ND33)*

Blum, A., Jiang, T., Li, M., Tromp, J., and Yannakakis, M. (1994), "Linear approximation of shortest superstrings", *J. ACM* **41**, 630–647. *(SR6)*

Blum, A., Konjevod, G., Ravi, R., and Vempala, S. (1998), "Semi-definite relaxations for minimum bandwidth and other vertex-ordering problems", *Proc. 30th ACM Symposium on Theory of Computing*, ACM, 100–105. *(GT42)*

Blum, A., Ravi, R., and Vempala, S. (1996), "A constant-factor approximation algorithm for the k-MST problem", *Proc. 28th ACM Symposium on Theory of Computing*, ACM, 442–448. *(ND8)*

Blum, M., and Kannan, S. (1989), "Designing programs that check their work", *Proc. 21st ACM Symposium on Theory of Computing*, ACM, 86–97.

Blundo, C., De Santis, A., and Vaccaro, U. (1994), "Randomness in distribution protocols", Unpublished manuscript. *(GT24)*

Bodlaender, H. L., Gilbert, J. R., Hafsteinsson, H., and Kloks, T. (1995), "Approximating treewidth, pathwidth, frontsize and shortest elimination tree", *J. Algorithms* **18**, 238–255. *(GT54)*

Bollobas, B. (1979), *Graph Theory, an introductory course*, Springer-Verlag, Berlin.

Bollobas, B. (1985), *Random Graphs*, Academic Press, New York.

Bollobas, B. (1988), "The chromatic number of random graphs", *Combinatorica* **8**, 49–55. *(GT12, GT23, LO12)*

Bollobas, B., and Erdös, P. (1976), "Cliques in random graphs", *Proc. Cambridge Philosophical Society* **80**, 419–427.

Bonizzoni, P., Duella, M., and Mauri, G. (1994), "Approximation complexity of longest common subsequence and shortest common supersequence over fixed alphabet", Technical Report 117/94, Dipartimento di Scienze dell'Informazione, Università degli Studi di Milano. *(SR5, SR7)*

Boppana, R., and Halldórsson, M. M. (1992), "Approximating maximum independent sets by excluding subgraphs", *BIT* **32**, 180–196. *(GT22)*

Bovet, D. P., and Crescenzi, P. (1994), *Introduction to the Theory of Complexity*, Prentice Hall, Englewood Cliffs NJ.

Boxma, O. J. (1985), "A probabilistic analysis of multiprocessor list scheduling: the Erlang case", *Stochastic Models* **1**, 209–220.

Brooks, R. L. (1941), "On coloring the nodes of a network", *Proc. Cambridge Philosophical Society* **37**, 194–197.

Bui, T. N., and Jones, C. (1992), "Finding good approximate vertex and edge partitions is NP-hard", *Information Processing Letters* **42**, 153–159. *(ND24, ND25)*

Călinescu, G., Fernandes, C. G., Finkler, U., and Karloff, H. (1996), "A better approximation algorithm for finding planar subgraphs", *Proc. 7th Annual ACM-SIAM Symposium on Discrete Algorithms*, ACM-SIAM, 16–25. *(GT31)*

Călinescu, G., Karloff, H., and Rabani, Y. (1998), "An improved approximation algorithm for multiway cut", *Proc. 30th ACM Symposium on Theory of Computing*, ACM, 48–52. *(ND21)*

Cerny, V. (1985), "A thermodynamical approach to the traveling salesman problem: An efficient simulation algorithm", *J. Optimization Theory and Applications* **45**, 41–51.

Bibliography

Chakrabati, S., Phillips, C. A., Schulz, A. S., Shmoys, D. B., Stein, C., and Wein, J. (1996), "Improved scheduling algorithms for minsum criteria", *Proc. 23rd International Colloquium on Automata, Languages and Programming*, Lecture Notes in Computer Science 1099, Springer-Verlag, Berlin, 646–657. *(SS12)*

Chandra, A. K., Hirschberg, D. S., and Wong, C. K. (1976), "Approximate algorithms for some generalized knapsack problems", *Theoretical Computer Science* **3**, 293–304. *(MP14, MP15)*

Chandra, B., Karloff, H., and Tovey, C. A. (1994), "New results on the old k-opt algorithm for the TSP", *Proc. 5th Annual ACM-SIAM Symposium on Discrete Algorithms*, ACM-SIAM, 150–159.

Charikar, M., Chekuri, C., Cheung, T., Dai, Z., Goel, A., Guha, S., and Li, M. (1998), "Approximation algorithms for directed Steiner problems", *Proc. 9th Annual ACM-SIAM Symposium on Discrete Algorithms*, ACM-SIAM, 192–200. *(ND8, ND10)*

Chaudhary, A., and Vishwanathan, S. (1997), "Approximation algorithms for the achromatic number", *Proc. 8th Annual ACM-SIAM Symposium on Discrete Algorithms*, ACM-SIAM, 558–563. *(GT6)*

Chekuri, C., Motwani, R., Natarajan, B., and Stein, C. (1997), "Approximation techniques for average completion time scheduling", *Proc. 8th Annual ACM-SIAM Symposium on Discrete Algorithms*, ACM-SIAM, 609–618. *(SS4)*

Chen, B. (1993), "A better heuristic for preemptive parallel machine scheduling with batch setup times", *SIAM J. Computing* **22**, 1303–1318. *(SS9)*

Chen, B. (1994), "Scheduling multiprocessor flow shops", in *Advances in Optimization and Approximation*, Kluwer, The Netherlands, 1–8. *(SS15)*

Chen, B., Glass, C. A., Potts, C. N., and Strusevich, V. A. (1996), "A new heuristic for three-machine flow shop scheduling", *Operations Research* **44**, 891–898. *(SS15)*

Chen, B., Potts, C. N., and Strusevich, V. A. (1998), "Approximation algorithms for two-machine flow shop scheduling with batch setup times", *Mathematical Programming* **82**, 255–271. *(SS16)*

Chen, B., and Strusevich, V. A. (1993a), "Approximation algorithms for three-machine open shop scheduling", *ORSA J. Computing* **5**, 321–326. *(SS14)*

Chen, B., and Strusevich, V. A. (1993b), "Worst-case analysis of heuristics for open shops with parallel machines", *European J. Operations Research* **70**, 379–390. *(SS14)*

Chen, J., Friesen, D.K., and Zheng, H. (1997), "Tight bound on johnson's algorithm for Max-SAT", *Proc. 12th Annual IEEE Conference on Computational Complexity*, 274–281.

Chen, Z.-Z. (1996), "Practical approximation schemes for maximum induced-subgraph problems on $k_{3,3}$-free or k_5-free graphs", *Proc. 23rd International Colloquium on Automata, Languages and Programming*, Lecture Notes in Computer Science 1099, Springer-Verlag, Berlin, 268–279. *(GT25)*

Cheriyan, J., and Thurimella, R. (1996), "Approximating minimum-size k-connected spanning subgraphs via matching", *Proc. 37th Annual IEEE Symposium on Foundations of Computer Science*, IEEE Computer Society, 292–301. *(ND27, ND28)*

Chiba, N., Nishizeki, T., and Saito, N. (1982), "An approximation algorithm for the maximum independent set problem on planar graphs", *SIAM J. Computing* **11**, 663–675.

Chlebikova, J. (1996), "Approximability of the maximally balanced connected partition problem in graphs", *Information Processing Letters* **60**, 225–230. *(GT16, ND4)*

Choi, J., Sellen, J., and Yap, C. K. (1997), "Approximate Euclidean shortest paths in 3-space", *International J. Computational Geometry and Applications* **7**, 271–295. *(MS13)*

Chor, B., and Sudan, M. (1995), "A geometric approach to betweenness", *Proc. 3rd Annual European Symposium on Algorithms*, Lecture Notes in Computer Science 979, Springer-Verlag, Berlin, 227–237. *(MS1)*

Christie, D. A. (1998), "A 3/2-approximation algorithm for sorting by reversals", *Proc. 9th Annual ACM-SIAM Symposium on Discrete Algorithms*, ACM-SIAM, 244–252. *(MS11)*

Christofides, N. (1976), "Worst-case analysis of a new heuristic for the travelling salesman problem", Technical Report 388, Graduate School of Industrial Administration, Carnegie-Mellon University, Pittsburgh, PA. *(ND33)*

Chudak, F. A., and Shmoys, D. B. (1997), "Approximation algorithms for precedence-constrained scheduling problems on parallel machines that run at different speed", *Proc. 8th Annual ACM-SIAM Symposium on Discrete Algorithms*, ACM-SIAM, 581–590. *(SS7)*

Chvátal, V. (1979), "A greedy heuristic for the set covering problem", *Mathematics of Operations Research* **4**, 233–235. *(SP4)*

Clementi, A., and Di Ianni, M. (1996), "On the hardness of approximating optimum schedule problems in store and forward networks", *IEEE/ACM Transactions on Networking* **4**, 272–280. *(SS19)*

Clementi, A., and Trevisan, L. (1996), "Improved non-approximability results for vertex cover problems with density constraints", *Proc. 2nd Annual International Conference on Computing and Combinatorics*, Lecture Notes in Computer Science 1090, Springer-Verlag, Berlin, 333–342. *(GT1)*

Cody, R. A., and Coffman Jr., E. G. (1976), "Record allocation for minimizing expected retrieval costs on drum-like storage devices", *J. ACM* **23**, 103–115. *(SP12)*

Bibliography

Coffman Jr., E. G., Flatto, L., and Lueker, G. S. (1984), "Expected makespan for largest–first multiprocessor scheduling", *Proc. 10th International Symposium on Models of Computer System Performance*, North-Holland, Amsterdam, 475–490.

Coffman Jr., E. G., Garey, M. R., and Johnson, D. S. (1997), "Approximation algorithms for bin-packing – a survey", in *Approximation Algorithms for NP-hard Problems*, PWS, Boston, Mass., 46–93. *(SR1)*

Coffman Jr., E. G., Garey, M. R., Johnson, D. S., and Lapaugh, A. S. (1985), "Scheduling file transfers", *SIAM J. Computing* **14**, 744–780. *(SS18)*

Coffman Jr., E. G., and Gilbert, E. N. (1985), "On the expected relative performance of list scheduling", *Operations Research* **33**, 548–561.

Coffman Jr., E. G., and Lueker, G. S. (1991), *Probabilistic Analysis of Packing and Partitioning Algorithms*, John Wiley & Sons, New York.

Cook, S. A. (1971), "The complexity of theorem-proving procedures", *Proc. 3rd ACM Symposium on Theory of Computing*, ACM, 151–158.

Cormen, T., Leiserson, C. E., and Rivest, R. L. (1990), *Introduction to Algorithms*, MIT Press and McGraw Hill.

Cornuejols, G., Fisher, M., and Nemhauser, G. (1977), "Location of bank accounts to optimize float: An analytic study of exact and approximate algorithms", *Management Science* **23**, 789–810. *(ND61)*

Cowen, L. J., Goddard, W., and Jesurum, C. E. (1997), "Coloring with defect", *Proc. 8th Annual ACM-SIAM Symposium on Discrete Algorithms*, ACM-SIAM, 548–557. *(GT5)*

Crama, Y., and Spieksma, F. C. R. (1992), "Approximation algorithms for three-dimensional assignment problems with triangle inequalities", *European J. Operations Research* **60**, 273–279. *(SP10)*

Crescenzi, P. (1997), "A short guide to approximation preserving reductions", *Proc. 12th Annual IEEE Conference on Computational Complexity*, IEEE Computer Society, 262–273.

Crescenzi, P., Fiorini, C., and Silvestri, R. (1991), "A note on the approximation of the MAX CLIQUE problem", *Information Processing Letters* **40**, 1–5.

Crescenzi, P., Kann, V., Silvestri, R., and Trevisan, L. (1999), "Structure in approximation classes", *SIAM J. Computing* **28**, 1759–1782. *(GT7, ND2, SR1)*

Crescenzi, P., and Panconesi, A. (1991), "Completeness in approximation classes", *Information and Computation* **93**, 241–262. *(LO8)*

Crescenzi, P., and Silvestri, R. (1990), "Relative complexity of evaluating the optimum cost and constructing the optimum for maximization problems", *Information Processing Letters* **33**, 221–226.

Crescenzi, P., Silvestri, R., and Trevisan, L. (1996), "To weight or not to weight: Where is the question?", *Proc. 4th Israel Symposium on Theory of Computing and Systems*, IEEE Computer Society, 68–77. *(GT1, ND14, ND16, LO1, LO2)*

Crescenzi, P., and Trevisan, L. (1994), "On approximation scheme preserving reducibility and its applications", *Proc. 14th Annual Conference on Foundations of Software Technology and Theoretical Computer Science*, Lecture Notes in Computer Science 880, Springer-Verlag, Berlin, 330–341.

Croes, G. A. (1958), "A method for solving traveling-salesman problems", *Operations Research* **6**, 791–812.

Dahlhaus, E., Johnson, D. S., Papadimitriou, C. H., Seymour, P. D., and Yannakakis, M. (1994), "The complexity of multiterminal cuts", *SIAM J. Computing* **23**, 864–894. *(ND21)*

D'Anzeo, C. (1996), "Optimization complexity of the SCS problem given a longest common subsequence", Unpublished manuscript. *(SR5)*

DasGupta, B., He, X., Jiang, T., Li, M., Tromp, J., and Zhang, L. (1997), "On distances between phylogenetic trees", *Proc. 8th Annual ACM-SIAM Symposium on Discrete Algorithms*, ACM-SIAM, 427–436. *(MS15)*

Datta, A. K., and Sen, R. K. (1995), "1-approximation algorithm for bottleneck disjoint path matching", *Information Processing Letters* **55**, 41–44. *(GT13)*

de Werra, D., and Hertz, A. (1989), "Tabu search techniques: A tutorial and an application to neural networks", *OR Spektrum* **11**, 131–141.

Demange, M., and Pascos, V.Th. (1996), "On an approximation measure founded on the links between optimization and polynomial approximation theories", *Theoretical Computer Science* **158**, 117–141.

Di Battista, G., Eades, P., Tamassia, R., and Tollis, I.G. (1999), *Graph Drawing. Algorithms for the Visualization of Graphs*, Prentice Hall, Englewood Cliffs NJ.

Di Ianni, M. (1998), "Efficient delay routing", *Theoretical Computer Science* **196**, 131–151. *(SS19)*

Doddi, S., Marathe, M. V., Mirzaian, A., Moret, B. M. E., and Zhu, B. (1997), "Map labeling and its generalizations", *Proc. 8th Annual ACM-SIAM Symposium on Discrete Algorithms*, ACM-SIAM, 148–157. *(MS14)*

Drangmeister, K. U., Krumke, S. O., Marathe, M. V., Noltemeier, H., and Ravi, S. S. (1998), "Modifying edges of a network to obtain short subgraphs", *Theoretical Computer Science* **203**, 91–121. *(ND8)*

Dreyfus, S. E., and Law, A. M. (1977), *The Art and Theory of Dynamic Programming*, Academic Press, New York.

Dudek, G., Romanik, K., and Whitesides, S. (1994), "Localizing a robot with minimum travel", Technical Report SOCS-94.5, McGill University, Canada. *(GP2)*

Bibliography

Dunham, B., Fridshal, D., Fridshal, R., and North, J. H. (1961), "Design by natural selection", Technical Report RC-476, IBM Res. Dept..

Eades, P., and Wormald, N. C. (1994), "Edge crossings in drawings of bipartite graphs", *Algorithmica* **11**, 379–403. *(ND15)*

Edmonds, J. (1971), "Matroids and the greedy algorithm", *Mathematical Programming* **1**, 127–136.

Engebretsen, L. (1999), "An explicit lower bound for TSP with distances one and two", *Proc. 16th Symposium on Theoretical Aspects of Computer Science*, Lecture Notes in Computer Science 1563, Springer-Verlag, Berlin, to appear. *(ND33)*

Eppstein, D. (1992), "Approximating the minimum weight triangulation", *Proc. 3rd Annual ACM-SIAM Symposium on Discrete Algorithms*, ACM-SIAM, 48–57. *(ND64)*

Errico, B., and Rosati, R. (1995), "Minimal models in propositional logics: approximation results", *Proc. 5th Italian Conference on Theoretical Computer Science*, World Scientific, Singapore, 547–562. *(LO7)*

Even, G., Naor, J., Rao, S., and Schieber, B. (1995), "Divide-and-conquer approximation algorithms via spreading metrics", *Proc. 36th Annual IEEE Symposium on Foundations of Computer Science*, IEEE Computer Society, 62–71. *(GT45)*

Even, G., Naor, J., Rao, S., and Schieber, B. (1997), "Fast approximate graph partitioning algorithms", *Proc. 8th Annual ACM-SIAM Symposium on Discrete Algorithms*, ACM-SIAM, 639–648. *(ND24)*

Even, G., Naor, J., Schieber, B., and Sudan, M. (1995), "Approximating minimum feedback sets and multi-cuts in directed graphs", *Proc. 4th International Conference on Integer Programming and Combinatorial Optimization*, Lecture Notes in Computer Science 920, Springer-Verlag, Berlin, 14–28. *(GT8, GT9)*

Even, G., Naor, J., and Zosin, L. (1996), "An 8-approximation algorithm for the subset feedback vertex set problem", *Proc. 37th Annual IEEE Symposium on Foundations of Computer Science*, IEEE Computer Society, 310–319. *(GT8)*

Farach, M., Kannan, S., and Warnow, T. (1993), "A robust model for finding optimal evolutionary trees", *Proc. 25th ACM Symposium on Theory of Computing*, ACM, 137–145. *(MS9)*

Farach, M., and Liberatore, V. (1998), "On local register allocation", *Proc. 9th Annual ACM-SIAM Symposium on Discrete Algorithms*, ACM-SIAM, 564–573. *(PO2)*

Feder, T., and Greene, D. H. (1988), "Optimal algorithms for approximate clustering", *Proc. 20th ACM Symposium on Theory of Computing*, ACM, 434–444. *(ND36, ND53, ND54)*

Feige, U. (1996), "A threshold of ln n for approximating set cover", *Proc. 28th ACM Symposium on Theory of Computing*, ACM, 314–318. *(GT2, SP4)*

Feige, U. (1998), "Approximating the bandwidth via volume respecting embedding", *Proc. 30th ACM Symposium on Theory of Computing*, ACM, 90–99. *(GT42)*

Feige, U., and Goemans, M. X. (1995), "Approximating the value of two prover proof systems, with applications to MAX 2SAT and MAX DICUT", *Proc. 3rd Israel Symposium on Theory of Computing and Systems*, IEEE Computer Society, 182–189. *(ND16, LO2)*

Feige, U., Goldwasser, S., Lovász, L., Safra, S., and Szegedy, M. (1991), "Approximating clique is almost NP-complete", *Proc. 32nd Annual IEEE Symposium on Foundations of Computer Science*, IEEE Computer Society, 2–12.

Feige, U., and Kilian, J. (1996), "Zero knowledge and the chromatic number", *Proc. 11th Annual IEEE Conference on Computational Complexity*, IEEE Computer Society, 278–287. *(GT5)*

Feige, U., Kortsarz, G., and Peleg, D. (1995), "The dense k-subgraph problem", Unpublished manuscript. *(GT35)*

Fekete, S., Khuller, S., Klemmstein, M., Raghavachari, B., and Young, N. (1996), "A network-flow technique for finding low-weight bounded-degree spanning trees", *Proc. 5th International Conference on Integer Programming and Combinatorial Optimization*, Lecture Notes in Computer Science 1084, Springer-Verlag, Berlin, 105–117. *(ND3)*

Feller, W. (1971), *An Introduction to Probability Theory and its Applications*, Vols. 1–2, John Wiley & Sons, New York.

Feo, T., Goldschmidt, O., and Khellaf, M. (1992), "One half approximation algorithms for the k-partition problem", *Operations Research* **40**, 170–172. *(ND55)*

Feo, T., and Khellaf, M. (1990), "A class of bounded approximation algorithms for graph partitioning", *Networks* **20**, 181–195. *(ND55)*

Fernandes, C. G. (1997), "A better approximation ratio for the minimum k-edge-connected spanning subgraph problem", *Proc. 8th Annual ACM-SIAM Symposium on Discrete Algorithms*, ACM-SIAM, 629–638. *(ND28)*

Fernandez de la Vega, W., and Karpinski, M. (1998), "Polynomial time approximation of dense weighted instances of MAX-CUT", Technical Report TR98-064, The Electronic Colloquium on Computational Complexity. *(ND14)*

Fernandez de la Vega, W., and Kenyon, C. (1998), "A randomized approximation scheme for metric MAX-CUT", *Proc. 39th Annual IEEE Symposium on Foundations of Computer Science*, IEEE Computer Society, 468–471. *(ND14)*

Bibliography

Fernandez de la Vega, W., and Lueker, G. S. (1981), "Bin packing can be solved within $1 + \varepsilon$ in linear time", *Combinatorica* **1**, 349–355.

Fernandez de la Vega, W., and Zissimopoulos, V. (1991), "An approximation scheme for strip-packing of rectangles with bounded dimensions", Technical Report 713, Laboratoire de Recherche en Informatique, Université de Paris, Orsay. *(SR2)*

Fialko, S., and Mutzel, P. (1998), "A new approximation algorithm for the Planar augmentation problem", *Proc. 9th Annual ACM-SIAM Symposium on Discrete Algorithms*, ACM-SIAM, 260–269. *(ND29)*

Fiduccia, C. M., and Mattheyses, R. M. (1982), "A linear-time heuristic for improving network partitions", *Proc. 19th ACM-IEEE Design Automation Conference*, ACM-IEEE Computer Society, 175–181.

Floréen, P., and Orponen, P. (1993), "Attraction radii in binary Hopfield nets are hard to compute", *Neural Computation* **5**, 812–821. *(MS6)*

Ford Jr., L. R., and Fulkerson, D. R. (1956), "Maximal flow through a network", *Canadian J. Mathematics* **8**, 399–404.

Formann, M., and Wagner, F. (1991), "A packing problem with applications to lettering of maps", *Proc. 7th Annual ACM Symposium on Computational Geometry*, ACM, 281–288. *(MS14)*

Fortnow, L., Rompel, J., and Sipser, M. (1988), "On the power of multi-prover interactive protocols", *Proc. 3rd Annual IEEE Conference on Structure in Complexity Theory*, IEEE Computer Society, 156–161.

Fraser, C. B., and Irving, R. W. (1995), "Approximation algorithms for the shortest common supersequence", *Nordic J. Computing* **2**, 303–325. *(SR5)*

Fraser, C. B., Irving, R. W., and Middendorf, M. (1996), "Maximal common subsequences and minimal common supersequences", *Information and Computation* **124**, 145–153. *(SR7)*

Frederickson, G. N. (1979), "Approximation algorithms for some postman problems", *J. ACM* **26**, 538–554. *(ND37)*

Frederickson, G. N. (1980), "Probabilistic analysis for simple one and two-dimensional bin packing algorithms", *Information Processing Letters* **11**, 256–161.

Frederickson, G. N., Hecht, M. S., and Kim, C. E. (1978), "Approximation algorithms for some routing problems", *SIAM J. Computing* **7**, 178–193. *(ND35, ND38, ND39, ND40)*

Frederickson, G. N., and Jájá, J. (1981), "Approximation algorithms for several graph augmentation problems", *SIAM J. Computing* **10**, 270–283. *(ND29, ND30)*

Frederickson, G. N., and Jájá, J. (1982), "On the relationship between the biconnectivity augmentation and traveling salesman problems", *Theoretical Computer Science* **19**, 189–201. *(ND27, ND29)*

Frederickson, G. N., and Solis-Oba, R. (1996), "Increasing the weight of minimum spanning trees", *Proc. 7th Annual ACM-SIAM Symposium on Discrete Algorithms*, ACM-SIAM, 539–546. *(ND12)*

Frenk, J. B. G., and Rinnooy Kan, A. H. G. (1987), "The asymptotic optimality of the LPT rule", *Mathematics of Operations Research* **12**, 241–254.

Frieze, A., Galbiati, G., and Maffioli, F. (1982), "On the worst-case performance of some algorithms for the asymmetric traveling salesman problem", *Networks* **12**, 23–39. *(ND33)*

Frieze, A., and Jerrum, M. (1997), "Improved approximation algorithms for MAX k-CUT and MAX BISECTION", *Algorithmica* **18**, 67–81. *(GT33, ND14, ND17)*

Fujito, T. (1996), "A unified local ratio approximation of node-deletion problems", *Proc. 4th Annual European Symposium on Algorithms*, Lecture Notes in Computer Science 1136, Springer-Verlag, Berlin, 167–178. *(GT26)*

Fürer, M., and Raghavachari, B. (1994), "Approximating the minimum-degree Steiner tree to within one of optimal", *J. Algorithms* **17**, 409–423. *(ND2)*

Gabow, H. N. (1990), "Data structures for weighted matching and nearest common ancestors with linking", *Proc. 1st Annual ACM-SIAM Symposium on Discrete Algorithms*, ACM-SIAM, 434–443.

Galbiati, G., Maffioli, F., and Morzenti, A. (1994), "A short note on the approximability of the maximum leaves spanning tree problem", *Information Processing Letters* **52**, 45–49. *(ND4)*

Galbiati, G., Maffioli, F., and Morzenti, A. (1995), "On the approximability of some maximum spanning tree problems", *Proc. 2nd International Symposium Latin American Theoretical Informatics*, Lecture Notes in Computer Science 911, Springer-Verlag, Berlin, 300–311. *(ND4)*

Garey, M. R., and Graham, R. L. (1975), "Bounds for multiprocessor scheduling with resource constraints", *SIAM J. Computing* **4**, 187–200. *(SS8)*

Garey, M. R., Graham, R. L., Johnson, D. S., and Yao, A. C. (1976), "Resource constrained scheduling as generalized bin-packing", *J. Combinatorial Theory, Series A* **21**, 257–298.

Garey, M. R., Graham, R. L., and Ullman, J. D. (1972), "Worst case analysis of memory allocation algorithms", *Proc. 4th ACM Symposium on Theory of Computing*, ACM, 143–150.

Garey, M. R., and Johnson, D. S. (1976a), "The complexity of near optimal graph coloring", *J. ACM* **23**, 43–49.

Garey, M. R., and Johnson, D. S. (1976b), "Approximation algorithms for combinatorial problems: an annotated bibliography", in *Algorithms and Complexity: New Directions and Recent Results*, Academic Press, New York, 41–52.

Bibliography

Garey, M. R., and Johnson, D. S. (1978), "Strong NP-completeness results: motivation, examples, and implications", *J. ACM* **25**, 499–508.

Garey, M. R., and Johnson, D. S. (1979), *Computers and Intractability: A Guide to the Theory of NP-completeness*, Freeman, San Francisco. *(ND2, SR1)*

Garey, M. R., and Johnson, D. S. (1981), "Approximation algorithms for bin packing problems – a survey", in *Analysis and Design of Algorithms in Combinatorial Optimization*, Springer-Verlag, Berlin, 147–172.

Garfinkel, R. S., and Nemhauser, G. L. (1972), *Integer Programming*, John Wiley & Sons, New York.

Garg, A., and Tamassia, R. (1994), "On the computational complexity of upward and rectilinear planarity testing", *Proc. 2nd DIMACS International Symposium on Graph Drawing*, Lecture Notes in Computer Science 894, Springer-Verlag, Berlin, 286–297. *(ND63)*

Garg, N. (1996), "A 3-approximation for the minimum tree spanning k vertices", *Proc. 37th Annual IEEE Symposium on Foundations of Computer Science*, IEEE Computer Society, 302–309. *(ND1)*

Garg, N., Konjevod, G., and Ravi, R. (1998), "A polylogarithmic approximation algorithm for the group Steiner tree problem", *Proc. 9th Annual ACM-SIAM Symposium on Discrete Algorithms*, ACM-SIAM, 253–259. *(ND8)*

Garg, N., Santosh, V. S., and Singla, A. (1993), "Improved approximation algorithms for biconnected subgraphs via better lower bounding techniques", *Proc. 4th Annual ACM-SIAM Symposium on Discrete Algorithms*, ACM-SIAM, 103–111. *(ND27)*

Garg, N., Saran, H., and Vazirani, V. (1994), "Finding separator cuts in planar graphs within twice the optimal", *Proc. 35th Annual IEEE Symposium on Foundations of Computer Science*, IEEE Computer Society, 14–23. *(ND24)*

Garg, N., Vazirani, V. V., and Yannakakis, M. (1994), "Multiway cuts in directed and node weighted graphs", *Proc. 21st International Colloquium on Automata, Languages and Programming*, Lecture Notes in Computer Science 820, Springer-Verlag, Berlin, 487–498. *(GT26, ND20, ND21, ND22)*

Garg, N., Vazirani, V. V., and Yannakakis, M. (1996), "Approximate max-flow min-(multi)cut theorems and their applications", *SIAM J. Computing* **25**, 235–251. *(GT32, ND22, LO11)*

Garg, N., Vazirani, V. V., and Yannakakis, M. (1997), "Primal-dual approximation algorithms for integral flow and multicut in trees", *Algorithmica* **18**, 3–20. *(ND22, ND47, SP4)*

Gens, G. V., and Levner, E. V. (1979), "Computational complexity of approximation algorithms for combinatorial problems", *Proc. 8th International Symposium on Mathematical Foundations of Computer Science*, Lecture Notes in Computer Science 74, Springer-Verlag, Berlin, 292–300. *(MP13, MP15)*

Bibliography

Gergov, J. (1996), "Approximation algorithms for dynamic storage allocation", *Proc. 4th Annual European Symposium on Algorithms*, Lecture Notes in Computer Science 1136, Springer-Verlag, Berlin, 52–61. *(SR4)*

Gil, J., and Itai, A. (1995), "Packing trees", *Proc. 3rd Annual European Symposium on Algorithms*, Lecture Notes in Computer Science 979, Springer-Verlag, Berlin, 113–127. *(SR11)*

Glover, F. (1989), "Tabu search (part I)", *ORSA J. Computing* **1**, 190–206.

Glover, F. (1990), "Tabu search (part II)", *ORSA J. Computing* **2**, 4–32.

Glover, F., Taillard, E., Laguna, M., and de Werra, D. (eds.) (1993), *Tabu Search*, volume 41 of *Annals of Operations Research*, , Annals of Operations Research, Baltzer, Basel.

Goemans, M. X., Goldberg, A. V., Plotkin, S., Shmoys, D. B., Tardos, É., and Williamson, D. P. (1994), "Improved approximation algorithms for network design problems", *Proc. 5th Annual ACM-SIAM Symposium on Discrete Algorithms*, ACM-SIAM, 223–232. *(ND10)*

Goemans, M. X., Queyranne, M., Schulz, A. S., Skutella, M., and Wang, Y. (1998), "Single machine scheduling with release dates", Unpublished manuscript. *(SS4)*

Goemans, M. X., and Williamson, D. P. (1994), "New 3/4–approximation algorithms for the maximum satisfiability problem", *SIAM J. Discrete Mathematics* **7**, 656–666.

Goemans, M. X., and Williamson, D. P. (1995a), "A general approximation technique for constrained forest problems", *SIAM J. Computing* **24**, 296–317. *(GT15, GT52, ND8, ND33)*

Goemans, M. X., and Williamson, D. P. (1995b), "Improved approximation algorithms for maximum cut and satisfiability problems using semidefinite programming", *J. ACM* **42**, 1115–1145. *(ND14)*

Goemans, M. X., and Williamson, D. P. (1996), "Primal-dual approximation algorithms for feedback problems in planar graphs", *Proc. 5th International Conference on Integer Programming and Combinatorial Optimization*, Lecture Notes in Computer Science 1084, Springer-Verlag, Berlin, 147–161. *(GT8, GT9, GT32)*

Goemans, M. X., and Williamson, D. P. (1997), "The primal-dual method for approximation algorithms and its applications to network design problems", in *Approximations Algorithms for NP-hard Problems*, PWS, Boston, Mass., 144–191.

Goldberg, D. E. (1989), *Genetic Algorithms in Search, Optimization and Machine Learning*, Addison-Wesley, Reading MA.

Goldberg, L. A., Paterson, M., Srinivasan, A., and Sweedyk, E. (1997), "Better approximation guarantees for job-shop scheduling", *Proc. 8th Annual ACM-SIAM Symposium on Discrete Algorithms*, ACM-SIAM, 599–608. *(SS17)*

Bibliography

Golden, B. L., and Stewart, W. R. (1985), "Empirical analysis of heuristics", in *The Traveling Salesman Problem*, John Wiley & Sons, New York, 207–249.

Goldreich, O. (1999), *Modern Cryptography, Probabilistic Proofs and Pseudo-Randomness*, Springer-Verlag, Berlin.

Goldschmidt, O., and Hochbaum, D. S. (1988), "Polynomial algorithm for the k-cut problem", *Proc. 29th Annual IEEE Symposium on Foundations of Computer Science*, IEEE Computer Society, 444–451. *(ND19)*

Goldwasser, S., Micali, S., and Rackoff, C. (1989), "The knowledge complexity of interactive proof-systems", *SIAM J. Computing* **18**, 186–208.

Gonzalez, T. F. (1985), "Clustering to minimize the maximum intercluster distance", *Theoretical Computer Science* **38**, 293–306. *(ND53, ND54)*

Gonzalez, T. F., and Zheng, S. (1989), "Improved bounds for rectangular and Guilhotine partitions", *J. Symbolic Computation* **7**, 591–610. *(MS10)*

Gonzalez, T. F., and Zheng, S. (1990), "Approximation algorithm for partitioning a rectangle with interior points", *Algorithmica* **5**, 11–42. *(MS10)*

Graham, R. L. (1966), "Bounds for certain multiprocessing anomalies", *Bell System Technical J.* **45**, 1563–1581.

Graham, R. L. (1969), "Bounds for multiprocessing timing anomalies", *SIAM J. Applied Mathematics* **17**, 263–269.

Graham, R. L., and Hell, P. (1985), "On the history of the spanning tree problem", *IEEE Annals of the History of Computing* **7**, 8–23.

Grigoriadis, M. D., and Khachiyan, L. G. (1994), "Fast approximation schemes for convex programs with many blocks and coupling constraints", *SIAM J. Optimization* **4**, 86–107. *(MP18)*

Grimmett, G. R., and Mc Diarmid, C. J. H. (1975), "On colouring random graphs", *Proc. Cambridge Philosophical Society* **77**, 313–324.

Grover, L. K. (1992), "Local search and the local structure of NP-complete problems", *Operations Research Letters* **12**, 235–243.

Gudmundsson, J., and Levcopoulos, C. (1997), "A linear-time heuristic for minimum rectangular coverings", *Proc. 15th International Conference on Fundamentals of Computation Theory,*, LNCS, 305–314. *(SR10)*

Gudmundsson, J., and Levcopoulos, C. (1999), "Close approximation of minimum rectangular coverings", *J. Combinatorial Optimization* **3**, 437–452. *(SR10)*

Guha, S., and Khuller, S. (1996), "Approximation algorithms for connected dominating sets", *Proc. 4th Annual European Symposium on Algorithms*, Lecture Notes in Computer Science 1136, Springer-Verlag, Berlin, 179–193. *(GT2)*

Guha, S., and Khuller, S. (1998), "Greedy strikes back: Improved facility location algorithms", *Proc. 9th Annual ACM-SIAM Symposium on Discrete Algorithms*, ACM-SIAM, 649–657. *(ND60)*

Gusfield, D., and Pitt, L. (1992), "A bounded approximation for the minimum cost 2-sat problem", *Algorithmica* **8**, 103–117. *(LO7)*

Guttmann-Beck, N., and Hassin, R. (1997), "Approximation algorithms for min-max tree partition", *J. Algorithms* **24**, 266–286. *(GT15)*

Guttmann-Beck, N., and Hassin, R. (1998a), "Approximation algorithms for minimum tree partition", *Discrete Applied Mathematics* **87**, 117–137. *(GT15)*

Guttmann-Beck, N., and Hassin, R. (1998b), "Approximation algorithms for minimum sum p-clustering", *Discrete Applied Mathematics* **89**, 125–142. *(ND55)*

Guttmann-Beck, N., and Hassin, R. (1999), "Approximation algorithms for minimum k-cut", Unpublished manuscript. *(ND19)*

Guttmann-Beck, N., Hassin, R., Khuller, S., and Raghavachari, B. (1999), "Approximation algorithms with bounded performance guarantees for the clustered traveling salesman problem", Unpublished manuscript. *(ND33)*

Hall, L. A., Schulz, A. S., Shmoys, D. B., and Wein, J. (1997), "On-line and off-line approximation algorithms", Unpublished manuscript. *(SS4)*

Hall, N. G., and Hochbaum, D. S. (1986), "A fast approximation algorithm for the multicovering problem", *Discrete Applied Mathematics* **15**, 35–40. *(MP1)*

Halldórsson, M. M. (1993a), "A still better performance guarantee for approximate graph coloring", *Information Processing Letters* **45**, 19–23. *(GT5, GT14)*

Halldórsson, M. M. (1993b), "Approximating the minimum maximal independence number", *Information Processing Letters* **46**, 169–172. *(GT4)*

Halldórsson, M. M. (1994), "Personal communication", Unpublished manuscript. *(GT17, GT18, GT37, GT46, SP1)*

Halldórsson, M. M. (1995a), "Approximating discrete collections via local improvements", *Proc. 6th Annual ACM-SIAM Symposium on Discrete Algorithms*, ACM-SIAM, 160–169. *(GT1, GT23, GT37)*

Halldórsson, M. M. (1995b), "Approximation via partitioning", Technical Report IS-RR-95-0003F, School of Information Science, Japan Advanced Institute of Science and Technology, Hokuriku. *(GT22, GT23, GT24, GT25, SR7, MP10)*

Halldórsson, M. M. (1996), "Approximating k-set cover and complementary graph coloring", *Proc. 5th International Conference on Integer Programming and Combinatorial Optimization*, Lecture Notes in Computer Science 1084, Springer-Verlag, Berlin, 118–131. *(GT5, GT14, GT17, SP4)*

Bibliography

Halldórsson, M. M., Iwano, K., Katoh, N., and Tokuyama, T. (1995), "Finding subsets maximizing minimum structures", *Proc. 6th Annual ACM-SIAM Symposium on Discrete Algorithms*, ACM-SIAM, 150–159. *(ND5)*

Halldórsson, M. M., Kratochvil, J., and Telle, J. A. (1998), "Independent sets with domination constraints", *Proc. 25th International Colloquium on Automata, Languages and Programming*, Lecture Notes in Computer Science 1443, Springer-Verlag, Berlin, 176–185. *(GT4)*

Halldórsson, M. M., and Radhakrishnan, J. (1994a), "Improved approximations of independent sets in bounded-degree graphs", *Nordic J. Computing* **1**, 475–492. *(GT23)*

Halldórsson, M. M., and Radhakrishnan, J. (1994b), "Greed is good: approximating independent set in sparse and bounded degree graphs", *Proc. 26th ACM Symposium on Theory of Computing*, ACM, 439–448.

Halldórsson, M. M., and Tanaka, K. (1996), "Approximation and special cases of common subtrees and editing distance", *Proc. 7th Annual International Symposium on Algorithms and Computation*, Lecture Notes in Computer Science 1178, Springer-Verlag, Berlin, 75–84. *(GT48)*

Halldórsson, M. M., Ueno, S., Nakao, H., and Kajitani, Y. (1992), "Approximating Steiner trees in graphs with restricted weights", *Proc. Asia-Pacific Conference on Circuits and Systems*, Sidney, Australia, 69–73. *(ND8)*

Hancock, T., Jiang, T., Li, M., and Tromp, J. (1996), "Lower bounds on learning decision lists and trees", *Information and Computation* **126**, 114–122. *(MS2)*

Haralambides, J., Makedon, F., and Monien, B. (1991), "Bandwidth minimization: an approximation algorithm for caterpillars", *Mathematical Systems Theory* **24**, 169–177. *(GT42)*

Hassin, R. (1992), "Approximation schemes for the restricted shortest path problem", *Mathematics of Operations Research* **17**, 36–42. *(ND43)*

Hassin, R., and Megiddo, N. (1991), "Approximation algorithms for hitting objects with straight lines", *Discrete Applied Mathematics* **30**, 29–42. *(SP7)*

Hassin, R., and Rubinstein, S. (1994), "Approximations for the maximum acyclic subgraph problem", *Information Processing Letters* **51**, 133–140. *(GT9)*

Hassin, R., and Rubinstein, S. (1998), "Robust matchings, maximum clustering, and maximum capacitated medians", *Proc. 1st International Conference on Fun with Algorithms*, 23–32. *(ND55)*

Hassin, R., Rubinstein, S., and Tamir, A. (1994), "Notes on dispersion problems", Unpublished manuscript. *(GT35)*

Håstad, J. (1996), "Clique is hard to approximate within $n^{1-\varepsilon}$", *Proc. 37th Annual IEEE Symposium on Foundations of Computer Science*, IEEE Computer Society, 627–636. *(GT22)*

Bibliography

Håstad, J. (1997), "Some optimal inapproximability results", *Proc. 29th ACM Symposium on Theory of Computing*, ACM, 1–10. *(GT1, GT32, ND14, ND16, MP10, LO2)*

Håstad, J., Phillips, S., and Safra, S. (1993), "A well-characterized approximation problem", *Information Processing Letters* **47**, 301–305. *(AN1)*

Hein, J., Jiang, T., Wang, L., and Zhang, K. (1996), "On the complexity of comparing evolutionary trees", *Discrete Applied Mathematics* **71**, 153–169. *(MS15)*

Hochbaum, D. S. (1982a), "Approximation algorithms for the set covering and vertex cover problems", *SIAM J. Computing* **11**, 555–556. *(GT1, SP4)*

Hochbaum, D. S. (1982b), "Heuristics for the fixed cost median problem", *Mathematical Programming* **22**, 148–162. *(ND60)*

Hochbaum, D. S. (1983), "Efficient bounds for the stable set, vertex cover and set packing problems", *Discrete Applied Mathematics* **6**, 243–254. *(GT23, SP2)*

Hochbaum, D. S. (1997), "Various notions of approximations: good, better, best, and more", in *Approximation Algorithms for NP-hard Problems*, PWS, Boston, Mass., 346–398. *(ND53)*

Hochbaum, D. S. (1998), "Approximating clique and biclique problems", *J. Algorithms* **29**, 174–200. *(GT27)*

Hochbaum, D. S., and Maass, W. (1985), "Approximation schemes for covering and packing problems in image processing and VLSI", *J. ACM* **32**, 130–136. *(SP8)*

Hochbaum, D. S., and Maass, W. (1987), "Fast approximation algorithms for a nonconvex covering problem", *J. Algorithms* **8**, 305–323. *(SP8)*

Hochbaum, D. S., Megiddo, N., Naor, J., and Tamir, A. (1993), "Tight bounds and 2-approximation algorithms for integer programs with two variables per inequality", *Mathematical Programming* **62**, 69–83. *(MP1)*

Hochbaum, D. S., and Shmoys, D. B. (1986), "A unified approach to approximation algorithms for bottleneck problems", *J. ACM* **33**, 533–550. *(ND36, ND53, ND54, ND56, ND62)*

Hochbaum, D. S., and Shmoys, D. B. (1987), "Using dual approximation algorithms for scheduling problems: theoretical and practical results", *J. ACM* **34**, 144–162. *(SS6)*

Hochbaum, D. S., and Shmoys, D. B. (1988), "A polynomial approximation scheme for machine scheduling on uniform processors: using the dual approach", *SIAM J. Computing* **17**, 539–551. *(SS10)*

Hofmeister, T., and Lefmann, H. (1998), "Approximating maximum independent sets in uniform hypergraphs", *Proc. 23rd International Symposium on Mathematical Foundations of Computer Science*, Lecture Notes in Computer Science 1450, Springer-Verlag, Berlin, 562–570. *(GT5, GT23)*

Bibliography

Hofri, M. (1987), *Probabilistic Analysis of Algorithms*, Springer-Verlag, Berlin.

Holyer, I. (1981), "The NP-completeness of edge-coloring", *SIAM J. Computing* **10**, 718–720. *(GT7)*

Hoogeveen, J. A. (1978), "Analysis of Christofides' heuristic: Some paths are more difficult than cycles", *Operations Research Letters* **10**, 178–193. *(ND33)*

Hoogeveen, J. A., Lenstra, J. K., and Veltman, B. (1995), "Three, four, five, six, or the complexity of scheduling with communication delays", *Operations Research Letters* **16**, 129–138. *(SS7)*

Horowitz, E., and Sahni, S. K. (1976), "Exact and approximate algorithms for scheduling nonidentical processors", *J. ACM* **23**, 317–327. *(SS6, SS10)*

Horowitz, E., and Sahni, S. K. (1978), *Fundamentals of Computer Algorithms*, Pitman. *(GT5)*

Hougardy, S., Prömel, H. J., and Steger, A. (1994), "Probabilistically checkable proofs and their consequences for approximation algorithms", *Discrete Mathematics* **136**, 175–223.

Hsu, W. L., and Nemhauser, G. L. (1979), "Easy and hard bottleneck location problems", *Discrete Applied Mathematics* **1**, 209–216. *(ND53)*

Hunt III, H. B., Marathe, M. V., Radhakrishnan, V., Ravi, S. S., Rosenkrantz, D. J., and Stearns, R. E. (1994b), "A unified approach to approximation schemes for NP- and PSPACE-hard problems for geometric graphs", *Proc. 2nd Annual European Symposium on Algorithms*, Lecture Notes in Computer Science 855, Springer-Verlag, Berlin, 424–435. *(GT1, GT2, GT3, GT11, GT12, GT23)*

Hunt III, H. B., Marathe, M. V., Radhakrishnan, V., Ravi, S. S., Rosenkrantz, D. J., and Stearns, R. E. (1994a), "Approximation schemes using L-reductions", *Proc. 14th Annual Conference on Foundations of Software Technology and Theoretical Computer Science*, Lecture Notes in Computer Science 880, Springer-Verlag, Berlin, 342–353. *(GT37)*

Hurkens, C. A. J., and Schrijver, A. (1989), "On the size of systems of sets every t of which have an SDR, with an application to the worst-case ratio of heuristics for packing problems", *SIAM J. Discrete Mathematics* **2**, 68–72. *(GT11, GT12, SP1, SP2)*

Ibarra, O. H., and Kim, C. E. (1975), "Fast approximation for the knapsack and sum of subset problems", *J. ACM* **22**, 463–468. *(MP13)*

Ihler, E. (1992), "The complexity of approximating the class Steiner tree problem", *Proc. 18th International Workshop on Graph-Theoretic Concepts in Computer Science*, Lecture Notes in Computer Science 570, Springer-Verlag, Berlin, 85–96. *(ND8)*

Jagota, A. (1993), "Constraint satisfaction and maximum clique", *Working Notes AAAI Spring Symposium on AI and NP-hard Problems*, Stanford University, 92–97. *(MS12)*

Jansen, K. (1992), "An approximation algorithm for the general routing problem", *Information Processing Letters* **41**, 333–339. *(ND41)*

Jansen, K. (1997), "Approximation results for the optimum cost chromatic partition problem", *Proc. 24th International Colloquium on Automata, Languages and Programming,* Lecture Notes in Computer Science 1256, Springer-Verlag, Berlin, 727–737. *(GT5)*

Jansen, K., and Öhring, S. (1997), "Approximation algorithms for time constrained scheduling", *Information and Computation* **132**, 85–108. *(SR1)*

Jiang, T., Kearney, P., and Li, M. (1998), "Orchestrating quartets: approximation and data correction", *Proc. 39th Annual IEEE Symposium on Foundations of Computer Science*, IEEE Computer Society, 416–425. *(MS4)*

Jiang, T., Lawler, E. L., and Wang, L. (1994), "Aligning sequences via an evolutionary tree: complexity and approximation", *Proc. 26th ACM Symposium on Theory of Computing*, ACM, 760–769. *(ND8, MS4)*

Jiang, T., and Li, M. (1994a), "Approximating shortest superstrings with constraints", *Theoretical Computer Science* **134**, 473–491. *(SR6)*

Jiang, T., and Li, M. (1994b), "On the approximation of shortest common supersequences and longest common subsequences", *Proc. 21st International Colloquium on Automata, Languages and Programming*, Lecture Notes in Computer Science 820, Springer-Verlag, Berlin, 191–202. *(SR5, SR7)*

Johnson, D. S. (1972), "Near-optimal bin packing algorithms", Technical Report MAC TR–109, Project MAC, MIT, Cambridge.

Johnson, D. S. (1974a), "Approximation algorithms for combinatorial problems", *J. Computer and System Sciences* **9**, 256–278. *(GT2, SP4, SP5, LO2)*

Johnson, D. S. (1974b), "Fast algorithms for bin packing", *J. Computer and System Sciences* **8**, 272–314.

Johnson, D. S. (1974c), "Worst case behavior of graph coloring algorithms", *5th Southeastern Conference on Combinatorics, Graph Theory and Computing*, 513–527.

Johnson, D. S. (1990), "A catalog of complexity classes", in *Algorithms and Complexity*, volume A of *Handbook of Theoretical Computer Science*, Elsevier Science, Amsterdam, 67–161.

Johnson, D. S., Aragon, C. R., McGeoch, L. A., and Schevon, C. (1989), "Optimization by simulated annealing: An experimental evaluation (part I)", *Operations Research* **37**, 865–892.

Johnson, D. S., Aragon, C. R., McGeoch, L. A., and Schevon, C. (1991), "Optimization by simulated annealing: An experimental evaluation (part II)", *Operations Research* **39**, 378–406.

Johnson, D. S, Demers, A., Ullman, J. D., Garey, M. R., and Graham, R. L. (1974), "Worst–case performance bounds for simple one–dimensional packing algorithms", *SIAM J. Computing* **3**, 299–325.

Bibliography

Johnson, D. S., and Garey, M. R. (1985), "A 71/60 theorem for bin-packing", *J. Complexity* **1**, 65–106. *(SR1)*

Johnson, D. S., Papadimitriou, C. H., and Yannakakis, M. (1988), "How easy is local search?", *J. Computer and System Sciences* **37**, 79–100.

Jonsson, P. (1997), "Tight lower bounds on the approximability of some NPO PB-complete problems", Technical Report 4, Department of Computer and Information Science, Linköping University, Sweden. *(MP2, LO6, LO7)*

Junger, M., Reinelt, G., and Rinaldi, G. (1994), "The traveling salesman problem", in *Handbook on Operations Research and Management Sciences: Network Models*, North-Holland, Amsterdam, 225–330.

Kahale, N. (1993), "On reducing the cut ratio to the multicut problem", Technical Report 93-78, DIMACS.

Kann, V. (1991), "Maximum bounded 3-dimensional matching is MAX SNP-complete", *Information Processing Letters* **37**, 27–35. *(GT11, SP1, SP2)*

Kann, V. (1992a), "On the approximability of the maximum common subgraph problem", *Proc. 9th Symposium on Theoretical Aspects of Computer Science*, Lecture Notes in Computer Science 577, Springer-Verlag, Berlin, 377–388. *(GT46, GT47)*

Kann, V. (1992b), *On the Approximability of NP-complete Optimization Problems*, PhD thesis, Department of Numerical Analysis and Computing Science, Royal Institute of Technology, Stockholm. *(GT2, GT4, GT8, GT9, GT46, SP6, LO6, LO9)*

Kann, V. (1994a), "Maximum bounded H-matching is MAX SNP-complete", *Information Processing Letters* **49**, 309–318. *(GT12)*

Kann, V. (1994b), "Polynomially bounded minimization problems that are hard to approximate", *Nordic J. Computing* **1**, 317–331. *(GT4, GT51, MP1, LO7, LO10, AL3)*

Kann, V. (1995), "Strong lower bounds on the approximability of some NPO PB-complete maximization problems", *Proc. 20th International Symposium on Mathematical Foundations of Computer Science*, Lecture Notes in Computer Science 969, Springer-Verlag, Berlin, 227–236. *(GT28, MP2, MP9)*

Kann, V., Khanna, S., Lagergren, J., and Panconesi, A. (1997), "Hardness of approximating MAX k-CUT and its dual", *Chicago J. Theoretical Computer Science* **1997**, 2. *(GT32, ND17)*

Kann, V., Lagergren, J., and Panconesi, A. (1996), "Approximability of maximum splitting of k-sets and some other APX-complete problems", *Information Processing Letters* **58**, 105–110. *(SP3, LO4)*

Kant, G., and Bodlaender, H. L. (1997), "Triangulating planar graphs while minimizing the maximum degree", *Information and Computation* **135**, 1–14. *(ND64)*

Karger, D., Motwani, R., and Ramkumar, G. D. S. (1997), "On approximating the longest path in a graph", *Algorithmica* **18**, 82–98. *(ND42)*

Karger, D., Motwani, R., and Sudan, M. (1998), "Approximate graph coloring by semidefinite programming", *J. ACM* **45**, 246–265. *(GT5)*

Karloff, H. (1991), *Linear Programming*, Birkhäuser.

Karloff, H., and Zwick, U. (1997), "A 7/8-approximation for MAX 3SAT?", *Proc. 38th Annual IEEE Symposium on Foundations of Computer Science*, IEEE Computer Society, 406–415. *(LO2)*

Karmarkar, N., and Karp, R. M. (1982), "An efficient approximation scheme for the one-dimensional bin packing problem", *Proc. 23rd Annual IEEE Symposium on Foundations of Computer Science*, IEEE Computer Society, 312–320. *(SR1)*

Karp, R. M. (1972), "Reducibility among combinatorial problems", in *Complexity of Computer Computations*, Plenum Press, 85–103.

Karp, R. M. (1976), "The probabilistic analysis of some combinatorial search problems", in *Algorithms and Complexity: New Directions and Recent Results*, Academic Press, New York, 1–19.

Karp, R. M. (1977), "Probabilistic analysis of partitioning algorithms for the traveling salesman in the plane", *Mathematics of Operations Research* **2**, 209–224.

Karp, R. M. (1982), "Lecture notes on the probabilistic analysis of algorithms", Technical report, Computer Science Division, University of California at Berkeley.

Karp, R. M., Luby, M., and Marchetti-Spaccamela, A. (1984), "A probabilistic analysis of multidimensional bin-packing problems", *Proc. 16th ACM Symposium on Theory of Computing*, ACM, 289–298.

Karp, R. M., McKellar, A. C., and Wong, C. K. (1975), "Near-optimal solutions to a 2-dimensional placement problem", *SIAM J. Computing* **4**, 271–286. *(MP8)*

Karp, R. M., and Steele, J. M. (1985), "Probabilistic analysis of heuristics", in *The Traveling Salesman Problem*, John Wiley & Sons, New York, 181–205.

Karpinski, M. (1997), "Polynomial time approximation schemes for some dense instances of NP-hard optimization problems", *Proc. 1st International Symposium on Randomization and Approximation Techniques in Computer Science*, Lecture Notes in Computer Science 1269, Springer-Verlag, Berlin, 1–14. *(GT1, ND8, SP4)*

Karpinski, M., and Wirtgen, J. (1997), "On approximation hardness of the bandwidth problem", Technical Report TR 97-041, The Electronic Colloquium on Computational Complexity. *(GT42)*

Bibliography

Karpinski, M., Wirtgen, J., and Zelikovsky, A. (1997), "An approximation algorithm for the bandwidth problem on dense graphs", Technical Report TR97-017, The Electronic Colloquium on Computational Complexity. *(GT42, GT43)*

Karpinski, M., and Zelikovsky, A. (1997a), "New approximation algorithms for the Steiner tree problem", *J. Combinatorial Optimization* **1**, 1–19. *(ND8)*

Karpinski, M., and Zelikovsky, A. (1997b), "Approximating dense cases of covering problems", Technical Report TR97-004, The Electronic Colloquium on Computational Complexity. *(GT1, ND8, SP4)*

Karuno, Y., Nagamochi, H., and Ibaraki, T. (1993), "Vehicle scheduling on a tree with release and handling times", *Proc. 4th Annual International Symposium on Algorithms and Computation*, Lecture Notes in Computer Science 762, Springer-Verlag, Berlin, 486–495. *(SS20)*

Kavvadias, D., Papadimitriou, C. H., and Sideri, M. (1993), "On Horn envelopes and hypergraph transversals", *Proc. 4th Annual International Symposium on Algorithms and Computation*, Lecture Notes in Computer Science 762, Springer-Verlag, Berlin, 399–405. *(LO14)*

Kellerer, H., Tautenhahn, T., and Woeginger, G. J. (1996), "Approximability and nonapproximability results for minimizing total flow time on a single machine", *Proc. 28th ACM Symposium on Theory of Computing*, ACM, 418–426. *(SS11)*

Kenyon, C., and Rémila, E. (1996), "Approximate strip packing", *Proc. 37th Annual IEEE Symposium on Foundations of Computer Science*, IEEE Computer Society, 31–36. *(SR2)*

Kern, W. (1989), "A probabilistic analysis of the switching algorithm for the Euclidean TSP", *Mathematical Programming* **44**, 213–219.

Kernighan, B. W., and Lin, S. (1970), "An efficient heuristic procedure for partitioning graphs", *Bell System Technical J.* **49**, 291–307.

Khachian, L.G. (1979), "A polynomial algorithm for linear programming", *Dokladi Akademiy Nauk SSSR* **244**, 1093–1096.

Khanna, S. (1997), "A polynomial time approximation scheme for the SONET ring loading problem", *Bell Labs Technical J.* **Spring**, 36–41. *(MS19)*

Khanna, S., Linial, N., and Safra, S. (1993), "On the hardness of approximating the chromatic number", *Proc. 2nd Israel Symposium on Theory of Computing and Systems*, IEEE Computer Society, 250–260. *(GT5)*

Khanna, S., and Motwani, R. (1996), "Toward a syntactic characterization of PTAS", *Proc. 28th ACM Symposium on Theory of Computing*, ACM, 329–337. *(LO1)*

Khanna, S., Motwani, R., Sudan, M., and Vazirani, U. (1999), "On syntactic versus computational views of approximability", *SIAM J. Computing* **28**, 164–191. *(GT2, SP4)*

Khanna, S., Muthukrishnan, S., and Paterson, M. (1998), "On approximating rectangle tiling and packing", *Proc. 9th Annual ACM-SIAM Symposium on Discrete Algorithms*, ACM-SIAM, 384–393. *(SP14)*

Khanna, S., Muthukrishnan, S., and Skiena, S. (1997), "Efficient array partitioning", *Proc. 24th International Colloquium on Automata, Languages and Programming*, Lecture Notes in Computer Science 1256, Springer-Verlag, Berlin, 616–626. *(SP13)*

Khanna, S., Sudan, M., and Trevisan, L. (1997), "Constraint satisfaction: the approximability of minimization problems", *Proc. 12th Annual IEEE Conference on Computational Complexity*, IEEE Computer Society, 282–296. *(LO12)*

Khanna, S., Sudan, M., and Williamson, D. P. (1997), "A complete classification of the approximability of maximization problems derived from Boolean constraint satisfaction", *Proc. 29th ACM Symposium on Theory of Computing*, ACM, 11–20. *(LO12)*

Khuller, S. (1997), "Approximation algorithms for finding highly connected subgraphs", in *Approximation Algorithms for NP-hard Problems*, PWS, Boston, Mass., 236–265. *(ND29)*

Khuller, S., Pless, R., and Sussmann, Y. J. (1997), "Fault tolerant k-center problems", *Proc. 3rd Italian Conference on Algorithms and Complexity*, Lecture Notes in Computer Science 1203, Springer-Verlag, Berlin, 37–48. *(ND53)*

Khuller, S., and Raghavachari, B. (1996), "Improved approximation algorithms for uniform connectivity problems", *J. Algorithms* **21**, 434–450. *(ND27, ND28)*

Khuller, S., Raghavachari, B., and Rosenfeld, A. (1994), "Localization in graphs", Technical Report UMIACS-TR-94-92, University of Maryland, UMIACS. *(GT53)*

Khuller, S., Raghavachari, B., and Young, N. (1993), "Maintaining directed reachability with few edges", Technical Report UMIACS-TR-93-87, University of Maryland, UMIACS. *(ND11)*

Khuller, S., Raghavachari, B., and Young, N. (1995), "Approximating the minimum equivalent digraph", *SIAM J. Computing* **24**, 859–872. *(GT38)*

Khuller, S., Raghavachari, B., and Young, N. (1996a), "Low degree spanning trees of small weight", *SIAM J. Computing* **25**, 355–368. *(ND3)*

Khuller, S., Raghavachari, B., and Young, N. (1996b), "On strongly connected digraphs with bounded cycle length", *Discrete Applied Mathematics* **69**, 281–289. *(ND27, ND28, ND30)*

Khuller, S., and Sussmann, Y. J. (1996), "The capacitated k-center problem", *Proc. 4th Annual European Symposium on Algorithms*, Lecture Notes in Computer Science 1136, Springer-Verlag, Berlin, 152–166. *(ND53)*

Bibliography

Khuller, S., and Thurimella, R. (1993), "Approximation algorithms for graph augmentation", *J. Algorithms* **14**, 214–225. *(ND29)*

Khuller, S., and Vishkin, U. (1994), "Biconnectivity approximations and graph carvings", *J. ACM* **41**, 214–235. *(ND10, ND28)*

Kim, S., and McNaughton, R. (1993), "Computing the order of a locally testable automaton", Unpublished manuscript. *(AL4)*

Kirkpatrick, S., Gelatt, C. D., Jr., and Vecchi, M. P. (1983), "Optimization by simulated annealing", *Science* **222**, 671–680.

Klein, P., Agrawal, A., Ravi, R., and Rao, S. (1990), "Approximation through multicommodity flow", *Proc. 31st Annual IEEE Symposium on Foundations of Computer Science*, IEEE Computer Society, 726–737. *(GT40, PO1)*

Klein, P., Plotkin, S. A., and Rao, S. (1993), "Excluded minors, network decomposition, and multicommodity flow", *Proc. 25th ACM Symposium on Theory of Computing*, ACM, 682–690. *(ND23)*

Klein, P., Rao, S., Agrawal, A., and Ravi, R. (1995), "An approximate max-flow min-cut relation for undirected multicommodity flow, with applications to routing and finding sparse cuts", *Combinatorica* **15**, 187–202.

Kleinberg, J. M. (1996), "Single-source unsplittable flow", *Proc. 37th Annual IEEE Symposium on Foundations of Computer Science*, IEEE Computer Society, 68–77. *(ND51)*

Kleinberg, J. M., and Tardos, É. (1995), "Approximations for the disjoint paths problem in high-diameter planar networks", *Proc. 27th ACM Symposium on Theory of Computing*, ACM, 26–35. *(ND48)*

Kloks, T., Kratsch, D., and Müller, H. (1995), "Approximating the bandwidth for asteroidal triple-free graphs", *Proc. 3rd Annual European Symposium on Algorithms*, Lecture Notes in Computer Science 979, Springer-Verlag, Berlin, 434–447. *(GT42)*

Knödel, W. (1981), "A bin packing algorithm with complexity $o(n \log n)$ and performance 1 in the stochastic limit", *Proc. 10th International Symposium on Mathematical Foundations of Computer Science*, Lecture Notes in Computer Science 118, Springer-Verlag, Berlin, 369–378.

Knuth, D. E. (1969), *The Art of Computer Programming: Fundamental Algorithms*, Addison-Wesley, Reading MA.

Knuth, D. E. (1971), *The Art of Computer Programming: Seminumerical Algorithms*, Addison-Wesley, Reading MA.

Knuth, D. E. (1973), *The Art of Computer Programming: Sorting and Searching*, Addison-Wesley, Reading MA.

Ko, M. T., Lee, R. C. T., and Chang, J. S. (1990), "An optimal approximation algorithm for the rectilinear m-center problem", *Algorithmica* **5**, 341–352. *(ND53)*

Kohli, R., Krishnamurti, R., and Mirchandani, P. (1994), "The minimum satisfiability problem", *SIAM J. Discrete Mathematics* **7**, 275–283. *(LO2, LO3)*

Kolaitis, P. G., and Thakur, M. N. (1994), "Logical definability of NP optimization problems", *Information and Computation* **115**, 321–353. *(LO5)*

Kolaitis, P. G., and Thakur, M. N. (1995), "Approximation properties of NP minimization classes", *J. Computer and System Sciences* **50**, 391–411. *(GT26, GT27)*

Kolliopoulos, S. G., and Stein, C. (1997), "Improved approximation algorithms for unsplittable flow problems", *Proc. 38th Annual IEEE Symposium on Foundations of Computer Science*, IEEE Computer Society, 426–435. *(ND51)*

Korobkov, V. K., and Krichevskii, R. E. (1966), *Algorithms for the Traveling Salesman Problem*, Nauka.

Korte, B., Lovasz, L., and Schrader, R. (1991), *Greedoids*, Springer-Verlag, Berlin.

Korte, B., and Schrader, R. (1981), "On the existence of fast approximation schemes", in *Non Linear Programming*, Academic Press, New York, 415–437.

Kortsarz, G., and Peleg, D. (1992), "Generating sparse 2-spanners", *Proc. 3rd Scandinavian Workshop on Algorithm Theory*, Lecture Notes in Computer Science 621, Springer-Verlag, Berlin, 73–82. *(GT36)*

Kortsarz, G., and Peleg, D. (1993), "On choosing a dense subgraph", *Proc. 34th Annual IEEE Symposium on Foundations of Computer Science*, IEEE Computer Society, 692–701. *(GT35)*

Kortsarz, G., and Peleg, D. (1994), "Generating low-degree 2-spanners", *Proc. 5th Annual ACM-SIAM Symposium on Discrete Algorithms*, ACM-SIAM, 556–563. *(GT36)*

Kortsarz, G., and Peleg, D. (1995), "Approximation algorithms for minimum time broadcast", *SIAM J. Discrete Mathematics* **8**, 401–427. *(ND52)*

Kortsarz, G., and Peleg, D. (1997), "Approximating shallow-light trees", *Proc. 8th Annual ACM-SIAM Symposium on Discrete Algorithms*, ACM-SIAM, 103–110. *(ND8)*

Kosaraju, S. R., Park, J. K., and Stein, C. (1994), "Long tours and short superstrings", *Proc. 35th Annual IEEE Symposium on Foundations of Computer Science*, IEEE Computer Society, 166–177. *(ND32)*

Kou, L. T., Stockmeyer, L. J., and Wong, C. K. (1978), "Covering edges by cliques with regard to keyword conflicts and intersection graphs", *Communications ACM* **21**, 135–139. *(GT17)*

Koutsoupias, E., Papadimitriou, C. H., and Yannakakis, M. (1996), "Searching a fixed graph", *Proc. 23rd International Colloquium on Automata, Languages and Programming*, Lecture Notes in Computer Science 1099, Springer-Verlag, Berlin, 280–289. *(ND45)*

Bibliography

Krentel, M. W. (1988), "The complexity of optimization problems", *J. Computer and System Sciences* **36**, 490–509.

Krentel, M. W. (1989), "Structure in locally optimal solutions", *Proc. 30th Annual IEEE Symposium on Foundations of Computer Science*, IEEE Computer Society, 216–222.

Krentel, M. W. (1990), "On finding and verifying locally optimal solutions", *SIAM J. Computing* **19**, 742–751.

Krumke, S. O. (1995), "On a generalization of the *p*-center problem", *Information Processing Letters* **56**, 67–71. *(ND53)*

Krumke, S. O., Marathe, M. V., Noltemeier, H., Ravi, R., Ravi, S. S., Sundaram, R., and Wirth, H. C. (1997), "Improving spanning trees by upgrading nodes", *Proc. 24th International Colloquium on Automata, Languages and Programming*, Lecture Notes in Computer Science 1256, Springer-Verlag, Berlin, 281–291. *(ND13)*

Lam, S., and Sethi, R. (1977), "Worst case analysis of two scheduling algorithms", *SIAM J. Computing* **6**, 518–536. *(SS7)*

Lawler, E. L., Lenstra, J. K., Kan, A. H. G. Rinnooy, and Shmoys, D. B. (1985), *The Traveling Salesman Problem*, John Wiley & Sons, New York.

Leighton, T., and Rao, S. (1988), "An approximate max-flow min-cut theorem for uniform multicommodity flow problems with applications to approximation algorithms", *Proc. 29th Annual IEEE Symposium on Foundations of Computer Science*, IEEE Computer Society, 422–431. *(ND26)*

Lenstra, J. K., and Kan, A. H. G. Rinnooy (1978), "Complexity of scheduling under precedence constraints", *Operations Research* **26**, 22–35. *(SS7)*

Lenstra, J. K., and Shmoys, D. B. (1995), "Computing near-optimal schedules", in *Scheduling Theory and its Applications*, John Wiley & Sons, New York, 1–14. *(SS14)*

Lenstra, J. K., Shmoys, D. B., and Tardos, É. (1990), "Approximation algorithms for scheduling unrelated parallel machines", *Mathematical Programming* **46**, 259–271. *(SS6)*

Leonardi, S., and Raz, D. (1997), "Approximating total flow time on parallel machines", *Proc. 29th ACM Symposium on Theory of Computing*, ACM, 110–119. *(SS11)*

Leung, J. Y.-T., and Wei, W.-D. (1995), "Tighter bounds on a heuristic for a partition problem", *Information Processing Letters* **56**, 51–57. *(SP12)*

Levcopoulos, C. (1987), "Improved bounds for covering general polygons with rectangles", *Proc. 7th Ann. Conf. on Foundations of Software Tech. and Theoret. Comput. Sci.*, , LNCS, 95–102. *(SR10)*

Levcopoulos, C., and Gudmundsson, J. (1997), "Approximation algorithms for covering polygons with squares and similar problems", *Proc. 1st International Symposium on Randomization and Approximation Techniques in Computer Science*, Lecture Notes in Computer Science 1269, Springer-Verlag, Berlin, 27–31. *(SR10)*

Levin, L. A. (1973), "Universal sorting problems", *Problems of Information Transmission* **9**, 265–266.

Li, C., McCormick, S. T., and Simchi-Levi, D. (1990), "The complexity of finding two disjoint paths with min-max objective function", *Discrete Applied Mathematics* **26**, 105–115. *(ND49)*

Li, C., McCormick, S. T., and Simchi-Levi, D. (1992), "On the minimum cardinality bounded diameter and the bounded cardinality minimum diameter edge addition problems", *Operations Research Letters* **11**, 303–308. *(ND31)*

Li, K., and Cheng, K. (1990), "On three-dimensional packing", *SIAM J. Computing* **19**, 847–867. *(SR2)*

Li, R., and Yue, M. (2000), "A tighter bound for FFD algorithm", *Acta Mathematica Applicatae Sinica* **4**, 337–347.

Li, M. (1990), "Towards a DNA sequencing theory", *Proc. 31st Annual IEEE Symposium on Foundations of Computer Science*, IEEE Computer Society, 125–134. *(SR6)*

Lin, C. (1994), "Hardness of approximating graph transformation problem", *Proc. 5th Annual International Symposium on Algorithms and Computation*, Lecture Notes in Computer Science 834, Springer-Verlag, Berlin, 74–82. *(GT49)*

Lin, J-H., and Vitter, J. S. (1992), "ε-approximations with minimum packing constraint violation", *Proc. 24th ACM Symposium on Theory of Computing*, ACM, 771–782. *(ND57)*

Lin, S. (1965), "Computer solutions for the traveling salesman problem", *Bell System Technical J.* **10**, 2245–2269.

Lin, S., and Kernighan, B. W. (1973), "An effective heuristic algorithm for the traveling salesman problem", *Operations Research* **21**, 498–516.

Linial, N., London, E., and Rabinovich, Y. (1995), "The geometry of graphs and some of its algorithmic applications", *Combinatorica* **15**, 215–246.

Lipton, R. J., and Tarjan, R. E. (1979), "A separator theorem for planar graphs", *SIAM J. Applied Mathematics* **36**, 177–189. *(ND25)*

Lipton, R. J., and Tarjan, R. E. (1980), "Applications of a planar separator theorem", *SIAM J. Computing* **9**, 615–627.

Lovász, L. (1975a), "On the ratio of optimal integral and fractional covers", *Discrete Mathematics* **13**, 383–390.

Bibliography

Lovász, L. (1975b), "Three short proofs in graph theory", *J. Combinatorial Theory, Series B* **19**, 269–271.

Lu, H., and Ravi, R. (1992), "The power of local optimization: Approximation algorithms for maximum-leaf spanning tree", *Proc. 30th Annual Allerton Conference on Communication, Control and Computing*, 533–542. *(ND4)*

Ludwig, W., and Tiwari, P. (1994), "Scheduling malleable and nonmalleable parallel tasks", *Proc. 5th Annual ACM-SIAM Symposium on Discrete Algorithms*, ACM-SIAM, 167–176. *(SS8)*

Lueker, G. S. (1980), "An average case analysis of bin packing with uniformly distributed item size", Technical report, Department of Information and Computer Science, University of California at Irvine.

Lund, C., Fortnow, L., Karloff, H., and Nisan, N. (1992), "Algebraic methods for interactive proof systems", *J. ACM* **39**, 859–868.

Lund, C., and Yannakakis, M. (1993), "The approximation of maximum subgraph problems", *Proc. 20th International Colloquium on Automata, Languages and Programming*, Lecture Notes in Computer Science 700, Springer-Verlag, Berlin, 40–51. *(GT25, GT26, GT28)*

Lund, C., and Yannakakis, M. (1994), "On the hardness of approximating minimization problems", *J. ACM* **41**, 960–981. *(GT2, GT5, GT14, GT17, GT18, SP4, SP5, SR1)*

Mahajan, S., and Ramesh, J. (1995), "Derandomizing semidefinite programming based approximation algorithms", *Proc. 36th Annual IEEE Symposium on Foundations of Computer Science*, IEEE Computer Society, 162–169. *(GT5, GT22, GT33, ND14, ND17)*

Makedon, F., and Tragoudas, S. (1990), "Approximating the minimum net expansion: near optimal solutions to circuit partitioning problems", *Proc. 16th International Workshop on Graph-Theoretic Concepts in Computer Science*, Lecture Notes in Computer Science 484, Springer-Verlag, Berlin, 140–153. *(ND26)*

Malesinska, E., and Panconesi, A. (1996), "On the hardness of frequency allocation for hybrid networks", *Proc. 22nd International Workshop on Graph-Theoretic Concepts in Computer Science*, Lecture Notes in Computer Science 1197, Springer-Verlag, Berlin, 308–322. *(MS16)*

Marathe, M. V., Breu, H., Hunt III, H. B., Ravi, S. S., and Rosenkrantz, D. J. (1995), "Simple heuristics for unit disk graphs", *Networks* **25**, 59–68. *(GT4, GT5)*

Marathe, M. V., Hunt III, H. B., and Ravi, S. S. (1996), "Efficient approximation algorithms for domatic partition and on-line coloring of circular arc graphs", *Discrete Applied Mathematics* **64**, 135–149. *(GT2)*

Marathe, M. V., Ravi, R., Sundaram, R., Ravi, S. S., Rosenkrantz, D. J., and Hunt III, H. B. (1995), "Bicriteria network design problems", *Proc. 22nd International Colloquium on Automata, Languages and Programming*, Lecture Notes in Computer Science 944, Springer-Verlag, Berlin, 487–498. *(ND52)*

Marchetti-Spaccamela, A., and Romano, S. (1985), "On different approximation criteria for subset product problems", *Information Processing Letters* **21**, 213–218.

Maruyama, O., and Miyano, S. (1995), "Graph inference from a walk for trees of bounded degree 3 is NP-complete", *Proc. 20th International Symposium on Mathematical Foundations of Computer Science*, Lecture Notes in Computer Science 969, Springer-Verlag, Berlin, 257–266. *(GT55)*

Mata, C. S., and Mitchell, J. S. B. (1995), "Approximation algorithms for geometric tour and network design problems", *Proc. 11th Annual ACM Symposium on Computational Geometry*, ACM, 360–369. *(ND34, GP2, MS17)*

Matula, D. W. (1970), "On the complete subgraph of a random graph", in *Combinatory Mathematics and its Applications*, Chapel Hill, North Carolina, 356–369.

Matula, D. W., and Beck, L. L. (1983), "Smallest last ordering and clustering and graph coloring algorithms", *J. ACM* **30**, 417–427.

Matula, D. W., Marble, G., and Isaacson, J. D. (1972), "Graph coloring algorithm", in *Graph Theory and Computing*, Academic Press, New York, 108–122.

Michel, C., Schroeter, H., and Srivastav, A. (1995), "TSP and matching in printed circuit board assembly", Unpublished manuscript. *(ND33)*

Middendorf, M. (1994), "On the approximation of finding various minimal, maximal, and consistent sequences", *Proc. 5th Annual International Symposium on Algorithms and Computation*, Lecture Notes in Computer Science 834, Springer-Verlag, Berlin, 306–314. *(SR5)*

Mitchell, J. S. B., Piatko, C., and Arkin, E. M. (1992), "Computing a shortest k-link path in a polygon", *Proc. 33rd Annual IEEE Symposium on Foundations of Computer Science*, IEEE Computer Society, 573–582. *(MS8)*

Mitchell, J. S. B., and Suri, S. (1992), "Separation and approximation of polyhedral objects", *Proc. 3rd Annual ACM-SIAM Symposium on Discrete Algorithms*, ACM-SIAM, 296–306. *(ND65)*

Möhring, R. H., Schäffter, M. W., and Schulz, A. S. (1996), "Scheduling jobs with communication delays: using infeasible solutions for approximation", *Proc. 4th Annual European Symposium on Algorithms*, Lecture Notes in Computer Science 1136, Springer-Verlag, Berlin, 76–90. *(SS7)*

Monien, B., and Speckenmeyer, E. (1985), "Ramsey numbers and an approximation algorithm for the vertex cover problem", *Acta Informatica* **22**, 115–123. *(GT1)*

Motwani, R. (1992), "Lecture notes on approximation algorithms–Volume I", Technical report, Department of Computer Science, Stanford University.

Motwani, R., and Naor, J. S. (1994), "On exact and approximate cut covers of graphs", Technical Report STAN-CS-TN-94-11, Department of Computer Science, Stanford University, California. *(GT21)*

Bibliography

Motwani, R., and Raghavan, P. (1995), *Randomized Algorithms*, Cambridge University Press, Cambridge.

Naor, J., and Zosin, L. (1997), "A 2-approximation algorithm for the directed multiway cut problem", *Proc. 38th Annual IEEE Symposium on Foundations of Computer Science*, IEEE Computer Society, 548–553. *(ND21)*

Nayak, A., Sinclair, A., and Zwick, U. (1998), "Spatial codes and the hardness of string folding problems", *Proc. 9th Annual ACM-SIAM Symposium on Discrete Algorithms*, ACM-SIAM, 639–648. *(MS5)*

Nigmatullin, R. G. (1976), "Approximate algorithms with bounded absolute errors for discrete extremal problems", *Kibernetika* **1**, 95–101.

Nishizeki, T., and Chiba, N. (1988), *Planar Graphs: Theory and Algorithms*, Annals of Discrete Mathematics 32, Elsevier Science, Amsterdam. *(GT25, SP1)*

Nishizeki, T., and Kashiwagi, K. (1990), "On the 1.1 edge-coloring of multigraphs", *SIAM J. Discrete Mathematics* **3**, 391–410. *(GT7)*

Ono, T., Hirata, T., and Asano, T. (1996), "Approximation algorithms for the maximum satisfiability problems", *Proc. 5th Scandinavian Workshop on Algortihm Theory*, Lecture Notes in Computer Science 1097, 100–111.

Orponen, P., and Mannila, H. (1987), "On approximation preserving reductions: Complete problems and robust measures", Technical Report C-1987-28, Department of Computer Science, University of Helsinki. *(ND32, MP1, LO7)*

Padberg, M. W., and Rinaldi, G. (1987), "Optimization of a 532 city symmetric traveling salesman problem by branch and cut", *Operations Research Letters* **6**, 1–7.

Padberg, M. W., and Rinaldi, G. (1991), "A branch and cut algorithm for the resolution of large scale symmetric traveling salesman problems", *SIAM Review* **33**, 60–100.

Panconesi, A., and Ranjan, D. (1993), "Quantifiers and approximation", *Theoretical Computer Science* **107**, 145–163. *(GT37, LO6)*

Papadimitriou, C. H. (1977), "The Euclidean traveling salesperson problem is NP-complete", *Theoretical Computer Science* **4**, 237–244.

Papadimitriou, C. H. (1985), "An algorithm for shortest-path motion in three dimensions", *Information Processing Letters* **20**, 259–263. *(MS13)*

Papadimitriou, C. H. (1992), "The complexity of the Lin-Kernighan heuristic for the traveling salesman problem", *SIAM J. Computing* **21**, 450–465.

Papadimitriou, C. H. (1994), *Computational Complexity*, Addison-Wesley, Reading MA.

Papadimitriou, C. H., Raghavan, P., Sudan, M., and Tamaki, H. (1994), "Motion planning on a graph", *Proc. 35th Annual IEEE Symposium on Foundations of Computer Science*, IEEE Computer Society, 511–520. *(GP1)*

Papadimitriou, C. H., and Steiglitz, K. (1977), "On the complexity of local search for the traveling salesman problem", *SIAM J. Computing* **6**, 76–83.

Papadimitriou, C. H., and Steiglitz, K. (1982), *Combinatorial Optimization: Algorithms and Complexity*, Prentice Hall, Englewood Cliffs NJ.

Papadimitriou, C. H., and Yannakakis, M. (1991), "Optimization, approximation, and complexity classes", *J. Computer and System Sciences* **43**, 425–440. *(GT1, GT2, GT9, GT23, GT33, ND14, ND16, SP4, LO1, LO2, LO4)*

Papadimitriou, C. H., and Yannakakis, M. (1993), "The traveling salesman problem with distances one and two", *Mathematics of Operations Research* **18**, 1–11. *(ND33)*

Park, J. K., and Phillips, C. A. (1993), "Finding minimum-quotient cuts in planar graphs", *Proc. 25th ACM Symposium on Theory of Computing*, ACM, 766–775. *(ND26)*

Paz, A., and Moran, S. (1981), "Non deterministic polynomial optimization problems and their approximations", *Theoretical Computer Science* **15**, 251–277. *(GT14)*

Peleg, D., and Reshef, E. (1998), "Deterministic polylog approximation for minimum communication spanning trees", *Proc. 25th International Colloquium on Automata, Languages and Programming*, Lecture Notes in Computer Science 1443, Springer-Verlag, Berlin, 670–679. *(ND7)*

Peleg, D., Schechtman, G., and Wool, A. (1993), "Approximating bounded 0-1 integer linear programs", *Proc. 2nd Israel Symposium on Theory of Computing and Systems*, IEEE Computer Society, 69–77. *(SP4)*

Petrank, E. (1992), "The hardness of approximation: gap location", Technical Report 754, Computer Science Department, Technion, Israel Institute of Technology, Haifa, Israel. *(MS3)*

Petrank, E. (1994), "The hardness of approximation: gap location", *Computational Complexity* **4**, 133–157. *(GT1, GT7, SP3)*

Pevzner, P. (1992), "Multiple aligning, communication cost, and graph matching", *SIAM J. Applied Mathematics* **52**, 1763–1779. *(MS4)*

Phillips, C., Stein, C., and Wein, J. (1995), "Scheduling jobs that arrive over time", *Proc. 4th Workshop on Algorithms and Data Structures*, Lecture Notes in Computer Science 955, Springer-Verlag, Berlin, 86–97. *(SS12)*

Phillips, C. A. (1993), "The network inhibition problem", *Proc. 25th ACM Symposium on Theory of Computing*, ACM, 776–785. *(ND18, ND43)*

Pitt, L. (1985), "A simple probabilistic approximation algorithm for vertex cover", Technical Report 404, Department of Computer Science, University of Yale.

Pitt, L., and Warmuth, M. K. (1993), "The minimum consistent DFA problem cannot be approximated within any polynomial", *J. ACM* **40**, 95–142. *(AL1)*

Bibliography

Plaisted, D. A., and Hong, J. (1987), "A heuristic triangulation algorithm", *J. Algorithms* **8**, 405–437. *(ND64)*

Plesník, J. (1980), "On the computational complexity of centers locating in a graph", *Aplikace Matematiky* **25**, 445–452. *(ND53)*

Plesník, J. (1981), "The complexity of designing a network with minimum diameter", *Networks* **11**, 77–85. *(ND6)*

Plesník, J. (1982), "Complexity of decomposing graphs into factors with given diameters or radii", *Math. Slovaca* **32**, 379–388. *(ND58)*

Plesník, J. (1986), "Bad examples of the metric traveling salesman problem for the 2-change heuristic", *Acta Mathematica Universitatis Comenianae* **55**, 203–207.

Plesník, J. (1987), "A heuristic for the p-center problem in graphs", *Discrete Applied Mathematics* **17**, 263–268. *(ND53)*

Plesník, J. (1988), "Two heuristics for the absolute p-center problem in graphs", *Math. Slovaca* **38**, 227–233. *(ND53)*

Plotkin, S., and Tardos, É. (1995), "Improved bounds on the max-flow min-cut ratio for multicommodity flows", *Combinatorica* **15**, 425–434.

Provan, J. S. (1988), "An approximation scheme for finding Steiner trees with obstacles", *SIAM J. Computing* **17**, 920–934. *(ND9)*

Queyranne, M. (1985), "Bounds for assembly line balancing heuristics", *Operations Research* **33**, 1353–1359. *(SR1)*

Queyranne, M. (1986), "Performance ratio of polynomial heuristics for triangle inequality quadratic assignment problems", *Operations Research Letters* **4**, 231–234. *(MP7)*

Rabani, Y. (1996), "Path coloring on the mesh", *Proc. 37th Annual IEEE Symposium on Foundations of Computer Science*, IEEE Computer Society, 400–409. *(ND48)*

Rabin, M. (1976), "Probabilistic algorithms", in *Algorithms and Complexity: New Directions and Recent Results*, Academic Press, New York, 21–39.

Raghavan, P. (1988), "Probabilistic construction of deterministic algorithms: approximating packing integer programs", *J. Computer and System Sciences* **37**, 130–143.

Raghavan, P., and Thompson, C. (1987), "Randomized rounding: a technique for provably good algorithms and algorithmic proofs", *Combinatorica* **7**, 356–374.

Raghavan, P., and Thompson, C. D. (1991), "Multiterminal global routing: A deterministic approximation scheme", *Algorithmica* **6**, 73–82. *(ND44)*

Raghavan, P., and Upfal, E. (1994), "Efficient routing in all-optical networks", *Proc. 26th ACM Symposium on Theory of Computing*, ACM, 134–143. *(ND48)*

Rao, S., and Richa, A. W. (1998), "New approximation techniques for some ordering problems", *Proc. 9th Annual ACM-SIAM Symposium on Discrete Algorithms*, ACM-SIAM, 211–218. *(GT39, GT44, SS2)*

Ravi, R. (1994a), "A primal-dual approximation algorithm for the Steiner forest problem", *Information Processing Letters* **50**, 185–190. *(ND8)*

Ravi, R. (1994b), "Rapid rumor ramification: approximating the minimum broadcast time", *Proc. 35th Annual IEEE Symposium on Foundations of Computer Science*, IEEE Computer Society, 202–213. *(ND52)*

Ravi, R., Sundaram, R., Marathe, M. V., Rosenkrantz, D. J., and Ravi, S. S. (1994), "Spanning trees short or small", *Proc. 5th Annual ACM-SIAM Symposium on Discrete Algorithms*, ACM-SIAM, 546–555. *(ND1)*

Ravi, R., and Williamson, D. (1997), "An approximation algorithm for minimum-cost vertex-connectivity problems", *Algorithmica* **18**, 21–43. *(ND27)*

Ravi, S. S., Rosenkrantz, D. J., and Tayi, G. K. (1991), "Facility dispersion problems: heuristics and special cases", *Proc. 2nd Workshop on Algorithms and Data Structures*, Lecture Notes in Computer Science 519, Springer-Verlag, Berlin, 355–366. *(ND59)*

Rayward-Smith, V. J. (1987), "Net scheduling with unit interprocessor communication delays", *Discrete Applied Mathematics* **18**, 55–71. *(SS7)*

Raz, R., and Safra, S. (1997), "A sub-constant error-probability low-degree test, and sub-constant error-probability PCP characterization of NP", *Proc. 29th ACM Symposium on Theory of Computing*, ACM, 475–484.

Reinelt, G. (1994a), *Contributions to Practical Traveling Salesman Problem Solving*, Springer-Verlag, Berlin.

Reinelt, G. (1994b), *The Traveling Salesman, Computational Solutions for TSP Applications*, Lecture Notes in Computer Science 840, Springer-Verlag, Berlin.

Rinnooy Kan, A. H. G., and Frenk, J. B. G. (1986), "On the rate of convergence to optimality of the LPT rule", *Discrete Applied Mathematics* **14**, 187–198.

Rogers Jr., H. (1987), *Theory of Recursive Functions and Effective Computability*, MIT Press, Cambridge MA.

Rosenkrantz, D. J., Stearns, R. E., and Lewis, P. M. (1977), "An analysis of several heuristics for the traveling salesman problem", *SIAM J. Computing* **6**, 563–581.

Sahni, S. (1973), *On the knapsack and other computationally related problems*, PhD thesis, Cornell University.

Sahni, S. K., and Gonzalez, T. F. (1976), "P-complete approximation problems", *J. ACM* **23**, 555–565. *(GT19, GT20, ND55, MP6, MP7)*

Bibliography

Salman, F. S., Cheriyan, J., Ravi, R., and Subramanian, S. (1997), "Buy-at-bulk network design: approximating the single-sink edge installation problem", *Proc. 8th Annual ACM-SIAM Symposium on Discrete Algorithms*, ACM-SIAM, 619–628. *(ND50)*

Saran, H., and Vazirani, V. (1991), "Finding k-cuts within twice the optimal", *Proc. 32nd Annual IEEE Symposium on Foundations of Computer Science*, IEEE Computer Society, 743–751. *(ND19)*

Schäffer, A. A., and Yannakakis, M. (1991), "Simple local search problems that are hard to solve", *SIAM J. Computing* **20**, 56–87.

Schiermeyer, I. (1994), "Reverse-fit: a 2-optimal algorithm for packing rectangles", *Proc. 2nd Annual European Symposium on Algorithms*, Lecture Notes in Computer Science 855, Springer-Verlag, Berlin, 290–299. *(SR2)*

Schrijver, A. (1986), *Theory of Linear and Integer Programming*, John Wiley & Sons, New York.

Schulz, A. S., and Skutella, M. (1996), "Scheduling–LPs bear probabilities: Randomized approximations for Min-sum criteria", Technical Report 533/1996, Technical University of Berlin, Department of Mathematics. *(SS4)*

Schulz, A. S., and Skutella, M. (1997a), "Scheduling-LPs bear probabilities: Randomized approximation for Min-sum criteria", *Proc. 5th Annual European Symposium on Algorithms*, Lecture Notes in Computer Science 1284, Springer-Verlag, Berlin, 416–429. *(SS12)*

Schulz, A. S., and Skutella, M. (1997b), "Random-based scheduling: New approximations and LP lower bounds", *Proc. 1st International Symposium on Randomization and Approximation Techniques in Computer Science*, Lecture Notes in Computer Science 1269, Springer-Verlag, Berlin, 119–133. *(SS4, SS12)*

Schwartz, J. T. (1980), "Fast probabilistic algorithms for verification of polynomial identities", *J. ACM* **27**, 701–717.

Sedgewick, R., and Flajolet, P. (1996), *An Introduction to the Analysis of Algorithms*, Addison-Wesley, Reading MA.

Seymour, P. D. (1995), "Packing directed circuits fractionally", *Combinatorica* **15**, 281–288. *(GT8)*

Seymour, P. D., and Thomas, R. (1994), "Call routing and the rat catcher", *Combinatorica* **14**, 217–241. *(ND11)*

Shachnai, H., and Tamir, T. (1998), "Noah's bagels – some combinatorial aspects", *Proc. 1st International Conference on Fun with Algorithms*, 65–78. *(MP16)*

Shamir, A. (1992), "IP=PSPACE", *J. ACM* **39**, 869–877.

Shamir, R., and Tsur, D. (1998), "The maximum subforest problem: Approximation and exact algorithms", *Proc. 9th Annual ACM-SIAM Symposium on Discrete Algorithms*, ACM-SIAM, 394–399. *(GT34)*

Shmoys, D. B. (1997), "Cut problems and their application to divide-and-conquer", in *Approximation Algorithms for NP-hard Problems*, PWS, Boston, Mass., 192–235. *(ND24)*

Shmoys, D. B., Stein, C., and Wein, J. (1994), "Improved approximation algorithms for shop scheduling problems", *SIAM J. Computing* **23**, 617–632. *(SS17)*

Shmoys, D. B., and Tardos, É. (1993), "Scheduling unrelated machines with costs", *Proc. 4th Annual ACM-SIAM Symposium on Discrete Algorithms*, ACM-SIAM, 448–454. *(SS6)*

Shor, P. W. (1986), "The average-case analysis of some on-line algorithms for bin packing", *Combinatorica* **6**, 179–200.

Simchi-Levi, D. (1994), "New worst-case results for the bin-packing problem", *Naval Research Logistics* **41**, 579–585. *(SR1)*

Simon, H. U. (1989), "Approximation algorithms for channel assignment in cellular radio networks", *Proc. 7th International Conference on Fundamentals of Computation Theory*, Lecture Notes in Computer Science 380, Springer-Verlag, Berlin, 405–416. *(MS7)*

Simon, H. U. (1990), "On approximate solutions for combinatorial optimization problems", *SIAM J. Discrete Mathematics* **3**, 294–310. *(GT17, GT18, AL1)*

Skutella, M. (1997), "Approximation algorithms for the discrete time-cost trade-off problem", *Proc. 8th Annual ACM-SIAM Symposium on Discrete Algorithms*, ACM-SIAM, 501–508. *(SS5)*

Solovay, R., and Strassen, V. (1977), "A fast Monte-Carlo test for primality", *SIAM J. Computing* **6**, 84–85.

Srinivasan, A. (1995), "Improved approximations of packing and covering problems", *Proc. 27th ACM Symposium on Theory of Computing*, ACM, 268–276. *(SP4, MP3, MP4)*

Srinivasan, A. (1996), "An extension of the Lovász local lemma and its applications to integer programming", *Proc. 7th Annual ACM-SIAM Symposium on Discrete Algorithms*, ACM-SIAM, 6–15.

Srinivasan, A. (1997), "Improved approximations for edge-disjoint paths, unsplittable flow, and related routing problems", *Proc. 38th Annual IEEE Symposium on Foundations of Computer Science*, IEEE Computer Society, 416–425. *(ND48)*

Srivastav, A., and Stangier, P. (1994), "Tight approximations for resource constrained scheduling problems", *Proc. 2nd Annual European Symposium on Algorithms*, Lecture Notes in Computer Science 855, Springer-Verlag, Berlin, 307–318. *(SS8)*

Bibliography

Sudan, M. (1995), *Efficient Checking of Polynomials and Proofs and the Hardness of Approximation Problems*, Lecture Notes in Computer Science 1001, Springer-Verlag, Berlin.

Tamir, A. (1991), "Obnoxious facility location on graphs", *SIAM J. Discrete Mathematics* **4**, 550–567. *(ND59)*

Tarhio, J., and Ukkonen, E. (1988), "A greedy approximation algorithm for constructing shortest common superstrings", *Theoretical Computer Science* **57**, 131–145. *(SR6)*

Trevisan, L. (1996), "Positive linear programming, parallel approximation and PCPs", *Proc. 4th Annual European Symposium on Algorithms*, Lecture Notes in Computer Science 1136, Springer-Verlag, Berlin, 62–75. *(LO12)*

Trevisan, L. (1997), "When Hamming meets Euclid: the approximability of geometric TSP and MST", *Proc. 29th ACM Symposium on Theory of Computing*, ACM, 21–29. *(ND34)*

Trevisan, L., Sorkin, G. B., Sudan, M., and Williamson, D. P. (1996), "Gadgets, approximation, and linear programming", *Proc. 37th Annual IEEE Symposium on Foundations of Computer Science*, IEEE Computer Society, 617–626. *(LO2)*

Turek, J., Schwiegelshohn, U., Wolf, J. L., and Yu, P. S. (1994), "Scheduling parallel tasks to minimize average response time", *Proc. 5th Annual ACM-SIAM Symposium on Discrete Algorithms*, ACM-SIAM, 112–121. *(SS8)*

Turner, J. S. (1989), "Approximation algorithms for the shortest common superstring problem", *Information and Computation* **83**, 1–20. *(SR6)*

Unger, W. (1998), "The complexity of the approximation of the bandwidth problem", *Proc. 39th Annual IEEE Symposium on Foundations of Computer Science*, 82–91. *(GT42)*

Vaidya, P. (1988), "Geometry helps in matching", *Proc. 20th ACM Symposium on Theory of Computing*, ACM, 422–425.

van Laarhoven, P. J. M., and Aarts, E. H. L. (1987), *Simulated Annealing: Theory and Applications*, Reidel, Dordrecht.

Verbitsky, O. (1994), "On the largest common subgraph problem", Unpublished manuscript. *(GT46)*

Verbitsky, O. (1995), "On the hardness of approximating some optimization problems that are supposedly easier than Max Clique", *Combinatorics, Probability and Computing* **4**, 167–180. *(GT12, GT23, LO12)*

Vishwanathan, S. (1992), "An approximation algorithm for the asymmetric travelling salesman problem with distances one and two", *Information Processing Letters* **44**, 297–302. *(ND33)*

Vishwanathan, S. (1996), "An $o(\log^* n)$ approximation algorithm for the asymmetric p-center problem", *Proc. 7th Annual ACM-SIAM Symposium on Discrete Algorithms*, ACM-SIAM, 1–5. *(ND53)*

Vizing, V. G. (1964), "On an estimate of the chromatic class of a p-graph", *Diskret. Analiz* **3**, 23–30. *(GT7)*

Wagner, F., and Wolff, A. (1995), "An efficient and effective approximation algorithm for the map labeling problem", *Proc. 3rd Annual European Symposium on Algorithms*, Lecture Notes in Computer Science 979, Springer-Verlag, Berlin, 420–433. *(MS14)*

Wang, Q., and Cheng, K. H. (1990), "A heuristic algorithm for the k-center problem with cost and usage weights", Technical Report TR #UH-CS-90-15, Computer Science Department, Houston University. *(ND56)*

Wee, T. S., and Magazine, M. J. (1982), "Assembly line balancing as generalized bin packing", *Operations Research Letters* **1**, 56–58. *(SR1)*

Wigderson, A. (1983), "Improving the performance guarantee for approximate graph coloring", *J. ACM* **30**, 729–735.

Williamson, D. P., Goemans, M. X., Mihail, M., and Vazirani, V. V. (1995), "A primal-dual approximation algorithm for generalized Steiner network problems", *Combinatorica* **15**, 435–454. *(ND10)*

Williamson, D. P., Hall, L. A., Hoogeveen, J. A., Hurkens, C. A. J., Lenstra, J. K., and Shmoys, D. B. (1994), "Short shop schedules", Unpublished manuscript. *(SS14, SS15, SS17)*

Wöginger, G. J., and Yu, Z. (1992), "A heuristic for preemptive scheduling with set-up times", *Computing* **49**, 151–158. *(SS9)*

Wolsey, L. A. (1980), "Heuristic analysis, linear programming and branch and bound", *Mathematical Programming Study* **13**, 121–134.

Wu, B. Y., Chao, K., and Tang, C. Y. (1998), "Exact and approximation algorithms for constructing ultrametric trees from distance metrics", *Proc. 4th Annual International Conference on Computing and Combinatorics*, Lecture Notes in Computer Science 1449, Springer-Verlag, Berlin, 299–308. *(MS9)*

Wu, B. Y., Lancia, G., Bafna, V., Chao, K., Ravi, R., and Tang, C. Y. (1998), "A polynomial time approximation scheme for minimum routing cost spanning tree", *Proc. 9th Annual ACM-SIAM Symposium on Discrete Algorithms*, ACM-SIAM, 21–32. *(ND7, MS4)*

Yannakakis, M. (1979), "The effect of a connectivity requirement on the complexity of maximum subgraph problems", *J. ACM* **26**, 618–630. *(GT29)*

Yannakakis, M. (1994), "On the approximation of maximum satisfiability", *J. Algorithms* **17**, 475–502.

Yannakakis, M., and Gavril, F. (1980), "Edge dominating sets in graphs", *SIAM J. Applied Mathematics* **38**, 364–372. *(GT10)*

Yu, B., and Cheriyan, J. (1995), "Approximation algorithms for feasible cut and multicut problems", *Proc. 3rd Annual European Symposium on Algorithms*, Lecture Notes in Computer Science 979, Springer-Verlag, Berlin, 394–408. *(ND22)*

Bibliography

Yue, M. (1991), "A simple proof of the inequality $MFFD(L) \leq \frac{71}{60}OPT(L) + \frac{78}{71}, \forall L$ for the $MFFD$ bin-pack algorithm", Technical Report RRR # 20-91, Rutcor, Rutgers Center for Operations Research, Rutgers University. *(SR1)*

Zhang, K., and Jiang, T. (1994), "Some MAX SNP-hard results concerning unordered labeled trees", *Information Processing Letters* **49**, 249–254. *(GT48)*

Zuckerman, D. (1993), "NP-complete problems have a version that's hard to approximate", *Proc. 8th Annual Conference on Structure in Complexity Theory*, IEEE Computer Society, 305–312. *(GT1, GT5, GT8, GT9, GT17, GT41, ND8, ND17, SP1, SP4, SP7, SP9, SS1)*

Zwick, U. (1998), "Approximation algorithms for constraint satisfaction problems involving at most three variables per constraint", *Proc. 9th Annual ACM-SIAM Symposium on Discrete Algorithms*, ACM-SIAM, 201. *(LO4, LO12)*

Index

k-opt Algorithm, 331–336

2-opt Algorithm, 65, 331

3-opt Algorithm, 331

Absolute
    Approximation Algorithm, 89
    Error, 89
achromatic number, 372
Algorithm
    k-opt, 331–336
    2-opt, 65, 331
    3-opt, 331
    Absolute Approximation, 89
    Approximation, 39, 88
    Best Fit Decreasing, 57, 299
    Cheapest Insertion, 327
    Christofides, 96
    Decreasing Degree, 59
    ε-approximate, 90
    Farthest Insertion, 327
    First Fit, 54, 146
    First Fit Decreasing, 55, 299
    Greedy, 40–50, 124, 127
    Greedy Sat, 92
    Independent Set Scheme, 108
    Integer Knapsack Scheme, 106
    List Scheduling, 51, 296
    Nearest Insertion, 327
    Nearest Neighbor, 48, 291, 327
    Next Fit, 54
    Nondeterministic, 14, 16
    PAIRMATCH, 299
    Partition PTAS, 103
    Polynomial-time, 21, 26
    Primal-dual, 67–69
    Pseudo-polynomial, 72
    r-approximate, 91
    Random Insertion, 327
    Randomized, 74–76, 153–171
    Sequential, 50–61
    Smallest Last, 60
    Smallest Sum Insertion, 328
    STRIP, 312
alignment, 465
Almost Transparent Proof, 221
Analysis
    Asymptotic, 6
    Worst Case, 4
AP-reducibility, 257, 260
Approximate
    Solution, 39
Approximation
    Algorithm, 39, 88, 254
    Classes, 88
    Performance Guarantee, 87
    Scheme, 72
        Asymptotic, 139
        Fully Polynomial-time, 73, 111
        Polynomial-time, 102
Arithmetization, 214–217, 231–237
array partition, 428
Asymptotic Analysis, 6
Asymptotic Approximation Scheme, 139

balanced connected partition, 377

Index

bandwidth, 387
Best Fit Decreasing Algorithm, 57, 299
betweenness, 464
bin packing, 430
binary constraints, 467
bipartite graphs, 378, 381–383, 400
bipartition, 383
bisection of graph, 399
Branch-and-bound, 322–325
Branch-and-cut, 324–325
broadcast time, 416
butterfly graphs, 414

caterpillar, 387, 392
channel assignment, 466
Cheapest Insertion Algorithm, 327
Chinese postman problem, 410
chordal graphs, 381–383, 386, 416
Christofides' Algorithm, 96
chromatic index, 373
chromatic number, 371
circular-arc graphs, 370, 381
Class
    APX, 93, 94, 111
    co–NP, 17
    EXPTIME, 11
    FPTAS, 111, 114
    NP, 16, 184, 191, 194
    NPO, 27, 94
    P, 11, 184
    PCP, 194
    PO, 29
    PSPACE, 11, 184
    PTAS, 105, 111, 114
    PTAS$^\infty$, 139
    RP, 192
claw free graphs, 369, 380
cliques, 376, 377, 379
Close Function, 209
Closure of Complexity Class, 20
clustering problems, 417, 418
coding theory, 464
coloring of graph, 371
common
    point set, 433
    sub-tree, 389
    subgraph, 388, 389
    subtree, 433
Complementary Problem, 17
Complete Problem, 20, 261
Complexity
    Lower Bound, 6
    Upper Bound, 6, 8, 182
Complexity Class, 11, 183
    Closure, 20
    Nondeterministic, 16
Composition Lemma, 245–248
Conditional Probabilities Method, 168–171
connectivity, 405, 406
Construction Heuristics, 325–329
Constructive Problem, 24, 31–32
Constructive Solution, 13
Convergence
    Almost Everywhere, 288
    in Expectation, 288
convex programming, 452
Cook-Levin Theorem, 185
Cost Measure
    Logarithmic, 3
    Uniform, 3
covering problems, 369–379, 424–426, 431, 434
covering with disks, 427
cut
    balanced, 403
    directed, 400
    flux, 405
    hypergraph, 424
    maximum, 399
    maximum k, 400
    minimum k, 401
    multi-cut, 402
    multiway, 402
    quotient, 404
    ratio, 403
    vertex, 402

data storage, 430–431
data storage problems, 431
Decision Problem, 9, 24, 28, 31
decision tree, 464
Decreasing Degree Algorithm, 59
digraph, equivalent, 385
disjoint paths, 414
Dominating Algorithm Technique, 299
dominating set, 370, 371
Dynamic Programming, 69–74

ε-approximable Problem, 92
ε-approximate Algorithm, 90
E3-Satisfiability, 188, 208–248
edge

Index

2-spanner, 385
coloring, 373
deletion, 382, 383
dominating set, 370
installation, 415
separator, 404
elimination tree height, 391
embedded sub-tree, 389
Encoding Scheme, 242
Equivalent Formulas, 20
Error
    Absolute, 89
    Relative, 90
Eulerian
    Graph, 95
Evaluation Problem, 24, 31
evolutionary trees, 465, 468
exact cover, 425
Expander Graph, 276

facility dispersion, 419
facility location, 420
Falsity, 17
Farthest Insertion Algorithm, 327
Feasible Solution, 22
feedback sets in graphs, 373, 374
finite automata, 461
First Fit Algorithm, 54, 146
First Fit Decreasing Algorithm, 55, 299
First Moment Method, 293
Fixed Partitioning Technique, 74
Fixed-depth Local Search, 330–336
flow problems, 413–416
folding of strings, 465
front size, 391
Fully Polynomial-time Approximation
    Scheme, 73, 111
Function
    Close, 209
    Linear, 210–221

Gap Technique, 100–102, 196
Genetic Algorithm, 344–347
geometric problems, 393, 395, 396,
    409, 421, 426, 448, 466
graph
    bipartite, 378, 381–383, 400
    bisection, 399
    butterfly, 414
    chordal, 381–383, 386, 416
    circular-arc, 370, 381
    claw free, 369, 380

coloring, 371
connectivity, 405, 406
covering with bipartite subgraphs,
    378
covering with cliques, 377
covering with cuts, 379
covering with cycles, 378, 379
diameter, 406, 419
drawing, 421
embedding, 400
Eulerian, 95
expander, 276
inference, 391
interval, 381, 382, 386
motion planning, 454
outerplanar, 108, 381–383, 416
partition into cliques, 376
planar, 60, 369–372, 374–376,
    380–384, 395, 401, 404,
    405, 408–410, 414, 417, 421
random, 302
strong connectivity, 406
transformation, 390
unit disk, 369–372, 375, 376, 380
Greedy Method, 40–50, 124, 127
Greedy Sat
    Algorithm, 92

Hamiltonian circuit, 386
Hard Problem, 261
heaviest subgraph, 385
Heuristics
    Kernighan-Lin, 338
    Lin-Kernighan, 339
hitting set, 426
Hopfield nets, 465
Horn clauses, 456
Horn core, 460
hypergraph
    cut, 424
    matching, 423
    quotient cut, 405
hyperplane consistency, 450

independence number, 371
independent
    dominating set, 371
    sequence of vertices, 380
    set, 380
Independent Set Scheme Algorithm,
    108
induced subgraph, 381, 382, 385, 389

Index

Input Size, 5
Insertion Sort, 7
Integer Knapsack Scheme Algorithm, 106
integer programming, 446, 447
interval graphs, 381, 382, 386
isomorphism problems, 388–390

Karp Reducibility, 17, 21
    Polynomial-time, 18, 184
Kernighan-Lin Heuristics, 338
knapsack problems, 450, 451

L-reducibility, 259
Language Recognized, 179, 182
lattices, 452
Lin-Kernighan Heuristics, 339
linear arrangement, 388
Linear Function, 210–221
Linear Programming, 65–69
linear systems of relations, 448–450
Linearity Test, 210–212
List Scheduling Algorithm, 51, 296
Local Optimum, 61, 330
Local Search, 61–65, 329–341
    Fixed-depth, 330–336
    Variable-depth, 336–341
Logarithmic Cost Measure, 3
logic problems, 455–460
Longest
    Common Subsequence, 432
    Computation, 461
    Induced Chordal Subgraph, 382
    Induced Cycle, 382
    Induced Path, 382
    Minimal Common
        Supersequence, 432
    Path, 411
    Path with Forbidden Pairs, 390
Low-degree Polynomial, 222–239
Low-degree Test, 225–229
Lower Bound
    Complexity, 6
LPT Rule, 52

map labeling, 468
matching problems, 375, 376, 423
mathematical programming, 446–452
Maximum
    d-Vector Covering, 431
    2-Satisfiability, 271, 456
    3-Dimensional Matching, 423

3-Satisfiability, 196–198, 260, 276, 278, 280, 455
Achromatic Number, 372
Acyclic Subgraph, 374
Agreement Subtree, 469
Balanced Connected Partition, 377
Betweenness, 464
Bisection, 399
Bounded 0-1 Programming, 446
Capacity Representatives, 428
Channel Assignment, 466
Class-Constrained Knapsack, 451
Clique, 31, 198–200, 256, 259, 302, 379
k-Clustering Sum, 418
k-Colorable Induced Subgraph, 385
k-Colorable Subgraph, 384
Common Embedded Sub-tree, 389
Common Induced Subgraph, 389
Common Point Set, 433
Common Subgraph, 388
Common Subtree, 433
Compatible Binary Constraint Satisfaction, 467
Compression, 432
Constrained Hamiltonian Circuit, 386
Constrained Sequencing to Minimize Tardy Task Weight, 435
k-Constraint Satisfaction, 459
Cut, 62–64, 260, 273, 399
k-Cut, 400
Degree-Bounded Connected Subgraph, 383
Directed Cut, 400
Disjoint Connecting Paths, 414
Distinguished Ones, 457
Domatic Partition, 370
Edge Subgraph, 384
k-Facility Dispersion, 419
k-Facility Location, 420
Flow, 130
Frequency Allocation, 469
Geometric
    Traveling Salesperson, 409
Geometric Square Packing, 427
Graph Transformation, 390
H-Matching, 375

Index

Horn Core, 460
Hypergraph Cut, 424
Hypergraph Matching, 423
Hyperplane Consistency, 450
Independent Sequence, 380
Independent Set, 43–47, 107–110, 259, 380
Independent Set of k-gons, 380
Induced Connected Subgraph with Property Π, 382
Induced Subgraph with Property Π, 381
Integer k-Choice Knapsack, 451
Integer Knapsack, 105–107
Integer m-Dimensional Knapsack, 451
Integral k-Multicommodity Flow on Trees, 413
Knapsack, 41–43, 69–74, 89, 450
Leaf Spanning Tree, 394
Map Labeling, 468
Matching of Consistent k-Cliques, 376
Metric
    Traveling Salesperson, 408
Minimum
    Metric k-Spanning Tree, 394
    Metric k-TSP, 395
    Spanning Tree Deleting k Edges, 398
    k-Steiner Tree, 395
Multi-commodity Flow, 131
Not-All-Equal 3-Satisfiability, 456
Not-All-Equal Satisfiability, 272
Number of Satisfiable Formulas, 458
Ones, 457
Outerplanar Subgraph, 383
Packing Integer Programming, 446
Planar Subgraph, 383
Priority Flow, 413
Product Knapsack, 112
Quadratic Programming, 447
Rectangle Packing, 429
Remote Minimum Spanning Tree, 395
Satisfiability, 74–76, 92–93, 259, 336, 455
k-Satisfiability, 455
Satisfiability of Horn Clauses, 456
Satisfiability of Quadratic Equations over GF[q], 453
Satisfying Linear Subsystem, 449
Scatter TSP, 410
k-Section, 401
Set Packing, 423
k-Set Packing, 423
Set Splitting, 424
String Folding, 465
Subforest, 384
Subset Sum, 451
Traveling Salesperson, 407
Triangle Packing, 375
Weighted 2-Satisfiability, 167–168
Weighted 3-Satisfiability, 265
Weighted Cut, 162–167
Weighted Satisfiability, 157–162, 169–171, 261, 262
Weighted Satisfiability with Bound, 458
Measure Function, 22
Membership Proof, 190
metric basis, 391
Minimum
    {0,1}-Linear Programming, 255, 265
    0-1 Programming, 446
    3-Dedicated Processor Scheduling, 441
    3-Dimensional Assignment, 427
    3DNF Satisfiability, 457
    Absolute k-Center, 417
    α-All-Neighbor k-Center, 417
    Array Partition, 428
    Assymmetric k-Center, 417
    Attraction Radius for Binary Hopfield Net, 465
    Balanced Cut, 338–339, 345
    b-Balanced Cut, 403
    Bandwidth, 387
    Bend Number, 421
    Biconnectivity Augmentation, 406
    Bin Packing, 53–58, 110, 143–148, 299, 430
    Bipartition, 383
    Bisection, 399
    Block-angular Convex Programming, 452
    Bottleneck Path Matching, 376
    Bounded Degree Spanning Tree, 394

Index

Bounded Diameter Augmentation, 406
Broadcast Time, 416
Capacitated k-Center, 417
k-Capacitated Tree Partition, 377
k-Center, 416
Chinese Postman for Mixed Graphs, 410
Chinese Postman Problem, 410
Chordal Graph Completion, 386
Chromatic Index, 373
Chromatic Number, 371
Clique Cover, 377
Clique Partition, 376
k-Clustering, 417
k-Clustering Sum, 418
Color Sum, 372
Coloring with Defect d, 372
Communication Cost Spanning Tree, 395
Complete Bipartite Subgraph Cover, 378
Consistent Finite Automaton, 461
Constrained Partition, 427
Covering Integer Programming, 447
Crossing Number, 400
Cut, 130
k-Cut, 401
Cut Cover, 379
Cut Linear Arrangement, 388
Decision Tree, 464
Degree Spanning Tree, 393
Degree Steiner Tree, 393
Diameter Spanning Subgraph, 395
Diameters Decomposition, 419
Directed Bandwidth, 387
Distinguished Ones, 457
Dominating Set, 370
Dynamic Storage Allocation, 431
Edge 2-Spanner, 385
Edge Coloring, 139–143, 373
k-Edge Connected Subgraph, 405
Edge Deletion
   Bipartition, 383
   k-Partition, 383
Edge Disjoint Cycle Cover, 379
Edge Dominating Set, 370
b-Edge Separator, 404
Edge-Deletion Subgraph with Property Π, 382

Elimination Tree Height, 391
Equivalence Deletion, 459
Equivalent Digraph, 385
Evolutionary Tree, 465
Exact Cover, 425
Facility Location, 420
Feedback Arc Set, 374
Feedback Vertex Set, 373
File Transfer Scheduling, 443
Flow-Shop Scheduling, 442
Flux Cut, 405
Fractional Chromatic Number, 372
Fractional Multi-Cut, 133
Front Size, 391
General Routing, 411
Generalized 0-1 Assignment, 447
Generalized Steiner Network, 397, 415
Generalized Tree Alignment, 465
Geometric
   3-Degree Spanning Tree, 393
   Angular Traveling Salesperson, 409
   Capacitated k-Center, 417
   k-Clustering, 418
   Disk Cover, 426
   k-Spanning Tree, 393
   Steiner Tree, 396
   Traveling Salesperson, 408
Graph Coloring, 25, 30, 58–61, 89, 102, 127–129, 311, 371
Graph Inference, 391
Graph Motion Planning, 454
Height Two Dimensional Packing, 430
Hitting Set, 426
Independent Dominating Set, 371
Interval Graph Completion, 386
Job Shop Scheduling, 443
Latency Problem, 413
Length Equivalent Frege Proof, 459
Length Triangulation, 421
Linear Arrangement, 388
k-Link Path in a Polygon, 466
Local Register Allocation, 463
Locally Testable Automaton Order, 461
Maximal Independence Number, 371

Index

Maximal Matching, 374
Maximum Disjoint Connecting Paths, 414
k-Median, 419
Metric
    Bottleneck Wandering Salesperson Problem, 409
    Traveling k-Salesperson Problem, 409
    Traveling Salesperson, 407
    Traveling Salesperson Path, 408
Metric Dimension, 391
Metric Traveling Salesperson, 48–50, 95–100
Multi-Cut, 129–138, 402
Multiprocessor Scheduling, 296, 437
Multiprocessor Scheduling with Speed Factors, 439
Multiway Cut, 402
k-Multiway Separator, 404
Net Expansion, 405
Network Inhibition on Planar Graphs, 401
Node-Del. Connected Subgraph with Prop. Π, 383
Nonplanar Edge Deletion, 383
Number of Satisfiable Formulas, 458
Numerical Taxonomy, 469
Ones, 458
Open-Shop Scheduling, 441
Parallel Processor Total Flow Time, 440
Partition, 103–104
k-Partition, 383
Partition of Rectangle with Interior Points, 467
Path, 26, 29
Path Coloring, 414
Path Width, 391
Permutation Group Base, 462
phylogenetic tree distance, 468
Planar Record Packing, 448
Point-To-Point Connection, 390
Precedence Constrained Scheduling, 437
Precedence Constrained Sequencing with Delays, 435

Preemptive Scheduling with Set-Up Times, 439
Quadratic 0-1 Assignment, 448
Quotient Cut, 404
Ratio-Cut, 403
Rectangle Cover, 434
Rectangle Tiling, 429
Rectilinear Global Routing, 412
Red-Blue Separation, 469
Register Sufficiency, 463
Relevant Variables in Linear System, 448
Resource Constrained Scheduling, 438
Ring Loading, 470
Routing Cost Spanning Tree, 395
Routing Tree Congestion, 397
Rural Postman Problem, 411
Satisfiability, 455
k-Satisfiability, 456
Satisfiability of Horn Clauses, 456
Schedule Length, 444
Scheduling on Identical Machines, 51–53
Separating Subdivision, 422
Sequence Alignment, 465
Sequencing with Release Times, 436
Set Cover, 124–127, 424
Single-Sink Edge Installation, 415
Size Ultrametric Tree, 466
Sorting by Reversals, 467
k-Spanning Tree, 393
Stacker Crane Problem, 410
k-Stacker Crane Problem, 411
Steiner Tree, 395
k-Steiner Tree, 396, 415
Steiner Trees with Obstacles, 397
Storage-Time Sequencing, 435
Strip Packing, 431
Strong Connectivity Augmentation, 406
Sum of Squares, 428
k-Supplier, 418
Survivable Network, 397
k-Switching Network, 421
Test Collection, 426
Time-Cost Tradeoff, 436
Travel Robot Localization, 454
Traveling Repairman, 412

Index

Traveling Salesperson, 24, 30, 47–50, 64–65, 94–100, 291, 312, 326–329, 331–336, 339–341, 346, 407
Tree Alignment, 465
Tree Compact Packing, 434
Tree Width, 391
Two-Processor Flow-Shop Scheduling with Batch Set-Up Times, 442
Unsatisfying Linear Subsystem, 450
Unsplittable Flow, 415
Upgrading Spanning Tree, 398
Vehicle Scheduling on Tree, 444
k-Vertex Connected Subgraph, 405
Vertex Cover, 23, 24, 28, 30, 190, 256, 348, 369
Vertex Disjoint Cycle Cover, 378
Vertex k-Cut, 402
b-Vertex Separator, 404
Vertex-Deletion Subgraph with Property Π, 381
Weighted Completion Time Scheduling, 440
Weighted Satisfiability, 262, 458
Weighted Vertex Cover, 66–69, 154–156
multicommodity flow, 413
multiprocessor scheduling problems, 437–441

Nearest
Codeword, 464
Lattice Vector, 452
Nearest Insertion Algorithm, 327
Nearest Neighbor Algorithm, 48, 291, 327
Neighborhood, 61, 330
network inhibition, 401
Next Fit Algorithm, 54
Non-Zero Polynomial, 192
Nondeterministic Algorithm, 14, 16
NP-complete Problem, 21, 185
NP-hard Problem, 30
numerical taxonomy, 469

Optimal Solution, 23
Optimization Problem, 22
Oracle, 19
Turing Machine, 189–190

outerplanar graphs, 108, 381–383, 416

packing problems, 375, 423, 427, 430, 448
PAIRMATCH Algorithm, 299
Partition PTAS Algorithm, 103
partitioning problems, 50–61, 369–379, 405, 428
path
longest, 382, 390, 411
matching, 376
shortest, 390, 412
width, 391
PCP
Normal Form, 243
Restricted, 242
Theorem, 195, 239
Performance Guarantee Approximation, 87
Performance Ratio, 39, 90
permutations, 462, 467
phylogenetic trees, 465, 468
planar graphs, 60, 369–372, 374–376, 380–384, 395, 401, 404, 405, 408–410, 414, 417, 421
polygon, link path, 466
polygon, subdivision, 422
Polynomial-time
Algorithm, 21
Karp Reducibility, 18, 184
Turing Reducibility, 20, 30
Polynomial-time Algorithm, 26
Polynomial-time Approximation Scheme, 102
Polynomially Bounded Optimization Problem, 113
preemptive scheduling, 439
Primal-dual Algorithm, 67–69
Primality, 5
Principle of Optimality, 70
printed circuit board assembly, 408
Probabilistic Analysis, 288
Probabilistic Turing Machine, 191–192
Probabilistically Checkable Proof, 190–195
Problem
APX-complete, 266–281
Complementary, 17
Complete, 20, 261
Constructive, 24, 31–32
Decision, 9, 24, 28, 31

Index

ε-approximable, 92
Evaluation, 24, 31
Hard, 261
NP-complete, 21, 185
NP-hard, 30
NPO-complete, 261–265
Optimization, 22
Partitioning, 50–61
Polynomially Bounded, 113
Pseudo-polynomial, 115–116
r-approximable, 92
Solution, 10, 14
Strongly NP-hard, 116
proof
    almost transparent, 221
    of tautology, 459
    transparent, 208
Pseudo-polynomial Algorithm, 72
Pseudo-polynomial Problem, 115–116

quadratic equations, 453
quadratic programming, 447
Quantified Boolean Formulas, 12, 16

r-approximable Problem, 92
r-approximate Algorithm, 91
Random Graph, 302
Random Insertion Algorithm, 327
Randomized Algorithm, 74–76, 153–171
Recognition of Language, 179, 182
rectangle cover, 434
rectangle packing, 429
rectangle partition, 467
rectangle tiling, 429
Reducibility, 17, 20
    AP, 257, 260
    Karp, 17, 21
        Polynomial-time, 18, 184
    L, 259
    Turing, 19
        Polynomial-time, 20, 30
register allocation, 463
register sufficiency, 463
Relative
    Error, 90
ring loading, 470
robot motion planning, 454, 468
Rounding, 66, 162
routing
    general, 411
    rectilinear, 412
    trees, 397
routing problems, 407–413
Running Space, 6
Running Time, 6–7
rural postman problem, 411

Satisfiability, 11, 13, 14, 18, 20, 21, 185, 191
satisfiability
    equivalence deletion, 459
    maximizing ones, 457
    minimizing ones, 457
    not-all-equal, 456
    of CNF clauses, 455, 456
    of CNF formulas, 458
    of DNF clauses, 457, 459
    weighted, 458
Satisfying Truth Assignment, 11
Scaling Technique, 74
scheduling problems, 437–445
Self-correcting
    Linear Functions, 213–214
    Low-degree Polynomials, 229–231
Self-improvability, 199
Semidefinite Programming, 162–168
separation of point sets, 469
sequence alignment, 465
sequencing problems, 435–437
Sequential Algorithm, 50–61
set
    cover, 424, 425
    dominating, 370
    hitting, 426
    independent, 380
    independent dominating, 371
    packing, 423
    partition, 427, 428
    splitting, 424
set problems, 423–429
shop scheduling problems, 441–443
Shortest
    Common Supersequence, 431
    Common Superstring, 432
    Computation, 461
    Maximal Common
        Non-Supersequence, 432
    Maximal Common Subsequence, 433
    Path Motion in 3 Dimensions, 468
    Path with Forbidden Pairs, 390

Index

Simulated Annealing, 341–344
Smallest Last Algorithm, 60
Smallest Sum Insertion Algorithm, 328
Solution
    Approximate, 39
    Constructive, 13
    Feasible, 22
    Measure, 23
    Optimal, 23
sorting by reversals, 467
spanning
    subgraphs, 395, 405
    trees, 393–395, 398
sparsest cut, 403
stacker crane problem, 410, 411
Steiner
    networks, 397
    trees, 393, 395, 396
storage-time sequencing, 435
string folding, 465
STRIP Algorithm, 312
strip packing, 431
strong connectivity, 406
Strongly NP-hard Problem, 116
sub-tree, 389
subgraph
    k-colorable, 384, 385
    common, 388, 389
    degree-bounded connected, 383
    heaviest, 384
    induced, 382
subgraph problems, 379–386
subsequences, 432
Sum-check Test, 234–237
supergraph problems, 386–387
supersequences, 431
superstrings, 432
survivable networks, 397
switching network, 421

Tabu Search, 347–349
test collection, 426
Transparent Proof, 208
traveling repairman, 412
traveling salesperson problems, 395, 407, 409
tree
    alignment, 465
    width, 391

Watchman Route, 454
Weight-Constrained Path, 412

triangle packing, 375
triangulation, 421
Turing Machine, 175–181
    Deterministic, 178–180
    Nondeterministic, 180–181
    Probabilistic, 191–192
Turing machines, 461
Turing Reducibility, 19
    Polynomial-time, 20, 30

ultrametric trees, 466
Underlying Language, 24, 30, 32
Uniform Cost Measure, 3
unit disk graphs, 369–372, 375, 376, 380
unsplittable flow, 415
Upper Bound
    Complexity, 6, 8

Variable Partitioning Technique, 112
Variable-depth Local Search, 336–341
Verifier, 193
    Restricted, 194
vertex
    cover, 369
    deletion, 381, 383
    separator, 404
Vertex Cover, 24

wandering salesperson problem, 409
Worst Case Analysis, 4

Zero-tester Polynomial, 232–234
$\{0,1\}$-Linear Programming, 18, 21